MONOGRAPHS ON THE PHYSICS AND CHEMISTRY OF MATERIALS

General Editors

H. FRÖHLICH
P. B. HIRSCH
N. F. MOTT

CHEMISTRY OF THE METAL - GAS INTERFACE

BY

M. W. ROBERTS

AND

C. S. McKEE

1478
1978

CLARENDON PRESS · OXFORD

Oxford University Press, Walton Street, Oxford OX2 6DP

OXFORD LONDON GLASGOW

NEW YORK TORONTO MELBOURNE WELLINGTON

IBADAN NAIROBI DAR ES SALAAM LUSAKA CAPE TOWN

KUALA LUMPUR SINGAPORE JAKARTA HONG KONG TOKYO

DELHI BOMBAY CALCUTTA MADRAS KARACHI

© Oxford University Press 1978

British Library Cataloguing in Publication Data

Roberts, M W
    Chemistry of the metal-gas interface.—(Monographs on the physics and
chemistry of materials).
    1. Gases in metals 2. Surface chemistry 3. Metallic surfaces
    I. Title II. McKee, C S III. Series
    546′.3   QD506   77-30413

ISBN 0-19-851339-9

Typeset by The Universities Press, Belfast
Printed in Great Britain
by J. W. Arrowsmith Ltd., Bristol

PREFACE

'Everything is incredible if you skim off the crust of obviousness our habits put on it.'

ALDOUS HUXLEY

IN THIS book we hope to provide a basic picture of gas-metal interactions which may be of use to all interested in the subject but which, in particular, will be relevant for those involved at either the graduate or undergraduate level. Of course, any teacher in such an area faces a major problem, namely, how to cover the most recent developments and at the same time relate the data to the substantial background information already available. A glance at some of the contemporary scientific papers in surface chemistry might even suggest that the subject was discovered no more than five years ago! One of the aims of this book is therefore to try to put recent developments in the context of earlier endeavour. Today's surface science is in a situation rather analogous to that of the physics of bulk atomic and condensed matter at the beginning of this century or the chemistry of inorganic complexes prior to X-ray diffraction analysis. We have much to learn.

No book can cover all aspects of the subject, but for reasons of either prejudice or ignorance (but not superstition!) we have chosen to divide the book into 13 chapters with approximately equal emphasis given to basic or fundamental aspects and specific, more specialized topics. We have not attempted to review the field in detail; for example we discuss physical adsorption in the narrow context of 'clean' surface interaction and in chemisorption we have chosen to focus our attention mainly on four molecules: oxygen, hydrogen, carbon monoxide, and nitrogen. Fundamental aspects of the determination of surface crystallography through low-energy electron diffraction and surface composition and bonding through spectroscopic studies are emphasized. How information from each of these approaches blends with more traditional studies and enhances our understanding of simple molecular processes is considered wherever possible. Also included are chapters dealing with the more traditional approaches adopted, kinetic, thermochemical, etc., and the possible implications of recent experimental developments on thinking in heterogeneous catalysis. In this way it is hoped that the reader will obtain a bird's eye view of the surface chemistry of metals through a blend of fundamental background information with current experimental data related to some specific molecules. The book should be relevant not only to those interested in surface chemistry *per se*, but also to anyone wishing to have a more basic understanding of interfacial phenomena and therefore including surface physicists and materials scientists.

The writing of this book owes much to the stimulus that we have received from the many research colleagues with whom both of us have been associated. One of us (M. W. R.) is also particularly conscious of the introduction he had to the subject through his association with K. W. Sykes, F. C. Tompkins, and C. Kemball, each of whom had a distinctive but different approach. We are grateful to our respective families for their patience and forbearance without which the book would not have been written. Finally, it is a pleasure to thank Mrs. D. J. Varley who so carefully typed the manuscript and Barry Firth who drew the diagrams.

'To find the truth is a matter of luck, the full value of which is only realized when we can prove that what we have found *is* true. Unfortunately, the certainty of our knowledge is at so low a level that all we can do is to follow along the lines of greatest probability.'

J. J. BERZELIUS

Bradford   M. W. R.
July 1977   C. S. M.

ACKNOWLEDGEMENTS

We wish to thank the many authors who have consented to the use of material from their published work. The permission of the following publishers for reproduction of copyright material is also gratefully acknowledged.

Academic Press Inc. (Figs 1.21, 3.28, 4.19, 5.1, 12.21; Table 11.1), Addison-Wesley (Figs 2.2, 2.5, 2.13, 2.14, 2.15, 2.16, 2.18, 2.19, 12.14, 12.18; Plate 1), The American Association for the Advancement of Science (Fig. 13.1; Table 6.1), American Institute of Physics (Figs 3.29, 4.21, 13.6; Table 3 1), The American Physical Society (Figs 1.10, 1.17, 4.22, 4.23, 5.6, 5.7, 5.9, 6.1, 6.7, 6.12, 6.15, 7.3, 8.2, 8.3, 8.4, 8.6, 8.8, 8.9a, 8.11, 8.12, 8.13, 9.2, 9.9, 10.1, 10.2, 10.3, 10.5b, 10.6, 10.7, 10.8, 10.9, 10.10, 10.11, 10.15, 10.16b, 10.17b, 11.10, 11.12, 12.6, 12.7, 12.15b, 13.4; Tables 7.3, 11.4), The Chemical Rubber Co. (Figs 9.7, 9.8), The Chemical Society (Figs 1.16, 4.17b, 4.18, 5.2, 5.10, 5.11, 7.2, 9.10, 9.11, 11.1a, 11.4, 11.6, 11.7, 11.8, 11.11, 11.14, 11.17, 11.18, 13.3; Table 4.2; Plates 3, 4d and e, 8, 9), Direttore del Nuovo Cimento (Figs 9.5, 9.6), Elsevier Scientific Publishing Company (Figs 4.10b, 11.13; Tables 9.3, 11.3), Gordon and Breach Ltd. (Figs 2.23, 2.28, 2.29, 2.30), Int. Atomic Energy Agency, Vienna (Fig. 4.2), The Institute of Physics (Figs 1.5a, 3.30, 4.16, 11.3a, 12.23), Japanese Journal of Applied Physics (Fig. 13.2), John Wiley and Sons Inc. (Figs 2.17, 4.9, 6.7, 9.4, 12.17), John Wiley Interscience (Fig. 2.17), Journal de Chimie Physique (Fig. 11.2), Macmillan Journals Ltd. (Fig. 4.20), McGraw-Hill Book Co. (Fig. 2.12), The Metals Society (Figs 2.8, 2.36b), National Bureau of Standards (Figs 4.3, 4.4), NATO Advanced Study Institutes Programme (Figs 1.13, 1.14, 1.20), The New York Academy of Sciences (Fig. 7.1; Plate 7), North-Holland Publishing Co. (Figs 1.9, 3.3, 3.27, 3.31, 3.35, 3.36, 3.37, 4.1b, 6.4, 6.11, 6.16, 6.17, 7.4, 8.5, 8.7, 9.3, 9.12, 9.13, 9.14, 10.4, 10.5a, 10.12, 10.13, 10.14, 10.17a, 10.18, 10.19, 10.20, 10.21, 10.23, 11.4c and d, 11.9, 11.15, 11.16, 12.5, 12.8, 12.9, 12.10, 12.11, 12.12, 12.13, 12.14, 12.15, 12.16, 12.18, 12.19, 12.22, 13.5; Tables 7.4, 7.5, 7.6, 12.4; Plates 4a and b, 5, 6), Pergamon Press Ltd. (Figs 2.24, 2.32, 2.33, 5.3, 5.4, 5.5, 6.13, 8.11a), Regiae Soc. Sci. Upsaliensis (Figs 4.7, 4.8), The Royal Society (Figs 4.10a, 4.12, 8.15, 8.16, 9.2, 10.23, 11.3b, 11.5; Table 8.2), Springer-Verlag (Fig. 1.5b), Taylor and Francis Ltd. (Figs 1.1, 1.2, 1.3, 1.4, 3.4, 12.3a).

CONTENTS

CONTENTS

1
SOME BASIC CONCEPTS

1.1. Introduction

A FEATURE of much of experimental surface chemistry over the last 25 years has been the acquisition of information on surface species from kinetic studies usually determined from pressure measurements or mass-spectrometric analysis of the gas phase. The approach is indirect and how the species are bonded to the surface is generally not known with any confidence, and frequently we resort to the asterisk to represent the surface site (*). For example, the dissociative chemisorption of ethylene to give a surface acetylenic species and two hydrogen adatoms might be described by

$$C_2H_4(g) + 4^* \rightarrow \underset{*}{\overset{|}{H}} - \underset{*}{\overset{|}{C}} = C - \underset{*}{\overset{|}{H}} + 2H$$

Clearly, unravelling the mysteries of the asterisk is of paramount significance in understanding the nature of the molecular events involved and there should in principle be no distinction drawn between those interested in the reactivity of atomically clean surfaces and those who are more interested in the characteristics or reactivity of commercial catalysts. Robert Gomer (1953) wrote

'The great practical importance of surfaces is equalled by their purely scientific interest. Although we have skilfully harnessed surface properties, much of our success is the result of luck and intuition and many fascinating problems remain unsolved.'

In spite of over 20 years of further very active research in the field of surface chemistry and catalysis, Gomer's statement is today still largely valid. That is not to suggest that no substantial progress has been made. On the contrary, we are now in a position to ask many more questions than was possible in 1953 and *that is* progress. The mysteries of the asterisk are beginning to unfold; extensive thermochemical data have become available for nearly all metals interacting with simple diatomic molecules; detailed kinetic data, information on the role of surface structure, the influence of surface steps on the reactivity of simple molecules, the ability to define the structure and chemical composition of surfaces, and direct information on the nature of surface bonding in some well-chosen systems are but some examples of the current position in surface chemistry. Coupled with these is a re-awakening of the theoretician who now is being supplied with suitable data for him to exercise his theoretical knowledge. That is indeed by any criterion progress, but we must be cautious as was suggested by Charles Kemball (1970) recently:

'... I would add here a plea to those who are fascinated by ultra-high vacua and by work with single crystals to remember that there is a danger they may become imprisoned in their own ivory towers...'

There is clearly a potential problem here but it is hoped that the more recent developments in research strategy will have a synergic rather than a restricting influence. In a review by Ahlborn Wheeler (1953) the 'clean-surface' and bulk-catalyst approaches are clearly delineated. We hope the distinction will become less apparent within the next decade.

It is not particularly easy to put the subject of surface chemistry, as it currently stands, in a proper perspective for an undergraduate or recent graduate student embarking on research in this area. The problem facing us is that it is important to give an indication of how the scientific frontiers of the subject are advancing and at the same time put these advances in the context of some 50 years of effort into understanding surface science. In this chapter emphasis is therefore laid on some of the more important fundamental aspects of the strategy adopted during this earlier period, and where possible their relevance to present day thinking is shown.

If we are to understand fully molecular events in surface chemistry and catalysis there are three aspects which ultimately have to be considered: *surface structure, chemical constitution of the surface,* and *electron distribution at the surface.* How are the answers to these questions to be achieved? Success or otherwise will depend on the experimental approach adopted and the theoretical framework for understanding the data obtained. We shall emphasize in later chapters these three aspects, the fundamental basis of the experimental approaches, and the measure of success achieved to date in each.

Following the advent of X-ray diffraction for studying structure, molecular models were developed where most of the atoms appeared to pack together like billiard balls with precise atomic radii. The radii were found to depend on the ionization state of the atom, its co-ordination number, and the nature of the interatomic bond, i.e. whether it was covalent, ionic, or metallic. Arising from such studies Goldschmidt and Pauling prepared tables of ionic, covalent, and metallic radii. Atom size and ratio of atomic radii have therefore dominated much of the thinking in structural studies; the role of polarization has frequently been given secondary consideration and this, somewhat ironically, is an area where quantum mechanics has made an important contribution. The gulf between quantum and structural theory is particularly obvious in the field of surface science and many problems still exist. We must therefore accept that, although it would be difficult to improve on Goldschmidt's model of the atom and atomic structure, there is little that is quantitative in the outlook adopted particularly when attempts are made in relating structure to other physical properties of solids. The problem (and this is a general one) therefore is of striking a compromise between the quantitative, rigorous approach on the one hand and the more general and perhaps intuitively attractive on the other. In developing models for surfaces and reactions at surfaces we shall frequently resort to using the billiard-ball description of atoms. It is, however, clear that this approach is pre-quantum mechanical, as in fact is most of our intuition of crystal structures in chemistry.

In view of the very significant impact that studies of electron interaction have had on the development of our understanding of solids and solid surfaces (electron microscopy, electron diffraction, and electron spectroscopy) over the last two decades it is worth reflecting first on two key 'sign-post' experiments. They were both carried out some 50 years ago and designed specifically to probe fundamental aspects of atomic structure. The first is due to Franck and Hertz (1911) and the second to Davisson and Germer (1927).

Franck and Hertz used an apparatus similar to that shown schematically in Fig. 1.1. Electrons were thermionically emitted from a hot tungsten wire W into a glass bulb containing mercury vapour at a low pressure. The electrons were then accelerated by

F I G. 1.1. The Franck–Hertz experiment (After Mott 1972.)

a voltage V_A to a gauze cylindrical grid G. If the voltage V_A is less than the first ionization potential (I_{Hg}) of Hg then no electrons, according to Franck and Hertz, will suffer any inelastic collisions with mercury atoms between the hot wire W and the grid G and the electrons will arrive at G with energy close to V_A. Between the grid and an earthed cylindrical electrode R another small potential, opposing the motion of the electrons, is applied. This is V_R and is the retarding potential (Fig. 1.1.). Provided $V_R < V_A$, i.e. that W is at a negative potential with respect to R, the electrons should arrive at the earthed electrode and be observed as a current. Furthermore, if $V_A > I_{Hg}$, emitted electrons from W will have sufficient energy to ionize the Hg atoms and therefore undergo inelastic collisions, thereby losing energy. Such electrons will not arrive at R. Thus if V_A is increased gradually a sudden decrease in the current is anticipated when $V_A = I_{Hg}$. The results of the experiments are shown in Fig. 1.2 where a drop occurs in the current at about 4·5 V. On increasing the voltage V_A further, sharp drops in current are observed at

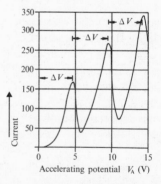

F I G. 1.2. Current–voltage curve in the Franck–Hertz experiment. The first drop occurs at a voltage ΔV and successive ones with further increments ΔV. (After Mott 1972.)

about 9 V and 14 V corresponding to the electrons undergoing two and three collisions with the mercury atoms respectively. In these experiments the pressure of the mercury vapour was such that the electrons undergo several collisions in traversing between W and G. A more sophisticated experimental design which provides more fundamental information on the mechanism of energy loss is the Dymond–Watson (1929) experiment where the electrons can only make one collision.

The Franck–Hertz and Dymond–Watson experiments gave some of the first direct experimental evidence of the existence of stationary states. They were superseded later by photons as impinging 'particles' rather than electrons, and we shall return to this later. We should also not lose sight of the fact that these were the first experimental manifestations of orbital structure and the forerunner of electron spectroscopy as we know it today.

Although the concept of particles having wave-like properties was due to the work of such theorists as de Broglie, Schrödinger, and Born, considerable impetus to their views was obtained from the experimental observations of Davisson and Germer in the U.S.A. and G. P. Thomson in the U.K.

In the experiments of Davisson and Germer (1927) a beam of electrons, accelerated through a potential of about 100 V, was reflected from the surface of a nickel crystal (Fig. 1.3). The electron beam was produced thermionically by the hot filament F; A is a cylinder with a hole which 'focuses' the electrons on to the crystal. The potential of the electrons can be varied by applying a voltage V_0 between A and F. After the crystal was heated (presumably to 'remove' surface contaminants) Davisson and Germer found that the scattered (reflected) electrons were

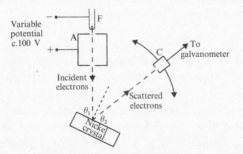

FIG. 1.3. Experimental arrangement of Davisson and Germer. (After Mott 1972.)

only observed when $\theta_1 \simeq \theta_2$. The electron beam was obviously behaving like an X-ray beam. Let us examine this in more detail. Now W. H. and W. L. Bragg had some 13 years previously formulated the Bragg Law (eqn (1.1)), i.e. the rules which govern the reflection of X-rays from solid surfaces:

$$n\lambda = 2d \sin \theta. \tag{1.1}$$

In other words, reflection only occurs when the angle of incidence θ is such that eqn (1.1) is obeyed, λ being the wavelength of the X-rays, d the distance between the crystal planes, and n an integer. Davisson and Germer showed that eqn (1.1) was also valid for the beam of electrons impinging on the crystal surface proving that they were wave-like in character with a wavelength λ given by $\lambda = h/P$, where P is the momentum of each electron and h is Planck's constant. Since the kinetic energy is given by $\frac{1}{2}m_e v^2$ and $P = m_e v$ then $P^2/2m_e$ is the kinetic energy of the electron. Now the kinetic energy is experimentally determined by the accelerating voltage V_0. Therefore $P^2/2m_e = eV_0$, and for V_0 values of 10, 100, 1000 V, λ has values of 0·4, 0·13, and 0·04 nm (4, 1·3, and 0·4 Å) respectively. We should in summary note two points: (a) these experiments showed that a beam of electrons interacting with a solid may be treated as waves (later we shall consider some characteristic properties of such waves and in particular the formulation of these properties in mathematical terms) and (b) that in the context of surface structural studies these results indicated the possible potential of electron diffraction using electrons of low energy (<100 eV) when wavelengths are comparable to interatomic distances (2–4 Å). The reasons why it took so long (some 30 years) for electron diffraction to become an accepted and useful tool in the study of solid surfaces will become clearer in later chapters.

1.2. Electrons in solids

Solids fall into three broad categories: metals which are good conductors of electricity with resistivities of 10^{-4} Ωm, insulators with a resistivity of 10^9 Ωm, and semiconductors with intermediate values, e.g. 1 Ωm. The methods used to describe electron behaviour in solids are analogous to those in gaseous molecules, the crystalline solid being treated as a 'giant molecule'. From the particular viewpoint of electrical conduction we would deduce that in metals the atoms have, as it were, lost their outer electrons and that these electrons are comparatively free to move in the metal. On the other hand, with insulators or semiconductors freedom of movement is restricted and they only move under the influence of considerable external stimulus.

Following Mott (1972), consider a one-dimensional model of a metal made up of a row of sodium or silver atoms which we can think of as 'one-electron atoms' since they have one electron outside the full shells (Fig. 1.4(a)). In order to avoid drawing 3s functions which would be correct for Na, Mott assumes them to be hydrogen atoms. Then the wavefunction with the lowest energy will be just like that for H but extended as in Fig. 1.4(b) throughout the lattice. There are also states which represent a wave moving through the lattice with some velocity V and wavelength λ (Fig. 1.4(c)). Because the electron is moving it must have an energy higher than that represented by the wavefunction shown in Fig. 1.4(b). If the array of atoms (Fig. 1.4) is bent around to form a ring of circumference L so that we can think of the electron as carrying a current, then λ can have values given by $n\lambda = L$ where n is zero or an integer. Obviously n cannot be greater than $N/2$ where N is the number of atoms; if it has this value λ is twice the distance between the atoms and provided the wavefunction is composed of atomic s functions it cannot be smaller.

Now for each value of n (other than zero and $N/2$) two solutions exist for the Schrödinger wave equation, one representing a wave travelling around the hypothetical circle of atoms in a clockwise direction and the other in an anticlockwise direction. There are therefore N states of the kind shown in Fig. 1.4, and

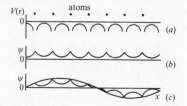

FIG. 1.4. (a) Potential energy of an electron in a crystal. (b) The wavefunction of an electron in the lowest state in a crystal. (c) The wavefunction of an electron moving through the crystal.

since according to the Pauli exclusion principle each state can accommodate *two* electrons, then in a solid composed of 'one-electron atoms' half the states will be occupied and half will be empty. In a solid which has 'two electrons per atom' all states are occupied. Now a characteristic feature of solids (compared with the free atom) is that the electrons have a range of energies; these are referred to as bands. The concept of bands leads naturally to the understanding of why some solids are insulators. If the band is half-filled one can have more electrons moving in one direction than another when a field is applied and therefore a current flows. Therefore 'one-electron atoms' invariably behave as metals in the solid state, with the exception of certain antiferromagnetic materials (such as $TiCl_3$), and one assumes that a band of energy levels is formed from the state of the outer electrons in the atom (e.g. 4 s in copper) and that this band is half-full. The energies within the band are extended over a wide range, several electron volts, up to a limiting energy which is referred to as the Fermi energy (level). Let us now consider how we may formulate the above views in terms of an energy model and what experimental evidence exists for it.

Undoubtedly the most obvious experimental support comes from soft-X-ray spectra, and Fig. 1.5(*a*) shows the spectra observed from the vapour and solid forms of aluminium (Skinner 1938). The spectrum of the vapour consists of lines while that of the solid consists of diffuse bands. How are these results to be interpreted and in what way do they provide support for the band structure of the metal? Let us consider the experimental basis of X-ray spectroscopy.

The specimen under investigation is bombarded by comparatively low-energy electrons usually produced thermionically. If the impinging electrons are of sufficient energy an electron will be ejected from a level in the solid and this vacancy will be filled by an electron falling from a higher level. The energy released will appear as a photon of comparatively long wavelength (low energy) in the region 10–30 nm. This is characteristic of 'soft' X-rays. The emitted radiation is analysed by a diffraction grating and the intensity measured by a photographic plate or a counter. Fig. 1.5(*b*) shows the energy-level diagram of magnesium metal determined in this manner (Tomboulian 1957). At the top (lowest energy) is the broad band (about 8 eV wide) of

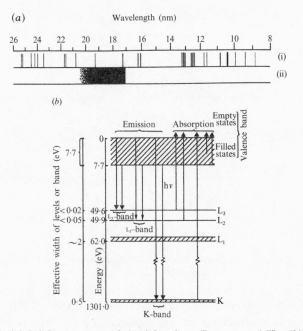

FIG. 1.5. (*a*) Soft X-ray spectra of aluminium from (i) vapour and (ii) solid metal. (After Skinner 1938.) (*b*) Energy-level diagram of magnesium metal. (After Tomboulian 1957.)

closely spaced levels occupied by the valence electrons, and at higher energy the core levels of the 'atomic-like ion–cores' of the lattice. These levels are comparatively narrow and discrete.

It is on the basis of such experiments as these that many of our current views of the metallic state are based. Furthermore they take on a further significance since they are complementary to X-ray-induced *electron-emission* studies. (ESCA or XPS, see Chapter 4). X-ray emission (Fig. 1.5) only enables *differences* between the energies involved to be measured whereas X-ray-induced electron emission maps out each individual atomic level. Comparison of the two sets of data is clearly valuable.

How electrons are distributed over possible energy values is described by $N(\varepsilon)$ curves, where $N(\varepsilon)$ is the number of allowed states in the range ε to $(\varepsilon + d\varepsilon)$. For example Fig. 1.6(*a*) shows that with increasing energy the number of electron states per unit volume of a particular metal increases to a maximum at A,

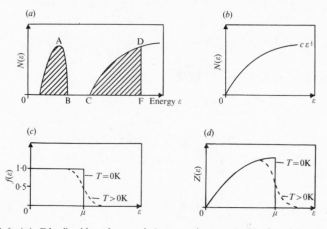

FIG. 1.6. (a) Distribution of occupied states in two bands in a metal. (b) Variation with energy of the number of allowed states in a free electron metal. (c) The Fermi function. (d) Distribution of occupied states in a free electron metal.

decrease to zero between B and C, and then increases again. In other words there are no states between B and C. The shaded area which terminates in the vertical line DF means that the number of electrons present is sufficient at 0 K to fill all energy states up to F. Such a sharp division between occupied and unoccupied states only occurs at the absolute zero of temperature.

The most elementary quantitative treatment of electrons in a simple metal (Sommerfeld 1928) considers the valence electrons within the uppermost, partially filled band to be completely free to move throughout the solid. The interior of the metal is assumed to be a region of constant potential bounded by a barrier (of height ~10 eV) at the surface; this restricts electron emission to a very low value at normal temperatures. For simplicity it will be assumed that the barrier takes the form of a step function, but in reality its shape is more complex owing, among other effects, to the existence of the image potential. The interaction between a metal and an electron at a distance x from its surface may be evaluated by assuming that the electron induces an imaginary positive charge $+e$ in the solid at a distance x below the surface plane. The attractive force F between the electron and its image is $-e^2/(2x)^2$. When the distance x is large compared with interatomic dimensions the surface may be considered homogeneous and the potential energy of an electron at x will be represented by the image potential $V(x)$. Remembering that in general $F = -\mathrm{d}V/\mathrm{d}x$ we have

$$V(x) = -\int_{\infty}^{x} \frac{e^2}{4x^2}\,\mathrm{d}x = \frac{e^2}{4x}. \tag{1.2}$$

This modifies the potential barrier at the surface in the way indicated in Fig. 1.7(a).

The distribution $N(\varepsilon)$ of allowed electron energy states for the free electrons is obtained by solving the Schrödinger equation for a particle in an equipotential box; taking the potential energy of an electron at rest inside the metal to be zero, the result (Fig. 1.6(b)) is $N(\varepsilon) = c\varepsilon^{\frac{1}{2}}$ where c is a constant. Next, the probability $f(\varepsilon)$ that a level of energy ε will be occupied is obtained by applying Fermi–Dirac statistics to the assembly of electrons. It is found that

$$f(\varepsilon) = \left\{1 + \exp\left(\frac{\varepsilon - \mu}{kT}\right)\right\}^{-1} \tag{1.3}$$

where $f(\varepsilon)$ is the Fermi function and μ is the Fermi energy. The number $Z(\varepsilon)$ of electrons actually occupying levels between ε and $(\varepsilon + \mathrm{d}\varepsilon)$ is thus

$$Z(\varepsilon) = N(\varepsilon)f(\varepsilon). \tag{1.4}$$

At a temperature of 0 K all levels of energy less than μ are occupied, since $f(\varepsilon) = 1$ for $\varepsilon \leqslant \mu$; all higher levels are empty. At temperatures above 0 K some electrons acquire thermal energy and are excited above the Fermi level ($\varepsilon_F = \mu$); some levels below this are therefore empty. The Fermi level is now identified as that level for which the probability of occupancy is one-half (Fig. 1.6(c)). Also, the Fermi energy μ represents the Helmholtz free energy of the system, that is the free-energy change on adding an electron to the solid. Therefore, when two or more solids are in equilibrium their Fermi levels must be coincident.

Various types of experiment lend support to this free-electron model, including those relating to thermionic emission, photo-electric emission, and field electron emission. If an electron within the metal conduction band acquires sufficient energy it may be raised to an empty level above the potential barrier at the surface and so escape from the solid. At 0 K the minimum barrier which must be overcome is that from the Fermi level to the vacuum level, the latter representing the energy of an electron at rest at infinity. This energy difference is known as the electronic work function ϕ (Fig. 1.7(a); for convenience, the vacuum level is here taken as energy zero). In the case of thermionic emission the number of electrons gaining sufficient thermal energy to escape may be calculated using the distribution given by the Richardson–Dushman equation

$$J = BT^2 \exp\left(\frac{-\phi}{kT}\right) \tag{1.5}$$

where J is the emission current density, B is a collection of universal constants, and k is the Boltzmann constant. This equation provides a reasonable description of experimental data. Note that the work function may be obtained from a plot of $\ln(J/T^2)$ *versus* $1/T$.

When photoemission occurs electrons surmount the surface barrier by absorbing photons of energy $h\nu$, where h is Planck's constant and ν is the radiation frequency. At 0 K no emission will take place for frequencies below the threshold ν_0, where $h\nu_0 = \phi$. The photocurrent at higher frequencies may be calculated on the basis, once again, of the free-electron picture of the metal; the result predicts the square root of the current to be proportional to ν (Fig. 1.7(b)). In the region of ν_0 the experimental data at $T > 0$ K deviate from this relationship owing to thermal excitation of electrons in the metal to levels above the Fermi level, but the work function of the system may be determined by extrapolating to obtain ν_0 (Fig. 1.7(b)).

Instead of electrons being removed from the solid by raising them above the potential barrier at the surface they may quantum mechanically tunnel through the barrier under suitable conditions. If a field F is applied to the metal the potential of an electron at a distance x outside the surface is no longer constant as in the case of zero field, but decreases as $-eFx$ (Fig. 1.7(c)). The surface barrier is then of finite width, and if the field is increased until the width d at a point opposite the Fermi level has decreased to the order of 1.0 nm, electrons at the Fermi level will tunnel through the barrier. The process is known as field electron

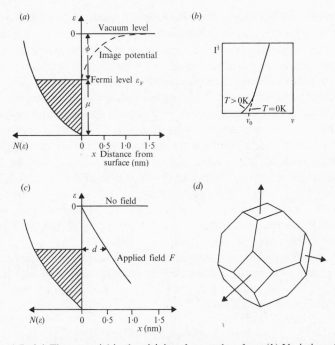

FIG. 1.7. (a) The potential in the vicinity of a metal surface. (b) Variation of the photoelectric yield from a metal with incident photon frequency in the region of the threshold frequency ν_0. (c) The potential in the vicinity of a metal surface in a field emission experiment. (d) The first Brillouin zone of the face-centred cubic structure.

emission. If the field is further increased tunnelling will occur from levels below the Fermi level and the emission current J will increase. On the basis of the free-electron model J is given by the Fowler–Nordheim equation

$$J = BF^2 \exp\left(-\frac{b\phi^{\frac{3}{2}}}{F}\right) \tag{1.6}$$

where B is the same constant as in eqn (1.5) and b is related to the radius of curvature of the emitting surface. In principle a plot of $\ln(J/F^2)$ *versus* $1/F$ will yield a value for ϕ, but in practice b is difficult to determine and so only changes in work function, caused for example by adsorption, can be measured by this method. Field electron emission is utilized in the field emission microscope, discussed in Chapter 6.

We have seen that the work function may be determined by the three techniques just described, but in surface chemistry changes in ϕ which occur as a result of adsorption are of greater interest than the absolute value of ϕ for the clean metal. (The negative of the work function change is also known as the surface potential.) Suppose the movement of charge on adsorption is such that the adsorbate becomes negatively charged with respect to the metal. An electron being removed from the metal must then pass through a double layer of charge at the surface. As long as it is outside this double layer the electron will not be influenced by it, but while between the sheets of charge it will be retarded by both the negative charge in front of it and the positive charge behind. The work necessary to remove the electron will therefore be greater in the presence of the adsorbed layer; adsorption increases the work function in this case. If on the other hand the adsorbate is positively charged with respect to the surface, the work function will be lower than that of the clean metal. The magnitude of the work-function change $\Delta\phi$ is given by the Helmholtz equation

$$\Delta\phi = 2\pi\sigma\mu \qquad (1.7)$$

where σ is the adsorbate concentration and μ is the moment of the dipole formed by the adsorbed particle and its imaginary image. Thus $\mu = 2dq$, where q is the charge on the adsorbate and d is the charge–surface separation. The Helmholtz equation is also often written as $\Delta\phi = 4\pi\sigma\mu$, with μ defined as dq; the relative merits of these two forms are discussed, for example, by Schmidt and Gomer (1965). Eqn (1.7) may be used to obtain values of q, although in practice difficulties can arise because of uncertainty in locating the charge sheets with respect to each other.

In addition to the three emission methods mentioned above, work-function changes can be determined in a number of other ways including the vibrating-capacitor and diode techniques; a comprehensive survey of the subject is given by Rivière (1969).

It was largely as a result of the elegant work of Skinner (1938) that experimental electron-distribution curves became available, and he derived these from the soft-X-ray spectra of the elements. The essential point to remember is that the frequency of the emitted radiation ν is directly related to the change in energy $\Delta\varepsilon$

by $\Delta\varepsilon = h\nu$ where h is Planck's constant. By studying the variation in intensity over the different frequencies of the emission band, it is feasible to derive $N(\varepsilon)$ curves, the intensity curve being a *crude* representation of the electron distribution over the emission band. The bands are usually 1–10 eV wide and naturally with increasing temperature the 'head' deviates more and more from the vertical. At room temperature the diffuse 'head of the band' is as much as a few tenths of an electron volt.

There are just two further points which may be noted. Firstly, each crystal structure gives rise to characteristically shaped $N(\varepsilon)$ curves, and one of the early triumphs of the 'electronic factor' in determining crystal structures was that $N(\varepsilon)$ curves lead to comparatively simple and straightforward explanations of why some particular alloy structure is stable rather than others. This approach, the rigid band model, is discussed in the context of more recent developments by Coles & Caplin (1976). The second point concerns how electrons behave in a *real crystal* of a metal. The free-electron theory described previously ignores the structure of the solid and assumes that the conduction electrons move in a uniform potential; in fact the potential must modulate with a period equal to the lattice spacing in the crystal (Fig. 1.4(a)). To take this into account we consider the reciprocal of the wavelength of the electron, $1/\lambda$, or for ease of notation $2\pi/\lambda$, i.e. the wave number k. The state of the electron can therefore be represented by a point in a k-space diagram in three dimensions. What do the wavefunctions and energy levels look like for the motion of electrons in a periodic one-dimensional potential? Kronig and Penney (1930) solved the Schrödinger equation for the motion of a single electron in this potential, and the wavefunction has the form

$$\psi(x) = \psi_0(x)u(x) \qquad (1.8)$$

where $\psi_0(x)$ is the solution in free space and $u(x)$ is another function which is periodic in the lattice potential. Functions of this kind are called Bloch functions and the solution is a running wave which may be written as a sine or cosine function and more generally as e^{ikx} where $k = 2\pi/\lambda$. This running wave therefore represents the behaviour of the free electron, but in the presence of the periodic potential the wave is modulated to give

$$\psi(x) = e^{ikx}u(x). \qquad (1.9)$$

This indicates that the electron wavefunction has the periodicity of the crystal lattice (cf. Fig. 1.4); this is a very important result. By analogy with Bragg's law of reflection ($n\lambda = 2d \sin \theta$), for each direction of the wavenumber there are critical wavenumbers at which an abrupt increase in the energy takes place. There are therefore forbidden ranges of energy that the electron cannot possess. In a two-dimensional wavenumber diagram the critical wavenumber at which the conditions for a Bragg reflection are satisfied lie on straight lines which form the boundaries of a polygon. Within it the electron energy varies continuously with wavenumber, but on passing through the boundary of the polygon there is an abrupt increase in energy. A polygon of this kind is called a two-dimensional Brillouin zone.

For a three-dimensional crystal a three-dimensional wave number diagram may be constructed with k_x, k_y, k_z representing the components of the wavenumbers in the crystal directions x, y, z. Fig. 1.7(d) shows the first Brillouin zone for the f.c.c. structure. There is of course a close connection between the actual crystallographic structure and the Brillouin zone structure since the latter depends on satisfying the Bragg equation for reflection of electrons from different kinds of atomic planes within the lattice.

Clearly there is an important difference between the electron energy relationship in the free-electron model and in the actual periodic field of a crystal. In the free-electron theory electrons may be added into the 'hypothetical box' with continuously increasing energies. With the real crystal electrons fill up the energy states in wavenumber space—2 per state—but on crossing the zone boundary there is a sudden increase in energy. The process is therefore discontinuous.

There is nothing strange about a gap of forbidden energies for an electron in a metal; similar situations exist for the hydrogen atom. In the latter case we have separate levels corresponding to energy values (eigenvalues) for which solutions ψ of the wave equation can be found. Further more detailed discussions of aspects of the band theory of solids are to be found elsewhere (Thomas 1975, Quinn 1973).

A concept worth reflecting on, since it is frequently referred to in discussions of surface chemistry and catalysis, is that of 'holes in the d band'. In the case of iron for example the electronic structure of the isolated atom is $\boxed{1s^2\ 2s^2\ 2p^6\ 3s^2\ 3p^6}\ 3d^6\ 4s^2$, and we would anticipate that as the atoms are brought together to form the solid a 3d band and a 4s band would be formed. The 3d electrons are nearer the nucleus than the 4s; hence 3d electrons of adjacent iron atoms will overlap other atoms less than the 4s electrons. In energy terms the consequence is that the 3d band is much narrower than the 4s band. The two bands overlap in energy, and when electrons are 'poured' into the bands they reach the same level (energy) in each and we find that 7·78 electrons per atom are in the 3d band and only 0·22 electrons in the 4s band. These values are calculated from the experimental magnetic properties of body-centred iron and we can refer to the 3d band as having 2·22 'electron holes'. This is clearly an over-simplified model, and one obvious point is that, since the 3d and 4s bands overlap in energy, an electron can be neither pure s nor pure d in character. The most appropriate wavefunction is more likely to be some hybrid of d and s states.

1.3. Adsorption

The preferential concentration of molecules at the surface of a solid is referred to as adsorption and must be distinguished from absorption which refers explicitly to penetration of these molecules below the surface of the solid. The relevance of adsorption to interfacial phenomena is obvious and in no field of endeavour is this more clear than heterogeneous catalysis. It is not our intention to give here a detailed account of adsorption, neither is it intended to give up-to-date details of the frontiers which the subject has attained during the last decade. We prefer to confine ourselves to some of the more conceptually important topics formulated over the last 30 or 40 years, interspersed by recent advances which tend to orientate current thinking in the field. In later chapters of the book specific adsorption systems will be considered in some detail, and it will become clear that our views are changing rather rapidly largely as a result of the experimental sophistication that is now possible.

It is invariable that in discussions of adsorption a distinction has been made between physical adsorption and chemisorption. A set of criteria are often used which hopefully allow a demarcation to be made. Physical adsorption is considered to be reversible and non-activated, but chemisorption, irreversible at one temperature, may also be non-activated and reversible at a higher temperature. Furthermore, when these criteria were first adopted there was little information on the crystallographic specificity of adsorption, i.e. how sensitive is adsorption (physical or chemisorption) to the atomic structure of the solid adsorbent. The criteria must therefore be viewed with considerable caution in the light of more recent experimental work and one or two examples will be discussed. We shall return to this later.

Central to much of the discussion of adsorption at solid surfaces during the period 1950 to 1970 has been the use of the Lennard-Jones potential-energy diagram. This is a plot of potential energy *versus* reaction co-ordinate originally conceived by Lennard-Jones (1932, 1937) to account for the complex shape of some adsorption isobars. Rather than showing a maximum adsorption at low temperature (physical adsorption) followed by a continuous decrease with increasing temperature several maxima appeared in the experimental data. Fig. 1.8 shows three Lennard-Jones curves proposed by Gundry and Tompkins (1956) to describe the $Fe + H_2$ adsorption system. Curve A represents the $H_2 + Fe$ interaction, curve B weak adsorption involving hydrogen adatoms, and curve C strong chemisorption. Clearly to go directly from $H_{2(g)}$ to strongly held chemisorbed hydrogen atoms would require, according to this model, a prohibitively large activation energy (E_a, Fig.1.8). On the other hand, by going through the intermediate state (curve B) the activation-energy is reduced. Furthermore Gundry and Tompkins proposed that the two chemisorbed states involved in the one case d orbitals of the surface atoms and in the other dsp hybrid orbitals. The heat of adsorption into the final chemisorbed state is ΔH (Fig. 1.8).

The activation energy for the adsorption of simple molecules such as hydrogen on atomically clean metal surfaces is very small and the $Fe + H_2$ system above is but one example. We note in passing that this is contrary to the early criterion for chemisorption, namely that it is 'an activated process analogous to a chemical reaction.' On the other hand, however, the chemisorption of simple hydrocarbons is frequently a 'slow process' which accelerates with increasing temperature, in other words it conforms to the early activated viewpoint of chemisorption. A variety of reasons have been proposed for the activated nature of the

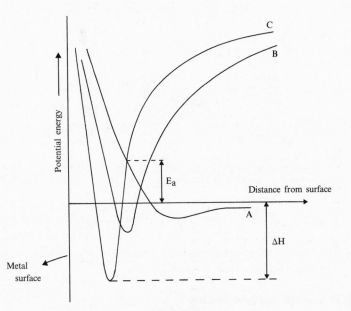

FIG. 1.8. Lennard–Jones potential energy diagram for chemisorption of H_2 on Fe (Gundry and Tompkins 1956).

process including the excitation of phonons and interband transitions in the solid, and excitation of electronic, vibrational, or rotational states of the gaseous molecule. Stewart and Ehrlich (1972, 1975) have recently explored the reasons for activated adsorption of methane on atomically clean rhodium by a combined molecular-beam–field-emission experiment. A schematic diagram of their apparatus is shown in Fig. 1.9(a). With the gas at 300 K no change in the work function of the rhodium crystal was observed (Fig. 1.9(b)) with increasing exposure; the rhodium was at 245 K. Raising the temperature of the rhodium to 400 K also produced no change.

However, on raising the temperature of the CH_4 to 680 K, with the crystal at 245 K, a continuous change in work function occurred (Fig. 1.9(b)). Adsorption has therefore been initiated by heating the CH_4 molecular beam to 500 K or higher. These experiments clearly indicate how excitation of the gas is sufficient to induce dissociative chemisorption of methane even on a 'cold' surface. Stewart and Erlich report an activation-energy barrier of

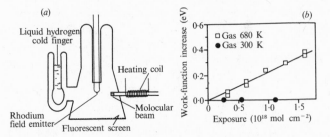

FIG. 1.9. (a) Schematic diagram of molecular-beam field emission microscope with liquid hydrogen cold finger. (b) Change in work function of rhodium crystal as a function of exposure to the molecular beam at two beam temperatures.

the order of $40 \, \text{kJ mol}^{-1}$ ($10 \, \text{kcal mol}^{-1}$), from which they deduce that both electronic and rotational excitation is unlikely leaving as the only possibilities excitation of the translational or vibrational modes of the molecule.

1.4. Surface characterization

There is little doubt that Langmuir (1960), even in his pioneering studies during the period 1912 to 1930, managed to work with 'clean' tungsten filaments. Somewhat later, in the mid-thirties, J. K. Roberts's (1935) very careful studies also deserve mention as furthering the 'clean-surface approach'. These are, however, two exceptions during this early period of development of surface chemistry when much greater effort had been put into understanding industrially relevant, and therefore catalytically interesting, solids. It is worth examining the review by Ahlborn Wheeler (1953) to obtain a perspective of the conflicting interests and viewpoints of the so-called 'clean-surface' and 'bulk-catalyst' schools. He commended the 'bulk-catalyst' approach, since it was indeed from studies with such surfaces that the qualitative chemistry of chemisorption and catalysis emerged. However, he did not lose sight of the unique contribution of Langmuir and Roberts, and from a comparison of the qualitative behaviour of the clean surface and bulk catalyst as adsorbents a set of criteria for atomic cleanliness was drawn up.

Let us examine the meaning of 'atomically clean' (Roberts 1967) by first recalling that there are two sources for potential contamination of a surface. The first is the ambient, and in this context the Hertz–Knudsen equation for relating the impact rate ν with the gas pressure P (eqn (1.10)) is important, m being the mass of the particular gas molecule, T the gas temperature, and k the Boltzmann constant:

$$\nu = \frac{P}{(2\pi mkT)^{\frac{1}{2}}}. \tag{1.10}$$

For nitrogen at $300\,\text{K}$ and a pressure of 10^{-6} torr $\nu \simeq 5 \times 10^{14}\,\text{cm}^{-2}\,\text{s}^{-1}$. It should be noted that ν is directly proportional to pressure, while the influence of temperature and the nature of the gas is less important since they appear in eqn (1.10) as $m^{-\frac{1}{2}}$ and $T^{-\frac{1}{2}}$. Since in general there are about 10^{15} atoms per cm^2 of surface, then assuming each molecular impact leads to adsorption, i.e. that the sticking probability S is unity, it would only take about 1 s to cover an initially 'clean' surface with a monolayer of foreign adatoms. We should recall, however, that for many surfaces S is significantly less than unity; furthermore it decreases with increase in surface coverage (Chapter 6). It has been found convenient to define exposure of a solid to a gas in terms of the Langmuir (L) where $1\,\text{L} = 1 \times 10^{-6}$ torr s.

The second source of surface contamination is the bulk of the solid adsorbent. In this context it is important to realise that a bulk concentration of 1 part per million of impurity when segregated at the surface of a solid of volume $1\,\text{cm}^3$ will give rise to about 20 monolayer equivalents of impurity. Thus the careful selection of high-purity materials is very important and specification to a purity of better than a few parts per million should be aimed at. We now have the facilities for providing 'on-tap' ultra-high vacuum ($\leqslant 10^{-9}$ torr); what is not available are comparably pure solid adsorbents.

Until the last decade a fundamental problem in the study of solid surfaces has been the inability to characterize their composition on the atomic scale. Frequently the only approach available has been to make comparisons of the adsorption characteristics of a metal-catalyst surface with that of a 'standard clean surface', usually the corresponding metal in the form of an evaporated film. By purposely adsorbing contaminants such as oxygen and sulphur the 'real-catalyst' surface could be simulated (Roberts and Sykes 1957, 1958). Clearly the arguments were circumstantial and deductive in nature and no unambiguous model could be proposed. It was nevertheless by this general approach that much of the descriptive surface chemistry was built up. The application of thermodynamic calculations frequently facilitated interpretation. For example in the reduction of NiO by H_2 the equilibrium constant P_{H_2O}/P_{H_2} is about 200 at 700 K. On these grounds it is thermodynamically feasible to reduce NiO to Ni. On the other hand, if the hydrogen had contained some hydrogen sulphide as an impurity, then it is likely that the solid nickel would have been covered by a surface sulphide. This suggestion is based on the fact that the reaction

$$\tfrac{1}{2}Ni_3S_2(s) + H_2(g) = \tfrac{3}{2}Ni(s) + H_2S(g) \tag{1.11}$$

is in equilibrium at about 700 K when the hydrogen contains 128 parts per million of H_2S. The surface sulphide, Ni_3S_2, is clearly much more thermodynamically stable than NiO. Whether it forms or not would be determined by the kinetics of the reaction.

One of the features of the development of improved sophistication of surface characterization has been the many false hopes that have been raised regarding the ultimate criterion. Among the techniques claiming the coveted prize have been work-function and photoelectric-threshold measurements, field emission, flash desorption, the evaluation of accommodation coefficients, and ellipsometry. The central problem has always been that a surface judged to be clean by one experimental approach may not necessarily be so when judged by another.

An interesting example is the observation that when oxygen was chemisorbed on a nickel film at 77 K the work function of the nickel increased by 1·5 eV. However, on warming *in vacuo* to 290 K the work function returned to that of the original value of the 'clean' nickel (Quinn and Roberts 1964, Roberts and Wells 1966). No oxygen was desorbed. Any argument for surface cleanliness, based on the measured work function, would be entirely fallacious and an interpretation of the molecular events involved is given in Chapter 11. Whetten (1965) observed, possibly facetiously, that there appears to be a correlation between the maximum secondary emission yield for MgO and the year in which the measurement was made (Fig. 1.10); clearly this is not the case and the correlation is likely to reflect the general improvement in vacuum techniques leading to cleaner surfaces and therefore higher yields between 1950 and 1958.

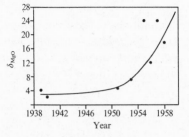

F<small>IG</small>. 1.10. Chronological history of the secondary-electron-emission yield δ of magnesium oxide. (From Whetten 1965.)

In recent years with the advent of Auger electron spectroscopy (AES) and X-ray photoelectron spectroscopy (XPS) it has for the first time been possible to probe directly the chemical composition of solid surfaces at the submonolayer level. There are still problems ranging from charging effects with XPS studies of 'insulator-like' adsorbents to electron-beam effects arising from the electron currents used in Auger electron spectroscopy. These topics are discussed in Chapter 4; it is sufficient to record here examples of the impurities observed by AES (Haas, Grant, and Dooley 1970) to be present on the surfaces of the elements listed in Table 1.1. The measure of the problem is obvious!

As to the choice of the form of the adsorbent, a powder, an evaporated metal film, a wire, or a single crystal, the investigator is restricted by (a) the experimental facilities available to him and (b) the question(s) he sets out to answer. He may be interested in comparing the catalytic activity of a range of metals for a given reaction; in this case he is prepared, rightly or wrongly, to forego the rather stringent experimental conditions inherent in elucidating the role of crystal structure and requiring diffraction and ultra-high vacuum techniques. Similarly he may decide that the electronic structure of a surface species is what is important, and again substrate structure is given secondary consideration and he concentrates on information obtained by various forms of electron spectroscopy. The ideal experiment is that designed to elucidate the influences of substrate structure on the electronic structure of surface species and how these in turn are manifested in catalytic activity. There are but a few experiments of this kind so far reported in the literature, and the gap between them and

TABLE 1.1

Some contaminants observed by Auger electron spectroscopy on various refractory metals after initial short outgassing in vacuum

Metal	S	C	Cl	O	N	Si
Sc	+	+	+	+	?	?
Ti	+	+	+	+	?	−
V	+	+	+	+	?	+
Cr	+	+	−	+	−	−
Fe	+	+	−	+	−	?
Co	+	+	−	+	−	?
Ni	+	+	−	+	−	−
Y	+	+	+	+	−	−
Zr	+	+	+	+	−	?
Nb	+	+	+	+	−	−
Mo	+	+	−	+	−	?
Ru	+	+	?	+	+	+
Rh	+	+	−	−	−	+
La	−	+	+	+	−	−
Hf†	+	+	+	+	−	?
Ta	+	+	+	+	?	−
W	+	+	−	+	−	−
Re‡	+	+	−	−	−	+
Ir	+	+	−	−	−	?
Pt	+	+	−	−	−	?
Au	+	+	−	−	−	−

+ Contaminant found; − not found; ? questionable.

† Zr also appeared at surface on heating Hf.

‡ Traces of Mo also apparently found and were removed by subsequent thermal treatments.

heterogeneous catalysis is still wide but the potential for bridging the gap is now considerable. Some examples illustrating the present optimistic climate in the field are discussed in Chapter 13.

Although there have been extensive studies of the role of defects in determining the reactivity of solids, the work of Thomas and his colleagues (Thomas, Evans, and Williams 1972) being outstanding, there have been few meaningful investigations of the significance of surface defects in influencing the chemical reactivity of metals. The reasons are obvious when cognisance is taken of the problems of experimentation with well-defined metal surfaces. Nevertheless there are copious literature references to the *possible* role of defects, largely on the basis of circumstantial evidence, and there are numerous examples where recourse has

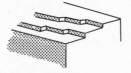

FIG. 1.11. A surface step.

been made to surface defects of one kind or another to explain experimental data. The concept of a surface step (Fig. 1.11) is an old one, the possibility of it being the site of enhanced chemical reactivity almost as old. What is new, however, is *proof* that surface steps are sites of enhanced reactivity. Evidence for such proof could only be achieved by a combination of sophisticated experimental techniques involving low-energy electron diffraction for definition of the surface and mass spectrometry for monitoring the gaseous products (Joyner and Somorjai 1973). We shall return to this later. What has also become clear from low-energy electron diffraction studies is that the adlayer itself may exhibit a number of interesting possibilities for catalysis. In particular, domain boundaries within an otherwise ordered layer may act as centres of enhanced reactivity (May, 1972). Such defects, rather interestingly, do not require a defective adsorbent surface but depend rather on the symmetry and atomic structure of both the adlayer and adsorbent.

1.5. Surface reactivity patterns and bonding

Chemisorption, the chemical 'reaction' between molecules and solid surfaces, must be intimately involved with the electron distribution of the solid adsorbent and the adsorbate, and therefore with the electronic and geometric configurations of the system as it moves from its initial state to its final state through some intermediate. In very simple terms we can describe this by the energy–distance relationship shown in Fig. 1.12, where ΔH is the enthalpy change, E_a the activation energy to form the transition state, and E_d the activation energy for desorption of the adsorbate. It follows that $E_d = \Delta H + E_a$. For a detailed understanding of the molecular events occurring during catalysis a knowledge of the structure and stabilities of the 'intermediates' involved is essential, but for many purposes a qualitative indication of the occurrence or otherwise of chemisorption is sufficient.

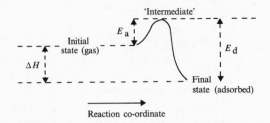

Fig. 1.12. Relationship between heat of adsorption (ΔH), activation energy of desorption (E_d) and activation energy of adsorption (E_a).

By comparing the activity of metal surfaces, usually prepared by evaporation in a high vacuum on to a glass substrate, Trapnell (1953) and his colleagues classified metals according to their ability to chemisorb O_2, CO, H_2, N_2, C_2H_4, and C_2H_2. The metals (Table 1.2) divide into seven groups designated in order of decreasing activity, A to E, and with few exceptions high activity is associated with transition metals. No transition metal is in the less active groups C to E, while groups A to B_2 contain exclusively transition metals.

TABLE 1.2

A classification of metals according to their abilities in chemisorption

Group	Metal	Gases						
		O_2	C_2H_2	C_2H_4	CO	H_2	CO_2	N_2
A	Ti, Zr, Hf, V, Nb, Ta, Cr, Mo, W, Fe, Ru, Os	+	+	+	+	+	+	+
B_1	Ni, Co	+	+	+	+	+	+	−
B_2	Rh, Pd, Pt, Ir	+	+	+	+	+	−	−
B_3	Mn, Cu	+	+	+	+	±	−	−
C	Al, Au	+	+	+	+	−	−	−
D	Li, Na, K	+	+	−	−	−	−	−
E	Mg, Ag, Zn, Cd, In, Si, Ge, Sn, Pb, As, Sb, Bi	+	−	−	−	−	−	−

+ Strong chemisorption occurs; ± Weak; − Unobservable.
After Bond 1974.

A direct consequence of this pattern of activity was the suggestion that chemisorption involves covalent bonding with the partly filled d band or with unpaired electrons in d orbitals (Dowden 1950, Trapnell 1953). With non-transition metals such d bands (orbitals) are completely filled and therefore inactive. It is therefore envisaged that the dissociative chemisorption of hydrogen, and there is good evidence for this, occurs by the formation of two covalent metal–hydrogen bonds. By simple thermodynamic arguments we can see that for the process

$$H_2(g) \rightarrow \begin{array}{cc} H_{(ads)} & + H_{(ads)} \\ | & | \\ M & M \end{array}$$

to be thermodynamically feasible (i.e. ΔH is exothermic if we neglect entropy changes) the condition is that $D_{M-H} \geqslant D_{H_2}/2$, i.e. $D_{M-H} \geqslant 214\,kJ$. In other words the strength of the adatom–substrate bond (D_{M-H}) must be at least half the bond energy of H_2. By such arguments we can rationalize why one of the most specific molecules in chemisorption is N_2. Table 1.2 shows that it is only chemisorbed (as N adatoms) with metals of Group A. The nitrogen bond energy is about 1000 kJ, and therefore the thermodynamic condition that dissociative chemisorption of nitrogen occurs is that $D_{M-N} \geqslant D_{N_2}/2$ and therefore $D_{M-N} \geqslant 500\,kJ$. Clearly the stringency for nitrogen adatom formation is much more severe than with hydrogen. In the case of nitrogen, dissociative chemisorption can only occur provided the metal–nitrogen bond formed is *at least* 500 kJ, whereas metal–hydrogen bonds can form when the M–H bond strength is as low as 214 kJ. Such arguments as these led to the development of the d-band theory of chemisorption. However, an alternative viewpoint is that the *activation energy* for dissociative chemisorption is smaller with transition metals than with sp metals. There is in fact some support for this in that when the activity of metals in the chemisorption of *hydrogen atoms* is considered (Roberts and Young 1970, Hayward, Herley, and Tompkins 1964) the distinction between transition and non-transition metals disappears. For example the sp metal lead chemisorbs hydrogen atoms (produced by the dissociation of $H_2(g)$ at a hot tungsten filament) at 77 K and at 290 K a solid hydride of overall stoichiometry $PbH_{0.2}$ is formed in the surface region (Roberts and Wells 1964).

The surface chemistry of solids therefore conforms to patterns clearly related to the periodicities inherent in Mendeléeff's periodic table and hence to the electronic structures of the substrate. In the chemistry of metal surfaces there is a clear distinction between transition and non-transition metals, and this is particularly obvious in Table 1.2 when we consider the reactivity pattern based on the dissociative chemisorption of hydrogen. The activity patterns of metals, oxides, and sulphides for the reactions of hydrogen show similar 'twin-peak' behaviour (Fig. 1.13). This is a strong indication of the importance, directly or indirectly, of electronic configuration and symmetry in surface chemistry. It is fair to state that, although the experimental evidence has been frequently impressive, support for the electronic theory is essentially qualitative. Nevertheless the correlation of activity and position in the periodic table leads naturally to a consideration of energy levels in metals. If only electrons in

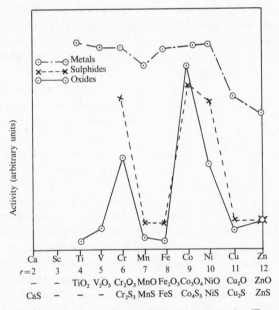

FIG. 1.13. The activity of transition metals and compounds. (From Dowden 1971.)

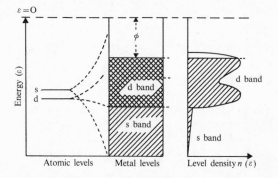

FIG. 1.14. Energy levels in atoms and metals (diagrammatic). (From Dowden 1971.)

excess of the inert gas configurations are considered (the 'valence electrons') and counted in the electron/atom ratios r, then at 0 K these electrons occupy levels in pairs up to the Fermi level (ε_F). The properties of the metals are therefore dependent on the energy density of levels $N(\varepsilon)$ as a function of energy (ε) and upon the extent to which the levels are filled (Fig. 1.14). Both theory and experiment indicate a sharp drop in the density of levels at $r \simeq 10\cdot5$ in each long period and particularly for the f.c.c. series of binary solid solutions formed between Ni, Pd, and Pt with Cu, Ag, or Au. Also, there has been some indication of a minimum in the middle of the long periods so that the density-of-states (levels) curve is frequently shown as twin peaked (Fig. 1.14). The analogy with Fig. 1.13 is clear.

In the field of heterogeneous catalysis as opposed to chemisorption a number of other correlations have been recognized. These have been summarized by Bond (1974) and two are shown here in Fig. 1.15. The 'volcano curve' is particularly interesting since it emphasizes that for efficient catalysis the adsorption should neither be too weak nor too strong.

Dowden (1950) one of the earliest protagonists of the electron theory, has recently reviewed (1972) progress over the last 25 years. In this he lays particular emphasis on the concept of 'ensembles' and discusses it in the context of electron-band filling. He concludes that the surface chemistry of completely random binary alloys, on which much of chemisorption theory is

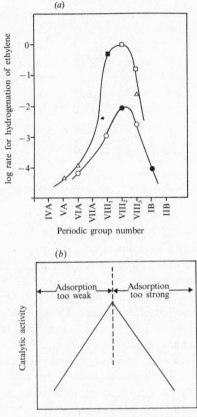

FIG. 1.15. (a) Periodic variation of log rate for hydrogenation of ethylene expressed relative to rhodium: Open symbols, evaporated metal films: filled symbols, silica-supported metals: $\bigcirc$, $\square$, $\triangle$, first, second and third row transition metals, respectively. (b) Diagrammatic representation of a 'volcano' curve. (From Bond 1974.)

based, requires detailed investigation. These conclusions are particularly apposite in view of the fact that recent electron spectroscopic studies have shown conclusively that in many alloy systems the composition at the surface is quite different from the bulk. In general it is found that it is the component of lowest sublimation energy that segregates to the surface. This is discussed in Chapter 13. The implication that this recent finding has for the interpretation of experiments with alloys, from which many of our current views stem, is obvious. It would therefore seem particularly appropriate with the experimental facilities now available to re-investigate some of these novel ideas which have played such an important role in the development of research in catalysis.

There have therefore been two extreme approaches to the problem of attempting to link up electronic structure and surface reactivity: the collective-electron or band model and the localized-bonding or valence-bond model. Bond (1966), following the work of Trost (1959) and Goodenough (1963), appreciated the possibility of formulating surface bonding in terms of molecular orbitals which have directional properties because of the crystal field. Confining attention to the f.c.c. metals he showed how it was possible to predict the direction of the emergence of two kinds of orbital at various faces of an f.c.c. crystal. Fig. 1.16 shows Bond's diagramatic representation of the emergence of orbitals from the (100) face of an f.c.c. metal. The application of this approach to the chemisorption of hydrocarbon molecules is appealing since it brings together thinking in two distinct fields: metal organic chemistry and heterogeneous catalysis with consequent links with homogeneous catalysis.

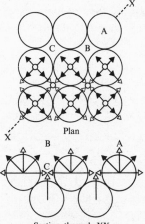

Plan

Section through XX

FIG. 1.16. Representation of the emergence of orbital lobes in the (100) face of f.c.c. metals: open circles, atomic orbitals emerging at right angles to plane of paper; open arrowheads, atomic orbitals within plane of paper; solid arrowheads, bonding orbitals within plane of paper in elevation but emerging at 45° to it in the plan. (After Bond 1966.)

Again experimental information has been lacking, but with the recent application of ultra-violet photoelectron spectroscopy to surface studies, and in particular angular distribution measurements, new and very exciting possibilities emerge. The future is clear from the work of Grimley (1974), Plummer and Liebsch (1974), Smith, Traum, and Di Salvo (1974), Gadzuk (1974), and Lloyd, Quinn, and Richardson (1975). Already experimental results of high quality are emerging, and the next five years are likely to bring very valuable and hitherto unattainable information on the surface bond. Such experimental work will undoubtedly go hand in hand with, and also give direction to, theoretical work. One of the earliest reported examples of the power of angular photoemission studies was that by Smith and Traum (1974) using TaSe$_2$. The results (Fig. 1.17) are particularly appealing to the surface chemist since the directional properties of the bonding are clearly to be seen.

Valence-band spectra of metals and alloys obtained by electron spectroscopy provide information on the density of states, particularly near to the Fermi level $N(\varepsilon_F)$. One interesting study is that of Bennett *et al.* (1974) who explored whether the similarity between tungsten carbide and platinum as a catalyst for the isomerization of 2,2-dimethyl propane is explicable in terms of their respective $N(\varepsilon_F)$ data. The density of states at ε_F for WC is

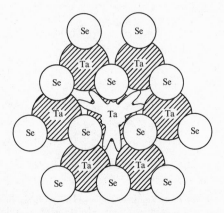

FIG. 1.17. Angular dependence of photoemission from TaSe$_2$ interpreted in terms of its structure. (From Smith and Traum 1974 and references therein.)

intermediate between that for W(111) where ε_F is in a region of low density of d-like states and Pt where ε_F is in a high-density region of d states. The authors conclude that a high $N(\varepsilon_F)$ is a necessary but not a sufficient condition for high catalytic activity. For example, the density-of-states argument alone would suggest that such metals as V or Ta would be efficient catalysts whereas they are not in practice.

1.6. Thermochemistry of surface processes

The adsorption of a gas on to the surface of a solid results in a change in the thermodynamic functions which characterize the system. In all physical adsorptions and most chemisorptions the processes are exothermic. This is understandable, since for any spontaneous process the free-energy change ΔG is negative and, since there is a loss of rotational and translational entropy of the adsorbate on adsorption, ΔH must also be negative according to the Gibbs–Helmholtz relationship

$$\Delta G = \Delta H - T \Delta S.$$

This is not to suggest that endothermic adsorption processes are ruled out; on the contrary there is evidence that hydrogen chemisorption on glass leads to hydrogen adatoms with complete two-dimensional mobility. There is in this case a gain in entropy, i.e. ΔS is positive and $T \Delta S$ large and negative, so that a positive ΔH term (i.e. endothermic adsorption) can exist with an overall negative change in the free energy. Such cases of endothermic adsorption are rare and have been discussed by Schwab (1957) and De Boer (1957). We should, however, not lose sight of the fact that the H adatoms are only endothermic with respect to $H_2(g)$ but are exothermic with respect to $H(g)$. This point appears to be frequently overlooked.

We therefore have established that the enthalpy change is an important quantity in defining an adsorbate–adsorbent system; furthermore it is related to the energy of the bonds formed as a result of adsorption. It must be emphasized, however, that the relationship between bond strength and enthalpy is not a simple one, but this has not deterred the evolution of empirical relationships (Roberts 1960, Tanaka and Tamaru 1963) which enable estimates to be made of the enthalpy of surface processes by

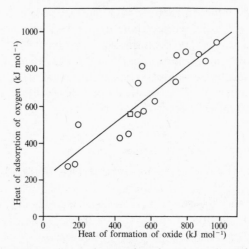

FIG. 1.18. Dependence of heat of adsorption of oxygen on heat of formation of most stable oxide at 25 °C. The square point is for the nitrogen–tantalum system. (From Bond 1974.)

recourse to, and analogy with, thermodynamic data for the corresponding bulk compound (Fig. 1.18). More recently Miyazaki and Yasumori (1976) have modified the Pauling–Eley formulation for estimating heats of chemisorption; they have achieved good agreement between calculated and experimental data for oxygen and nitrogen chemisorption.

The two most useful thermodynamic definitions of the enthalpy of adsorption are the differential heat of adsorption q_d defined by

$$q_d = \left(\frac{\partial q_{int}}{\partial n_A}\right)_T$$

and the isosteric heat of adsorption q_{st} defined by

$$\left(\frac{\partial \ln P}{\partial T}\right)_\theta = \frac{q_{st}}{RT^2}$$

where q_{int} is the integral heat of adsorption for n_A moles of gas and P is the equilibrium pressure for a coverage θ at temperature T. It can be shown that

$$q_{st} = q_d + RT$$

so that at room temperature the difference between isosteric and differential heats of adsorption is about $2 \cdot 5 \, kJ \, mol^{-1}$. This is no more than the error usually involved in the measurement of q_{st} and q_d.

Isosteric heats of adsorption may be calculated from a series of adsorption isotherms at different temperatures. A prerequisite is, however, that the adsorbate–adsorbent system has attained a thermodynamically reversible equilibrium. In such cases it is then possible to apply the Clausius–Clapeyron equation for determining the heat of adsorption:

$$q_{st} = RT^2 \left(\frac{\partial \ln P}{\partial T} \right)_\theta = -R \left(\frac{\partial \ln P}{\partial (1/T)} \right)_\theta$$

Fig. 1.19 shows schematically adsorption isotherms drawn at two temperatures T_1 and T_2, where $T_1 > T_2$. If at coverage θ_1 the equilibrium pressures at T_1 and T_2 are P_1 and P_2, then the isosteric heat of adsorption is q_{st} where

$$\ln \frac{P_1}{P_2} = \frac{q_{st}}{R} \left(\frac{1}{T_1} - \frac{1}{T_2} \right).$$

Thus the variation of the heat of adsorption with coverage can be determined. One point to note is that, since the isosteric heat of adsorption varies with temperature as required by Kirchhoff's Law

$$q_{st} = q_{st(0)} + \int_0^T \Delta C_p \, dT \tag{1.12}$$

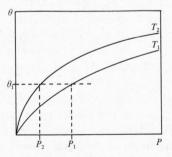

FIG. 1.19. Schematic diagram of adsorption isotherms at two temperatures T_1 and T_2.

where ΔC_p is the difference in the heat capacities of the adsorbate in the gas phase and in the adsorbed state at a constant pressure, the temperature range of the isotherms used for the calculation of isosteric heats should be small. Another obvious restriction is where there is a change in the nature of the bonding over the temperature range used, although as a qualitative guide to molecular processes at surfaces the information would be useful.

The simplest method of determining the heat of adsorption via isotherm data is by the classical volumetric method, but other methods have been developed particularly where problems arise from accurate pressure measurement when vapours are used. In this context gas chromatographic methods have been particularly helpful, the column being packed with the adsorbent and the adsorbate introduced into the carrier gas. The shape of the adsorbate peak emerging from the column is dependent on the way in which the adsorbate interacts with the stationary phase and therefore provides information on the form of the adsorption isotherm. Methods of peak-shape analysis in the context of adsorption isotherms have been proposed by Wilson (1944) and Glueckauf (1949) who followed arguments used in equilibrium chromatography where the column is considered to be ideal and the adsorption process reversible.

It is frequently the case, however, that equilibrium conditions are not maintained, so that the thermodynamic approach to determining enthalpy changes cannot be applied. In such cases two further methods are open to us: (a) the kinetic method and (b) the application of calorimetric techniques. Two modifications of the kinetic method are the flash-filament desorption technique and the temperature-programmed desorption method; both are inherently identical in philosophy, the former being confined to filaments and the latter to metallic powders. The fundamental basis of these two experimental methods is given in detail in Chapter 6.

The relationship between the enthalpy of adsorption ΔH and the surface coverage θ is one of the most important in surface chemistry, and calorimetric determination of the dependence of ΔH on θ has been central to the development of the subject. We shall not consider here the details of calorimeter design but draw attention to a number of distinct calorimetric approaches.

1.6.1. *Isothermal calorimeters*

The design is such that the adsorption cell and the surrounding isothermal shield are coupled so that a temperature differential between them exists. In the constant-heat-exchange calorimeter described by Kiselev, Dzhigit, and Muttik (1962) the adsorbent is heated by pulses of current in a heater, this maintaining a constant temperature difference between the adsorption cell and the isothermal outer jacket. For an exothermic adsorption (or reaction) the frequency of the heating pulses is decreased. The variation of the heating current during adsorption enables the enthalpy change to be determined.

1.6.2. *Adiabatic calorimeters*

In this case the temperature of the (adiabatic) shield follows closely that of the calorimeter itself. Central to the determination of enthalpy changes by this approach is knowledge of the heat capacity of the calorimeter and the adsorbent. Kington and Smith (1964) have drawn particular attention to the problem arising from the variation of the heat capacity during adsorption, i.e. as a function of coverage. Electrical calibration is therefore essential at all stages of adsorption; this is not easy since it can effect desorption. A typical design for an adiabatic calorimeter has been described by Morrison and Los (1950).

Most so-called adiabatic calorimeters are, however, not truly adiabatic, since as soon as adsorption commences a temperature gradient will exist between the inner vessel and the surrounding shield leading to heat losses. The decrease in the calorimeter temperature will follow Newton's law. Provided, however, that the change in enthalpy on adsorption is virtually instantaneous the heat evolved at time zero can be determined to a good degree of accuracy.

The most frequently used calorimeter design for adsorption studies has been that due to Otto Beeck (Beeck, Cole, and Wheeler 1950). The particular point of interest is that the adsorbent is an evaporated metal film which is deposited on the inner surface of the glass calorimeter by evaporation from a filament. The temperature increase during adsorption is usually followed by a resistance thermometer wound around the calorimeter vessel. Calibration is again achieved by passing a known current through a heater wound around the calorimeter. The main problem with the Beeck-type calorimeter is the non-uniform distribution of adsorbate over the film adsorbent so that ΔH versus θ curves may be misleading. The calorimeters are frequently cylindrical in shape and the gas introduced from the top; clearly, banding of adsorbate may occur along the length of the cylindrical adsorbent (film). The extent of this banding will depend on the sticking probability of the adsorbate. Extensive application of the Beeck-type calorimeter has been described by Brennan, Hayward, and Trapnell (1960), Wedler and Brocker (1966), and Czerny and Ponec (1968), while Gale, Haber, and Stone (1962) have described a modification for studying granulated catalysts.

One of the earliest attempts to determine heats of adsorption on non-porous adsorbents was that of J. K. Roberts (1935) who used a tungsten-filament adsorbent as the calorimeter itself. The rise in temperature of the wire on adsorption and its subsequent cooling are followed by monitoring the resistance of the filament which forms one arm of a Wheatstone bridge. More recently Kisliuk (1959) and Eley and Norton (1970) have studied nitrogen chemisorption on tungsten filaments and hydrogen chemisorption on nickel respectively. This method has the advantage of being able to clean the adsorbent by high-temperature heating and also of compatibility with ultra-high vacuum techniques. A further development in this area has been the application of Auger electron spectroscopy to monitor the surface species directly and recently the $W+N_2$ system has been studied in this way (Joyner, Rickman, and Roberts 1974). A particular advantage of the approach used was that both kinetic and thermodynamic parameters of the adsorbed phase could be evaluated. The data may be compared with the calorimetric results of Kisliuk (1959) determined under non-equilibrium conditions.

In the context of heterogeneous catalysis we are not only interested in the heat of adsorption of a single adsorbate, but also the interaction between two or more adsorbed species and possibly the interaction between a surface species and a gaseous molecule. Stone (1962) has given a detailed discussion of the application of adsorption calorimetry for such studies. In particular, Tian-Calvet microcalorimeters are very useful as described by Gravelle and Teichner (1969). Fig. 1.20 shows the differential heats of reaction for mixtures $(CO+\frac{1}{2}O_2)$ over anatase at 723 K

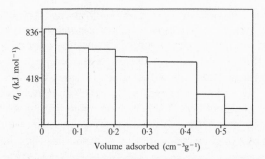

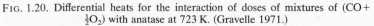

Volume adsorbed $(cm^{-3}g^{-1})$

FIG. 1.20. Differential heats for the interaction of doses of mixtures of $(CO + \frac{1}{2}O_2)$ with anatase at 723 K. (Gravelle 1971.)

which has been previously reduced by carbon monoxide. The heats are very much higher than the normal heat of combustion of CO (285 kJ mol^{-1} of CO) but they are similar to the differential heats of adsorption of $O_2(g)$ on anatase at the same temperature (723 K).

Czerny and Ponec (1968) and Gravelle (1971) have reviewed in detail the application of thermochemistry to problems in surface chemistry.

1.7 Kinetic studies

The kinetic approach, although a valuable one, must be regarded as indirect in the sense that reaction mechanisms are largely derived by inference, there being no direct monitoring of the molecular events occurring *at the surface*. Nevertheless, by exploring in detail the influence of pressure and temperature on the observed rate valuable conclusions may be drawn. The general rate equation for the catalytic reaction

$$A(g) + B(g) \xrightarrow{\text{catalyst}} AB(g)$$

would in very simple terms be of the form

$$\frac{dP_{AB}}{dt} = k(P_A)^x(P_B)^y \exp\left(\frac{-E}{RT}\right) \qquad (1.13)$$

where x and y are the pressure dependences of the reaction, E the observed activation energy, T the temperature, R the gas constant, and k a term frequently regarded as a constant but which need not be so. We shall comment later on possible interpretations of k. Clearly such expressions do not include the concentration of *surface reactants* but only the dependence of the observed rate of reaction on the gas phase concentration.

The simplest heterogeneous reaction may be thought of as occurring by either a Langmuir–Hinshelwood mechanism (eqn (1.14)) or an Eley–Rideal mechanism (eqn (1.15)). Central to a detailed analysis of the kinetic data is a precise knowledge of the terms B(ads) and A(ads) and in general this has been rarely attainable:

$$A(g) + B(g) \xrightarrow{\text{catalyst}} A(ads) + B(ads) \longrightarrow AB(g) \qquad (1.14)$$

$$A(g) + B(g) \xrightarrow{\text{catalyst}} A(g) + B(ads) \longrightarrow AB(g) \qquad (1.15)$$

The approach adopted has been analogous to that used in gas-phase homogeneous kinetics. If, for example, the rate of formation of AB was found to be independent of the pressure of B and directly proportional to the pressure of A, we could conclude that the *surface concentration* of B is independent of pressure, over the particular range investigated, while that of A varies linearly with pressure. A possible conclusion is that B is strongly adsorbed while A is weakly adsorbed. We might then proceed to enquire whether k (eqn (1.13)) is dependent on the surface coverage of either reactants A and B.

The essence of the above approach is, however, too simple, as emphasized by Tompkins (1974) recently. If, for example, B is a diatomic molecule which is chemisorbed dissociatively and A is molecularly adsorbed, the former would require two adjacent 'catalytic sites', whereas the adsorption of A is less stringent and may require only a single 'catalytic site'. Depending on the sticking probabilities of A and B it is conceivable that the concentration of A could therefore increase substantially during the course of the reaction and put in jeopardy the simple analysis inherent in eqn (1.13). Clearly we can adopt this approach (eqn (1.13)) under steady-state conditions only. There is in fact substantial experimental evidence for variation in the terms k, x, y, and E occurring during the course of surface reactions. Such variations offer important clues regarding mechanism and have also implications for the industrialist interested in the formulation and performance of catalysts. A decrease in k and/or an increase in E would be manifested as catalyst poisoning and would be reflected in chemical and/or physical changes in the nature of the catalyst.

An interesting and particularly illustrative example of this point is the recent work of Van Herwignen and De Jong (1974), who showed that the thermal treatment of a copper/zinc oxide CO-shift catalyst in a reducing gas causes a decrease in the measured surface area as well as partial reduction of zinc oxide followed by the formation of α-brass. The formation of the brass leads to poisoning of the catalyst, and this is reflected in the observed correlation between the rate of the CO conversion and the concentration of zinc in the copper-rich phase. The authors in developing their argument considered first the thermodynamics of the reduction of CuO and ZnO:

$$CuO + H_2 = Cu + H_2O \qquad (1.16)$$

$$\Delta G_{473\ K} = -107\ kJ\ mol^{-1}$$

$$ZnO + H_2 = Zn + H_2O \qquad (1.17)$$

$$\Delta G_{473\ K} = +80\cdot 6\ kJ\ mol^{-1}.$$

For CuO, reduction is thermodynamically favourable whereas this is not the case for ZnO at 473 K. They also considered the thermodynamics of the formation of the α-phase of brass:

$$y\,Cu + ZnO + H_2 \rightleftharpoons \alpha\ Cu_yZn + H_2O. \qquad (1.18)$$

The results in Fig. 1.21 show that the formation of brass is thermodynamically feasible when applying conditions similar to those prevailing during the reduction of the CO-shift catalyst where values of up to 1000 for the pressure ratio P_{H_2}/P_{H_2O} can occur.

We therefore have a good example of how back-up thermodynamic data can give a clear lead as to the molecular interpretation of catalyst poisoning.

1.7.1. *Absolute rate theory*

There are a number of different approaches to the interpretation of the pre-exponential factor in an Arrhenius-type rate expression (rate $= k \exp(-E/RT)$) for a surface reaction. One of the most successful was the absolute rate theory (Laidler 1954), the keystone of the theory being the assumption that the reactant molecules are in a transition (activated) state in statistical equilibrium with molecules in the gas phase. It should be emphasized that even when equilibrium is not established between reactants and products, the transition complex is assumed to be in equilibrium with the reactants (Fig. 1.12). For a simple process involving reaction of a gaseous molecule A at a single site

$$A(g) + solid \rightarrow activated\ complex \rightarrow products$$

the equilibrium between the initial and activated states is given by

$$\frac{C^{\neq}}{C_g C_s} = \frac{Q^{\neq}}{Q_g Q_s} \exp\left(\frac{-E_0^0}{RT}\right) \qquad (1.19)$$

The concentrations C_g, C_s and $C^{\neq}$ refer to gaseous reactant, vacant surface sites and activated complexes, respectively, while the Q's are the corresponding partition functions; E_0^0 is the energy at 0 K of the activated complex with reference to the reactants, all in their standard states, and it is therefore the activation energy at that temperature. Using the normal reasoning of absolute rate theory it follows from eqn (1.19) that the rate of reaction is given by

$$Rate = C_g C_s \frac{kT}{h} \frac{Q^{\neq}}{Q_g Q_s} \exp\left(\frac{-E_0^0}{RT}\right) \qquad (1.20)$$

There are situations in which eqn (1.20) may be simplified. For example, if the surface is only sparsely covered by A the concent-

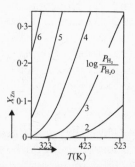

FIG. 1.21. Equilibrium composition for formation of α-brass, reaction (1.18). X_{Zn} is the mole fraction of zinc.

ration of vacant sites C_s will be approximately equal to the total site concentration and so will be virtually independent of C_g. The reaction will then be first order with respect to A(g). On the other hand, if the surface is almost completely covered by adsorbed molecules the concentration of vacant sites will vary with gas pressure. To obtain a value for C_s in this case we may consider that equilibrium exists between A in the gas phase and on the surface and write an adsorption isotherm in the form

$$\frac{C_a}{C_g C_s} = \frac{Q_a}{Q_g Q_s} \exp\left(\frac{-\Delta U_0^0}{RT}\right) \tag{1.21}$$

where C_a, Q_a are the concentration and partition function for the adsorbed species and ΔU_0^0 is the standard internal energy change for adsorption at 0 K. From eqns (1.20) and (1.21) it follows that the rate is given by

$$\text{Rate} = C_a \frac{kT}{h} \frac{Q^{\neq}}{Q_a} \exp -\left\{\frac{E_0^0 - \Delta U_0^0}{RT}\right\}$$

$$= C_a \frac{kT}{h} \frac{Q^{\neq}}{Q_a} \exp\left(\frac{-E_0^0(\text{ad})}{RT}\right) \tag{1.22}$$

where $E_0^0(\text{ad})$ and E_0^0 are the activation energies at 0 K for formation of the activated complex from the adsorbed state and from the gas phase, respectively. If we restrict attention to the case in which both adsorbed particles and activated complexes are immobile on the surface then $Q^{\neq} = Q_a = 1$ and eqn (1.22) becomes

$$\text{Rate} = C_a \frac{kT}{h} \exp\left(\frac{-E_0^0(\text{ad})}{RT}\right) \tag{1.23}$$

This is similar to the Polanyi-Wigner equation frequently found to describe, reasonably accurately, rates of desorption with a pre-exponential factor of $10^{13} \, \text{s}^{-1}$ (see Section 6.5). Further restrictions or conditions of adsorption such as dissociation, requiring dual sites, are considered in detail by Laidler (1954).

Although the absolute rate approach to heterogeneous reaction kinetics has been frequently successful and always stimulating, it has serious drawbacks as a platform for the future development of this aspect of surface chemistry. We also draw general attention to the problem of ascertaining pressure dependence and activation-energy data for surface processes (e.g. eqn (1.24))

$$\frac{dP_{AB}}{dt} = k_\theta P_{\theta_A}^x P_{\theta_B}^y \exp\left(\frac{-E_\theta}{RT}\right) \qquad (1.24)$$

when the activation energy E and the pre-exponential factor k are both coverage dependent. In the case of H_2S dissociation on iron surfaces the initial rate when θ is changing rapidly appears not to be first order. However, when θ is changing only slowly so that the pressure dependence can be determined at virtually constant θ the kinetics are accurately first order (Roberts and Ross 1966). Other examples include H_2S interaction with Pd (Roberts and Ross 1969) and CO desorption from W as discussed by Goymour and King (1973).

When considering the rates of heterogeneous reactions, and particularly of chemisorption, it is helpful to compare the observed rate with the molecular impingement rate ν. The magnitude of ν is related to the gas pressure P, the molecular weight of the gas, and the temperature T by eqn (1.10). For oxygen at 295 K and a pressure of 10^{-6} torr, $\nu \simeq 5 \times 10^{14}$ impacts $cm^{-2}\, s^{-1}$. Since the surface atom density is usually about $10^{15}\, cm^{-2}$, it is clear that the surface would be covered completely in 1–2 s if every impacting molecule was adsorbed successfully. The probability of a molecule being adsorbed on impact with the surface is termed the sticking probability S, and depending on the nature of the surface and the temperature S can approach unity (CO + W at 77 K) or be as small as 10^{-14}(!) for O_2 on $(000l)$ surfaces of graphite (Barber, Evans, and Thomas 1973). The determination of the sticking probability is discussed in Chapter 6, attention being given to the situation when desorption is also a significant factor. Although it is easy to envisage from the above discussion that surface contamination may be very serious under some experimental conditions, less attention has been given to the role of surface segregation in perturbing surface composition. We should not lose sight of the fact that a bulk impurity concentration of 1 in 10^6 when segregated at the surface would be equivalent to some 20 monolayers on a typical single-crystal surface.

REFERENCES

BARBER, M., EVANS, E. L., and THOMAS, J. M. (1973). *Chem. Phys. Lett.* **18**, 423.

BEECK, O., COLE, W. A., and WHEELER, A. (1950). *Disc. Faraday Soc.* **8**, 314.

BENNETT, L. H., CUTHILL, J. R., McALISTER, A. J., ERICKSON, N. E., and WATSON, R. E. (1974). *Science* **184**, 563.

BOND, G. C. (1966). *Disc. Faraday Soc.* **41**, 200.

—— (1974). *Heterogeneous catalysis: principles and applications.* Clarendon Press, Oxford.

BRENNAN, D., HAYWARD, D. O., and TRAPNELL, B. M. W. (1960). *Proc. R. Soc. A* **256**, 81.

COLES, B. R. and CAPLIN, A. D. (1976). *The Electronic Structures of Solids.* Edward Arnold, London.

CZERNY, S., and PONEC, V. (1968). *Catalysis Rev.* **2**, 249.

DAVISSON, C. J., and GERMER, L. H. (1927). *Phys. Rev.* **30**, 705.

DE BOER, J. H. (1957). *Adv. Catalysis* **9**, 472.

DOWDEN, D. A. (1950). *J. chem. Soc.* p. 242.

—— (1971). *Fundamental principles of heterogeneous catalysis.* NATO Advanced Study Institute, Italy.

—— (1972). *Proc. Vth. int. congr. catalysis,* paper 41. Florida.

DYMOND, E. G., and WATSON, R. (1929). *Proc. R. Soc. A.* **122**, 571.

ELEY, D. D., and NORTON, P. R. (1970). *Proc. R. Soc. A.* **314**, 301.

FRANCK, J., and HERTZ, G. (1911). *Ver. dt. Phys. ges.* **16**, 12.

GADZUK, J. W. (1974). *Solid st. Commun.* **15**, 1011.

GALE, R. L., HABER, J., and STONE, F. S. (1962) *J. Catalysis* **1**, 32.

GLUECKAUF, E. (1949). *Disc. Faraday Soc.* **7**, 199.

GOMER, R. (1953). In *Structure and properties of solid surfaces* (Eds R. Gomer and C. S. Smith), preface. University of Chicago Press, Chicago, Ill.

GOODENOUGH, J. B. (1963). *Magnetism and the chemical bond.* John Wiley Interscience, New York.

GOYMOUR, C. G., and KING, D. A. (1973). *J. chem. Soc. Faraday I,* **69**, 736, 749.

GRAVELLE, P. C. (1971). *Fundamental principles of heterogeneous catalysis.* NATO Advanced Study Institute, Italy.

GRAVELLE, P. C., and TEICHNER, S. J. (1969). *Adv. Catalysis,* **20**, 167.

GRIMLEY, T. B. (1974). *Disc. Faraday Soc.* **58**, 7.

GUNDRY, P. M., and TOMPKINS, F. C. (1956). *Trans. Faraday Soc.* **52**, 1609.

HAAS, T. W., GRANT, J. T., and DOOLEY, G. J. (1970). *J. vac. Sci. Technol.* **7**, 43.

HAYWARD, D. O., HERLEY, P. J., and TOMPKINS, F. C. (1964). *Surf. Sci.* **2**, 156.

JOYNER, R. W., RICKMAN, J., and ROBERTS, M. W. (1974). *J. chem. Soc. Faraday I* **70**, 1825.

JOYNER, R. W., and SOMORJAI, G. A. (1973). *Spec. period. rep: surface and defect properties of solids* (Senior Reporters, M. W. Roberts and J. M. Thomas), Vol. 2, p. 1. Chemical Society, London.

KEMBALL, C. (1970). In *Chemisorption and catalysis* (Ed. P. Hepple). Institute of Petroleum, London.

KINGTON, G. L., and SMITH, P. S. (1964). *J. sci. Instrum.* **41**, 145.

KISELEV, A. V., DZHIGIT, O. M., and MUTTIK, G. G. (1962). *J. phys. Chem.* **66**, 2127.

KISLIUK, P. (1959). *J. chem. Phys.* **31**, 1605.

KRONIG, R. L. and PENNEY, W. G. (1930). *Proc. R. Soc. A* **130**, 499.

LAIDLER, K. J. (1954). *Catalysis* (Ed. P. H. Emmett), Vol. 1, p. 195. Rheinhold, New York.

LANGMUIR, I. (1960). *Collected works of Irving Langmuir* (Ed. G. Suits). Pergamon Press, Oxford.

LENNARD-JONES, J. E. (1932). *Trans. Faraday Soc.* **28**, 333.

—— (1937). *Physica* **4**, 941.

LLOYD, D. R., QUINN, C. M., and RICHARDSON, N. V. (1975). *J. Phys. C Solid st. Phys.* **8**, L 371.

MAY, J. W. (1972). *Proc. R. Soc. A* **331**, 185.

MIYAZAKI, E., and YASUMORI, I. (1976). *Surf. Sci.* **55**, 747.

MORRISON, J. A., and LOS, L. M. (1950). *Disc. Faraday Soc.* **8**, 321.

MOTT, N. F. (1972). *Elementary quantum mechanics*, Wykeham, London.

PLUMMER, E. W., and LIEBSCH, A. (1974). *Disc. Faraday Soc.* **58**, 19.

QUINN, C. M.(1973). *An introduction to the quantum chemistry of solids.* Clarendon Press, Oxford.

QUINN, C. M., and ROBERTS, M. W. (1964). *Trans. Faraday Soc.* **60**, 899.

RIVIÈRE, J. C. (1969). In *Solid state surface science* (Ed. M. Green), vol. 1, p. 179. Marcel Dekker, New York.

ROBERTS, J. K. (1935). *Proc. R. Soc. A* **152**, 445. See also A. R. Miller, *The adsorption of gases on solids*, Cambridge University Press, London (1949).

ROBERTS, M. W. (1960). *Nature, Lond.* **188**, 1020.

ROBERTS, M. W., and ROSS, J. R. H. (1966). *Trans. Faraday Soc.* **62**, 2301.

—— (1969). *Reactivity of solids* (Ed. J. W. Mitchell, R. C. De, Vries, R. W. Roberts, and P. Cannon). John Wiley, New York.

ROBERTS, M. W., and SYKES, K. W. (1957). *Proc. R. Soc. A* **242**, 534.

—— (1958). *Trans. Faraday Soc.* **54**, 548.

ROBERTS, M. W. and WELLS, B. R. (1964). *Proc. Chem. Soc.* p. 173.

—— (1966). *Disc. Faraday Soc.* **41**, 162.

ROBERTS, M. W., and YOUNG. N. J. (1970). *Trans. Faraday Soc.* **66**, 2636.

ROBERTS, R. W. (1967). *Adhesion or cold welding of materials in space environments.* Rep. No. ASTM STP 431, American Society of Testing Materials.

SCHMIDT, L. D., and GOMER, R. (1965). *J. chem. Phys.* **42**, 3573.

SCHWAB, G. M. (1957). *Adv. Catalysis,* **9**, 496.

SKINNER, R. W. B. (1938). *Rep. Prog. Phys.* **5,** 257.

SMITH, N. V. and TRAUM, M. M. (1974). *Surf. Sci.* **45,** 745.

SMITH, N. V., TRAUM, M. M., and DI SALVO, F. J. (1974). *Phys. Rev. Lett.* **32,** 1241; *Solid st. Commun.* **15,** 211.

SOMMERFELD, A. (1928). *Z. Phys.* **47,** 1.

STEWART, C. N., and EHRLICH, G. (1972). *Chem. phys. Lett.* **16,** 203.

—— (1975). *J. Chem. Phys.* **62,** 4672.

STONE, F. S. (1962). *Adv. Catalysis* **13,** 1.

TANAKA, K., and TAMARU, K. (1963). *J. Catalysis* **2,** 366.

THOMAS, J. M., EVANS, E. L., WILLIAMS, J. O. (1972). *Proc. R. Soc. A* **331,** 417.

THOMAS, J. M. (1975). In *Electronic states of inorganic compounds* (Ed. P. Day), p. 27. Reidel, New York.

TOMBOULAIN, D. H. (1957). *Hand. Phys.* **30,** 246. Springer-Verlag, Berlin.

TOMPKINS, F. C. (1974). *CRC crit. rev. solid st. Sci.* **4,** 279.

TRAPNELL, B. M. W. (1953). *Proc. R. Soc. A* **218,** 566.

TROST, W. R. (1959). *Can. J. Chem.* **37,** 460.

VAN HERWIGNEN, T., and DE JONG, W. A. (1974). *J. Catalysis* **34,** 209.

WEDLER, G., and BROCKER, F. J. (1966). *Disc. Faraday Soc.* **41,** 87.

WHEELER, A. (1953). In *Structure and properties of solid surfaces* (Eds. R. Gomer and C. S. Smith). University of Chicago Press, Chicago, Ill.

WHETTEN, N. R. (1965). *J. vac. Sci. Technol.* **2,** 84.

WILSON, J. N. (1944). *J. am. chem. Soc.* **62,** 1583.

GENERAL REFERENCES FOR FURTHER READING

ANDERSON, J. R. (Ed.) (1971). *Chemisorption and reactions on metallic films,* vols. 1, 2. Academic Press, New York, London.

—— (1975). *Structure of metallic catalysts.* Academic Press, New York, London.

BLAKELY, J. M. (1973). *Introduction to the properties of crystal surfaces.* Pergamon Press, Oxford.

BOND, G. C. (1962). *Catalysis by metals.* Academic Press, New York, London.

—— (1974). *Heterogeneous catalysis: principles and applications.* Clarendon Press, Oxford.

CLARK, A. (1974). *The chemisorptive bond.* Academic Press, New York, London.

ERTL, G., and KÜPPERS, J. (1974). *Low energy electrons and surface chemistry.* Verlag Chemie, Weinheim.

GASSER, R. P. H. (1970). Ann. Repts. (Chem. Soc.), **67,** 213.

GREENWOOD, N. N. (1968). *Ionic crystals, lattice defects, and non-stoichiometry.* Butterworths, London.

HAYWARD, D. O., and TRAPNELL, B. M. W. (1964). *Chemisorption.* Butterworths, London.

HOBSON, J. P. (1974). Vacuum and the solid-gas interface. *Adv. Colloid Interface Sci.* **4,** 79.

KITTEL, C. (1976). *Introduction to solid state physics.* John Wiley, New York.

MOORE, W. J. (1967). *Seven solid states.* W. A. Benjamin, New York.

OUDAR, J. (1975). *Physics and chemistry of surfaces.* Blackie, Edinburgh.

PONEC, V., KNOR, Z., and CZERNY, S. (1974). *Adsorption on solids.* Butterworths, London.

PRUTTON, M. (1975). *Surfaces physics.* Clarendon Press, Oxford.

RIDEAL, E. K. (1968). *Concepts in catalysis.* Academic Press, New York, London.

ROBERTS, R. W. (1967). *Adhesion or cold welding of materials in space enviroments.* Rep. No. ASTM STP 431, American Society of Testing Materials.

ROBERTSON, A. J. B. (1970). *Catalysis of gas reactions by metals.* Logos Press, London.

SACHTLER, W. M. H. (1966). Sorption on solids. *Proc. 3rd Int. Vacuum Congress.* Pergamon Press, Oxford.

SOMORJAI, G. A. (1972). *Principles of surface chemistry.* Prentice-Hall, Englewood Cliffs N. J.

THOMAS, J. M., and THOMAS, W. J. (1967). *Introduction to the principles of heterogeneous catalysis.* Academic Press, New York, London.

THOMSON, S. J., and WEBB, G. (1968). *Heterogeneous catalysis.* Oliver and Boyd, Edinburgh.

WEDLER, G. (1976). *Chemisorption: An Experimental Approach.* Translated by D. F. Klemperer, Butterworths, London.

2

CRYSTALLOGRAPHY OF METALS

2.1. Introduction

IN CHAPTER 1 it was suggested that an understanding of surface chemistry at the molecular level must involve consideration of three fundamental properties of surface layers—electron distribution, chemical constitution, and structure. A description of the surface structure of the clean solid, prior to adsorption, probably represents the least difficult of these problems, especially since future discussion is to be restricted mainly to metals. It therefore seems appropriate to examine this subject at the outset of our detailed consideration of adsorption.

The structure of a metal surface can be readily determined on the assumption that in this region the arrangement of atoms which occurs in the bulk of the solid is maintained; the structures of an ideal single-crystal surface and of the corresponding planes in the bulk will then be the same. (Evidence derived from low-energy electron diffraction experiments (see Chapter 3) indicates that in many cases this assumption is valid, although in some instances distortion of the surface layer has been detected.) The bulk structures of ideal metals, described in terms of the close packing of identical spheres, are well established and many excellent descriptions of the subject exist; see, for example, the appropriate sections in the texts by Barrett (1966), Cullity (1956), Hume-Rothery (1966), and Azaroff (1969). Also, the crystallography of a wide range of single-crystal surfaces is superbly illustrated in a work by Nicholas (1965) which should be prescribed reading (and viewing) for anyone involved in surface chemistry.

No adsorbent possesses an ideal single-crystal surface. All contain defects, many are polycrystalline, and in the case of supported catalysts the metal particles may be so small that their surface properties no longer correspond to those of more extensive surfaces. So in the latter part of this chapter we briefly introduce the topics of surface stability, structural defects, and also diffusion in solids since this represents a mechanism by which surface structure may be modified.

2.2. The basic point lattices

A crystal is by definition a periodic arrangement of identical units and so it is convenient to relate its structure to a periodic collection of points in space, the surroundings of every point being the same. This point lattice (or space lattice in the three-dimensional case) is a geometrical abstraction, and a real crystal is obtained by populating it with an identical group of atoms, known as the basis, at each point:

LATTICE + BASIS = CRYSTAL STRUCTURE.

As is evident from Fig. 2.1(a) the point lattice may be built up by stacking identical unit cells side by side in three dimensions, each cell being defined by vectors **a**, **b**, and **c**. The position vector of any lattice point relative to another such point as origin is then given by

$$\mathbf{r} = n_1\mathbf{a} + n_2\mathbf{b} + n_3\mathbf{c} \qquad (2.1)$$

where the n_i are arbitrary integers. Since there are no natural limits to the magnitudes and relative orientations of the vectors **a**, **b**, and **c** an infinite number of different lattices may exist, but when symmetry properties are taken into consideration it is possible to classify all of them in terms of 14 different types, referred to as the Bravais lattices.

Crystallographic axes are chosen to correspond with the rotation axes of the lattice. (An n-fold rotation axis exists if the structure is repeated after rotation of the system through an angle

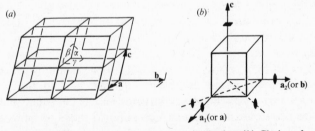

FIG. 2.1. (a) Stacking of unit cells to form a point lattice. (b) Choice of axes in the tetragonal system: ⧫ two-fold axis; ■ four-fold axis.

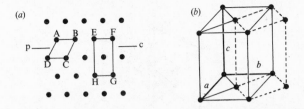

FIG. 2.2. (*a*) Retention of symmetry by a centred cell. (*b*) Relation of tetragonal C lattice (solid lines) to tetragonal P lattice (broken lines). (After Cullity 1956.)

of $2\pi/n°$ about this axis.) The *c* axis is taken in the direction of the unique symmetry axis, if such exists; for example in the tetragonal system *c* is parallel to the four-fold axis (Fig. 2.1(*b*)). Conventionally the *b* axis is drawn horizontally to the right and the *c* axis upwards; α is the angle between *b* and *c*, β that between *a* and *c*, and γ ($\geqslant90°$) that between *a* and *b* (Fig. 2.1(*a*)). If any two, or three, of the axes are equivalent by symmetry they may be alternatively labelled a_1, a_2 (and a_3).

If lattice points occur only at the corners of the unit cell, then, since each point is shared between eight adjacent cells, the total number of points in any particular cell is one. Such cells are referred to as primitive (symbol P, except for the trigonal cell which is labelled R). In order to emphasize the overall symmetry of a lattice it is also convenient in a number of cases to consider non-primitive cells, each containing more than one lattice point. This is most easily illustrated in two dimensions, as in Fig. 2.2(*a*). The primitive unit ABCD does not show the rectangular symmetry of the lattice, but this becomes apparent on consideration of the centred orthogonal unit EFGH. In three dimensions only three kinds of non-primitive cell need be considered—body centred (symbol I), face centred (F), and side centred (A, B, or C, depending on which pair of opposite faces are centred). In certain cases a given type of centred cell may reduce to a primitive cell of the same symmetry, and so nothing is gained by use of the more complex version; a tetragonal C cell, reducing to a primitive cell which is also tetragonal, is shown in Fig. 2.2(*b*). In other cases the non-primitive cell has distinct advantages, the most familiar examples being the face-centred (f.c.c.) and body-centred (b.c.c.) variations of the cubic system. The primitive cell in these cases is a rhombohedron (Fig. 2.3) which effectively obscures the higher

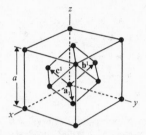

FIG. 2.3. The rhombohedral primitive cell of an f.c.c. crystal.

cubic symmetry actually present. In 7 of the 14 Bravais lattices the unit cells are primitive, the remainder being centred.

In two dimensions the point lattice is known as a plane lattice, or net, and is constructed from repeating *unit meshes*. There are five different lattice types (Fig. 2.4), of which the parallelogram is the most general. In the case of the diamond lattice the symmetry is emphasized by considering a non-primitive centred rectangular mesh. (Note that, as in three dimensions, the *b* axis is drawn horizontally to the right and the *a* axis downwards, either vertically or pointing to the left).

2.3. Metal crystals

The crystal structures of a large number of metals are of one of three types, f.c.c., b.c.c., or hexagonal close-packed (h.c.p.) (Table 2.1). In the first two cases the structure is obtained by placing a single atom at each point in the corresponding Bravais lattice, but the h.c.p. case is rather more complicated. In the first

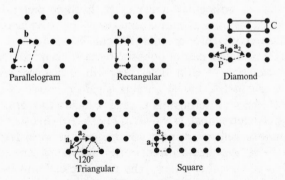

FIG. 2.4. The five plane lattices.

TABLE 2.1
Crystal structure of metals

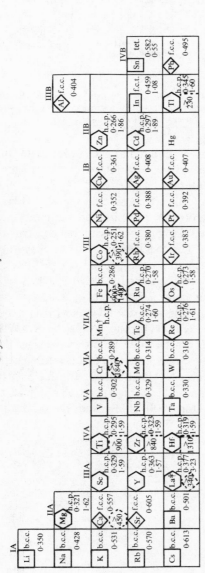

□ body-centred cubic (b.c.c.); ◇ face-centred cubic (f.c.c.); ⬡ hexagonal close-packed (h.c.p.); ⬠ tetragonal (tet); □ face-centred tetragonal (f.c.t.). Solid-line symbols represent the forms stable at room temperature. Other forms are indicated by broken-line symbols, with the temperature range of their stability shown in degrees Celsius.

Values of the unit cell parameter *a* (nm) and of the axial ratio *c/a* (where appropriate) are given in each case.

† Two complicated structures, α and β, with 58 and 20 atoms per unit cell respectively, both cubic.

‡ Structure possesses a doubled axial ratio; stacking sequence ABACABAC.....

(Data taken from Smithells 1976 and Hume-Rothery 1966).

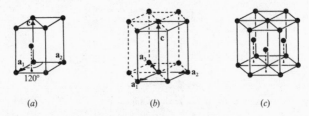

place a basis of two atoms is required at each point in the Bravais lattice. Secondly, the hexagonal symmetry of the system, although implicit in the interaxial angle $\gamma = 120°$, is visually emphasized only when three cells are considered together (Fig. 2.5.).

The f.c.c. and h.c.p. structures differ in their symmetry properties but they are related in that both can be constructed by packing identical spheres in the closest possible way; f.c.c. is sometimes referred to as cubic close-packed (c.c.p.). If we consider a single layer of atoms (A) the most closely packed arrangement has hexagonal symmetry (Fig. 2.6(a)). A second similar layer (B) is now placed on the first, the centres of its atoms lying over the positions marked B in the lower layer. When a third layer is added this can be done in two ways (Fig. 2.6(b)). If the atoms rest at the positions in the second layer marked A so that the stacking sequence is ABABAB . . . the h.c.p. arrangement results. The close-packed plane is known as the basal plane; its indices (see Section 2.4) are (0001). Alternatively, the third-layer

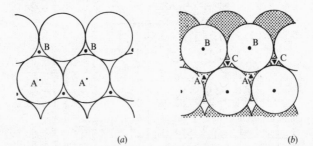

Fig. 2.6. Alternative stacking sequences for h.c.p. layers. (a) Single layer (A), showing sites (B) for second layer. (b) Two layers, showing possible sites (A or C) for third layer.

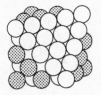

FIG. 2.7. F.c.c. structure showing the close-packed (111) plane.

atoms may be placed in positions C in the second layer (Fig. 2.6(*b*)) giving the stacking sequence ABCABC.... This arrangement produces the f.c.c. structure. As shown in Fig. 2.7 the close-packed plane is (111) in this case, and, since there are four equivalent sets of planes of this type in the structure, in general the physical and mechanical properties of f.c.c. metals are more isotropic than those of h.c.p. metals which contain a single set of close-packed planes.

In an ideal h.c.p. system the axial ratio is $c/a = 1\cdot633$ and the co-ordination number (the number of nearest neighbours surrounding any atom) is 12. In many h.c.p. metals the axial ratio actually lies in the range $1\cdot57$–$1\cdot60$ (Table 2.1) and the six neighbours in the basal plane are then a little further away than the three in the plane above and the three in the plane below. Zinc and cadmium, however, have ratios of approximately $1\cdot9$, and the nearest neighbours are then the six in the basal plane.

In the f.c.c. system the co-ordination number is also 12, the nearest-neighbour distance expressed in terms of the unit-cell dimension a being $a/\sqrt{2} = 0\cdot707a$. The six next-nearest neighbours lie at a distance a. The b.c.c. structure is more open, having eight-fold co-ordination and a nearest-neighbour distance of $a\sqrt{3}/2 = 0\cdot866a$. The six next-nearest neighbours, however, are situated only slightly further away, at a distance a.

The interstitial positions in a metal lattice are important as possible sites for small impurity atoms, for example carbon, and also in relation to penetration into the bulk of species adsorbed at a surface. In the three structures under consideration the interstitial sites are composed of tetrahedral holes (surrounded by four atoms in a tetrahedral arrangement) and octahedral holes (surrounded by six atoms at the corners of an octahedron). In the b.c.c. lattice these tetra- and octahedra are not regular (Fig. 2.8). Assuming the metals to be constructed from hard spheres of

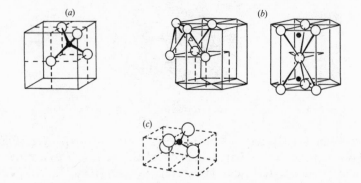

FIG. 2.8. (a) Tetrahedral hole (●) in the f.c.c. structure; (b) octahedral hole (△) and tetrahedral holes (●) in the close-packed hexagonal structure; (c) tetrahedral hole (●) in the b.c.c. structure. (From Hume-Rothery 1966.)

radius r, the maximum radius of a sphere which can be accommodated interstitially in each case is shown in Table 2.2. It is interesting to note that the octahedral hole in the close-packed structures is larger than either of the holes in the more open b.c.c. lattice.

2.4. Miller indices

Within a space lattice, or within a crystal formed by populating the lattice with groups of atoms, the orientations of planes relative to the axes of the system are defined in terms of Miller indices, although these do not specify the position of a plane relative to the origin, since the position of the origin itself is arbitrary. The indices for a particular plane are determined in three steps: (i) the intercepts of the plane with the three axes are found and expressed relative to the unit cell dimensions a, b, c; (ii) the reciprocals of these numbers are determined; (iii) if fractions result, they are reduced to the three smallest integers

TABLE 2.2
Interstices in metal lattices

Structure	h.c.p. $(c/a = 1{\cdot}633)$	f.c.c.	b.c.c.
Tetrahedral hole	$0{\cdot}225r$	$0{\cdot}225r$	$0{\cdot}291r$
Octahedral hole	$0{\cdot}410r$	$0{\cdot}410r$	$0{\cdot}154r$

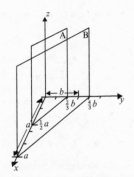

FIG. 2.9. Determination of the Miller indices of a plane.

having the same ratio, and in general form these integers are represented by the letters h, k, and l. The procedure is illustrated in Fig. 2.9. It is irrelevant which plane from an infinite stack of parallel planes is selected; consider for example plane A.

Step (i)	Intercepts	$\frac{1}{2}a/a = \frac{1}{2}$	$\frac{2}{3}b/b = \frac{2}{3}$	$\infty c/c = \infty$
Step (ii)	Reciprocals	2	$\frac{3}{2}$	0
Step (iii)	Indices	4	3	0.

(A plane parallel to an axis cuts that axis at infinity and the corresponding index is zero. A plane cutting an axis in the negative direction produces a negative index, e.g. $-h$, written $\bar{h}$.) The indices are enclosed within round brackets, so that A is a (430) plane, as also is any parallel plane such as B (Fig. 2.9).

Any set of equivalent lattice planes which are related by symmetry are called planes of a form and are represented by the indices of any one enclosed in braces $\{hkl\}$. Thus all the faces of a cube are of the form $\{100\}$ (Fig. 2.10(a)), but this is a simple case.

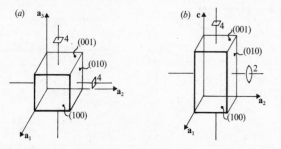

FIG. 2.10. (a) Cube faces of form $\{100\}$. (b) The $\{100\}$ and $\{001\}$ forms in the tertagonal system.

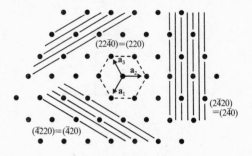

FIG. 2.11. Some planes in the hexagonal system related to three symmetry-equivalent axes a_1, a_2, a_3.

In the tetragonal system for example the planes (100), (010), ($\bar{1}$00), and (0$\bar{1}$0) are of the form {100} related by the 4-axis (i.e. axis of four-fold rotational symmetry), while (001) and (00$\bar{1}$) are of the form {001} related by the 2-axis (Fig. 2.10(b)). All planes of a given form have similar physical properties.

In the hexagonal system it is possible to use indices based on just three axes, a_1, a_2, and c, but symmetry-related planes such as (220), (2$\bar{4}$0), and ($\bar{4}$20) (Fig. 2.11) do not then have similar indices. Such difficulties are avoided by the use of three symmetry-equivalent a axes, 120° apart, and a fourth c axis normal to these. The corresponding indices, known as the Bravais–Miller indices, are ($hkil$) and the first three always obey

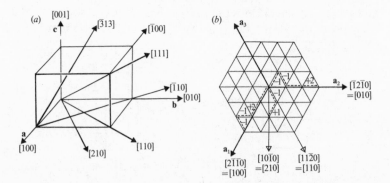

FIG. 2.12. (a) Indices of directions. (b) Indices of directions in the hexagonal system. (From Barrett 1966.)

the relation

$$h + k = -i. \tag{2.2}$$

The three planes in Fig. 2.11 are now clearly seen to be of the form $\{22\overline{4}0\}$.

To represent a direction in a lattice a line parallel to this direction is drawn through the origin. If this line passes through a point with co-ordinates x, y, z these are first converted, if necessary, to the set of smallest integers in the same ratio, u, v, w. Written in square brackets $[uvw]$ these represent the indices of the given direction or of any parallel direction (Fig. 2.12(a)). It is useful to remember that the direction $[hkl]$ is normal to the plane (hkl) in the following cases: (i) when the plane cuts equal, orthogonal axes, i.e. any plane in the cubic system and $(hk0)$ planes in the tetragonal system; (ii) when the plane is perpendicular to a single axis and parallel to two others. The relationship is not valid in any other case. Directions which are symmetry related are known as directions of a form, and a set of these is described by the indices of any of them enclosed in angular brackets $\langle uvw \rangle$; for example in the cubic system $[100]$, $[010]$, and $[001]$ are directions of the form $\langle 100 \rangle$.

To define a direction in the hexagonal system a line parallel to this direction is drawn through the origin, as before. The direction indices are then the translations parallel to each of the four axes which are required to reach a point on this line. It is essential to note that the sum of the first two indices must be the negative of the third index. Some examples are shown in Fig. 2.12(b).

For the specification of net rows and of directions in a two-dimensional net a similar system is adopted based on two indices (hk) or $[hk]$. With a hexagonal unit mesh some confusion with three dimensions could be introduced by using the three indices (h, k, i), and so use of the i index is not recommended (Wood 1964); it may always be determined from eqn (2.2).

On certain occasions it may be convenient to change the axes of a lattice from $\mathbf{a}, \mathbf{b}, \mathbf{c}$ to a new set $\mathbf{A}, \mathbf{B}, \mathbf{C}$; an example would be the transformation from a primitive cell to a non-primitive one which emphasizes the lattice symmetry. Any translation in the original lattice may be represented by

$$\mathbf{t} = u\mathbf{a} + v\mathbf{b} + w\mathbf{c} \tag{2.3}$$

and so the new axes can be defined by the relationships

$$\mathbf{A} = u_1\mathbf{a} + v_1\mathbf{b} + w_1\mathbf{c}$$
$$\mathbf{B} = u_2\mathbf{a} + v_2\mathbf{b} + w_2\mathbf{c} \qquad (2.4)$$
$$\mathbf{C} = u_3\mathbf{a} + v_3\mathbf{b} + w_3\mathbf{c}.$$

The constants u_n, v_n, w_n completely specify the transformation, and are conveniently written as a transformation matrix $\mathbf{T}$, where

$$\mathbf{T} = \begin{bmatrix} u_1 v_1 w_1 \\ u_2 v_2 w_2 \\ u_3 v_3 w_3 \end{bmatrix}. \qquad (2.5)$$

The same matrix governs the transformation of the indices (hkl) of any plane in the original lattice to the values (HKL) appropriate to the new one.

$$\mathbf{T} \begin{bmatrix} h \\ k \\ l \end{bmatrix} = \begin{bmatrix} H \\ K \\ L \end{bmatrix}. \qquad (2.6)$$

Any set of planes which have a common line of intersection are known as planes of a zone, the direction of the common line $[uvw]$ being the zone axis (Fig. 2.13). The condition for any plane (hkl) to be parallel to the direction $[uvw]$ is

$$hu + kv + lw = 0 \qquad (2.7)$$

and so this must be satisfied by each plane in the zone (see Section 2.7).

The zone to which two non-parallel planes $(h_1 k_1 l_1)$ and $(h_2 k_2 l_2)$ belong is generally determined using the following

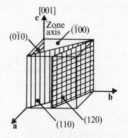

FIG. 2.13. Some planes in the cubic lattice belonging to the [001] zone. (From Cullity 1956.)

mnemonic. The indices of the faces are written in duplicate, one below the other, as follows.

$$\begin{matrix} h_1 \\ h_2 \end{matrix} \left| \begin{matrix} k_1 l_1 h_1 k_1 \\ k_2 l_2 h_2 k_2 \end{matrix} \right| \begin{matrix} l_1 \\ l_2 \end{matrix}.$$

The first and last columns are rejected and from the resulting matrix three determinants are formed by taking two columns at a time.

$$\left| \begin{matrix} k_1 l_1 \\ k_2 l_2 \end{matrix} \right| \left| \begin{matrix} l_1 h_1 \\ l_2 h_2 \end{matrix} \right| \left| \begin{matrix} h_1 k_1 \\ h_2 k_2 \end{matrix} \right|.$$

Evaluation of matrices gives the indices of the zone axis:

$$u = k_1 l_2 - k_2 l_1$$
$$v = h_2 l_1 - h_1 l_2 \qquad (2.8)$$
$$w = h_1 k_2 - h_2 k_1.$$

A relationship equivalent to (2.8) holds between the indices $[u_1 v_1 w_1]$ and $[u_2 v_2 w_2]$ of two directions and the indices (hkl) of a plane which is parallel to both directions.

Three planes belong to the same zone, or have a common line of intersection, when the determinant of their indices is zero, i.e.

$$\left| \begin{matrix} h_1 k_1 l_1 \\ h_2 k_2 l_2 \\ h_3 k_3 l_3 \end{matrix} \right| = 0. \qquad (2.9)$$

Similarly, three directions $[u_1 v_1 w_1]$, $[u_2 v_2 w_2]$, $[u_3 v_3 w_3]$ are coplanar when their indices satisfy a condition equivalent to (2.9).

2.5. Projections of crystals

Crystallographic problems can be solved by analytical methods, some of which have already been outlined, but in other cases it is convenient to resort to graphical procedures. If a crystal is located at the centre of a large sphere and the planes in the crystal are extended to cut the sphere they will do so in a series of great circles. These circles can be used to represent various properties of the planes, such as the angles between them, but the spherical surface becomes rather congested if a large number of planes is involved. This difficulty may be avoided and the same properties illustrated if each plane is represented instead by its pole, which is the intersection of the normal to the plane with the surrounding sphere (Fig. 2.14); the resulting construction is a spherical projection of the crystal. It is more convenient, however, to use a planar projection, and this can be prepared in a number of ways, each of which preserves certain features of the original spherical map; a geographer for example might choose a projection to maintain equality of area.

For crystallography the stereographic projection is the most suitable and its construction is illustrated in Fig. 2.14. The plane of projection is placed at A, normal to a chosen diameter AB of the sphere, and the other end of this diameter (B) is used as the point of projection. The projection P′ of a pole lying on the sphere at P is obtained by extending the line BP to cut the projection plane at P′. The great circle NESW projects to form the basic circle N′E′S′W′ and all poles on the 'A side' of this circle

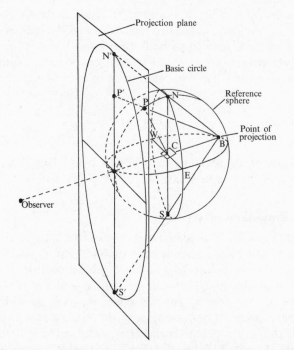

FIG. 2.14. The stereographic projection; P is the pole of plane C. (From Cullity 1956.)

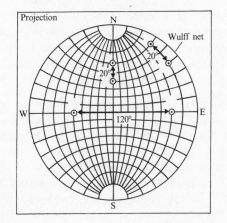

FIG. 2.15. Stereographic projection of three pairs of poles superimposed on the Wulff net for measurement of the angles between the poles. (From Cullity 1956.)

may be represented in the projection; those on the other side may be obtained by moving the projection plane to B and the point of projection to A.

The projection preserves the symmetry properties of the crystal direction (AB) lying normal to the plane of projection and also the angular relationships between poles, which are identical to the relationships between the corresponding planes. (Actually an angle χ on the sphere projects as the angle $\chi/2$, but this is allowed for by a scaling factor when making measurements on the projection.) In addition great circles on the reference sphere project as circular arcs, passing through two diametrically opposed points on the basic circle; if they pass through the points AB (Fig. 2.14) they appear as straight lines through the centre of the projection.

Measurements on a stereographic projection are most easily carried out using a Wulff net, which is a projection of a sphere ruled with lines of latitude and longitude, onto a plane parallel to the N–S axis of the sphere (Fig. 2.15). The stereographic projection is drawn on transparent paper with a basic circle of the same diameter as the Wulff net and is then superimposed on the net, so that angular relationships can be read off directly.

The angle between any two planes in a crystal is equal to the angle between their poles in the projection, provided these poles

lie on a great circle (Fig. 2.15). If they do not the projection must
be rotated relative to the Wulff net until the poles do lie on such
a circle whereupon the angular measurement can be made. This
procedure is permissible since angular relationships are not
changed by rotation of poles about the axis of projection.

Standard projections. If some important crystal plane of low
indices (e.g. (100), (110), (111), or (0001)) is chosen as the plane
of projection a standard projection results. Its construction requires of course a knowledge of the angles between the various
planes and these depend in general not only on the system to
which the crystal belongs, but also on the particular axial ratio
(c/a) involved. In the cubic system, however, the interplanar
angles ϕ are fixed by symmetry and for all cubic crystals are given
by the relationship

$$\cos \phi = \frac{h_1 h_2 + k_1 k_2 + l_1 l_2}{\{(h_1^2 + k_1^2 + l_1^2)(h_2^2 + k_2^2 + l_2^2)\}^{\frac{1}{2}}}. \tag{2.10}$$

It is thus possible to prepare standard projections valid for all
such crystals. Three examples, based on (001), (011), and (111),
are shown in Figs 2.16 and 2.17; a further large selection is given
by Johari and Thomas (1969).

The normals to all planes in a given zone are coplanar and
perpendicular to the zone axis (see Fig. 2.13). Hence the poles of
these planes will all lie on a single great circle in the stereographic projection and the zone axis will be at 90° to this circle.
Also, important planes generally belong to more than one zone
and so their poles are located at the intersection of zone circles.

Projections such as those shown in Fig. 2.16 consist of 24
similar and equivalent unit stereographic triangles, each having

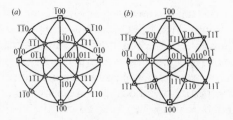

FIG. 2.16. Standard projections of cubic crystals on (*a*) (001) and (*b*) (011).
(From Cullity 1956.)

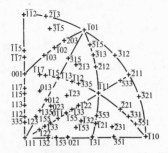

FIG. 2.17. One quadrant of a detailed standard cubic (111) projection. (From Johari and Thomas 1969.)

{100}, {110}, and {111} at its vertices. If we are concerned only with the orientation of a single-crystal surface, then we need consider only one such triangle and not the complete projection since the pole of the surface considered must fall within one of these equivalent triangles. This leaves some ambiguity as to the orientation of the specimen as a whole, since it allows one rotational degree of freedom about the surface normal, but this is unimportant.

2.6. Orientation of single crystals

Since many experiments in surface chemistry and physics are now performed on single-crystal surfaces it is essential to be able to determine the orientation of such surfaces to within at least 1°. This is generally carried out by the back-reflection Laue method of X-ray diffractometry. A polychromatic X-ray beam is used and the orientation of the specimen to the beam is fixed. Since the wavelength is unknown, determination of the Bragg angle for a given spot does not enable the set of planes producing the spot to be identified. The orientation with respect to the incident beam of the normal to the plane can be established, however, since the normal will bisect the incident and diffracted directions. When a white X-ray beam strikes the specimen the beams diffracted by all planes in a given zone will be on a conical surface centred on the zone axis (Fig. 2.18(a)). The semi-vertical angle ζ of the cone is equal to the angle between the zone axis and the transmitted beam. The cone intersects the film in a hyperbola HK and so all spots from one zone will lie on such a curve. If $\zeta = 90°$ the intersection generates a straight line.

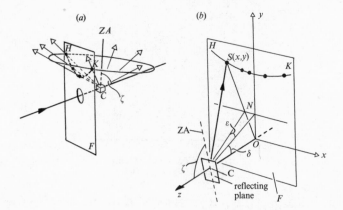

FIG. 2.18. The back-reflection Laue method. (a) Intersection (HK) of a conical array of diffracted beams with the film; C, crystal; F, film; ZA, zone axis. (b) Location of a back-reflection spot S, with angular co-ordinates δ and ε. (After Cullity 1956.)

In Fig. 2.18(b) suppose a reflecting plane C with normal CN produces a spot on the film at S. (The incident beam, diffracted beam, and normal are coplanar). By measuring the spot co-ordinates (x, y) it is possible, using the appropriate geometrical relationships, to derive the orientation of the normal in terms of its angular co-ordinates δ and ε. Alternatively, this calculation may be avoided by using a Greninger chart which, when placed on the film, gives δ and ε directly for any spot. Knowing these co-ordinates the pole of the reflecting plane can be plotted on a stereographic projection (Fig. 2.19). The Wulff net must in this case be orientated with its meridians running E–W, since the Laue spots lying on curves of constant δ originated from planes of a zone, the poles of which must lie on a great circle in the projection.

Having plotted all the poles great circles are drawn through the sets of poles corresponding to the different hyperbolae on the film. The individual poles are then indexed by a trial-and-error method, remembering that poles lying at the intersections of great circles generally belong to planes of low indices such as {100}, {110}, {111}, and {112} in cubic systems. This also applies to the axis of a zone lying at 90° to the related great circle. By comparison of the angles between important poles with tabulated

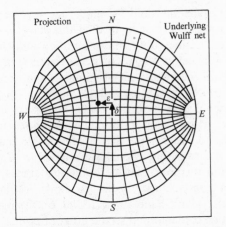

FIG. 2.19. Use of the Greninger chart to plot the pole of a reflecting plane with angular co-ordinates δ and ε on a stereographic projection.

interplanar angles the orientation of the crystal may be established absolutely.

In the applications of interest in surface chemistry, which concern the surface of a single crystal, the orientation is often known approximately and the procedure is then considerably simplified. Only the unit stereographic triangle containing the normal to the surface need be considered. Assuming that the crystal was mounted with its face perpendicular to the X-ray beam the centre of the projection corresponds to the position of the surface normal and so the deviation of the normal from the desired pole may be measured. If necessary the orientation may be adjusted by regrinding or recutting the crystal.

2.7. Reciprocal space

For the various sets of parallel planes in a three-dimensional structure the stereographic projection discussed in Section 2.5 enables the relative plane orientations to be conveniently represented by a set of points in two dimensions, but in certain circumstances, such as problems involving diffraction (see Chapter 3), this information is insufficient. When constructing a projection a given set of planes (hkl) is first represented by its normal, i.e. by a line with direction but of no fixed magnitude. If this line is now converted to a vector by specifying that its magnitude σ_{hkl} be proportional to the reciprocal of the spacing of the (hkl) planes ($\sigma_{hkl} \propto 1/d_{hkl}$), then the vector defines both the orientation and the spacing of these planes. Further, if the vectors associated with the different sets of planes in the crystal all originate at the same point then each vector can be replaced by a single point at its terminus. This three-dimensional collection of points, known as the reciprocal lattice of the crystal, thus presents two important parameters relating to the planes compared with the single parameter represented by the two-dimensional stereographic projection. The locations of the reciprocal lattice points for some planes in the [001] zone of a monoclinic crystal are shown in Fig. 2.20 (since all planes in a given zone are parallel to a single line their normals must be coplanar and so the array of points in this case is two dimensional).

As the name reciprocal lattice implies, all points such as those shown in Fig. 2.20 possess the important property that they form a lattice. Thus any point h, k, l can be located by a reciprocal lattice vector

$$\sigma_{hkl} = h\mathbf{a}^* + k\mathbf{b}^* + l\mathbf{c}^* \qquad (2.11)$$

where $\mathbf{a}^*$, $\mathbf{b}^*$, $\mathbf{c}^*$ are vectors defining a unit cell in the reciprocal lattice and h, k, and l are integers. This may be proved by construction as in Fig. 2.20 but an analytical proof is most easily

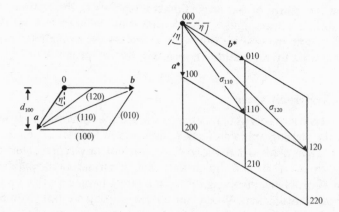

FIG. 2.20. Reciprocal lattice points for some planes in the [001] zone of a monoclinic crystal.

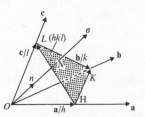

FIG. 2.21. Relation between reciprocal-lattice vector σ_{hkl} and crystal plane (hkl).

given by reversing the course we have just taken; we shall define a lattice with certain properties and show that it corresponds to the reciprocal lattice.

If **a**, **b**, **c** are the unit-cell vectors of the direct, or real-space, lattice, consider a second lattice with cell vectors $\mathbf{a}^*$, $\mathbf{b}^*$, $\mathbf{c}^*$ such that

$$\mathbf{a}\cdot\mathbf{b}^* = \mathbf{a}\cdot\mathbf{c}^* = 0; \qquad \mathbf{b}\cdot\mathbf{a}^* = \mathbf{b}\cdot\mathbf{c}^* = 0; \qquad \mathbf{c}\cdot\mathbf{a}^* = \mathbf{c}\cdot\mathbf{b}^* = 0$$

$$(2.12)$$

$$\mathbf{a}\cdot\mathbf{a}^* = \mathbf{b}\cdot\mathbf{b}^* = \mathbf{c}\cdot\mathbf{c}^* = 1. \qquad (2.13)$$

From eqns (2.12) and (2.13) it follows that **a** is perpendicular to $\mathbf{b}^*$ and $\mathbf{c}^*$, **b** is perpendicular to $\mathbf{a}^*$ and $\mathbf{c}^*$, and so on. This is illustrated in two dimensions in Fig. 2.20 for a crystal viewed along the direction of the c axis.

Let us first prove that the reciprocal lattice vector σ_{hkl} (eqn (2.11)) is normal to the set of planes (hkl) in the direct lattice. In Fig. 2.21 HKL represents part of the plane which is nearest the origin in this set and so it cuts the crystal axes at points H, K, and L defined by the vectors $\mathbf{a}/h$, $\mathbf{b}/k$, $\mathbf{c}/l$. By vector addition

$$\mathbf{b}/k - \mathbf{a}/h = \mathbf{HK}. \qquad (2.14)$$

Then

$$\sigma_{hkl}\cdot\mathbf{HK} = (h\mathbf{a}^* + k\mathbf{b}^* + l\mathbf{c}^*)\cdot(\mathbf{b}/k - \mathbf{a}/h)$$
$$= 1 - 1 = 0. \qquad (2.15)$$

When the scalar product of two vectors is zero they must be perpendicular to each other and it can similarly be shown that σ_{hkl} is perpendicular to **KL**. Since **HK** and **KL** together define the plane HKL then σ_{hkl} must be normal to this plane.

We must now consider the magnitude of σ_{hkl}. Let **n** be a unit vector in the direction of σ_{hkl}, i.e.

$$n = \frac{\sigma_{hkl}}{|\sigma_{hkl}|}. \tag{2.16}$$

In Fig. 2.21 the interplanar spacing d_{hkl} corresponds to ON, and so

$$\frac{1}{d} = \frac{1}{ON} = \frac{1}{(a/h) \cdot n}$$

$$\left(= \frac{1}{(b/k) \cdot n} \text{ etc.} \right). \tag{2.17}$$

From eqns (2.16) and (2.17)

$$\frac{1}{d} = \frac{|\sigma_{hkl}|}{(a/h) \cdot \sigma_{hkl}} = \frac{|\sigma_{hkl}|}{(a/h) \cdot (ha^* + kb^* + lc^*)}$$

$$= |\sigma_{hkl}|. \tag{2.18}$$

The magnitude of σ_{hkl} is thus equal to the reciprocal of the spacing of the planes (hkl) in the direct lattice. This relationship is in fact expressed by eqns (2.12) and (2.13) with which the lattice a^*, b^*, c^* was in part defined; in Fig. 2.20 for example the spacing of the (100) planes is seen to be

$$d_{100} = |a| \cos \eta \tag{2.19}$$

where η is the angle between a and a^*. From eqn (2.13), however,

$$a \cdot a^* = aa^* \cos \eta = 1. \tag{2.20}$$

Thus $1/d_{100} = |a^*|$, which is identical to $|\sigma_{100}|$ (see Fig. 2.20).

We have now shown that the lattice a^*, b^*, c^* defined by eqns (2.12) and (2.13) satisfies the conditions required for a lattice reciprocal to a, b, c. The reciprocal lattice vector σ_{hkl} has the following properties: (i) its direction is normal to planes (hkl) in the real-space lattice; (ii) its magnitude is the reciprocal of the spacing of the (hkl) planes.

It is also important to note the relationships between axes in direct and in reciprocal space. For example, in the three-dimensional case a^* is normal to b and c; in two dimensions a^* is normal to b, and b^* to a (Fig. 2.20). If the axes are orthogonal the situation is particularly simple since a^* is then parallel to a and b^* to b; also, $a^* = 1/a$ and $b^* = 1/b$.

As mentioned previously the concept of reciprocal space is extremely useful when discussing problems in diffraction but its application is not limited to this case. Numerous relationships involving direct and reciprocal lattice parameters may be readily derived (Azaroff 1969). As a specific example we may consider the planes lying in a particular zone $[uvw]$ in a crystal; the zone axis will be given by the equation

$$\text{zone axis} = u\mathbf{a} + v\mathbf{b} + w\mathbf{c}. \qquad (2.21)$$

while the normal to any plane (hkl) is defined, in the reciprocal lattice, by the equation

$$\boldsymbol{\sigma}_{hkl} = h\mathbf{a}^* + k\mathbf{b}^* + l\mathbf{c}^*. \qquad (2.22)$$

If (hkl) is to lie parallel to the zone axis, then $\boldsymbol{\sigma}_{hkl}$ must be perpendicular to this axis and the scalar product

$$(u\mathbf{a} + v\mathbf{b} + w\mathbf{c}) \cdot (h\mathbf{a}^* + k\mathbf{b}^* + l\mathbf{c}^*) \qquad (2.23)$$

must be zero. Since $\mathbf{a} \cdot \mathbf{a}^* = 1$, $\mathbf{a} \cdot \mathbf{b}^* = 0$, etc. (eqns (2.12) and (2.13)), this reduces to the requirement

$$hu + kv + lw = 0 \qquad (2.24)$$

which was given without proof in Section 2.4 as the condition for the plane (hkl) to be in the $[uvw]$ zone.

If a distance in a particular direction is large in the direct lattice it will be small in the corresponding reciprocal lattice (see for example Fig. 2.20). We can imagine a two-dimensional structure in real space as being formed by gradual separation of the lattice planes in a three-dimensional structure to infinity, say along the $\mathbf{c}$ direction. In the reciprocal lattice this will result in a continual decrease in the separation of the lattice points in the $\mathbf{c}^*$ direction until they finally fuse together. Therefore in a lattice reciprocal to a two-dimensional structure the lattice elements are rods of infinite length, normal to the plane of the direct lattice. If the process is continued to produce a one-dimensional structure by separating the net rows to infinity in the $\mathbf{b}$ direction, the reciprocal lattice rods will fuse in the $\mathbf{b}^*$ direction. Thus in a lattice reciprocal to a one-dimensional structure the lattice elements are planes of infinite extent, normal to the line of the direct lattice.

In the two-dimensional case conversion from direct to reciprocal unit meshes and the reverse operation can be very neatly carried out using a matrix method first introduced by Brillouin (1946) (see also Park and Madden 1968). If the origin of a pair of orthogonal axes x and y is taken to coincide with the origin of the direct lattice the basis vectors **a** and **b** may be written in terms of their Cartesian components as

$$\mathbf{a} = a_x\mathbf{i} + a_y\mathbf{j} \tag{2.25}$$

$$\mathbf{b} = b_x\mathbf{i} + b_y\mathbf{j} \tag{2.26}$$

where **i** and **j** are unit vectors in the x and y directions respectively. The direct unit mesh may thus be represented by a matrix **A**:

$$\mathbf{A} = \begin{bmatrix} a_x & a_y \\ b_x & b_y \end{bmatrix}. \tag{2.27}$$

This is converted to the matrix $\mathbf{A}^*$ representing the reciprocal unit mesh by the operation of matrix inversion

$$\mathbf{A}^* = \frac{1}{a_x b_y - a_y b_x} \begin{bmatrix} b_y & -a_y \\ -b_x & a_x \end{bmatrix}$$

$$= \frac{1}{|A|} \begin{bmatrix} b_y & -a_y \\ -b_x & a_x \end{bmatrix} \tag{2.28}$$

where $|A|$ is the determinant of **A**. The components of the

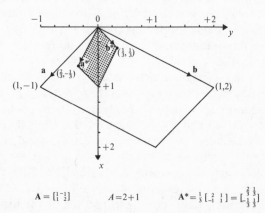

$$\mathbf{A} = \begin{bmatrix} 1 & -1 \\ 1 & 2 \end{bmatrix} \qquad A = 2+1 \qquad \mathbf{A}^* = \tfrac{1}{3}\begin{bmatrix} 2 & 1 \\ -1 & 1 \end{bmatrix} = \begin{bmatrix} \tfrac{2}{3} & \tfrac{1}{3} \\ -\tfrac{1}{3} & \tfrac{1}{3} \end{bmatrix}$$

FIG. 2.22. The use of the matrix method for conversion between real-space and reciprocal-space.

reciprocal mesh vectors are given by the *columns* of $\mathbf{A}^*$, i.e.

$$\mathbf{a}^* = \frac{b_y}{|A|}\mathbf{i} - \frac{b_x}{|A|}\mathbf{j} \tag{2.29}$$

$$\mathbf{b}^* = \frac{-a_y}{|A|}\mathbf{i} + \frac{a_x}{|A|}\mathbf{j}. \tag{2.30}$$

The process is much easier to carry out than may be apparent from the above outline and it provides the most reliable means of conversion between the two types of net. A numerical example is given in Fig. 2.22.

2.8. Single-crystal metal surfaces

2.8.1. *Surface models*

The configurations of atoms in ideally flat single-crystal surfaces have been considered in detail by Nicholas (Moore and Nicholas 1961, Nicholas 1961) and a comprehensive collection of photographs of ball models of various types of surface has been published (Nicholas 1965). Fig. 2.23 shows some of the low-index faces of the b.c.c., f.c.c., and h.c.p. structures. These can be visualized easily by examining the three-dimensional unit cells, but this is often difficult for faces of higher indices and so it is instructive to investigate the rules for building the corresponding ball models.

As a simple but important example the construction of an (hkl) surface of an f.c.c. or b.c.c. crystal from a stack of (100) planes will be outlined (Moore and Nicholas 1961). Because of the cubic symmetry, discussion can be limited to planes with normals lying in the unit stereographic triangle in which $h \geqslant k \geqslant l \geqslant 0$; this includes planes representative of all possible forms. The first problem is to determine the configuration of the atoms along the edge of each (100) layer. It is convenient to choose crystal axes x, y lying in the (100) plane to coincide with the directions of the square unit mesh (Fig. 2.24). The repeat distance along the edge can then be characterized by a total number, A, of displacements in the x direction combined with B in the y direction, each displacement being through one atomic diameter. It can be

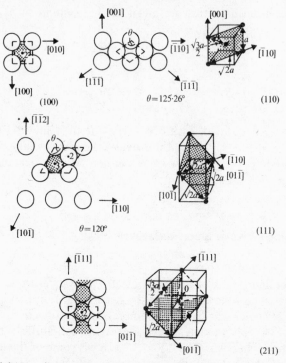

FIG. 2.23(a). Low-index planes in the b.c.c. system. The primitive surface unit mesh is indicated by broken lines. Atoms labelled 2 (3) lie in the first (second) layer below the surface.

shown (Moore and Nicholas 1961) that the following relationships apply:

$$\frac{B}{A} = \frac{k+l}{k-l} \quad \text{f.c.c. case}$$

$$\frac{B}{A} = \frac{l}{k} \quad \text{b.c.c. case.} \tag{2.31}$$

A and B are the smallest positive integers which satisfy this equation and which have no common factor. The A and B displacements occur in groups; in the f.c.c. structure for example the A displacements occur one at a time, the B in groups of I or $(I+1)$ where I is the integral part of B/A. The order of occurrence of these groups can be specified precisely (Nicholas 1961) but is obvious in most cases. If a straight edge touches atoms at equivalent positions along the mean direction of the edge of the layer there must be no vacant sites behind the straight edge which could accommodate a complete atom.

It is interesting to note that the same edge structure is possessed by any surface (*hkl*) having a given value of *B/A*. All such surfaces in fact lie in the same zone. When building the structure from (100) planes this zone must also contain (100) and so its

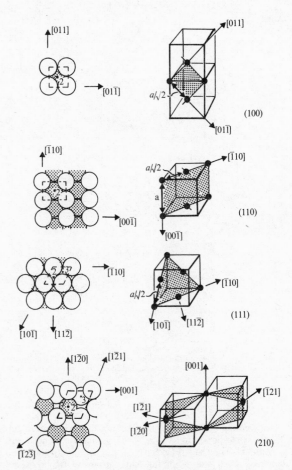

Fig. 2.23(*b*). Low-index planes in the f.c.c. system. For details see Fig. 2.23(*a*).

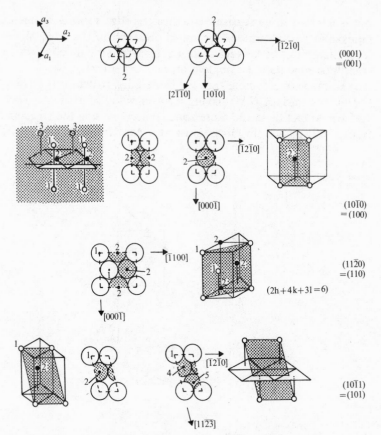

FIG. 2.23(c). Low-index planes in the h.c.p. system. For details see Fig. 2.23(a).

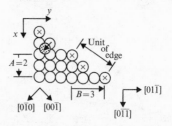

FIG. 2.24. Arrangement of balls in a (100) plane of the face-centred cubic model. The edge shown corresponds to $A = 2$, $B = 3$, and balls marked with a cross represent kink atoms. The shaded circle shows the position in the layer above of the ball separated from the adjoining kink atom by the chosen interlayer translation. (After Moore and Nicholas 1961.)

indices will be $[0vw]$ (eqn (2.7)). Further, as (hkl) lies in the zone

$$0 + kv + lw = 0$$

or

$$v/w = -l/k.$$

Therefore the zone is $[0\bar{l}k]$. Some planes lying in various zones in the cubic system are shown in Fig. 2.25 and the structures of the (100) layer edges for f.c.c. planes in a number of different zones are shown in Fig. 2.26; the b.c.c. zones which have these (100) edges are also identified. We might have chosen to construct models by stacking (111) rather than (100) planes and therefore the (111) layer edges for a number of zones are shown in Fig. 2.27. Note that if the plane upon which the construction is based is not a plane of a given zone then the edge structure will vary from plane to plane in that zone. For example, all planes in the $[0\bar{1}2]$ zone have the same (100) edge but each plane in the $[\bar{1}2\bar{1}]$ zone has a different (100) edge.

Having established the structure of a single (100) layer, the way in which the layers are stacked to generate the desired surface can then be specified by two further translations. An interlayer translation, which will carry an atom from a site in the first layer to a site in the second layer can be defined in an infinite number of ways, but one of the simplest involves a vertical movement followed by displacement in the $[0\bar{1}0]$ direction (Fig. 2.24).

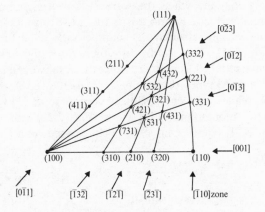

FIG. 2.25. Some planes lying in various zones in the cubic system.

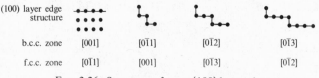

(100) layer edge structure			
b.c.c. zone	[001]	[0$\bar{1}$1]	[0$\bar{1}$2]	[0$\bar{1}$3]
f.c.c. zone	[0$\bar{1}$1]	[001]	[0$\bar{1}$3]	[0$\bar{1}$2]

FIG. 2.26. Structures of some {100} layer edges.

Alternatively, it may be viewed as a displacement through one atom diameter in the [1$\bar{1}$0] direction. This effectively generates an atomic step. It may then be necessary to move the edge perpendicular to itself in the plane of the second layer, this being equivalent to removal of rows of atoms lying parallel to the edge. The complete movement from an edge in one layer to that in the next layer will consist of an interlayer translation followed by a number (possibly zero) of row removals, the latter determining the widths of the terraces on the surface. The sequence of such combined translations will vary periodically as successive layers are added to the structure, each period containing a total of C row removals and D steps, where C and D are integers without a common factor. The relationships by which they are determined are listed in Table 2.3. Again, the row removals and the steps will occur in groups in an order which may be specified exactly but which is generally obvious from the 'straight-edge' criterion.

The foregoing discussion has considered f.c.c. and b.c.c. surfaces to be composed of terraces of {100} configuration separated by atomic steps, and when the terraces are of significant width and the steps are monatomic this is the most satisfactory way of viewing the various structures; examples would be f.c.c. (411), (310), and (721) (Fig. 2.28). In other cases, however, {100} 'terraces' are reduced to very small width, or may even be absent, and other terrace types predominate. Thus the (332), (221), and (331) f.c.c. planes are very obviously composed of {111} terraces separated by monatomic {111} steps (Fig. 2.29) and the b.c.c. (210) and (320) planes consist of {110} terraces and {100} steps

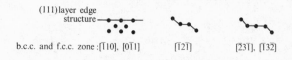

(111)layer edge structure

b.c.c. and f.c.c. zone :[$\bar{1}$10], [0$\bar{1}$1]   [$\bar{1}$2$\bar{1}$]   [$\bar{2}$3$\bar{1}$], [$\bar{1}$3$\bar{2}$]

FIG. 2.27. Structures of some {111} layer edges.

TABLE 2.3

	Based on (100) layers	Based on (111) layers
f.c.c.	$B/A = (k+l)/(k-l)$ $C/D = A\{(h-k)/(k-l)\}$	$B/A = (k-l)/(h-k)$ $C/D = A\{(k+l)/(h-k)\}$
b.c.c.	$B/A = l/k$ $C/D = A(h-k+l)/2k$	$B/A = (k-l)/(h-k)$ $C/D = A(-h+k+l)/2(h-k)$

(Fig. 2.30). Finally, there are many cases in which several low-index configurations make important contributions to the structure of a particular plane. In the f.c.c. system (311) may be considered to consist of {100} terraces two atom rows wide and monatomic {111} steps or of {111} terraces and {100} steps, while f.c.c. (210) exhibits elements of all three low-index faces, {100}, {110}, and {111} (Fig. 2.31).

In the case of h.c.p. structures the situation is complicated by the fact that two classes of atom can be identified on the basis of their environments. If the structure is considered to be built from {0001} layers all atoms in any layer are of the same class, distinguished by whether the layer occupies an A or a B position in the stacking order ABABAB .... As seen from Fig. 2.6 an atom (such as A) in one class sits on three atoms in the underlying plane which are in the configuration $\triangle$, while for an atom of the other class (such as B) the corresponding configuration is $\triangledown$. Thus

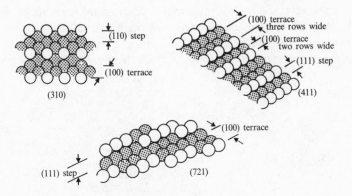

FIG. 2.28. Some face-centred cubic planes viewed as {100} terraces and mon-atomic steps. (After Nicholas 1965.)

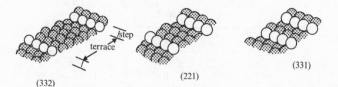

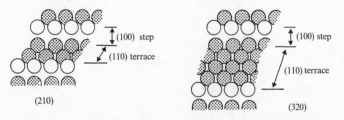

FIG. 2.29. Some face-centred cubic planes viewed as {111} terraces and {111} monatomic steps. (After Nicholas 1965.)

two types of (0001) surface are possible, each with the same unit mesh but differing in the atom positions in the underlying layers, and this is generally true for any h.c.p. surface (*hkil*) such that $(2h + 4k + 3l)$ is not a multiple of 6. The (0001) surfaces differ only by a rotation through 180°, but in other cases the difference is not trivial (Fig. 2.23(*c*)).

At this stage it will be obvious that the full subtlety of a given structure can often be appreciated only after examination of an appropriate model. In most cases this can easily be constructed by sticking half-inch diameter polystyrene balls onto a rigid base in a {100} or {111} array and arranging upon this layer additional loose balls in the proper configuration. Data necessary for the generation of various surfaces are collected in Table 2.3 and further information is given by Nicholas (1965).

2.8.2. *The density and co-ordination of surface atoms*
When discussing chemisorption on solid surfaces the adsorbate coverage θ at various stages of the process is often related to the 'monolayer capacity' of the surface. If the solid is polycrystalline the monolayer can be defined only in terms of the number of molecules adsorbed at 'saturation' either of the gas in question or of some reference gas such as krypton or xenon. The 'monolayer'

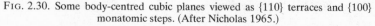

FIG. 2.30. Some body-centred cubic planes viewed as {110} terraces and {100} monatomic steps. (After Nicholas 1965.)

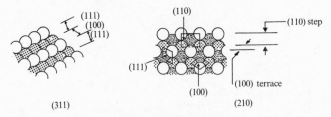

FIG. 2.31. Contributions of low-index planes to the f.c.c. (311) and (210) surfaces.

therefore varies in magnitude with the gas used and also possibly with the adsorption temperature. In the case of single-crystal surfaces it is more satisfactory to express adsorbate concentrations relative to the atom density in the surface layer of the solid, since the monolayer concentration ($\theta = 1$) then has a fixed value for a given surface irrespective of the gas involved. For low-index planes the definition of surface-atom density is obvious, but on high-index surfaces the unit mesh is large and the choice of a location for the surface plane becomes somewhat arbitrary. Atoms which are technically in the second or third layers below the surface may be exposed to an extent similar to those in the top layer (see, for example, Fig. 2.29). Nevertheless it is still often convenient in the case of more open planes to equate the monolayer with the atom density in the first surface layer, although this may result in saturation adsorbate concentrations which are greater than one monolayer ($\theta > 1$).

The surface-atom density as just defined will be equal to the inverse of the area S_{hkl} of a primitive unit mesh in the surface; this latter quantity is given by the following relationships (Nicholas 1965):

$$S_{hkl} = \tfrac{1}{4}Qa^2(h^2 + k^2 + l^2)^{\frac{1}{2}} \quad \text{b.c.c. and f.c.c.}$$
$$S_{hkl} = \tfrac{1}{2}a^2[4r^2(h^2 + hk + k^2) + 3l^2]^{\frac{1}{2}} \quad \text{h.c.p.} \tag{2.32}$$

where for b.c.c. $Q = 2$ if $(h+k+l)$ is even and $Q = 4$ if $(h+k+l)$ is odd, for f.c.c. $Q = 1$ if $h, k, l,$ are all odd and $Q = 2$ if h, k, l are of mixed parity, a is the cubic lattice parameter, and r is the h.c.p. axial ratio c/a. Atom densities calculated using these equations are given in Table 2.4 for a selection of planes of various metals.

TABLE 2.4
Surface-atom densities

f.c.c. structure

Plane	(100)	(110)	(111)	(210)	(211)	(221)	(310)	(311)	(320)
Density relative to (111)	0·866	0·612	1·000	0·387	0·354	0·289	0·274	0·522	0·240
Metal	Al	Rh	Ir	Ni	Pd	Pt	Cu	Ag	Au
Density of (111) (atom cm^{-2} × 10^{-15})	1·415	1·599	1·574	1·864	1·534	1·503	1·772	1·387	1·394

b.c.c. structure

Plane	(100)	(110)	(111)	(210)	(211)	(221)	(310)	(311)	(320)
Density relative to (110)	0·707	1·000	0·409	0·316	0·578	0·236	0·447	0·213	0·196
Metal	V	Nb	Ta	Cr	Mo	W	Fe		
Density of (110) (atom cm^{-2} × 10^{-15})	1·547	1·303	1·299	1·693	1·434	1·416	1·729		

h.c.p. structure

Plane	(0001)	(10$\bar{1}$0)	(10$\bar{1}$1)	(10$\bar{1}$2)	(11$\bar{2}$0)	(11$\bar{2}$2)		
Density relative to (0001)	1·000	$\dfrac{3}{2r}$	$\dfrac{\sqrt{3}}{(4r^2+3)^{\frac{1}{2}}}$	$\dfrac{\sqrt{3}}{(4r^2+12)^{\frac{1}{2}}}$	$\dfrac{1}{r}$	$\dfrac{1}{2(r^2+1)^{\frac{1}{2}}}$		
Metal	Zr	Hf	Re	Ru	Os	Co	Zn	Cd
Density of (0001) (atom cm^{-2} × 10^{-15})	1·110	1·130	1·514	1·582	1·546	1·830	1·630	1·308
axial ratio $r = c/a$	1·59	1·59	1·61	1·58	1·58	1·62	1·86	1·89

An atom in a surface layer has fewer nearest neighbours, or a lower co-ordination number z, than an atom in the bulk of the solid, and this feature may of course be regarded as being basically responsible for the various unique properties of surfaces. It is therefore of some interest to determine, for a given surface, the densities of atoms having various values of z, and a generalized method for doing so has been given by Nicholas (Mackenzie, Moore, and Nicholas 1962, Mackenzie and Nicholas 1962). Results for the b.c.c. structure are shown in Fig. 2.32; $n(z)$ represents the number of atoms with co-ordination number z lying within an area a^2 of the surface, where a is the three-dimensional unit-cell parameter. The number of such atoms within a primitive surface unit mesh of area S_{hkl} can be easily calculated using eqns (2.32). A given surface atom in the b.c.c. structure will have $(8-z)$ fewer nearest neighbours than an atom in the bulk; for the f.c.c. structure the value is $(12-z)$. Mackenzie, Moore, and Nicholas have also estimated the densities of atoms with various co-ordination number in a polycrystalline b.c.c. surface in which the grains are assumed to be oriented at random. It is interesting to note that the result is not the same as that obtained by averaging results from the three low index faces, (100), (110), and (111), an approximation which has been used in the past in discussions of polycrystalline surfaces.

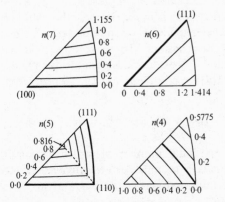

FIG. 2.32. Contour plots for a b.c.c. crystal of the density $n(z)$ of atoms with co-ordination number z. The density is referred to area a^2 of surface. Zero contours are plotted as thick lines. The broken line shows where corners occur in the contours. (From MacKenzie, Moore and Nicholas 1962.)

2.8.3. *The stability of crystal surfaces*

When we consider a solid immersed in a fluid (liquid or gas) at constant temperature it is found that the equilibrium shape of the solid particle and the geometrical relationships between its intersecting surfaces are determined by a quantity γ, properly known as the specific surface work. (A variety of names for γ appear in the literature, including surface energy and surface tension; for a discussion of nomenclature, together with a comprehensive review of surface thermodynamics, see Linford (1973). The present simplified treatment largely follows Blakely (1973).) Physically, γ is the reversible work required to form unit area of new solid surface by cleavage at constant temperature, volume, and number of moles of each component present.

Any extensive property of the solid–fluid system, for example the Helmholtz free energy F, may be written in the form

$$F_{\text{total}} - F_\pi = F_\alpha + F_\beta. \tag{2.33}$$

In this equation F_α and F_β are the values F would have for the solid and fluid phases, if these phases maintained their bulk properties up to a dividing plane located at the solid surface. In reality, of course, the change from solid to fluid is not discontinuous but occurs across a 'surface phase' π of finite thickness with which we associate the term F_π. This term is thus a thermodynamic excess quantity, representing the amount by which F for the actual system is in excess over the value $(F_\alpha + F_\beta)$ for the hypothetical system with a sharp boundary. For the creation of an area dA of new surface the work done will be $\gamma\,dA$ and at constant V and T work done is equal to the change in Helmholtz free energy for the complete system. Hence, from eqn (2.33) we obtain

$$[\gamma\,\partial A = \partial F_{\text{tot}} = \partial F_\pi + \partial(F_\alpha + F_\beta)]_{V,T,n_i} \tag{2.34}$$

where n_i is the total number of moles of component i in the system. Formation of new surface involves transfer of material between the bulk and surface phases and so

$$\partial(F_\alpha + F_\beta) = \sum_i \mu_i\,dn_{i,\alpha+\beta} = -\sum_i \mu_i\,dn_{i,\pi} \tag{2.35}$$

where μ_i is the chemical potential of component i. Combining eqns (2.34) and (2.35) finally gives

$$\gamma = f_\pi - \sum_i \mu_i \Gamma_i \tag{2.36}$$

where $f_\pi = \mathrm{d}F_\pi/\mathrm{d}A$ is the specific (i.e. per unit area) surface Helmholtz free energy and $\Gamma_i = \mathrm{d}n_{i,\pi}/\mathrm{d}A$ is the surface excess (or surface coverage) of component i. The effects on γ of temperature changes and of adsorption of species i are given by the Gibbs adsorption equation

$$\mathrm{d}\gamma = -s_\pi \,\mathrm{d}T - \sum_i \Gamma_i \,\mathrm{d}\mu_i \tag{2.37}$$

where s_π is the specific surface entropy.

The equilibrium shape of the solid will be determined by the condition that F_{tot} should have a minimum value. If γ is independent of orientation, as in a liquid, this condition becomes (using eqn (2.34))

$$F_{\mathrm{tot}} = \int_A \gamma \,\mathrm{d}A = \text{minimum} \tag{2.38}$$

where the integral is taken over the complete surface. The equilibrium state corresponds to minimum surface area and the particle will be spherical. In solids, however, γ is generally a function of orientation, $\gamma = \gamma(\mathbf{n})$, where $\mathbf{n}$ is a unit vector in the direction of the surface normal. Equilibrium now requires that

$$\sum_m \int_{A_m} \gamma(\mathbf{n}_m) \,\mathrm{d}A = \text{minimum} \tag{2.39}$$

and the equilibrium shape will be a polyhedron with surfaces of low γ preferentially exposed; the actual shape can be determined from a polar plot of γ as a function of orientation known as the Wulff plot (see Blakely 1973). Such polyhedra have been observed experimentally with small metal particles and are of obvious interest in relation to supported metal catalysts.

In generating a surface by cleavage a greater number of bonds must be broken in the case of an atom located at the edge of a terrace than for an atom in the middle of a low-index plane. Thus low-index planes hkl (also called singular surfaces) should have relatively small values of γ. Orientations a few degrees away from

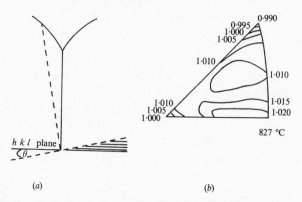

FIG. 2.33. (a) A cusp in the γ plot; (b) γ plot for clean copper. (After McLean 1971.)

a low-index plane (i.e. vicinal surfaces) will consist of *hkl* terraces separated by monatomic steps and because of the presence of the steps will have γ values higher than γ_{hkl}. Thus in a plot of γ as a function of orientation, minima, or cusps, will occur in directions normal to the singular surfaces (Fig. 2.33(a)). Experimental results for clean copper (McLean 1971) are shown in Fig. 2.33(b), plotted as contours within the stereographic triangle.

Adsorption on a clean surface lowers γ. That this should be so can be seen by considering eqn (2.37) under constant temperature conditions (in which case it is known as the Gibbs adsorption isotherm):

$$d\gamma_T = -\sum \Gamma_i \, d\mu_i. \tag{2.40}$$

To develop this equation a little further let us consider (c.f. Linford 1973) a solid phase (component 1) with negligible vapour pressure in contact with an ideal gas (component 2) which is insoluble in the solid. The dividing surface between the two phases can be chosen so that $\Gamma_1 = 0$, and so, at constant T,

$$d\gamma = -\Gamma_2 \, d\mu_2. \tag{2.41}$$

For an ideal gas

$$\mu = \mu_0 + kT \ln P$$

and

$$d\mu_T = kT \, d(\ln P).$$

Hence

$$d\gamma_T = -kT \, \Gamma_2 \, d(\ln P_2). \tag{2.42}$$

Suppose that it is found that Γ_2 (coverage) is proportional to P_2. For example, this might be the case with physical adsorption at low coverages. Then

$$d\gamma_T = -kT \tag{2.43}$$

and the reduction in γ resulting from adsorption is directly proportional to T at a given coverage. The effect of physical adsorption on γ is relatively small, but in the case of chemisorption the reduction in γ may be considerable. In general, adsorption also accentuates the variation of γ with surface orientation.

It is quite often observed that a flat surface will decompose into a hill-and-valley structure consisting of facets of two or more orientations, at least one of these being of low index. Although the total surface area is increased in the process, the formation of surfaces of low γ results in an overall decrease in Helmholtz free energy (c.f. eqn (2.39)). Any flat surface which is not part of the equilibrium shape for the solid under a given set of conditions will be unstable, and so all faceting behaviour could be predicted if the complete Wulff plot (and hence the equilibrium shape) for the solid was known. Herring (1951) has shown that the criterion for stability of a plane represented by some point A on a γ plot (Fig. 2.34(a)) can be defined as follows. A sphere is constructed which passes through the origin O of the plot and is tangent to the plot at A. If the plot nowhere passes inside this sphere (Fig. 2.34(a)) then the surface corresponding to A will not facet to the low index orientation S; in contrast, the surface corresponding to A′ (Fig. 2.34(b)) would facet. This criterion must be applied using

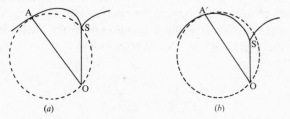

(a) (b)

FIG. 2.34. Criteria for stability of a plane A (or A′) with respect to faceting to a low index pole S; 0 is the origin of the γ plot.

the complete three-dimensional Wulff plot, but in many cases, particularly for surface orientations far from a low index pole, sufficient experimental data are not available for this to be done. In general, however, faceting will be favoured by the presence of extensive cusps in the γ plot at the low index poles.

For clean metal surfaces the variation of γ with orientation is of the order of only 1 per cent (Fig. 2.33(b)) and all orientations should be stable, a prediction which is supported by LEED observations. Faceting therefore only occurs in the presence of adsorbed impurities and is the result of an increase in the orientational anisotropy of γ which is brought about by variations in both the coverage and the binding energy of the adsorbate with crystal plane. The process is discussed further in Chapter 12.

2.9. Crystal defects

The structures discussed so far have been those of ideal solids consisting of perfectly repeating arrays of unit cells, but in real materials imperfections inevitably occur. These may be point, line, surface, or volume defects, the simplest being the point defect. In a metal this can arise in the following ways.

(i) By removal to the surface, to a dislocation, etc. of an atom from a point in an otherwise perfect lattice producing a vacancy or Schottky defect.

(ii) By removal of an atom to an interstitial position, i.e. a position not coinciding with a point in the perfect lattice; this produces a Frenkel defect.

(iii) By introduction of an impurity to an interstitial position.

(iv) By replacement of a parent metal atom by an impurity; this is termed a substitutional impurity.

Schottky and Frenkel defects are always present in any solid at temperatures greater than 0 K, since they result in an increase in the configurational entropy of the system. In the Schottky case, if the energy required to remove an atom from the lattice is E_s and the change in entropy is ΔS, the equilibrium number of vacancies n is given by

$$n = (N-n) \exp\left(\frac{-E_s}{RT} + \frac{\Delta S}{R}\right) \approx N \exp\left(\frac{-E_s}{RT}\right) \qquad (2.44)$$

where N is the total number of atoms in the solid. Taking copper as an example, $E_s \approx 110 \text{ kJ mol}^{-1}$. If the entropy term in eqn (2.44)

TABLE 2.5

Approximate values for the ratio of the equilibrium number of Schottky defects to the total number of atoms in solid Cu at various temperatures.

T/K	0	300	600	900	1200	1350	(m.p. 1356)
n/N	0	7×10^{-20}	3×10^{-10}	4×10^{-7}	2×10^{-5}	6×10^{-5}	

is neglected, the temperature variation of the equilibrium ratio (n/N) is as shown in Table 2.5. The annihilation of Schottky and Frenkel defects involves diffusion in the solid and so a high concentration of these defects can be 'frozen' into the structure by rapid cooling from a high temperature.

The concept of the type of line (i.e. one dimensional) imperfection known as a dislocation was introduced to account for the ease with which metals can be plastically, or permanently, deformed. In many cases this deformation occurs by a process of slip, part of the crystal on one side of a slip plane sliding over the other part (Fig. 2.35(a)). If it is assumed that simultaneous movement of one complete plane of atoms over another is

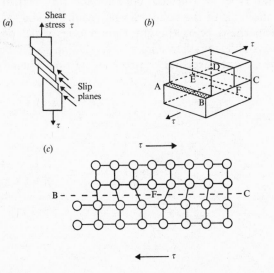

FIG. 2.35. (a) The process of slip under an applied shear stress τ. (b) Production of an edge dislocation by slip. (c) Atomic arrangement in the region of an edge dislocation.

involved, the calculated value of the shear stress required for plastic deformation is much higher than that observed experimentally. It is therefore proposed that the slip process progresses gradually across the slip plane, involving successive movements of individual atoms; this requires a much lower stress. The boundary between the slipped and unslipped regions of the crystal represents a dislocation line. If slip has occured in the manner shown in Fig. 2.35(*b*) over the left half of a slip plane ABCD but not over the right half, the boundary EF represents an *edge* dislocation. At points remote from the dislocation line the perfect structure remains unperturbed, while the detailed atomic arrangement around the line is as shown in Fig. 2.35(*c*); effectively it is equivalent to the introduction of an extra half-plane of atoms into the crystal above the slip plane. The other basic type of dislocation, the *screw* dislocation, is generated if displacement of the crystal on either side of the slip plane takes place as shown in Fig. 2.36(*a*); in this case the atoms around the dislocation line E′F′ are arranged in a spiral (Fig. 2.36(*b*)). The structures of real dislocations are generally more complex than those of the two simple cases just outlined.

Several types of 'surface' imperfection (i.e. two dimensional, within the bulk of the solid) can be readily envisaged, the most obvious perhaps being the grain boundaries which separate the single crystallites making up a polycrystalline specimen. Stacking faults represent a second example, occurring when the regular

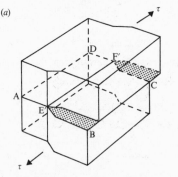

FIG. 2.36. (*a*) Production of a screw dislocation by slip. (*b*) Arrangement of atoms around a screw dislocation. (After Hume-Rothery 1966.)

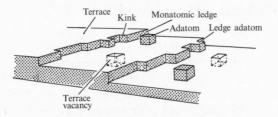

FIG. 2.37. The terrace–ledge–kink model for a surface.

sequence in which atomic planes are stacked to form the crystal is interrupted. For example, in a perfect h.c.p. structure (0001) planes are arranged in the order ... ABABAB ... (Section 2.3); the sequence ... ABABABCAB ... would represent a stacking fault.

Defects on the surface of a solid arise when the surface is intersected by line or surface defects from the bulk. A number of point defects will also be present at any temperature above 0 K. The simplest picture of a defective single-crystal surface is provided by the terrace–ledge–kink (TLK) model originally suggested by Kossel (1927) (see Gjostein 1963, 1967). A surface of low index such as f.c.c. (100) or (111) is essentially planar, but at high temperatures adatom–vacancy pairs will be formed (Fig. 2.37); their concentration, however, will be low. On a vicinal surface, that is one with an orientation comparatively close to a low-index pole, steps must occur to maintain the inclination with respect to the low-index plane. These steps may or may not contain kink sites in the perfect structure (see Fig. 2.28), but in any case defect kink sites and jog sites will be formed by atoms moving away from a ledge onto the adjoining terrace or by terrace adatoms moving to the ledge (Fig. 2.37). The concentration of ledge defects will be much greater than that of adatom–vacancy pairs on terraces.

Defects in specific materials have been experimentally detected by a number of techniques, including electron microscopy, but the most powerful is undoubtedly field ion microscopy since this allows direct observation of defects in metals on the atomic scale. The field ion microscope is discussed in Chapter 6; for more comprehensive discussion of defects in bulk solids see, for example, Hume-Rothery (1966), Cottrell (1967), Hull (1968), and Henderson (1972).

2.10. Diffusion in solids and on solid surfaces

If we consider diffusion within a solid a number of different situations can be envisaged. For example, an impurity atom small enough to occupy an interstitial position may hop to a neighbouring site during a temporary local expansion of the lattice (Fig. 2.38(a)); a mechanism of this type would be typical of diffusion of carbon in iron and also is discussed in Chapter 11 in connection with the oxidation of nickel. In the cases of a large impurity atom occupying a substitutional position or of self-diffusion of an atom in its own lattice the lattice expansion required for the above mechanism would be excessive and diffusion occurs predominantly by way of vacancies (Fig. 2.38(b)). Note that this does involve some degree of lattice distortion. Whatever the mechanism, the moving atom will experience a potential involving an activation energy barrier of magnitude E_m. A similar situation will pertain for diffusion on a surface, although as might be expected motion here is much less restricted than in the bulk.

For the purpose of illustration we shall consider bulk diffusion of an impurity (concentration C) along a direction x under the influence of an impurity concentration gradient. Suppose that in a solid of unit cross-sectional area the value of C in an atomic plane 1 is C_1 (atoms per unit volume) and in an adjacent plane 2 is C_2, with $C_1 > C_2$. The number of impurity atoms in these planes will be $N_1 = C_1 a$ and $N_2 = C_2 a$, where a is the atomic diameter of the lattice planes. Each impurity atom will vibrate with a frequency ν of the order of $10^{13}\,\mathrm{s}^{-1}$ and on any given

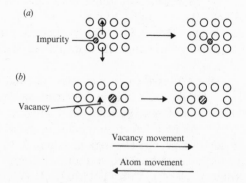

FIG. 2.38. (a) Diffusion of an interstitial atom. (b) Vacancy diffusion mechanism.

vibration will jump to a site in a neighbouring plane provided it can overcome the activation energy barrier E_m. Thus the actual jump frequency will be

$$f = \nu \exp\left(\frac{-E_m}{kT}\right) s^{-1}. \tag{2.45}$$

An atom in any plane may jump either 'forwards' or 'backwards' and so the rate of movement of atoms from plane 1 to plane 2 will be $\frac{1}{2}fN_1$; there will also be movement from 2 to 1 at a rate $\frac{1}{2}fN_2$. Hence the *net* rate of movement from the plane of higher impurity concentration (1) to that of lower concentration (2) will be

$$\frac{dN}{dt} = \frac{1}{2}f(N_1 - N_2) = \frac{1}{2}\nu \exp\left(-\frac{E_m}{kT}\right)a(C_1 - C_2).$$

We can write for the concentration gradient

$$\frac{dC}{dx} = \frac{C_2 - C_1}{a} \tag{2.46}$$

and so

$$\frac{dN}{dt} = -\frac{1}{2}\nu a^2 \exp\left(-\frac{E_m}{kT}\right)\frac{dC}{dx}. \tag{2.47}$$

This last equation may be abbreviated to

$$\frac{dN}{dt} = -D\frac{dC}{dx} \tag{2.48}$$

where

$$D = D_0 \exp\left(-\frac{E_m}{kT}\right)$$

or

$$D = \frac{1}{2}fa^2$$

and

$$D_0 = \frac{1}{2}\nu a^2.$$

Eqn (2.48) is an expression of Fick's first law of diffusion, the negative sign implying that flow is *down* the concentration gradient; D is known as the *diffusion coefficient*. The exact form of the pre-exponential factor D_0 depends on the structure of the system. In a three-dimensional simple-cubic solid, for example, an atom may move to any one of six equivalent positions on a given jump, but only one of these will carry it down the concentration gradient; hence $D_0 = \frac{1}{6}va^2$. Again, on a surface of four-fold symmetry we would expect

$$D_0 = \tfrac{1}{4}va^2. \tag{2.49}$$

Activation energies for interstitial diffusion in the bulk are of the order of ·100 kJ mol^{-1} or less, while those associated with the vacancy mechanism are somewhat higher, generally greater than about 200 kJ mol^{-1} (see Table 2.6). In this second case, however, the observed jump frequency will depend not only on the probability that during any of its vibrations an atom will acquire the activation energy E_m required to move into an adjacent vacancy, but also on the probability that a vacancy actually exists adjacent to the atom considered. This latter probability will be $\exp(-E_v/kT)$ where E_v is the energy of formation of a vacancy. The vacancy diffusion coefficient is therefore of the form

$$D = D_0 \exp\left\{-\frac{(E_m + E_v)}{kT}\right\} \tag{2.50}$$

where E_m and E_v are both generally of the order of 100 kJ mol^{-1} or greater.

In connection with diffusion on surfaces (Chapter 8) we shall find that it is important to estimate the distance x a diffusing particle may travel in some time t. If a_p is the distance moved on

TABLE 2.6
Bulk diffusion data

System	$D_0(\text{m}^2\,\text{s}^{-1})$	$E_m(\text{kJ mol}^{-1})$†
Li in Ge	$2\cdot0 \times 10^{-7}$	50
N in Fe (b.c.c.)	$5\cdot0 \times 10^{-7}$	77
C in Fe (b.c.c.)	$2\cdot0 \times 10^{-6}$	86
Fe in Fe (b.c.c.)	$1\cdot4 \times 10^{-4}$	237
Cu in Cu	$6\cdot9 \times 10^{-5}$	210
Ta in Ta	$2\cdot0 \times 10^{-4}$	460
Au in Ni	$2\cdot0 \times 10^{-6}$	147

† In the older metallurgical literature energies are often quoted in electron volts: $1\text{ eV} = 1\cdot9 \times 10^{-19}\text{ J}$, $1\text{ eV mol}^{-1} \simeq 100\text{ kJ mol}^{-1}$.
From Wert and Thompson 1964.

any one jump then

$$x = \sum_{p=1}^{P} a_p \qquad (2.51)$$

where $P = ft$ is the total number of jumps executed in time t. If we consider a large value of P, then the *net* value of x will be zero since the probability of a jump in one direction is identical to that of a jump in the reverse direction. It is more useful to calculate the root-mean-square average of x, $[\langle x^2 \rangle]^{\frac{1}{2}}$ since this is non-zero. It is a measure of the total distance travelled in time t, regardless of direction. Now

$$\begin{aligned} x^2 = \sum^{P} a_p^2 &+ 2a_1(a_2 + a_3 + \ldots) \\ &+ 2a_2(a_3 + a_4 + \ldots) \\ &+ \ldots. \end{aligned} \qquad (2.52)$$

If attention is restricted to situations in which all jump distances are identical then $a_p = a_q = a$. Since each a_p has an equal chance of being positive or negative then all terms such as $2a_1a_2$, $2a_2a_3$, etc. will be zero when averaged over a large number of jumps. Thus

$$[\langle x^2 \rangle]^{\frac{1}{2}} = (Pa^2)^{\frac{1}{2}} \qquad (2.53)$$

or

$$[\langle x^2 \rangle]^{\frac{1}{2}} = (fta^2)^{\frac{1}{2}} = (2Dt)^{\frac{1}{2}}. \qquad (2.54)$$

A considerable amount of work has been done on bulk diffusion in solids using macroscopic methods, for example radioactive-tracer techniques, while the various spectroscopies of the surface region, such as AES and XPS (see Chapter 4), now offer a sensitive means of investigating diffusion from bulk to surface and *vice versa*. Diffusion *on* the surfaces of metals may be studied in the field emission and field ion microscopes and is discussed more fully in Chapters 6, 8, and 12.

REFERENCES

AZAROFF, L. V. (1969). *Elements of X-ray crystallography*. McGraw-Hill, New York.

BARRETT, C. S. (1966). *Structure of metals*. McGraw-Hill, New York.

BLAKELY, J. M. (1973). *Introduction to the properties of crystal surfaces*. Pergamon Press, Oxford.

BRILLOUIN, L. (1946). *Wave propagation in periodic structures*. Dover Publications, New York.

COTTRELL, A. H. (1967). In *Materials, a Scientific American book*, p. 39. W. A. Freeman, San Francisco.

CULLITY, B. D. (1956). *Elements of X-ray diffraction*. Addison-Wesley, Reading, Mass.

GJOSTEIN, N. A. (1963). In *Metal surfaces: structure, energetics and kinetics*, ASM/AIME Seminar, Cleveland, 1962, p. 99. American Society of Metals, Ohio.

—— (1967). In *Surfaces and interfaces* (eds J. J. Burke, N. L. Reed, and V. Weiss), Syracuse University Press, Syracuse, N.Y.

HENDERSON, B. (1972). *Defects in crystalline solids*. Edward Arnold, London.

HERRING, C. (1951). *Phys. Rev.* **82**, 87.

HULL, D. (1968). *Introduction to dislocations*. Pergamon Press, Oxford.

HUME-ROTHERY, W. (1966). *Elements of structural metallurgy*. Institute of Metals, London.

JOHARI, O., and THOMAS, G. (1969). *The stereographic projection and its applications. Techniques of Metals Research*, Vol. 2a (ed. R. F. Bunshah). John Wiley Interscience, New York.

KOSSEL, W. (1927). *Nachr. Akad. Wiss. Goettingen* 135.

LINFORD, R. G. (1973). In *Solid state surface science* (ed. M. Green), Vol. 2, p. 1. Marcel Dekker, New York.

McLEAN, M. (1971). *Acta Metall.* **19**, 387.

MACKENZIE, J. K., MOORE, A. J. W., and NICHOLAS, J. F. (1962). *J. Phys. Chem. Solids* **23**, 185.

MACKENZIE, J. K., and NICHOLAS, J. F. (1962). *J. Phys. Chem. Solids* **23**, 197.

MOORE, A. J. W., and NICHOLAS, J. F. (1961). *J. Phys. Chem. Solids* **20**, 222.

NICHOLAS, J. F. (1961). *J. Phys. Chem. Solids* **20**, 230.

—— (1965). *An Atlas of models of crystal structures*. Gordon and Breach, New York.

PARK, R. L., and MADDEN, H. H. (1968). *Surf. Sci.* **11**, 188.

SMITHELLS, C. J. (1976). *Metals reference book* (5th ed). Butterworth, London.

WERT, C. A., and THOMSON, R. M. (1970). *Physics of solids* (2nd ed). McGraw-Hill, New York.

WOOD, E. A. (1964). *J. appl. Phys.* **35**, 1306.

THE DETERMINATION OF
SURFACE STRUCTURE

3.1. Introduction

WHEN an obstacle is placed in the path of a light wave a shadow
is formed. To a first approximation the edges of this shadow are
quite sharp, but close examination shows that across the boundary region brightness falls to zero gradually and, if monochromatic radiation is used, the shadow edges are seen to be surrounded
by a set of bright and dark bands known as fringes. Similar effects
are observed if the light passes through an aperture in an opaque
screen. These observations illustrate the phenomenon of diffraction, which is quite generally associated with any wave motion.
Since the effects are a function of wavelength, they are very
easily detected in cases such as sound and water waves, but with
light and electron waves and X-rays detection requires more
refined techniques.

In the diffraction process information about the structure of the
obstacle is transferred to the diffracted waves, as is obvious from
a superficial examination of the patterns produced by a circular
and by a rectangular aperture (Plate 1). Diffraction should thus
provide a means for structure determination, but unfortunately it
is also obvious from Plate 1 that the relation between pattern
and obstacle is not simple. For determining structure the ideal
instrument is one which produces directly an accurate image of
the object, i.e. a microscope. In the microscope the first stage of
operation does involve diffraction of the illuminating radiation by
the object, but the diffracted rays are then gathered together by a
lens in such a way that an image of the object is synthesized (Fig.
3.1). The greater the number of diffracted rays collected by the
lens the greater is the resolution of the instrument, or the
accuracy with which the image reflects the object. For a given
wavelength λ the angular spread of the diffraction pattern increases as the size of the object decreases; with a circular
aperture of diameter D, for example, the first dark fringe occurs

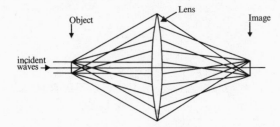

FIG. 3.1. Formation of image of diffraction grating. Five orders of diffraction are shown.

at $\sin \theta = 1 \cdot 22 \lambda / D$ $(\theta \le \pi/2$ (Plate 1)). If $D < 1 \cdot 22 \lambda$ this fringe cannot be observed, regardless of the lens aperture, and it becomes impossible to collect sufficient of the diffraction pattern to produce an image, even of the poorest quality. Thus to examine very small objects low-wavelength radiation must be used, and for atomic dimensions (a few tenths of one nanometer) only X-rays and electrons are suitable in this respect. For copper K_α X-radiation, for example, $\lambda = 0 \cdot 154$ nm, while an electron which has been accelerated through a potential of V V acquires energy E_p $(=eV)$ and has wavelength

$$\lambda \text{ (nm)} = (1 \cdot 5/V)^{\frac{1}{2}}. \tag{3.1}$$

If $V = 20$ V, $\lambda \approx 0 \cdot 27$ nm; if $V = 100$ kV, $\lambda \approx 0 \cdot 004$ nm.

At this point a further problem arises—the construction of suitable lenses. In the case of X-rays none are available, and although electrons can be focused by electric and magnetic fields the resolution which can be achieved in an electron microscope is limited by fundamental features of lens design to about 50 times the wavelength, for example $\sim 0 \cdot 2$ nm with 100 keV electrons under ideal conditions. In a few cases single heavy-metal atoms have been detected in scanning electron microscope images (see e.g. Crewe 1971), but in general when using electrons, or X-rays, to obtain structural data on the *atomic* scale one is at present denied the luxury of a direct image and must be content with extracting information from the diffraction pattern of the object. (Individual atoms may in fact be viewed directly in the field ion microscope, but wave phenomena are not involved; this technique is discussed in Chapter 6.)

High-energy electrons, and to an even greater degree X-rays, can easily penetrate solids of considerable thickness. For example 44 per cent of Cu K_α radiation (energy 8040 eV) incident on a nickel foil 0·02 mm (or approximately 10^5 atom layers) thick will be transmitted, while transmission electron microscopes using energies of 100 keV can examine solids with thickness of the order of 100 nm. In such a specimen, however, the surface atoms ($\sim 10^{13}$ mm^{-2}) constitute only a few per cent of the total number of atoms present and so the contribution of the structure of the surface to diffraction patterns produced by either of these radiations will be negligible. When *surface* structure is to be determined an alternative is obviously required; two possibilities exist. In reflectance high-energy electron diffraction (RHEED) electrons of moderately high energies (30–50 keV) are used, but are made to strike the solid at a glancing angle so that they are scattered predominantly in the surface layer. In the second technique—low-energy electron diffraction (LEED)—the electron energies lie in a range, approximately 10–300 eV, in which the scattering amplitudes of atoms are high. Therefore even at normal incidence these low-energy electrons are completely scattered within the first few atom layers at the surface. Both LEED and RHEED thus produce information on the atomic structure of surfaces which is essentially free of contribution from the bulk of the material; in both cases obviously only electrons back-scattered from the specimen can be detected. Of the two LEED has been much more extensively applied, but in many respects RHEED provides important complementary data; in the remainder of this chapter we shall concentrate exclusively on LEED, however, since the fundamental principles of all diffraction techniques are similar. Further information on RHEED can be obtained, among other sources, from a comprehensive review by Bauer (1969*a*) and a paper by Nielsen (1973). LEED has been the subject of a considerable number of reviews, for example by Lander (1965), Bauer (1969*b*), Estrup (1970), May (1970), Estrup and McRae (1971), Joyner and Somorjai (1973) and Mason and Textor (1976). The subject in general is also discussed in books by Somorjai (1972), Prutton (1975), and Ertl and Küppers (1974), while intensity theory has been treated in considerable and lucid detail in a book by Pendry (1974).

3.2. LEED: basic experimental observations

The most widely used type of LEED apparatus, the 'post-acceleration display system' is shown in Fig. 3.2. Primary electrons (10–300 eV; 0·388–0·071 nm) are generated in an electron gun and strike the specimen surface at normal incidence. The beam current is generally 1–2 μA and the beam diameter approximately 1 mm; the current density may therefore be sufficient to cause changes in adsorbed surface layers (cf. Chapter 4) and this possibility must always be borne in mind.

Various processes occur in the surface region of the solid resulting in back-scattering of electrons with energies covering the complete spectrum from zero to the primary energy E_p (see Chapter 4). Only those which have been elastically scattered (i.e. are of energy E_p) carry diffraction information, however; all others must be filtered out since they would produce a background intensity in the diffraction pattern which would tend to obscure the diffraction spots. This is achieved by passing the electrons through a series of suitably biased hemispherical mesh grids which are concentric with the specimen. These grids may be two, or more usually three or four, in number, the last being preferred if the system is also to be used as a retarding field analyser for Auger electron spectroscopy (see Chapter 4). The exit shield of the electron gun, the specimen, and the outermost grid are at ground potential so that the incident and reflected electrons move in a field-free region. The second grid is at a potential just slightly less than $-E_p$, and therefore only the elastically scattered electrons proceed beyond this point to be finally accelerated through a potential of several kilovolts onto a hemispherical fluorescent screen where they produce a diffraction

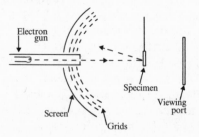

FIG. 3.2. LEED display system.

pattern consisting of bright spots on a dark background. This may be viewed and photographed through a window in the vacuum chamber. Examples are shown in Plate 2; the dark object in the upper half of the photograph is part of the specimen manipulator, carrying a circular specimen at its lower end. The bright spot in the centre is not to be confused with a diffraction feature; it is produced by light from the filament of the electron gun. Using normal incidence there will in fact always be a diffraction spot at the centre of the pattern due to specular reflection of the primary beam, but it is obscured by the specimen.

As will be discussed below, the relative intensities of the spots in the pattern plus the variation of intensity as a function of electron wavelength are of central importance in diffraction theory. When using a display system these intensity features are generally measured by means of an external photometer which can be focused on individual spots through the window. In the principal alternative form of LEED apparatus (Park and Farnsworth 1964) intensities can be determined directly using a movable Faraday cup. At any instant this collects and filters the electrons scattered in a particular direction, and by rotating the specimen and varying the co-altitude of the cup, all directions may be rapidly scanned to give the complete diffraction pattern, which is presented on a storage oscilloscope. Alternatively, the electron wavelength may be varied while the cup remains stationary in the path of a diffracted beam.

LEED experiments may be carried out on metals, semiconductors, and insulators, although in the last case a problem arises if the specimen accumulates negative charge, since the primary beam will then be deflected. Fortunately the secondary-electron yield for many insulators has a value greater than 1 for primary energies above about 30 eV and therefore initially they emit more electrons than they receive. The resulting build-up of positive charge is accompanied by an increase in space charge and eventually the two come to equilibrium, the secondary yield adjusting to unity. While there is little restriction on the chemical constitution of a specimen, the same is not true of its crystallography—in general only single-crystal surfaces can be successfully examined. In a LEED pattern any given spot is observed at all electron wavelengths above some threshold value (see Section 3.3). Hence in diffraction by a polycrystalline specimen the observed pattern will be a superposition of the complete patterns generated by each individual crystallite and the multiplicity of spots usually defies interpretation. This contrasts with X-ray diffraction in which case a given spot is observed only at certain wavelengths or angles of incidence so that the number of spots in a powder pattern, for example, lies within a managable range.

When various LEED patterns are compared two general features are immediately obvious.

(i) The spot *positions* depend on the nature of the surface involved (Plate 2). Also, if the electron wavelength is varied by adjusting the accelerating potential in the electron gun (cf. eqn (3.1)), all spots in the pattern, with the exception of the central one, change position, moving inwards to (or outwards from) the centre as the wavelength decreases (or increases).

(ii) For certain surfaces, for example Cu(210) (Plate 2(c)), there is considerable variation in the relative *intensities* of adjacent spots in the pattern; in other cases (e.g. Plate 2(a, b)) this variation is very much less pronounced. For *all* surfaces, however, the intensity of a particular spot is modulated quite strongly if the wavelength is changed (Fig. 3.3).

On the basis of these observations we may expect spot positions and intensities to be related to structural features of the

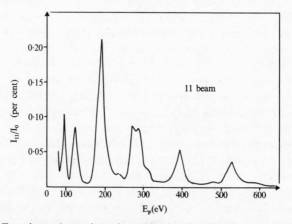

FIG. 3.3. Experimental spot intensity profile for Cu (100) 11 spot. (After Andersson 1969.)

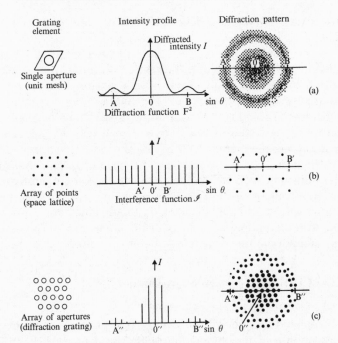

FIG. 3.4. Elements of a diffraction pattern. (After Lipson and Lee 1970.)

surface, and the nature of the relationship is illustrated most simply by means of an optical analogue (Lipson and Lee 1970). Suppose a two-dimensional grating consists of a space lattice which is an array of points populated by unit meshes each of which is a circular aperture in a parallelogram of opaque material (Fig. 3.4). The diffraction pattern which would be produced by a single unit mesh is known as the diffraction function F^2; in this case it would be a series of alternating bright and dark rings (Fig. 3.4(a)). The pattern which would be produced by the space lattice alone, known as the interference function $\mathscr{I}$, would be an array of spots of equal intensity (Fig. 3.4(b)). When these two grating elements are superimposed to produce the complete grating the actual diffraction pattern is obtained by superimposing F^2 on $\mathscr{I}$. It is seen (Fig. 3.4(c)) that the spot *positions* are determined, through the interference function, by the space lattice, that is by the *size and shape of the unit mesh*. The modulation in relative spot *intensities* on the other hand is determined, through the diffraction function, by the *structure within the unit mesh*.

The problem of deciphering the structural information contained in a diffraction pattern therefore falls into two main parts, the first involving consideration of spot positions to reveal the dimensions and symmetry of the mesh. This is relatively straightforward and is discussed in the next section. Secondly, an analysis of spot intensities is required to reveal the contents of the mesh, i.e. the number and positions of atoms within it. This, in its rigorous form, is by contrast a problem of great complexity; it is outlined in Section 3.5. An additional category of information becomes available if certain of the spots in a pattern are diffuse or otherwise deformed, since such abnormalities are caused by defects in the perfect periodicity of the lattice array. The extraction of information on defect structures is discussed in Section 3.7.

3.3. LEED: basic pattern interpretation

3.3.1. *General physics of diffraction*

A complete description of any diffraction process would involve consideration of the interaction of the incident wave field with the entire diffraction grating to produce a scattered field which is detected experimentally in the form of a diffraction pattern. This is an extremely complex problem, as can be seen by examining the case of light striking an aperture in an opaque screen. In the surface layers of the screen, both facing the source and in a narrow region around the edges of the aperture, the electrons are induced to oscillate by the incident field. Each oscillating electron gives rise to a secondary wave and it is interference between these waves which produces the diffraction effects. A rigorous calculation of the diffraction pattern would involve an exact solution of Maxwell's equations and has been obtained in a few simple cases only, such as that of a semi-infinite screen with a straight edge made of a perfectly reflecting metal of zero thickness. In general an approximate model must therefore be adopted which has its basis in ideas first put forward by Huygens in the seventeenth century, developed by Fresnel, and given quantitative mathematical treatment by Kirchoff. As waves propagate from a source the surface they reach at any instant is known as the wavefront. Huygens suggested that the wavefront at any later time may be found by assuming that each point on the original front acts as a point source from which spherical waves spread out and that the new wavefront is given by the envelope of these wavelets. In the diffraction situation it is assumed that as the incident wave passes through an aperture the points on the wavefront act as secondary sources. The pattern at some point of observation is then derived by considering the interference between the secondary wavelets, the net disturbance being obtained by superimposing all the individual wavelets. Thus the oscillating electrons in the screen have been replaced by a set of fictitious oscillators in the aperture. This is a scalar model, in that the waves are represented by a single scalar field rather than by coupled electric and magnetic vector fields as required in the rigorous treatment. The one quantity produced by the model which can be compared with experimental data is the intensity at a given point in the diffraction pattern. In the theory of electromagnetic waves intensity is proportional to the square of the amplitude of the electric field vector, and by analogy the square of the amplitude of the scalar is taken to be the relevant quantity. The proportionality constant cannot be determined, however, and so only the ratios of intensities at various points in the pattern can be obtained.

3.3.2. *Spot positions*

The model outlined above satisfactorily describes a number of features common to all diffraction processes, including LEED. The fundamental concept involved is that the total scattered wave may be obtained by superimposing the wavelets scattered by individual elements of the grating. As we shall see in Section 3.4.1 this leads readily to an equation which in principle describes the intensity at all points in space and is of the type used by Laue in the first comprehensive analysis of X-ray diffraction patterns. In this section, however, we wish to concentrate on spot positions. We then need enquire only as to the conditions which give rise to *principal maxima* in the scattered intensity, i.e. to spots, and can ignore any weak scattering to all other points in space. We may also ignore the fact that low-energy electrons penetrate beyond the top layer of atoms at the solid surface; as will be seen later penetration affects the relative intensities of spots but not their positions.

A. *Diffraction by a one-dimensional lattice.* The simplest situation to consider is that of diffraction by a row of identical atoms, a one-dimensional periodic structure with a single atom within each unit mesh. Rather than consider the interaction of an incident wave front with the system as a whole, we assume that individual wavelets are scattered by individual atoms and that the diffracted wave is obtained by applying the principle of superposition to the scattered wavelets. Thus the diffracted intensity will be maximum in those directions in space in which all scattered wavelets are exactly in phase, i.e. when their points of maximum displacement coincide. This situation is illustrated in Fig. 3.5(a) for a given angle of incidence θ_i. In the direction at angle θ_k from the surface the intensity will be maximum, but if we looked in the direction at θ'_k (Fig. 3.5(b)) we would see nothing. The wavelets scattered at θ'_k are exactly out of phase and when superimposed cancel each other completely. The condition for maxima is the familiar one—the path difference Δ for waves scattered by adjacent atoms must be an integral number of wavelengths. From Fig.

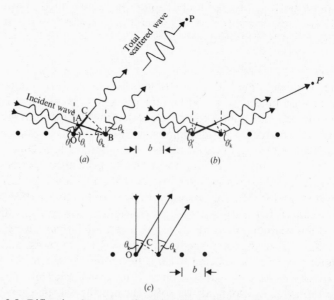

FIG. 3.5. Diffraction by a one-dimensional row of point scatterers. (a) Non-normal incidence, scattered waves in phase. (b) Non-normal incidence, scattered waves exactly out of phase. (c) Normal incidence.

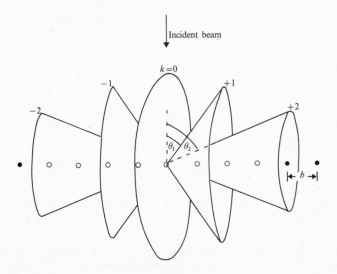

FIG. 3.6. Cones of scattered radiation produced by a one-dimensional grating of point scatterers.

$3.5(a)$ it can be seen that this is of the form

$$\Delta = OC - AB$$

$$k\lambda = b(\sin \theta_k - \sin \theta_i) \quad k = 0, \pm 1, \pm 2, \ldots. \tag{3.2}$$

The integer k is known as the order of diffraction. Note that the angles of incidence (θ_i) and scattering (θ_k) are measured with respect to the surface normal. In LEED the angle of incidence is generally 0° so that condition (3.2) becomes (see Fig 3.5(c))

$$k\lambda = b \sin \theta_k. \tag{3.3}$$

Fig. 3.5 might imply that the intensity maxima radiate from the grating along particular lines in space, but since the incident wave is scattered in all directions the maxima actually lie on the surfaces of cones (Fig. 3.6); these correspond to the various values of k and have semi-vertical angles θ_k (eqn (3.3)).

If a fluorescent screen which forms part of a spherical surface is placed with the grating as its centre of curvature then the diffraction cones will intersect the screen so as to give a pattern consisting of a set of parallel lines. These lines lie in the direction

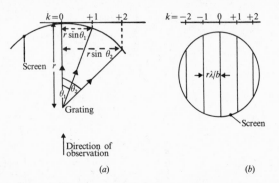

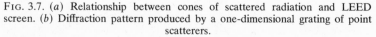

FIG. 3.7. (a) Relationship between cones of scattered radiation and LEED screen. (b) Diffraction pattern produced by a one-dimensional grating of point scatterers.

normal to the line of the lattice, and when viewed or photographed they will appear with equal spacings (Fig. 3.7) of magnitude $r(\lambda/b)$, where r is the radius of curvature of the screen. Note that this spacing is proportional to the reciprocal of the lattice spacing b. The case of the one-dimensional lattice is important when considering LEED patterns produced by surfaces which contain elements of structural disorder lying in a preferred direction.

B. *Diffraction by a two-dimensional lattice.* When the treatment is extended to a two-dimensional point lattice the situation may be viewed in two ways. If the two directions **a** and **b** defining the unit mesh are the only ones considered, the lattice may be constructed from parallel rows of points running in the **b** direction, the separation between these rows being **a**; Fig. 3.8 illustrates the simple case of a rectangular lattice. For diffraction by any row the condition represented by eqn (3.3) must be satisfied. A similar condition will be required for the production of maxima in the radiation intensity scattered by the various rows, namely

$$h\lambda = a \sin \theta_h \qquad h = 0, \pm 1, \pm 2. \qquad (3.4)$$

This defines a second family of cones lying along the direction normal to the rows. (Note the implication that in determining the *positions* of intensity maxima only the *spacing* of the grating elements is important, their *structure* being irrelevant. A row of points and an array of parallel lines produce similar sets of cones.) Intensity maxima in the resulting diffraction pattern can occur only when (3.3) and (3.4) are simultaneously satisfied. Geometrically this condition is represented by the lines along which cones from the two families h and k intersect (Fig. 3.8). When these lines of radiation strike the screen a pattern of spots is produced with spacings $r(\lambda/a)$ in the direction normal to the lattice rows and $r(\lambda/b)$ parallel to the rows (Fig. 3.9).

A spot is generally identified by the corresponding values of h and k, while the values of θ_h and θ_k locate its position on the screen. For instance the specularly reflected beam, for which $\theta_h = \theta_k = 0°$, is referred to as the 00 spot; other examples are shown in Fig. 3.9. When a clean surface constitutes the diffraction grating, h and k in eqns (3.3) and (3.4) are always integers and the spots in the diffraction pattern from such a surface are therefore often referred to as 'integral-order' spots. When an overlayer is present we shall see below that additional 'fractional-order' spots may appear.

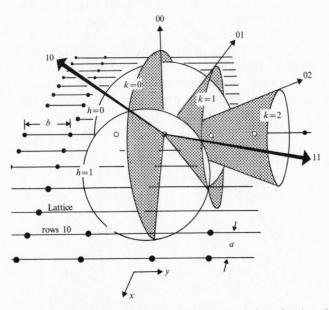

FIG. 3.8. Cones of scattered radiation produced by a two-dimensional grating of point scatterers. Diffraction maxima occur in directions defined by the lines along which the cones intersect.

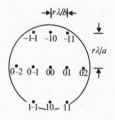

FIG. 3.9. Diffraction pattern produced by a two-dimensional grating of point scatterers.

In Fig. 3.8 the lattice was considered to be constructed from rows of points lying in the [01] direction, parallel to the y-axis, but many other possible constructions exist, for example the sets of rows (11) or (21) (Fig. 3.10). The alternative view of two-dimensional diffraction is equivalent to the Bragg method in three dimensions and proceeds as follows. Consider any set of lattice rows with inter-row spacing d_{hk}. Once again the diffraction process can be imagined as giving rise to two families of cones but, of that family with axis parallel to the lattice rows, we now ignore all except the cone of zero order; this is actually a disc lying in the azimuthal plane (Fig. 3.10). The only diffraction maxima then defined are those resulting from the intersections of this zero-order disc with the cones of the family with axis normal to the rows. These maxima all lie in the azimuthal plane normal to the lattice rows. They correspond to the diffraction conditions

$$n\lambda = d_{hk} \sin \theta_{nh,nk} \tag{3.5}$$

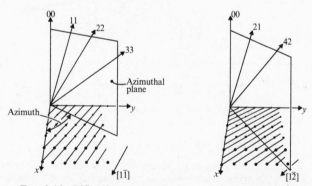

FIG. 3.10. Diffraction maxima lying in various azimuthal planes.

and may be indexed as nh, nk spots ($n = 0, \pm1, \pm2 \ldots$). For example, the (11) rows produce spots 00, $\pm1 \pm 1$, $\pm2 \pm 2$, etc., and the (21) rows 00, $\pm2 \pm 1$, $\pm4 \pm 2$, etc. (Fig. 3.10). The entire diffraction pattern can thus be constructed by considering in turn all possible sets of lattice rows.

From eqn (3.5) and Figs 3.7 and 3.10 it follows that the spacing between the spots in any azimuth is inversely related to the spacing of the lattice rows lying normal to this azimuth. This is a quite general and fundamental relationship.

Features in a diffraction pattern in any given direction reflect the structure of the diffraction grating in the direction normal to this, the periodic spacings in pattern and grating being inversely proportional to each other.

This inverse relationship immediately suggests that the concept of reciprocal space, discussed in Chapter 2, should be useful in connection with diffraction problems, and of course these repres- ent one of the main areas of its application. Consider for example the (01) lattice rows in a surface with an oblique unit mesh and the corresponding diffraction maxima lying in the $[uv]$ azimuth. The reciprocal space of the two-dimensional surface in this azimuth consists of rods of infinite length lying parallel to the z-axis and passing through the reciprocal net points, i.e. sepa- rated by the distance $1/d_{01}$ (Fig. 3.11). The relationship between reciprocal lattice and diffraction pattern is most readily ap- preciated if we now proceed with a construction first proposed by Ewald. Starting at the origin O of the real lattice draw a vector OO' of length $1/\lambda$ lying in the direction of the incident beam (Fig. 3.11). Let the terminus O' of this vector be the origin of the reciprocal lattice. Next, with the origin O of the real lattice as centre draw a circle of radius $1/\lambda$. Vectors from the centre of this circle to its points of intersection with the reciprocal lattice rods then represent the directions of possible diffraction maxima in the $[uv]$ azimuth. This may be confirmed by examination of Fig. 3.11; the direction OA for example satisfies the condition for the 02 maximum (cf. eqn (3.5)):

$$\sin \theta_{02} = \frac{2/d_{01}}{1/\lambda}.$$

The construction of Fig. 3.11 can be generalized to interpret the complete diffraction pattern from a two-dimensional grating

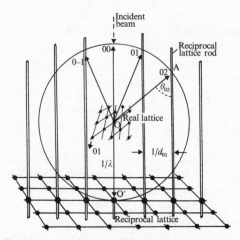

FIG. 3.11. The Ewald construction for diffraction maxima in the $[uv]$ azimuth produced by a two-dimensional grating of point scatterers. The real and reciprocal lattices both lie normal to the plane of the paper.

by considering the whole of reciprocal space instead of a single azimuth. The circle of radius $1/\lambda$ then becomes a sphere of the same radius known as the *Ewald sphere* or the *sphere of reflection*. Diffraction by one- and three-dimensional structures may also be investigated by drawing the appropriate reciprocal-space elements (planes and points respectively, see for example Lander (1965)).

A number of important general conclusions emerge from examination of Fig. 3.11.

(i) Since the screen in a LEED apparatus is usually constructed to be part of a spherical surface, with the specimen at its centre of curvature, it follows that the screen and the Ewald sphere are concentric. By considering Fig. 3.11 we are then led to the conclusion that *the spot pattern on the screen provides a direct picture of the reciprocal space elements of the surface*, as viewed in the direction of the incident beam.

(ii) As the wavelength of the radiation used is varied the Ewald sphere expands (or contracts). In the case of diffraction by a three-dimensional structure the condition for a maximum is satisfied only when a reciprocal lattice point lies on the sphere and therefore a given reflection will occur at certain wavelengths only (assuming that the angle of incidence is fixed). This is the situation in X-ray diffraction where different maxima are observed by varying, for example, the angle of incidence at fixed wavelength.

(iii) With one- or two-dimensional structures a given reciprocal space element h (or hk) will intersect the sphere for all wavelengths such that

$$0 \leqslant \sin \theta_{h(k)} \leqslant 1,$$

that is

$$0 \leqslant h\lambda \leqslant d_h \qquad \text{(or } d_{hk}\text{).} \tag{3.6}$$

Therefore the diffraction pattern always contains a 00 spot and the reflection h (or hk) will be present at *all* wavelengths satisfying (3.6) (although the *intensity* of a given reflection generally varies with wavelength, cf. Plate 2).

(iv) As the wavelength increases the observed reflections diverge from the 00 spot and fewer of them will be intercepted by the viewing screen.

3.3.3. *Clean-surface patterns*

We have just seen that the diffraction pattern is a direct picture of the reciprocal lattice and relationships established in connection with the latter can therefore be applied directly to extract structural information from the pattern. In Chapter 2 it was seen that the reciprocal lattice is a device for conveniently representing the direction and separation of planes of atoms in a three-dimensional crystal. In the case of a two-dimensional structure it fulfils the same purpose with respect to the atom rows from which the structure can be considered to be constructed. Thus a two-dimensional reciprocal mesh vector σ_{hk} has the following properties: (i) its direction is normal to the hk atom rows; (ii) its magnitude is inversely related to the spacing of these rows.

It may then be deduced that the vectors $\mathbf{a}$ and $\mathbf{b}$ defining the unit mesh of the structure in real space are related to the corresponding reciprocal space vectors $\mathbf{a}^*$ and $\mathbf{b}^*$ by the following equations (cf. eqns (2.12) and (2.13)):

$$\mathbf{a} \cdot \mathbf{a}^* = \mathbf{b} \cdot \mathbf{b}^* = 1 \tag{3.7}$$

$$\mathbf{a} \cdot \mathbf{b}^* = \mathbf{b} \cdot \mathbf{a}^* = 0. \tag{3.8}$$

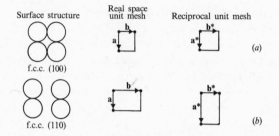

FIG. 3.12. Relationships between real-space and reciprocal-space unit meshes for f.c.c. (100) and (110) surfaces.

(For those familiar with the Kronecker delta symbol these may be succinctly combined in the form $\mathbf{a}_i \cdot \mathbf{a}_j^* = \delta_{ij}$.) Thus $\mathbf{a}^*$ for example is normal to $\mathbf{b}$, i.e. to the atom rows lying in the $\mathbf{b}$ direction, and the magnitude of $\mathbf{a}^*$ is inversely proportional to $a \cos \eta$, which is the spacing of the $\mathbf{b}$ rows (η is the angle between $\mathbf{a}$ and $\mathbf{a}^*$, or between $\mathbf{b}$ and $\mathbf{b}^*$).

Let us now use these facts to interpret some simple LEED patterns. Fig. 3.12(a) shows the (100) face of an f.c.c. crystal such as copper and the corresponding real- and reciprocal-space unit meshes. This is a particularly simple case since $\mathbf{a}$ and $\mathbf{b}$ are normal to each other and of equal magnitude. Eqns (3.7) and (3.8) then demonstrate that the real and reciprocal meshes are identical except for a scaling factor. If this was an 'unknown' surface, simple observation of the diffraction pattern would thus immediately tell us that its unit mesh was square. If we knew the sample-to-screen distance and the magnification of the photographic process, by measuring the distances $\mathbf{a}^*$ and $\mathbf{b}^*$ on a photograph of the pattern we could calculate the unit mesh dimensions $\mathbf{a}$ and $\mathbf{b}$ absolutely. This is rarely done, however, since the identity of the substrate is generally known and these dimensions can be more accurately derived from X-ray data.

In Fig. 3.12(b) the structure of a Cu(110) surface is compared with its reciprocal mesh. Since in this case the unit mesh is rectangular the reciprocal relationship is more apparent; in a given direction the smaller the real-space vector (e.g. $\mathbf{a}$) the larger is the corresponding reciprocal vector ($\mathbf{a}^*$). In fact, as the mesh is orthogonal, $\mathbf{a}^*$ is simply the direct reciprocal of $\mathbf{a}$ (cf. eqn (3.7)). Note in this and in the previous example that while $\mathbf{a}^*$ is parallel to $\mathbf{a}$, and $\mathbf{b}^*$ to $\mathbf{b}$, this is a trivial consequence of the meshes being orthogonal; the fundamental relationship is $\mathbf{a}^*$ perpendicular to $\mathbf{b}$ and $\mathbf{b}^*$ to $\mathbf{a}$.

The general case involving non-equal, non-orthogonal mesh vectors is illustrated by the Cu(210) surface (Fig. 3.13). Interconversion between real and reciprocal space becomes more complex than in the two examples previously discussed since now $|\mathbf{a}^*|$ is the reciprocal of $|\mathbf{a}| \cos \eta$. As emphasized in Chapter 2 the easiest and most reliable means of transforming one space into the other, in all but the simplest cases, is to use the matrix method (see caption for Fig. 3.13).

3.3.4. Overlayer patterns

It has already been pointed out that in surface chemistry the primary interest is not in the structures of clean surfaces, but in changes in structure brought about by processes such as adsorption. Let us expose a clean Cu(210) crystal to oxygen at 10^{-6} torr, 298 K. The LEED pattern is found to change from that of Fig. 3.14(a) to that of 3.14(b). We shall temporarily assume that no reconstruction of the surface occurs so that the top surface layer now consists of adsorbed oxygen rather than of copper. Since it is the symmetry of the top layer which determines spot positions in the diffraction pattern, the fact that adsorption has altered the pattern shows that the structure (i.e. the unit mesh) of the oxygen layer must be different from that of the original copper surface. Spots do appear in the new pattern at positions identical to those of spots in the copper pattern and are therefore sometimes referred to as 'clean-surface' spots. Similarly, the spots in the new pattern which are additional to those in the original may be

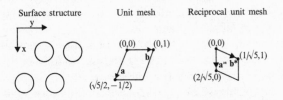

FIG. 3.13. Relationship between real and reciprocal-space unit meshes for an f.c.c. (210) surface. The matrix method for real-reciprocal space transformation in this case is

$$\mathbf{A} = \begin{bmatrix} \sqrt{5}/2 & -1/2 \\ 0 & 1 \end{bmatrix} \qquad \mathbf{A}^* = \frac{1}{\sqrt{5}/2 - 0}\begin{bmatrix} 1 & 1/2 \\ 0 & \sqrt{5}/2 \end{bmatrix} = \begin{bmatrix} 2/\sqrt{5} & 1/\sqrt{5} \\ 0 & 1 \end{bmatrix}$$

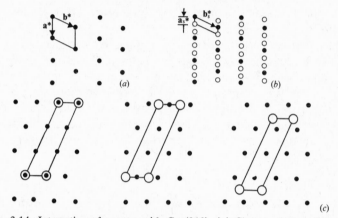

FIG. 3.14. Interaction of oxygen with Cu (210). (a) Clean surface pattern. (b) (3×1) pattern formed after exposure to O_2 at 1×10^{-6} torr. (●) Integral order spots; (○) fractional order spots. (c) Alternative locations for a (3×1) over-layer mesh on an f.c.c. (210) surface.

classed as 'extra' features. It must be emphasized, however, that the *positions of all spots* in Fig. 3.14(b) are determined by the periodicity of the top (oxygen) surface layer alone and not by the underlying copper; 'clean surface' and 'extra' are merely terms of convenience. From the point of view of diffraction theory it is more appropriate to refer to 'integral order' and 'fractional order' spots (see Section 3.4.1).

Comparison of Figs 3.14(a,b) shows that in the a^* direction oxygen adsorption has tripled the number of spots; in other words the reciprocal mesh vector in this direction for the ad-sorbed layer, a_s^*, has magnitude $a_s^* = \frac{1}{3}a^*$. In the b^* direction the spacing remains unchanged. Therefore in the **a** direction in real space the size of the oxygen mesh is three times that of the clean surface mesh, while in the **b** direction the dimensions of both meshes are identical. We have therefore established the size and symmetry of the unit mesh of the adsorbed layer. This is in fact as far as an analysis of spot positions alone will take us, but determination of the structure remains incomplete since (a) the atoms in this layer have not been positively identified as oxygen (they could be displaced copper) and (b) the position of the top-layer mesh with respect to atom positions in the second layer is entirely unknown. Thus the three (out of many) alternative structures shown in Fig. 3.14(*c*) would give rise to diffraction patterns in which the spot *positions* were identical to those in Fig. 3.14(*b*). The relative spot *intensities*, however, would be expected to differ in each case and it is by an analysis of intensities that the complete solution of the structural problem is to be obtained (see Sections 3.4 and 3.5).

A further point arises from the above discussion; it was emphasized that spot positions are determined by the periodicity of the top surface layer. This statement introduces no conceptual problems provided the top layer has a reasonably close-packed structure (e.g. Fig. 3.15(*a*)), but if the adsorbate concentration is low, or if planes of high index are considered (Fig. 3.15(*b*)), then atoms in the layers immediately below the top may be quite directly exposed to the incident beam and must therefore have a strong influence on the diffraction process. This is true, but it is an influence on spot intensities only, since the periodicity of the overall structure in these cases still corresponds to the atomic spacing in the outermost layer. In certain circumstances, exemplified by the system W(100)–H$_2$, the situation may be more complex, however. The exact radius of the adsorbate will depend on its charge state, but it may be so small that it sits *below* the tungsten surface plane (Fig. 3.15(*c*)) and the periodicity of the outermost layer, viewed in isolation, remains unchanged by adsorption. There is nevertheless a modification in the periodicity of the quasi-three-dimensional 'surface region' and in the most general case it is this which governs spot positions. Extra spots therefore appear in the W(100)–H$_2$ pattern. The effects of underlying layers are further discussed in Sections 3.4, 3.5, and 3.6.

Some nomenclature for adsorbate or overlayer structures is obviously required. The one most frequently used is due to Wood (1964). The clean substrate net is taken as a reference and the vectors of the overlayer mesh ($\mathbf{a}_s$, $\mathbf{b}_s$) are defined in terms of those

(*a*)   (*b*)   (*c*)

FIG. 3.15. Definition of surface periodicity in the cases of (*a*) a close-packed plane, (*b*) a high index plane, and (*c*) an adsorbed layer such as W(100)-H$_2$, in which the adsorbate is located below the top metal layer.

of the substrate $(\mathbf{a}, \mathbf{b})$. In the simplest cases $\mathbf{a}_s$ will be parallel to $\mathbf{a}$, and $\mathbf{b}_s$ to $\mathbf{b}$, so that

$$\mathbf{a}_s = P\mathbf{a} \qquad \mathbf{b}_s = Q\mathbf{b}$$

where P and Q are integers. The overlayer is then said to have a $(P \times Q)$ mesh; for example in the Cu(210)–oxygen case just discussed a (3×1) ('three-by-one') structure is formed (Fig. 3.14(c)). Its full designation would be Cu(210)$-(3 \times 1)-$O$-\theta$ where the penultimate letter indicates the chemical nature of the adsorbate producing the new structure and θ is its coverage expressed as a fraction of a monolayer. In spite of its wide application this system of notation breaks down in certain cases. A more comprehensive, but also rather more cumbersome, alternative involves designation of a structure in terms of its transformation matrix $\mathbf{A}$ (see Chapter 2). For example, in the case of Fig. 3.14(c) this would be

$$\text{Cu(210)} - \begin{vmatrix} 3 & 0 \\ 0 & 1 \end{vmatrix} - \text{O} - \theta.$$

Let us now examine some other patterns which commonly result from adsorption. Fig. 3.16 illustrates the case of an oxygen layer on a W(100) surface. It is seen that

$$\mathbf{a}_s = 2\mathbf{a} \quad \text{and} \quad \mathbf{b}_s = 2\mathbf{b}.$$

Therefore the overlayer has a 'two-by-two' structure, designated W(100)–$(2 \times 2)-$O. A slightly more complicated case is shown in Fig. 3.17. Here one tends intuitively to consider the centred, non-primitive mesh outlined on the left-hand side of the diagram since this has the same orientation as the substrate mesh, and to term the structure 'centred two by two', $c(2 \times 2)$. Noting that the mesh dimensions in the $\mathbf{a}_s$ and $\mathbf{b}_s$ directions are twice those in the

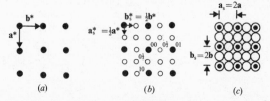

FIG. 3.16. The (2×2) structure. (a) Diffraction pattern from clean W(100); (b) Diffraction pattern from W(100)$+$O$_2$. (O) Fractional order spots. (c) Possible (2×2) surface structure for W(100)$+$O$_2$. (O) W atom; (●) O atom.

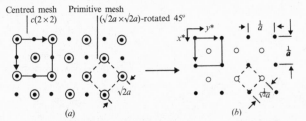

FIG. 3.17. (a) The c(2×2) surface structure (real space). (●) Substrate atom: (○) adsorbate atom. (b) Diffraction pattern (reciprocal space). (●) Integral-order spots hk; (○) fractional-order spots h/2, k/2.

substrate **a** and **b** directions it is at first sight surprising to find that in reciprocal space (Fig. 3.17(b)) no 'extra' spots appear in the $\mathbf{x}^*$ and $\mathbf{y}^*$ directions. This is equivalent to systemmatic absences in X-ray diffraction. The diffracted beams which pro-duce spots such as $0\frac{1}{2}$ and $\frac{1}{2}0$ in the (2×2) structure (Fig. 3.16) are eliminated in the present case by destructive interference with beams diffracted by the centre atoms in each mesh (this point is pursued further in Section 3.5). The relationship between real and reciprocal space becomes obvious on considering the primitive mesh shown on the right-hand side of Fig. 3.17(a); in terms of this the Wood notation for the structure is $(\sqrt{2}a \times \sqrt{2}a)$ R−45° where R represents 'rotated with respect to the substrate mesh'. This is perhaps the simplest case in which matrix transformation may be useful for anyone unfamiliar with the structure, but it is trivial in the sense that the c(2×2) is one of the most commonly observed LEED patterns.

Based on an analysis of a large number of LEED observations Somorjai and Szalkowski (1971) have suggested that the follow-ing rules may be used to predict the type of structure to be expected in various situations.

(i) 'Adsorbed species pack as closely as possible, consistent with adsorbate dimensions and adsorbate–adsorbate and adsorbate–substrate interactions.' Large reciprocal unit meshes are therefore uncommon and are in any case generally rationalized in terms of close-packed structures by involving multiple scattering (see Section 3.6). The most frequently observed meshes are of the same size as the substrate mesh, i.e. (1×1), or are approximately twice as large, e.g. (2×2), c(2×2), (2×1), ($\sqrt{3}\times\sqrt{3}$) (Fig. 3.18).

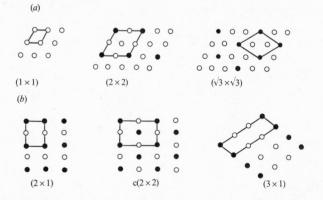

FIG. 3.18. (a) Most common surface structures on substrates with six-fold rotational symmetry, e.g. f.c.c. (111). (b) Most common surface structures on substrates with two-fold rotational symmetry, e.g. f.c.c. (110), b.c.c. (211), b.c.c. (110). (O) Substrate atom; (●) adsorbate atom, on top of substrate atom.

(ii) 'Adsorbed layers tend to form ordered surface structures with unit mesh vectors closely related to the substrate unit cell vectors, rather than to those of the bulk condensate.' Exceptions to this rule occur in the case of highly charged adsorbate layers, such as alkali metals on Ni (Gerlach and Rhodin 1969) in which adatom–adatom repulsions predominate over adatom–substrate interactions.

(iii) 'Adsorbed structures have the same rotational symmetry as the substrate.' If the surface unit mesh has a lower symmetry than the substrate, then domains of the various possible mesh orientations are to be expected on different areas of the surface with a resulting increase in symmetry. For example, Fig. 3.19(b) shows a LEED pattern from an overlayer on a four-fold substrate; $\frac{1}{5}$th-order reflections are present, but since they are limited to $h+n/5$, k and h, $k+n/5$ the structure is not (5×5). The mesh is in fact of the type (5×1), but on its own (5×1) would produce the spots $h+n/5$, k only. At low coverage, however, while (5×1) meshes form on some parts of the surface, (1×5) meshes should form with equal probability on other parts and these will give rise to the h, $k+n/5$ spots. As coverage increases domains of the two orientations will develop. On

its own each has two-fold rotational symmetry but together they give a structure with the four-fold symmetry of
the substrate; a specific example would be the reconstructed (5×1)-Au surface (Fedak and Gjostein 1967).

3.4. LEED: kinematic diffraction theory

In the last section we discussed the way in which the positions of
spots in a diffraction pattern are related to the size and symmetry
of the surface unit mesh. For this only the points of maximum
intensity in the distribution of diffracted intensity over all space
need be considered, since, provided the surface structure is
perfect, the main diffraction spots will be the only observable
features in the pattern. If the grating is imperfect, however,
additional features will appear and for an analysis of these it is
essential to have an expression for the diffracted intensity I at all
points in space, in other words for reciprocal space intensity
profiles. We have also seen that to determine the structure factor,
or the number and positions of atoms within the unit mesh, we
must analyse spot intensities. It is a simple matter to change the
incident electron wavelength in LEED and so a record of the
intensity of a particular spot as the electron accelerating voltage
V is varied proves to be the most convenient form in which to
collect data; this we may call the spot intensity profile or the $I–V$
curve. We therefore have the additional theoretical requirement
for an expression for I as a function of V for a given spot. In this
section it will be shown that a suitable equation can be established on the basis of the kinematic, or single-scattering, assumption. As its name implies, this supposes that the incident wave is

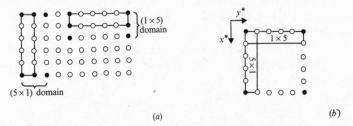

(a) (b)

FIG. 3.19. (a) Domains of (5×1) and (1×5) structure occurring on different
parts of a surface. (O) Substrate atom; (●) adsorbate atom. (b) The corresponding
diffraction pattern. (O) Fractional order spots.

scattered once only during its interaction with the diffraction grating. The assumption is not unreasonable in the cases of optical and X-ray diffraction, but in general it proves to be invalid for LEED. Nevertheless, the kinematic intensity equation is useful for qualitative interpretation of patterns from defect structures and familiarity with it is essential before discussing more realistic intensity theories. Consultation of the extensive literature on the kinematic theory of optical diffraction may be found helpful; see, for example, Towne (1967) and Stone (1963).

3.4.1. *Reciprocal-space intensity profiles*

A. Diffraction by a one-dimensional grating. Consider a screen containing a row of M apertures of arbitrary but identical shape, all oriented in the same way, and let the distance between representative points Q in any two apertures be d (Fig. 3.20). As we have seen, the diffraction pattern produced by this, or any other, grating has certain features which depend only on the number and relative positions of the apertures, and other features which depend only on the shape of the individual apertures; the former will now be considered.

Choose some arbitrary point O as origin; for convenience this will be taken to coincide with the representative point of the first aperture (Fig. 3.20). Let radiation of wavelength λ from a monochromatic point source S fall on the screen, and let the diffraction pattern be observed at a point P. The waves reaching the screen from S and those reaching P from the screen will in fact be spherical, but if the distance R from both S and P to the screen is sufficiently great these waves may be taken to be plane, and Fraunhöfer diffraction conditions are established. (In the case of a two-dimensional grating, if the apertures occupy any approximately circular area of radius r, the condition for Fraunhofer diffraction is $R \geqslant 5r^2/\lambda$.)

Let the plane wave of amplitude A arriving at the point of observation P by way of the aperture at the origin O be represented (cf. eqn (A 12)) by

$$\psi_0 = A \cos(\alpha - \omega t) = A \exp(i\alpha) . \exp(-i\omega t) \qquad (3.9)$$

where

$$\alpha = kX + \delta = \frac{2\pi X}{\lambda} + \delta,$$

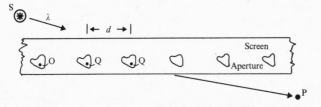

FIG. 3.20. Interaction of radiation from a point source S with a screen containing identical apertures spaced distance d apart. P is the point of observation.

k is the wave number, X is the distance SOP, δ is the phase angle, and ω is the angular frequency. The wave arriving at P by way of any other aperture m will have travelled a distance Δ_m further than the reference wave given by (cf. Fig. 3.5 and eqn (3.2))

$$\Delta_m = (m-1)d(\sin \theta_i - \sin \theta) = (m-1)\Delta$$

where θ_i, θ are the angles between the normal to the grating and the incident and diffracted wave directions respectively. The wave ψ_m arriving at P will therefore be represented by

$$\psi_m = A \cos\left[\frac{2\pi}{\lambda}\{X + (m-1)\Delta\} + \delta - \omega t\right]$$

$$= A \exp i\left\{\alpha + \frac{2\pi}{\lambda}(m-1)\Delta\right\}\exp -i\omega t.$$

Since the waves incident on each aperture come from a single source, each will have the same value of A and δ. The resultant disturbance at P due to interference between the waves from the M apertures will, according to the principle of superposition, then be

$$\Psi = A \exp(-i\omega t)\exp(i\alpha) \sum_{m=0}^{M-1} \exp i\left(\frac{2\pi}{\lambda} m\Delta\right). \qquad (3.10)$$

It is generally more convenient to define the path difference Δ in an alternative way as follows. Let $\mathbf{S}_0$ and $\mathbf{S}$ be unit vectors defining the directions of the incident and scattered waves, respectively (Fig. 3.21). Then

$$\mathbf{d} \cdot \mathbf{S}_0 = d \sin \theta_i$$

$$\mathbf{d} \cdot \mathbf{S} = d \sin \theta.$$

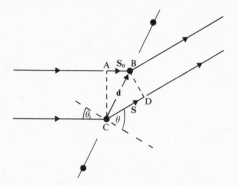

FIG. 3.21. Interaction of a wave with a row of point scatterers. The incident and scattered directions are defined by unit vectors S_0 and S, respectively.

Hence

$$\Delta = -\mathbf{d} \cdot (\mathbf{S} - \mathbf{S}_0).$$

The phase difference ϕ between waves from two adjacent apertures is thus

$$\phi = \frac{2\pi\Delta}{\lambda} = \frac{-2\pi}{\lambda} \mathbf{d} \cdot (\mathbf{S} - \mathbf{S}_0). \tag{3.11}$$

A *scattering vector* $\mathbf{K}$ may now be defined as

$$\mathbf{K} = \frac{2\pi}{\lambda} (\mathbf{S} - \mathbf{S}_0) \tag{3.12}$$

so that the phase factor takes the particularly simple form

$$\phi = \frac{2\pi\Delta}{\lambda} = -\mathbf{K} \cdot \mathbf{d}. \tag{3.13}$$

The scattering vector has magnitude $|\mathbf{K}| = 4\pi/\lambda \sin\frac{1}{2}(\theta - \theta_i)$ and lies in the direction bisecting the angle between $\mathbf{S}$ and $-\mathbf{S}_0$. This is in fact a quite general and very important result. The phase difference between any two scatterers separated by a vector $\mathbf{r}$ is $\phi = -\mathbf{K} \cdot \mathbf{r}$.

Omission of the time factor in (3.10) and substitution of (3.13) gives the following expression for the complex amplitude of the resultant wave at P:

$$A_c = A \exp(i\alpha) \sum_{m=0}^{M-1} \exp(i\phi_m)$$

$$= A \exp(i\alpha) \sum_{m=0}^{M-1} \exp(-im\mathbf{K} \cdot \mathbf{d}). \qquad (3.14)$$

The summation in this equation is a geometric progression of the form

$$1 + x + x^2 + \ldots + x^{M-1} = \frac{1 - x^M}{1 - x}$$

where $x = \exp(-i\mathbf{K} \cdot \mathbf{d})$. Therefore

$$\sum_{m=0}^{M-1} \exp(-im\mathbf{K} \cdot \mathbf{d}) = \left\{ \frac{1 - \exp(-iM\mathbf{K} \cdot \mathbf{d})}{1 - \exp(-i\mathbf{K} \cdot \mathbf{d})} \right\}$$

$$= \frac{\exp(-\tfrac{1}{2}iM\mathbf{K} \cdot \mathbf{d})\{\exp(\tfrac{1}{2}iM\mathbf{K} \cdot \mathbf{d}) - \exp(-\tfrac{1}{2}iM\mathbf{K} \cdot \mathbf{d})\}}{\exp(-\tfrac{1}{2}i\mathbf{K} \cdot \mathbf{d})\{\exp(\tfrac{1}{2}i\mathbf{K} \cdot \mathbf{d}) - \exp(-\tfrac{1}{2}i\mathbf{K} \cdot \mathbf{d})\}}$$

$$= \exp\{-\tfrac{1}{2}i(M-1)\mathbf{K} \cdot \mathbf{d}\} \frac{\sin(\tfrac{1}{2}M\mathbf{K} \cdot \mathbf{d})}{\sin(\tfrac{1}{2}\mathbf{K} \cdot \mathbf{d})} .$$

Hence (3.14) may be written as

$$A_c = A \exp i\{\alpha - \tfrac{1}{2}(M-1)\mathbf{K} \cdot \mathbf{d}\} \frac{\sin(\tfrac{1}{2}M\mathbf{K} \cdot \mathbf{d})}{\sin(\tfrac{1}{2}\mathbf{K} \cdot \mathbf{d})} . \qquad (3.15)$$

The amplitude of the resultant wave is thus

$$|A_c| = A \frac{\sin(\tfrac{1}{2}M\mathbf{K} \cdot \mathbf{d})}{\sin(\tfrac{1}{2}\mathbf{K} \cdot \mathbf{d})} \qquad (3.16)$$

and its phase is $\tfrac{1}{2}(M-1)\mathbf{K} \cdot \mathbf{d}$ relative to that of the reference wave at the origin (cf. eqn (3.9)). The significant quantity is the intensity I obtained by squaring $|A_c|$:

$$I(\mathbf{K}) = \frac{A^2 \sin^2(\tfrac{1}{2}M\mathbf{K} \cdot \mathbf{d})}{\sin^2(\tfrac{1}{2}\mathbf{K} \cdot \mathbf{d})} . \qquad (3.17)$$

Note that to compare intensities at various points in the diffraction pattern only the *relative* phases $\phi = -\mathbf{K} \cdot \mathbf{d}$ of the waves scattered by adjacent components of the grating need to be known. Absolute phases and amplitudes are not required. Also the calculations do not depend explicitly on the quantities $\mathbf{S}_0$, $\mathbf{S}$, and λ, but only on the combination $\mathbf{K} = (2\pi/\lambda)(\mathbf{S} - \mathbf{S}_0)$.

The intensity $I(\mathbf{K})$ is shown as a function of $\mathbf{K} \cdot \mathbf{d}$ in Fig. 3.22 for various values of M. Principal maxima occur when

$$\mathbf{K} \cdot \mathbf{d} = 2\pi q \qquad \text{(equivalent to } q\lambda = d(\sin\theta_i - \sin\theta))$$

(3.18)

where q is zero or any integer. This relationship is known as the Laue condition. From (3.13) it is equivalent to the familiar condition

$$\Delta = q\lambda$$

i.e. that the path difference for scattering by different elements in the grating is an integral number of wavelengths. Since under these conditions the numerator and denominator of (3.17) are both zero, the function must be evaluated as

$$\lim_{\gamma \to 0} \frac{\sin M\gamma}{\sin\gamma} = \lim_{\gamma \to 0}\left[M\left\{\frac{\sin M\gamma}{M\gamma}\frac{\gamma}{\sin\gamma}\right\}\right].$$

Now

$$\lim_{\theta \to 0}\left(\frac{\sin\theta}{\theta}\right) = \lim_{\theta \to 0}\left\{\frac{1}{\theta}\left(\theta - \frac{\theta^3}{\underline{|3}} + \frac{\theta^5}{\underline{|5}} - \cdots\right)\right\} = 1.$$

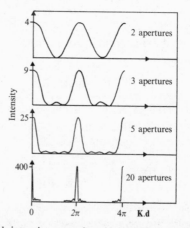

FIG. 3.22. Diffracted intensity as a function of $\mathbf{K} \cdot \mathbf{d}$ for gratings containing various numbers of apertures. The aperture spacing is $\mathbf{d}$, and $\mathbf{K}$ is the scattering vector.

Hence

$$\lim_{\gamma \to 0} \frac{\sin M\gamma}{\sin \gamma} = M$$

and so the principal maxima of $I(\mathbf{K})$ (eqn (3.17)) take the value $A^2 M^2$. Over any 2π interval of $\mathbf{K} \cdot \mathbf{d}$ the curve $I(\mathbf{K})$ has $(M-2)$ subsidiary maxima. When M is large these are too small to be observed, and the diffraction pattern consists of a line of sharp reflections produced by the principal maxima alone. (We must remember that for a one-dimensional grating intensity in reciprocal space is not limited by a second Bragg condition. This means that the maxima drawn in Fig. 3.22 actually extend in the direction normal to the plane of the paper and so would give rise to a series of *lines* on a LEED screen (cf. Section 3.3.2).) When, however, the grating consists of a small number of elements (M small), or when defective gratings are considered, the subsidiary maxima in eqn (3.17) may not be of negligible intensity and streaks rather than sharp spots may be observed. Also, the width at half height of the principal maxima is proportional to $1/M$, so that as M decreases these maxima broaden; this is equivalent to the 'particle size' broadening observed in X-ray diffraction.

B. Diffraction by a two-dimensional grating. It is now necessary to consider diffraction by a two-dimensional structure which may have more than a single atom in the unit mesh. If $\mathbf{a}$, $\mathbf{b}$, are the vectors defining the unit mesh, then the origin of any given mesh in the structure may be located by a vector $(m_1\mathbf{a} + m_2\mathbf{b})$ where m_1, m_2 are integers. The position of any atom within the mesh may then be defined with respect to the mesh origin by a vector $\mathbf{r}_n$, and hence with respect to the origin of the entire structure by the vector

$$\mathbf{R}_m^n = m_1\mathbf{a} + m_2\mathbf{b} + \mathbf{r}_n. \qquad (3.19)$$

As was seen above (eqn (3.13)) the phase difference between waves scattered at the origin and at the distance $\mathbf{R}_m^n$ from the origin is simply

$$\phi_m^n = -\mathbf{K} \cdot \mathbf{R}_m^n. \qquad (3.20)$$

Substitution of (3.19) and (3.20) into (3.14) gives the amplitude of the resultant wave scattered by the complete structure

$$A_c = A \exp(i\alpha) \sum_{m_1=0}^{(M_1-1)} \sum_{m_2=0}^{(M_2-1)} \sum_{n=0}^{(N-1)} f_n$$
$$\times \exp -\{i\mathbf{K} \cdot (m_1\mathbf{a} + m_2\mathbf{b} + \mathbf{r}_n)\}, \quad (3.21)$$

where f_n is the scattering factor for atom n. The crystal dimensions are M_1a, M_2b, and N is the number of atoms in the unit mesh. The resultant amplitude scattered by a *single unit mesh* is known as the structure factor; its square is the diffraction function. These terms are defined by the relationships

$$F = \sum_{n=0}^{(N-1)} f_n \exp(-i\mathbf{K} \cdot \mathbf{r}_n) \qquad (3.22a)$$

$$F^2 = \sum_{m=0}^{(N-1)} \sum_{n=0}^{(N-1)} f_m f_n^* \exp\{-i\mathbf{K} \cdot (\mathbf{r}_m - \mathbf{r}_n)\} \qquad (3.22b)$$

$$= \left\{\sum_{n=0}^{(N-1)} f_n \cos(-\mathbf{K} \cdot \mathbf{r}_n)\right\}^2 + \left\{\sum_{n=0}^{(N-1)} f_n \sin(-\mathbf{K} \cdot \mathbf{r}_n)\right\}^2 \qquad (3.22c)$$

$$= \sum_{n=0}^{(N-1)} f_n^2 + \sum_{\substack{m=0 \\ m \neq n}}^{(N-1)} \sum_{n=0}^{(N-1)} f_m f_n^* \cos\{\mathbf{K} \cdot (\mathbf{r}_m - \mathbf{r}_n)\}. \qquad (3.22d)$$

The quantity F^2 will be required for calculation of the diffracted intensity; since, in general, F is complex, F^2 must be evaluated as FF^*, where F^* is the complex conjugate of F. Using eqn (3.22a) eqn (3.21) may be written as

$$A_c = A \exp(i\alpha) F \sum_{m_1=0}^{(M_1-1)} \exp(-im_1\mathbf{K} \cdot \mathbf{a}) \sum_{m_2=0}^{(M_2-1)} \exp(-im_2\mathbf{K} \cdot \mathbf{b}).$$
$$(3.23)$$

The term F in (3.23) represents the resultant scattering by one unit mesh in terms of a single scatterer at the mesh origin. The two summation terms represent the resultant scattering by a two-dimensional lattice of point scatterers each located at a unit mesh origin; these terms are each equivalent to the summation in eqn (3.14) for the scattering by a one-dimensional grating.

From (3.23) the scattered intensity is given by

$$I(\mathbf{K}) = A^2 F^2 \frac{\sin^2(\frac{1}{2}M_1\mathbf{K}\cdot\mathbf{a})}{\sin^2(\frac{1}{2}\mathbf{K}\cdot\mathbf{a})} \frac{\sin^2(\frac{1}{2}M_2\mathbf{K}\cdot\mathbf{b})}{\sin^2(\frac{1}{2}\mathbf{K}\cdot\mathbf{b})} \qquad (3.24a)$$

$$I(\mathbf{K}) \propto F^2 \mathscr{I} \qquad (3.24b)$$

where $\mathscr{I}$ is the *interference function*. It consists of two functions of the form of Fig. 3.22 and will have maxima of magnitude $M_1^2 M_2^2$ when the two Laue conditions

$$\begin{aligned}\mathbf{K}\cdot\mathbf{a} &= 2\pi h &\quad (\text{or } h\lambda = a\sin\theta_h)\\ \mathbf{K}\cdot\mathbf{b} &= 2\pi k &\quad (\text{or } k\lambda = b\sin\theta_k)\end{aligned} \qquad h, k = \text{zero or integer}$$

$$(3.25)$$

are simultaneously satisfied. Eqn (3.24) may be compared with the elements of the optical diffraction process illustrated in Fig. 3.4.

In the presence of an adsorbate with mesh dimensions $\mathbf{a}_s = P\mathbf{a}$, $\mathbf{b}_s = Q\mathbf{b}$ (P and Q integers) eqns (3.23) and (3.24) are modified by the replacement of $\mathbf{a}$, $\mathbf{b}$ by $\mathbf{a}_s$, $\mathbf{b}_s$ and eqns (3.25) become

$$\mathbf{K}\cdot\mathbf{a}_s = 2\pi h \qquad \mathbf{K}\cdot\mathbf{b}_s = 2\pi k$$

or

$$\left.\begin{aligned}\mathbf{K}\cdot\mathbf{a} &= 2\pi h/P = 2\pi h'\\ \mathbf{K}\cdot\mathbf{b} &= 2\pi k/Q = 2\pi k'\end{aligned}\right\}. \qquad (3.26)$$

When $\mathbf{K}\cdot\mathbf{a}/2\pi = h'$ and $\mathbf{K}\cdot\mathbf{b}/2\pi = k'$ are both integral, spots occur at positions identical to those occupied in the clean surface pattern; these are integral order spots. In contrast to the clean surface case, however, an overlayer can also generate non-integral values of $\mathbf{K}\cdot\mathbf{a}/2\pi$ and $\mathbf{K}\cdot\mathbf{b}/2\pi$. These values give rise to fractional order (or 'extra') spots, which occur at positions not occupied in the clean surface pattern.

The structure factor F in eqns (3.24) for diffraction by a clean surface is given by (3.22). The vector $\mathbf{r}_n$ locating an atom within the unit mesh with respect to the mesh origin may be written

$$\mathbf{r}_n = x_n\mathbf{a} + y_n\mathbf{b} \qquad (3.27)$$

where $x_n = X_n/a$, $y_n = Y_n/b$ are the *fractional co-ordinates* of the atom; its Cartesian co-ordinates are (X_n, Y_n). Thus

$$F = \sum_{n=0}^{(N-1)} f_n \exp\{-i\mathbf{K}\cdot(x_n\mathbf{a} + y_n\mathbf{b})\}. \qquad (3.28)$$

The principal intensity maxima occur only when (3.25) are satisfied, and under these conditions (3.28) becomes

$$F = \sum_{n=0}^{(N-1)} f_n \exp\{-2\pi i(hx_n + ky_n)\}. \tag{3.29}$$

Note that this summation does not lead to the same type of expression as do the summations in eqn (3.23), since in general x_n, y_n are not periodic within the unit mesh, but its value can be easily obtained using a computer. In the case of an overlayer eqn (3.29) is modified only by the replacement of h, k by $h' = h/P$, $k' = k/Q$ respectively.

Eqn (3.24), with an appropriate substitution for F (e.g. eqn (3.28)), enables us to calculate reciprocal-space intensity profiles since variation of $\mathbf{K}\cdot\mathbf{a}$ and $\mathbf{K}\cdot\mathbf{b}$ corresponds to examination of scattering in different directions from the surface (cf. Fig. 3.21). For example, if $\mathbf{K}\cdot\mathbf{b}$ is fixed at zero and $\mathbf{K}\cdot\mathbf{a}$ is varied between zero and 2π, we generate the profile along a line in the pattern between the 00 ($\mathbf{K}\cdot\mathbf{a} = \mathbf{K}\cdot\mathbf{b} = 0$) and 10 ($\mathbf{K}\cdot\mathbf{a} = 2\pi$, $\mathbf{K}\cdot\mathbf{b} = 0$) spots. If there is only one atom in the unit mesh the profile would be of the form shown previously in Fig. 3.22.

Eqn (3.24) also confirms two conclusions established qualitatively in Section 3.2. The *positions* of points of maximum intensity (spots in the pattern) are determined by the interference function $\mathscr{I}$ alone and hence yield the size and symmetry of the unit mesh (i.e. $\mathbf{a}$ and $\mathbf{b}$). The relative spot *intensities* are determined by the square of the structure factor $|F|^2$, and F is related through its phase factors (the exponential terms in eqn (3.28)) to the *positions of atoms within the unit mesh*. The experimentally determined intensities are, however, proportional to $F^2 = FF^*$ and so give the amplitude of F, but not its phase. This constitutes the *phase problem* in structure analysis. In the X-ray case various means have been devised by which it may be overcome and complete structures may be determined. In LEED, however, an additional problem is introduced by the fact that the diffraction process is not strictly two-dimensional so that the structure factor is not accurately represented by (3.28).

3.4.2. *Spot intensity profiles—qualitative observations*

The positions of spots in the pattern enable us to determine the size and shape of the surface unit mesh, in other words the space-group symmetry of the two-dimensional surface lattice. Having done this, a number of important questions remain unanswered.

   (i) What is the site symmetry of the adsorbate species: one-fold, two-fold, etc. (Fig. 3.23(a,b))? This is equivalent to asking where the overlayer is placed relative to the top surface layer.

  (ii) Is there more than a single atom in the unit mesh (Fig. 3.23(c))? Note that both the above questions should be considered in relation to clean adsorbent surfaces, as well as to adsorbed layers, since the termination of the bulk lattice may cause some rearrangement of surface atoms

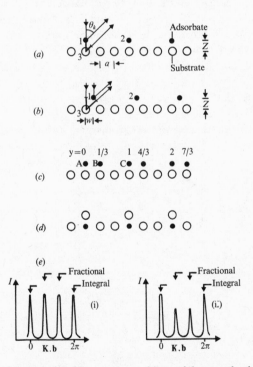

FIG. 3.23. (a) Section through a surface with a triple-spaced adsorbed layer containing a single adsorbate atom within the unit mesh. (b) As in (a), but with the adsorbate located at sites of different symmetry. (c) As in (a) but with two adsorbate atoms within the unit mesh. (d) As in (a) but with the surface reconstructed. (e) The reciprocal-space intensity profiles predicted by simple kinematic theory for the structures in (i), case (a) and (ii), case (b) above.

from the periodicity expected on the basis of the bulk structure.

(iii) Does adsorption lead to reconstruction, that is, are atoms of the substrate shifted from their original sites; do adsorbate species penetrate the substrate lattice (Fig. 3.23(d))?

Suppose low-energy electrons were in fact scattered entirely by the surface layer, as we have thus far assumed. The structure factor for a two-dimensional unit mesh is given by eqn (3.29) as

$$F = \sum_{n=0}^{(N-1)} f_n \exp\{-2\pi i(hx_n + ky_n)\}. \qquad (3.29)$$

In the case of a single atom per unit mesh (Fig. 3.23(a)) $x_n = y_n = 0$, and F is identical with the atomic scattering factor f. Therefore, since F is independent of h and k, all spots in the pattern will be of equal intensity, at least in the first approximation; (there will be some variation, however, since f is a function of the scattering vector $\mathbf{K}$ and hence of the direction of the reflection in reciprocal space (see Section 3.5)). In terms of the Ewald construction that is equivalent to all reciprocal-space rods being of identical and uniform diameter (Fig. 3.24(a)).

If we now take a case of two atoms per mesh in a triple-spaced structure, such as Fig. 3.23(c), we have $h = 0$, $y_1 = 0$, $y_2 = \frac{1}{3}$, with

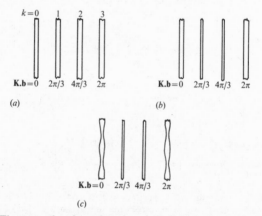

FIG. 3.24. Elements of reciprocal lattices according to simple kinematic theory. (a) The case of scattering by the adsorbate layer in Fig. 3.23(a). (b) Scattering by the adsorbate layer in Fig. 3.23(c). (c) Scattering by both the adsorbate and top substrate layers in Fig. 3.23(c).

respect to the substrate. Then

$$F = 1 + \exp(-2\pi i k/3) \qquad\qquad (3.30a)$$

$$F^2 = 2(1 + \cos 2\pi k/3) \qquad k = 0, \pm 1, \ldots . \qquad (3.30b)$$

Note that we included only atoms A and B (Fig. 3.23(c)) in the calculation, assigning to each a scattering factor f of unity, although A is shared with a neighbouring mesh and an additional atom (C) also contributes to the mesh being considered. If, however, we put $f_A = f_C = \frac{1}{2}$ and write F as

$$F = \tfrac{1}{2} + \exp(-2\pi i k/3) + \tfrac{1}{2}\exp(-2\pi i k)$$

we arrive at an expression identical to (3.30a). From this we may draw the general conclusion that the scattering contribution from atoms at the corners of a unit mesh is equivalent to that of a single scatterer at the mesh origin.

Returning to (3.30b) we see that for integral-order spots $k/3 = 0$ or an integer and $F^2 = 4$. For fractional order spots $k/3$ is non-integral, the cosine term in eqn (3.30b) will be less than unity, and so the intensities of these spots will be less than those of the integral orders. In this case each reciprocal lattice rod still has a uniform diameter, but diameters vary from rod to rod, depending on the indices of the rod (Fig. 3.24(b)). The distinction between the structures in Fig. 3.23(a,c) on the basis of kinematic intensities is illustrated by Fig. 3.23(e).

Centred lattices represent an extreme example of intensity modulation by atoms within the unit mesh. The structure factor for a c(2×2) mesh for example is

$$F = 1 + \exp\{-i2\pi(\tfrac{1}{2}h + \tfrac{1}{2}k)\}$$

and

$$F^2 = 2\{1 + \cos 2\pi(\tfrac{1}{2}h + \tfrac{1}{2}k)\}.$$

For spots h, k or $h/2$, $k/2$ the value of F^2 is 4, but for $h/2$, k or h, $k/2$ it is zero, hence the systematic absences in the c(2×2) compared with the p(2×2) pattern (Figs 3.16 and 3.17).

According to eqns (3.30) it would appear that the intensities of diffraction beams produced by a two-dimensional grating are independent of the wavelength of the radiation used, since the expression for the structure factor does not contain a term in λ. This is not quite true, since we have omitted to observe that the atomic scattering factor f is rather weakly wavelength dependent and as a result the beams from a two-dimensional grating would decrease slowly and *smoothly* as the primary electron energy was increased. It was seen in Section 3.2, however, that fairly rapid modulation of spot intensities is always observed experimentally as a consequence of wavelength variation. Let us first establish in a most rudimentary way that this modulation would result if low-energy electrons penetrated below the top surface layer, i.e. if the grating had some degree of three-dimensional character.

We shall consider the structure shown in Fig. 3.23(a). The position of a given spot of index k will be determined by scattering from overlayer atoms such as 1 and 2 through the relationship $k\lambda = 3a \sin \theta_k$. If interference between the waves scattered by overlayer atom 1 and by substrate atom 3, for example, is introduced, this will be constructive only if the path difference Δ_{13} is an integral number of wavelengths, i.e.

$$\Delta_{13} = Z(1 + \cos \theta_k) = n\lambda = n\left(\frac{1 \cdot 504}{V}\right)^{\frac{1}{2}}. \tag{3.31}$$

Provided this condition is fulfilled 1–3 scattering will *reinforce* the intensity of spot k, otherwise it will be diminished. Therefore if the electron wavelength λ, or the accelerating potential V, is varied, the intensity I of a given spot will be modulated owing to variation in the contribution to the intensity from scattering by the underlying layers. A plot of I against V for a particular spot (e.g. Fig. 3.3) is the spot intensity profile; according to the oversimplified picture just considered, intensity maxima, the so-called Bragg maxima, should occur at those values of V for which eqn (3.31) is satisfied. The observation of such intensity variation confirms that low-energy electrons do indeed penetrate at least some distance into the solid.

Intensity criteria can now be used in principle to distinguish between structures such as Fig. 3.23(a,b). For the case of Fig. 3.23(b)

$$\Delta_{13} = Z(1 + \cos \theta_k) + w \sin \theta_k \tag{3.32}$$

where w measures the relative displacement of the overlayer parallel to the surface. Obviously if condition (3.31) holds at a particular electron energy then condition (3.32) cannot. Therefore at this energy 1–3 scattering in the case of the structure in

Fig. 3.23(a) would add to the intensity of spot k, but would detract from it in the case in Fig. 3.23(b). Hence the two different site symmetries could in principle be distinguished on the basis of spot intensities. Note incidentally that the structures in Fig. 3.23(a,b) could both be differentiated from Fig. 3.23(c) if the adsorbate coverages were accurately known. For example, when an Ni(110) surface interacts with oxygen the following patterns are observed as time of exposure to the gas increases: (3×1), (2×1), (3×1). In the two (3×1) patterns the spot *positions* are identical but, because of the order in which they are formed, the corresponding surface structures are presumably different. Such structural differences would also be reflected in the corresponding intensity profiles, and, although a complete analysis of these is not a routine matter at the present time, they may nevertheless prove useful as 'fingerprints' for distinguishing different structures.

3.5. Spot-intensity profiles—quantitative considerations

3.5.1. *Kinematic theory*

If a rigorous analysis of spot intensities is to be attempted the various processes which a primary electron may experience within the solid must be established. As a low-energy (say 10–500 eV) electron approaches a surface it first comes under the influence of the long-range image potential, but on close approach it will encounter the valence-band electrons contained within the solid by the inner potential barrier V_0 at the surface (Fig. 3.25). As it crosses this barrier the electron will experience a change in energy from E_p to $(E_p - V_0)$. This is equivalent to

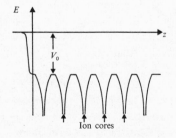

FIG. 3.25. A cross-section of the potential in a crystal.

refraction since it leads to a change in wavelength from $\lambda = (1\cdot504/V)^{\frac{1}{2}}$ to λ' where

$$\lambda' = \left(\frac{1\cdot504}{E_{\mathrm{p}} - V_0}\right)^{\frac{1}{2}}. \tag{3.33}$$

In the region of V_0 the incident electron can begin to undergo various *inelastic* interactions with the valence electrons, which scatter it predominantly in the forward direction. These include the excitation of inter- and intra-band transitions, in which the energy loss will be rather less than ~10 eV, but the most important is the production of surface and bulk plasmons, for which the energy loss is ~10–20 eV. If the electron penetrates a few tenths of a nanometer further into the solid it reaches the first plane of ion cores from which it may be *elastically* scattered, both in the forward and backward directions. Since the ion cores are arranged in a periodic array, this elastic scattering constitutes a diffraction process and it is the back-scattered component which leaves the solid and is allowed to filter through the grids of the LEED optics to generate the spot pattern.

The total cross-section σ_{e} for elastic ion-core scattering is of the order of 10^{-19}–10^{-21} m^2, decreasing slowly with increasing electron energy (see e.g. Kambe 1970); for inelastic loss by plasmon excitation the value is ~10^{-20} m^2. The electron mean free path λ_{e} is related to σ by the expression

$$\lambda_{\mathrm{e}} = (\rho\sigma)^{-1}$$

where ρ is the atom density in the solid, i.e. ~10^{28}–10^{29} atoms m^{-3}. It follows that for both elastic and inelastic scattering the mean free path is ~1 nm (see also Chapter 4). Therefore an incident electron will undergo both types of scattering somewhere within the top one to five layers at the surface.

The inelastically scattered electrons are eliminated from the LEED experiment by the grid system, yet it is the high probability of inelastic scattering which makes the technique surface sensitive. This, however, also implies a strong and hence complex interaction between the incident electron and the solid, and is responsible for the fact that the interpretation of experimental data is much more difficult with LEED than in the case of weakly scattered (highly penetrating) radiation such as X-rays.

Having found that a LEED pattern will have contributions

from events occurring perhaps five or more atom layers below the surface the first step in developing a description of intensities might be to start with the expression used in the previous discussion of a two-dimensional grating, namely (eqn (3.24))

$$I(\mathbf{K}) = A^2 F^2 \mathcal{I}. \tag{3.24}$$

We might think of modifying the interference function $\mathcal{I}$ to take account of the extension of the structure into the third dimension and write

$$\mathcal{I} = \frac{\sin^2(\tfrac{1}{2}M_1\mathbf{K} \cdot \mathbf{a})}{\sin^2(\tfrac{1}{2}\mathbf{K} \cdot \mathbf{a})} \frac{\sin^2(\tfrac{1}{2}M_2\mathbf{K} \cdot \mathbf{b})}{\sin^2(\tfrac{1}{2}\mathbf{K} \cdot \mathbf{b})} \frac{\sin^2(\tfrac{1}{2}M_3\mathbf{K} \cdot \mathbf{c})}{\sin^2(\tfrac{1}{2}\mathbf{K} \cdot \mathbf{c})}. \tag{3.34}$$

In the z-direction (normal to the surface) c is the translational periodicity and M_3 is the number of unit cells falling within the region of depth $M_3 c$ at the surface concerned in the diffraction of low-energy electrons. As with a two-dimensional grating the M_1 and M_2 terms will produce sharp maxima when $\mathbf{K} \cdot \mathbf{a}/2\pi$ and $\mathbf{K} \cdot \mathbf{b}/2\pi$ take integral values, and will be effectively zero elsewhere. On the other hand, since M_3 is small (2–10), the final sine term in eqn (3.34) will have appreciable magnitude over a considerable range of $\mathbf{K} \cdot \mathbf{c}$ (Fig. 3.22) and therefore will not impose any very strict conditions as to the values of $\mathbf{K} \cdot \mathbf{c}$ required for production of a spot. An alternative way of expressing this is to say that the periodicity in the z-direction is not sufficiently extensive to enable a third Laue condition, $\mathbf{K} \cdot \mathbf{c}/2\pi = l$, to be established. Spot *positions* are therefore still determined, as with a two-dimensional grating, by terms involving only parameters related to the surface unit mesh (i.e. $\mathbf{a}$ and $\mathbf{b}$).

In eqn (3.34) the third sine term effectively plays the part of a structure factor and it is therefore more convenient to cast the intensity equation in a form which specifically recognizes this fact. The crystal is considered to be constructed from the two-dimensional surface lattice, each point in which is populated by a three-dimensional cell extending to the depth of electron penetration below the surface and containing N scatterers. Then, since we are considering spot positions only, F will be given by the equation

$$F = \sum_{n=0}^{(N-1)} f_n \exp\{-i\mathbf{K} \cdot (x_n\mathbf{a} + y_n\mathbf{b} + Z_n)\} \tag{3.35}$$

where, for scatterer n, x_n, and y_n are fractional co-ordinates in directions parallel to the surface, and Z_n is the absolute co-ordinate normal to the surface. For a particular spot $\mathbf{K} \cdot \mathbf{a} = 2\pi h$, $\mathbf{K} \cdot \mathbf{b} = 2\pi k$ and also, from eqn (3.13)

$$-\mathbf{K} \cdot \mathbf{Z}_n = 2\pi\Delta/\lambda' \qquad (3.36)$$

where Δ is the path difference in the z-direction for scattering by atom n and by an atom at the origin. So, using eqn (3.31), we can write

$$-\mathbf{K} \cdot \mathbf{Z}_n = \frac{2\pi Z_n(1 + \cos\theta)}{\lambda'}. \qquad (3.37)$$

Hence

$$F = \sum_{n=0}^{(N-1)} f_n \exp\left[-2\pi i\left\{hx_n + ky_n + \frac{Z_n(1 + \cos\theta_{hk})}{\lambda'}\right\}\right]. \qquad (3.38)$$

The full intensity eqn (3.24) will now consist of the square of this structure factor multiplied by an interference function made up of the first two sine terms in eqn (3.34). Note that the wavelength required in eqn (3.38) is that of the electron *inside* the solid, and so the value of the inner potential must be taken into account in its calculation (eqn (3.33)). Eqn (3.38) conforms with experimental observation in that both the presence of underlying atoms and wavelength variation will influence F and hence spot intensity.

In reciprocal space the interference function corresponding to a two-dimensional lattice generates a set of lattice rods of uniform diameter (Fig. 3.24(a) and (b)), but when a structure factor of type (3.38) is introduced its cosine term may modulate the diameter of a rod. The area of intersection of the Ewald sphere with the rod (i.e. the spot intensity) therefore changes as the electron energy varies and the sphere expands and contracts. It is an interesting exercise to show that simple kinematic theory predicts modulation in the case of integral order rods only; the diameter of fractional order rods should be independent of wavelength (Fig. 3.24(c)). This point can be demonstrated by writing the structure factor as the sum of two terms, one due to scattering by the overlayer and the other to scattering by the top substrate layer (cf. eqn (3.46), Section 3.6.1).

For integral order spots simple kinematic theory thus predicts that the intensity profile will consist of a series of peaks, the

Bragg maxima, centred at values V_{max} of the electron accelerating potential V for which the third Laue condition $\mathbf{K} \cdot \mathbf{c} = 2\pi l$ is satisfied.

Eqn (3.38) represents what is generally known as the *kinematic approximation* for the structure factor and is based on the assumption that the probability of an electron being scattered by the ion cores is small. The possibility of it being scattered for a second, third, etc. time before eventually leaving the solid may therefore be neglected, and so this is a *single-scattering* approximation, as opposed to more complex *multiple-scattering*, or *dynamical*, theories.

3.5.2. Modified kinematic theory

In general eqn (3.38) does not lead to realistic intensity profiles. While in certain cases it may reproduce the positions of intensity maxima with moderate success, it very rarely succeeds in predicting the correct magnitudes of peaks and the finer details of the profile. Its most obvious shortcomings are that it does not take into account (*a*) any dependence of the scattering factor f_n on electron energy or on the incident and diffracted directions and (*b*) inelastic scattering.

The simplest attempt to incorporate any of these features is due to Lander and Morrison (1962, 1963), who suggested that allowance could be made for the strong elastic scattering by considering atoms in underlying layers to be 'shadowed' by atoms above them. This effectively reduces the scattering factor f_n for a shadowed atom by a factor t_n. In certain cases limited success is apparently obtained simply by taking t_n to be either 1 or 0 depending on whether or not atom n is directly 'visible' to the incident beam but it is unfortunately found that a number of possible structures lead to equally satisfactory results.

The more sophisticated attempts to devise an acceptable kinematic theory introduce detailed consideration of both elastic and inelastic scattering. Elastic scattering occurs at the ion cores, and the corresponding scattering factors f_n depend on the electron–ion-core potential. It is conveniently found that the lineshapes of observed I–V curves are quite adequately described by rather simple models such as the 'muffin-tin' approximation. This model assumes that the potential is spherically symmetrical within the largest possible non-overlapping spheres drawn about

FIG. 3.26. Cross-section of a crystal divided into muffin tins.

each nucleus and constant elsewhere (Fig. 3.26). Its use leads to realistic scattering factors, but in experiments carried out at room temperature these would be perturbed by the thermal displacements of the ion cores. Correction for this effect is difficult, since the magnitudes of the displacements are not well known, and it is therefore desirable to collect intensity data at temperatures below about 100 K where the influence of lattice vibration is negligible.

Some examples of scattering factors calculated by Fink, Martin, and Somorjai (1972) are shown in Fig. 3.27; they are found to agree well with experimental results for energies above about 50 eV. It is found, as stated earlier, that forward-scattering predominates over back-scattering. There is no systematic variation

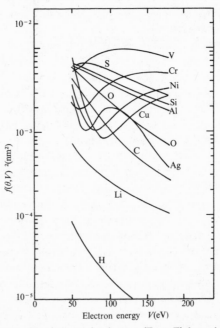

FIG. 3.27. Elastic scattering factors. (From Fink *et al.* 1972.)

of the shape or magnitude of f with atomic number, and scattering from elements such as oxygen may be larger than that from heavier elements under certain conditions. This is important in relation to the earlier development of LEED, since it was originally thought that light-element scattering would be very weak and that the presence of, for example, adsorbed oxygen at a surface could not produce extra features in the diffraction pattern.

Inelastic electron–electron scattering leading to absorption of the incident beam is accounted for by introducing an imaginary component V_{0i} into the inner potential V_0, the real component of which (V_{0r}) arises, as we have already seen, from the interaction of the incident electron with the valence electrons. The calculation of V_{0i} is one of the main areas of difficulty in LEED theory. At the present time it is introduced using empirical formulae. These may have a physical basis but in general both V_{0i} and V_{0r} must be treated as parameters which are to be adjusted to achieve the best (multiple-scattering) description of the experimental curve for a clean surface. The value of V_{0r} is derived from the work function and band width, while V_{0i} is obtained by fitting the experimental peak widths. Uncertainties in V_{0r} produce errors in surface lattice parameters, especially for overlayers, but similar uncertainty in V_{0i} is less significant since V_{0i} mainly affects peak widths and intensities, while it is peak positions which are of primary importance for structure determination.

Having obtained realistic values for f_n and V_0 these may be used in a kinematic intensity calculation (eqn (3.38)) but the results are still almost always unsatisfactory, except in the following two situations.

   (a) When the primary energy is high, somewhere in the range 100–500 eV, the electron mean free path increases. This change decreases the probability of multiple scattering and hence favours the kinematic approach. A comparison of the latter theory with results for carbon shows good agreement above 100 eV (Hirabayashi 1968), but in a case such as a tungsten (100) surface, kinematic behaviour would not be expected below about 450 eV. At this energy, however, not only is surface sensitivity severely reduced, but background intensity in the diffraction pattern is high and quantitative measurements are difficult.

(b) If absorption due to inelastic processes is high, then after the first scattering event the electron wave will be strongly attenuated and the probability of it undergoing scattering for a second time will be greatly reduced.

The requirement for elastic scattering to be weak compared with absorption is best fulfilled when the bulk unit cell of the solid is large (Pendry 1974), as it is in crystals of some inert gases for example. Fig. 3.28(a) compares the experimental and calculated (eqn (3.38)) intensity profile of the 00 reflection from a (111) surface of xenon, for which $a_0 = 0.6154$ nm (Ignatjev, Pendry, and Rhodin 1971), and it is seen that agreement is excellent. For a 00 reflection, at normal incidence, eqn (3.38)

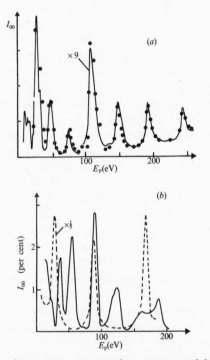

FIG. 3.28. (a) Intensity-energy spectrum taken near normal incidence from a xenon (100) surface. Black dots show points calculated in the kinematic approximation with scattering fitted to experiment. Close agreement of peak shape and position indicates a kinematic experimental spectrum. (b) Comparisons with experiment (solid curve) of kinematic (broken curve) intensities reflected from a (001) nickel surface at normal incidence, 00 beam. (From Pendry 1974.)

takes the particularly simple form

$$F = \sum_{n=0}^{(N-1)} f \exp\left\{-2\pi i\left(\frac{2Z_n}{\lambda'}\right)\right\} \tag{3.39}$$

since $h = k = 0$ and $\theta_{00} = 0°$. As the interlayer spacing c is uniform throughout the crystal, for any atom

$$Z_n = mc \qquad (m = 0, 1, 2 \ldots). \tag{3.40}$$

Each exponential term in eqn (3.39) will have its maximum value of unity when

$$2c = l\lambda' \qquad (l = 0, 1, 2 \ldots) \tag{3.41}$$

because under this condition

$$\exp\left(-\frac{4\pi i Z_n}{\lambda'}\right) = \exp(-2\pi i m l) = 1.$$

Eqn (3.41) thus represents the conditions for principal maxima in F and hence for Bragg maxima in the I–E_p curve. It is equivalent to the third Laue condition $\mathbf{K} \cdot \mathbf{c} = 2\pi l$:

$$\mathbf{K} \cdot \mathbf{c} = \frac{-2\pi\Delta}{\lambda'} \qquad \text{(eqn (3.13))}$$

$$= 2\pi l \qquad l = 0, \pm 1, \pm 2 \ldots$$

since in this case $\Delta = 2c = l\lambda'$. The prediction is that the Bragg maxima will occur at wavelengths corresponding to those of reflections from the bulk of the crystal, but this is strictly true only when the spacing between *all* layers in the crystal is constant (as it is with xenon). If the overlayer–top substrate layer spacing for example is different from that in the bulk the maxima would no longer be accurately determined by an equation of type (3.41).

Eqn (3.41) may be rewritten (cf. eqn (3.33))

$$2c = l\left(\frac{1\cdot504}{E_{p,max} - V_0}\right)^{\frac{1}{2}} \tag{3.42}$$

or

$$E_{p,max} = \frac{1\cdot504}{4}\frac{l^2}{c^2} + V_0. \tag{3.43}$$

where $E_{p,max}$ is the primary electron energy at which a maximum occurs in the I–E_p curve for a given value of l. A plot of $E_{p,max}$

against l^2 for the peaks in the intensity profile thus provides in principle a method for determination of c and V_0.

Eqn (3.43) shows that the main effect of the inner potential is to shift all the peaks in the intensity profile to higher energies by a constant amount V_0 eV. Actually, this is oversimplified since V_0 is a function of both wavelength and angle of incidence, but in practice a constant value often appears to be satisfactory. Using the data shown in Fig. 3.28(a) it is found that the inner potential for xenon is $10 \cdot 0$ eV.

In the xenon case a complete kinematic calculation reproduces the experimental curve almost exactly, but this is quite exceptional. Generally ion-core scattering is strong relative to absorption and a kinematic theory predicts neither peak positions nor shapes with any certainty; Fig. 3.28(b) shows results for the 00 reflection from a (100) nickel surface. Disagreement with experiment is particularly poor at low incident energies, $E_p \leqslant 100$ eV, when many more peaks appear than predicted by eqn (3.43); these are known as secondary Bragg maxima.

To sum up at this point, we find that kinematic, or single-scattering theory, provides an inadequate description of LEED intensities except in a few cases. One reason for its failure lies in the strength of the elastic electron-scattering process in the *forward* direction. Because of this, when an electron incident on the solid is scattered for the first time there is a high probability that it will give rise to diffracted beams travelling further into the solid, and each of these can act as an incident beam for a second scattering event. The result will be back-scattered beams greater in number than those produced by a single-scattering process, and the appearance of 'secondary Bragg maxima' (Fig. 3.28(b)) is always taken as evidence for strong elastic forward-scattering, which leads to double or *multiple diffraction*.

3.5.3. *Data-reduction methods*

The next stage in increasing complexity of intensity analysis recognizes that, although the experimental I–E_p curves do show strong dynamical behaviour, their major peaks generally lie close to the positions expected for Bragg maxima. It has been suggested that the curves can be considered as an array of

randomly positioned individual multiple-scattering peaks modulated by a kinematic 'Bragg envelope' which is determined only
by the geometry of the surface. On this basis attempts have been
made to take a large body of experimental intensity data and
treat it in such a way as to filter out the dynamical features,
leaving what is hopefully a kinematic intensity profile. One way
in which this may be done (Lagally, Ngoc, and Webb 1971, 1972;
Webb and Lagally 1973) is to record profiles for a particular spot
at various azimuthal angles and then take the average over all
profiles. This should remove the randomly positioned multiple-
scattering features and does appear to produce a profile consisting of Bragg peaks only which may then be analysed in a
straightforward way using eqn (3.38). It has been established
(Duke and Smith 1972) that this type of averaging is capable of
determining layer spacings and registry of the surface layers of
clean metals to within about 5 per cent accuracy, but its validity
for analysis of adsorbate structures has not yet been demonstrated. An alternative and rather more simple energy-averaging
technique has been proposed for overlayers (Tucker and Duke
1972), and is based on the observation that for fractional-order
beams multiple scattering between the overlayer and the substrate will mainly redistribute kinematic intensity within a given
beam rather than transfer it from one beam to another. Thus if
the dynamical intensity is integrated over an energy range of $\frac{1}{2}\Delta E$
on either side of the position of a kinematic peak, the result
should be directly proportional to the integrated kinematic intensity over this range, i.e. to F^2, the square of the structure factor.
Very recently a third technique, Fourier-transform deconvolu-
tion, has been used by Landman and Adams (1974, 1975),
following the earlier work of Mason (see Mason and Textor
1976).

A major attraction of these data-reduction methods is their use
of kinematic theory with its associated computational simplicity,
but this is offset by the experimental effort required to collect
data in the volume required. Also, they are based on the assumption that all variations in the I–E_p curve are geometrical in origin,
and this may not be correct. The validity of the first two techniques has been investigated for a limited number of systems; the
conclusions in relation to Tucker and Duke's scheme were un-
favourable (Mitchell, Woodruff, and Vernon 1974), but Webb's

method led to a unique solution for the structure of Cu(100)-c(2×2)-O (McDonnell, Woodruff, and Mitchell 1975) which was in reasonable agreement with preliminary dynamical calculations. On the other hand, Lagally, Buchholz, and Wang (1975) found in the case of W(110)-(2×1)-O that it was difficult to average out multiple-scattering contributions at low energies, while at higher energies the decrease in surface sensitivity due to increased beam penetration reduced the accuracy with which oxygen could be located.

3.5.4. *Dynamical theories*

The most complete description of intensities is obtained by bowing before the weight of multiple-scattering evidence and specifically including this feature in the calculations to produce a dynamical theory; details of the methods involved may be obtained, for example, in the book by Pendry (1974) and in various reviews, such as that by Strozier, Jepsen, and Jona (1975).

One set of experimental systems on which considerable attention has been focused is the c(2×2) structures formed by the chalcogens oxygen, sulphur, selenium, and tellurium on Ni(100) (for a summary of recent work see Rhodin and Tong 1975). Initially, three types of approach were employed, due to Duke (e.g. Duke *et al.* 1973), Marcus and Jepsen (e.g. Demuth, Jepsen, and Marcus 1973), and Pendry (e.g. Andersson *et al.* 1973). Essentially these are all equivalent, using elastic- and inelastic-scattering potentials of the types discussed above in a multiple-scattering context. Differences arise in the mathematical techniques adopted (Marcus and Jepsen's method requires very large computing facilities for example), and also, of more significance, in details of the models used for the electron–solid interactions. The various uncertainties involved are greatest at electron energies of less than about 50 eV and the most satisfactory results are obtained by analysing non-specular beams over wide ranges of energy above 50 eV. In the case of adsorbed layers one might foresee an additional complication due to the fact that the nature of the adsorbate–surface bond will influence the ion-core scattering potentials. The exact characteristics of the bond will depend in turn on the co-ordination of the adsorbed species, which is one of the main parameters to be determined. The following question therefore arises: if we calculate the intensity profiles for a given species in two alternative positions on a given surface will differences in the profiles result mainly from differences (a) in coordination of the adsorbate or (b) in its electronic properties? Very fortunately, it appears that the dominant influence is that of co-ordination, and so realistic structural information can be obtained even if factors determining electronic interactions are known, at best, semiquantitatively.

The three computational approaches mentioned above successfully reproduced the intensity profiles for clean metals and in addition there was general agreement that O and S occupied the sites of four-fold co-ordination on Ni(100). There was disagreement, however, as to the adatom–surface spacing z; for example, for O–Ni Pendry's group obtained a value of $0·15 \pm 0·01$ nm while Demuth *et al.* reported a value of $0·09 \pm 0·01$ nm. Further work has shown that such differences have arisen either because of the choice of substrate potential, or when accuracy has had to be sacrificed because of excessive computing requirements, and there now appears to be a consensus in favour of the values shown in Table 3.1. This shows a satisfying correspondence between the bond lengths of Ni–X on the surface (Fig. 3.29(a)) and in bulk complexes and compounds; a comparison of experimental and calculated intensity profiles is made in Fig. 3.29(b). In an interesting extension of this work Andersson and Pendry (1976) have investigated mixed sulphur–sodium adsorption on Ni(100). If both S and Na atoms occupied identical positions to those which they would separately occupy on this surface, the structure would be as shown in Fig. 3.30(a), with a Ni–Na spacing of $0·23$ nm. The result of the intensity analysis reveals that in fact the structure is that given in Fig. 3.30(b), with the Na at $0·25$ nm from the surface (the accuracy of the calculation is $0·01$ nm). This spacing suggests that sodium is essentially bonding to sulphur rather than to nickel and may be compared with the S–Na spacing in sodium sulphide of $0·264$ nm.

A firm basis has obviously been established for the dynamical analysis of LEED patterns generated by simple structures, but exact calculations involve very considerable computing resources. Perturbation methods, which are still dynamical in nature but which are much more economical, offer an attractive alternative. These are based on the fact that the incident beam penetrates only the first few atom layers and therefore only the first few

TABLE 3.1

Summary of results of structure analyses on centred and primitive (2×2) overlayers of chalcogen atoms at a four-fold site on Ni(100). (From Rhodin and Tong 1975.)

| | Theoretical | | | | Experimental |
| | c(2×2) structure† | | p(2×2) structure‡ | | |
	z/nm	Bond length/nm	z/nm	Bond length/nm	Bond length/nm
Ni–O	0·090±0·01	0·198±0·005	0·090±0·01	0·198±0·005	0·184–0·206 Ni-chelate complexes
Ni–S	0·130±0·01	0·219±0·006	0·130±0·01	0·219±0·006	0·210–0·223 Ni-chelate complexes
Ni–Se	0·145±0·01	0·228±0·006	0·155±0·01	0·234±0·007	0·232 Ni-chelate complexes
Ni–Te	0·190±0·01	0·259±0·007	0·180±0·01	0·252±0·007	0·264 Ni-Te bulk compounds

† Demuth *et al.* 1973, Marcus *et al.* 1975.   ‡ Van Hove and Tong 1975.

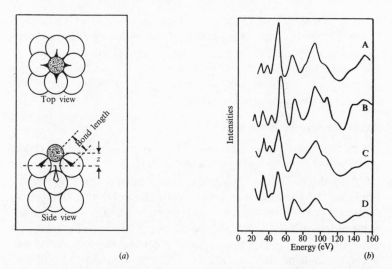

(a)   (b)

FIG. 3.29. (a) The adsorption site for a chalcogen atom on Ni(100). (b) Comparison between theory and experiment for a centred 2×2 sulphur overlayer on Ni(001), the $(\frac{1}{2} \ \frac{1}{2})$ beam at $\theta = 0°$. Experiment: A, Demuth and Rhodin (1974), $T = 300$ K; B, Duke et al. (1973), $T = 30$ K. Theory: C, Demuth et al. (1973), $z = 0.13$ nm; D, Van Hove and Tong (1975), $z = 0.13$ nm. (From Rhodin and Tong 1975.)

scattering events can be large. The electron wave inside the solid can then be expanded in terms of perturbation theory, with each order including an additional set of multiple-scattering events. In most cases a convergent solution of high accuracy can be obtained with less than about five perturbation iterations. This type of method was originally introduced by Pendry (1971) in a treatment of clean Cu(100) and has since been extended to overlayer systems; for example the Ni(100)–p(2×2)–chalcogen

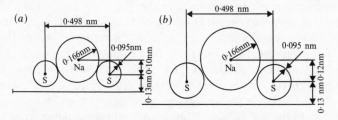

FIG. 3.30. Mixed sulphur–sodium adsorption on Ni(100). (a) Positions which S and Na would occupy separately on the surface. (b) Actual structure determined by intensity analysis. (From Andersson and Pendry 1976.)

data in Table 3.1 were calculated in this way (Van Hove and Tong 1975) and also using full dynamical theory (Marcus, Demuth, and Jepsen 1975) with identical results. Obviously one of the main hopes for the immediate future must be that rapid perturbation methods will be improved to a point where the calculations become quite routine.

3.6. Coincidence lattices

Up to this point we have restricted discussion to simple overlayer structures, i.e. those for which $a_s = Qa$, where Q is an integer. In addition the following assumptions have been made (the second implicitly): (i) during interaction with the solid the electron is scattered once only (the kinematic approximation) and (ii) there is no modulation of atom parameters, such as scattering factor or co-ordinates, with a period greater than the dimensions of the overlayer unit mesh.

Under these conditions it has been seen that the presence of the underlying layers will modulate spot intensities but in general is unlikely to lead to the total extinction of any particular spot. Determination of the unit mesh *geometry* therefore appears to be free of ambiguity, but on closer examination the situation is found to be less simple than this. The possibility of complication is suggested when we consider certain experimentally observed LEED patterns exhibiting high-order 'extra' spots.

The interaction of ammonia with the (211) plane of tungsten provides a suitable example (May, Szostak, and Germer 1969); at one stage during the process a (7×2) pattern is observed, the adsorbed layer consisting of NH_2 species. This could of course be interpreted in terms of an overlayer unit mesh of dimensions seven times that of the substrate mesh in the $[\bar{1}11]$ or x-direction and twice that of the substrate in the $[0\bar{1}1]$ or y-direction. The overlayer mesh may contain only one NH_2 species (Fig. 3.31(a)) or it may contain several (Fig. 3.31(b)). Structures such as these present problems, however, since each implies the existence of an unusual condition at the surface. The first would be that the ordering forces between adatoms should be effective over considerable distances, in the specific case of Fig. 3.31(a) $7 \times \frac{1}{2}\sqrt{(3)}a_0 = 1\cdot91$ nm in the x-direction. The second, which would apply to the structure in Fig. 3.31(b), is that a range of different adatom–adatom and/or adatom–substrate interactions should operate

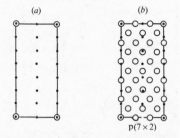

FIG. 3.31. Possible structures of W(211)-p(7×2)-NH$_3$: (a) one adsorbed species per unit mesh; (b) coincidence mesh containing 30 adsorbed species. (●) W surface atoms; (○) NH$_2$ adsorbed species. (From May, Szostak and Germer 1969).

within a single unit mesh. Neither of these conditions is intuitively appealing, and the following question therefore arises: can an alternative structure, which is physically more reasonable, be found which will also give rise to the diffraction pattern? Such a structure, one of a general class known as coincidence lattices, can indeed be postulated and will produce the desired pattern provided either that electrons undergo *multiple* scattering within the solid or that there is modulation of some parameter relating to individual scatterers in the overlayer. The concept of the coincidence lattice was first introduced by Tucker (1966) in order to rationalize high-fractional-order spot patterns.

In the case of a simple (one-dimensional) overlayer the mesh vector $\mathbf{a}_s$ is an integral multiple of the substrate mesh vector $\mathbf{a}$. This type of structure will be preferred when the adsorbate–surface interaction is stronger than adsorbate–adsorbate interactions, but if the reverse is true the overlayer will adopt a periodicity of its own regardless of that of the substrate. The two lattices should, nevertheless, coincide at certain points, forming an apparent superstructure or coincidence lattice (Fig. 3.32(a)). The mesh vectors will then be in the ratio of two integers P and Q, i.e.

$$P\mathbf{a}_s = Q\mathbf{a}. \qquad (3.44)$$

In fact all lattices are defined by this relationship. In the case of a simple overlayer $P = 1$, while if the distance Pa_s is greater than the coherence width of the incident electron beam, the diffraction process will not register the apparent superperiodicity and the structure will be effectively *incoherent*. In general, however, the

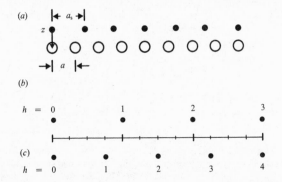

FIG. 3.32. (a) Coincidence mesh $3\mathbf{a}_s = 4\mathbf{a}$. (b) Pattern from clean surface corresponding to (a). (c) pattern from (a) if overlayer alone scatters.

term coincidence lattice is applied to those structures in which $2 < P \leqslant 10$.

3.6.1. Scattering by a coincidence lattice

We know that a diffraction pattern containing $1/Q$th fractional-order spots may be interpreted in terms of a simple overlayer structure of periodicity $\mathbf{a}_s = Q\mathbf{a}$. We now aim to show that this pattern could also arise from a coincidence lattice with parameters $P\mathbf{a}_s = Q\mathbf{a}$; for ease of explanation the case of $\frac{1}{4}$-order features will be considered in relation to a lattice for which $3\mathbf{a}_s = 4\mathbf{a}$ (Fig. 3.32(a)). If the incident electrons did not penetrate beyond the top surface layer the pattern observed would be due to the overlayer alone (Fig. 3.32(c)); note that this would contain spots at positions h corresponding to those of the clean surface pattern only when $h = Q \times$ integer. Experimentally however, it is almost invariably found that *all* 'clean-surface' positions are occupied in an overlayer pattern (provided that the overlayer is not so thick that electrons cannot penetrate to the substrate). Once again we are therefore forced to consider the effects of the substrate layers and to view diffraction as a process involving the crystal as a whole. It was first emphasized by Bauer (1967) that the *overall periodicity is then that of the coincidence lattice*. Pursuing our one-dimensional model this means that the structure must be considered to consist of a lattice of spacing $P\mathbf{a}_s$ ($= Q\mathbf{a}$) populated by unit 'cells', each of dimension $P\mathbf{a}_s \times z$, where z is the beam penetration depth. If there are M such units within the coherence width of the beam the scattered intensity is given by

$$I = F^2 \mathscr{I} = \frac{F^2 \sin^2\{\frac{1}{2} M \mathbf{K} \cdot (Q\mathbf{a})\}}{\sin^2\{\frac{1}{2}\mathbf{K} \cdot (Q\mathbf{a})\}} . \tag{3.45}$$

In this case the form of the interference function $\mathscr{I}$ specifically reflects the coincident periodicity, and will permit diffraction maxima when $\frac{1}{2} Q \mathbf{K} \cdot \mathbf{a} = h\pi$, i.e. when $\mathbf{K} \cdot \mathbf{a} = 2\pi h/Q$. Provided the structure factor F is always finite under these conditions, all Qth-order spots would therefore be produced.

The most rigorous way to proceed would be to use dynamical methods to calculate F, but given that this is effectively impossible it is instructive to use first simple kinematic theory. Considering the unit 'cell' to consist of the overlayer and top substrate layer only substitution of the appropriate atom co-ordinates into eqn (3.35) and rearrangement gives

$$F = f_s \frac{\sin[\frac{1}{2}P\mathbf{K} \cdot \{(Q/P)\mathbf{a}\}]}{\sin[\frac{1}{2}\mathbf{K} \cdot \{(Q/P)\mathbf{a}\}]} + f \exp(-i\mathbf{K} \cdot \mathbf{z}) \frac{\sin(\frac{1}{2}Q\mathbf{K} \cdot \mathbf{a})}{\sin(\frac{1}{2}\mathbf{K} \cdot \mathbf{a})} . \tag{3.46}$$

The first term on the right hand side of eqn (3.46) represents scattering by the overlayer and the second term scattering by the first substrate layer. The maxima in F will occur when

$$\frac{\mathbf{K} \cdot \mathbf{a}}{2\pi} = h \text{ or } \frac{P}{Q} h' \tag{3.47}$$

and since these positions correspond to maxima in the interference function in eqn (3.45) they will be occupied by spots. These spots are shown in Fig. 3.33 and obviously certain of the fractional-orders are absent. Since, however, P and Q are small numbers the sine terms in eqn (3.46) will vary relatively slowly with $\mathbf{K} \cdot \mathbf{a}$ and will have finite values at positions other than those

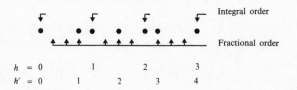

FIG. 3.33. Kinematic pattern produced by structure of Fig. 3.32(a).

corresponding to conditions (3.47). It might be suggested therefore that those $1/Q$th order spots not generated by this relation
could nevertheless be present; if this were so the $\frac{1}{4}$-order spots
missing in Fig. 3.33 for example could appear, although with
reduced intensity. This is not possible, however; when $\mathbf{K} \cdot \mathbf{a}/2\pi =$
h/Q the sine terms in the numerators in eqn (3.46) are zero, and
so F will disappear unless the sines in the denominators are also
zero, i.e. unless eqn (3.47) is satisfied. Thus, although we have
specifically recognized the periodicity of the coincidence lattice,
this is not reflected in the pattern predicted on a simple kinemati-
cal basis; the separation of overlayer and substrate terms in eqn
(3.46) implies that these elements of the structure act independently. Short of a dynamical calculation, there remain two relatively simple alternative approaches, multiple scattering and
parameter modulation, but it must be emphasized that like the
kinematic calculation these are approximations only.

3.6.2. *Multiple scattering*

Although the simple kinematic theory does not consider the
possibility, there is no reason to suppose that an electron scattered in the forward direction by the overlayer cannot be scattered a second time, for example, in the backward direction by
the substrate layer. Along with other double- and multiple-
scattering combinations which may be envisaged this event would
contribute to the observed diffraction pattern.

Using again the example of a one-dimensional coincidence
mesh ($P\mathbf{a}_s = Q\mathbf{a}$) and considering for the purpose of illustration a
double-diffraction event involving first the substrate layer and
then the overlayer (Fig. 3.34) we may write, for beams diffracted
by the substrate,

$$m\lambda = a \sin \phi_m \qquad (m = 0, \pm 1, \pm 2, \ldots). \qquad (3.48)$$

These beams will be incident at angle ϕ_m on the overlayer and
will be diffracted by it at angles θ_n given by the equation

$$n\lambda = a_s(\sin \theta_n - \sin \phi_m) \qquad (n = 0, \pm 1, \pm 2 \ldots)$$

$$= \frac{Q}{P} a\left(\sin \theta_n - \frac{m\lambda}{a}\right). \qquad (3.49)$$

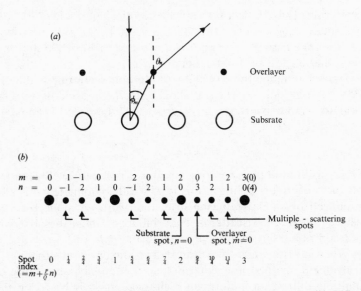

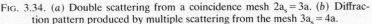

FIG. 3.34. (a) Double scattering from a coincidence mesh $2a_s = 3a$. (b) Diffraction pattern produced by multiple scattering from the mesh $3a_s = 4a$.

This rearranges to

$$(mQ + nP)\lambda = Qa \sin \theta_n = Pa_s \sin \theta_n \qquad (3.50)$$

or

$$\left(\frac{m}{a} + \frac{n}{a_s}\right)\lambda = \sin \theta_n. \qquad (3.51)$$

Eqn (3.51) may be easily generalized (Taylor 1966) to give the angle ϕ_J at which a beam will emerge from the surface after undergoing J scattering events:

$$\lambda \sum_{j=1}^{J} \frac{m_j}{a_j} = \sin \theta_J. \qquad (3.52)$$

The m_j are the intermediate diffraction orders and the a_j the intermediate lattice constants.

Examination of eqn (3.51) shows that values of $m = 0$ or $n = 0$ give respectively the diffraction patterns of the overlayer and of the substrate, as if each was behaving as an independent kinematic scatterer. When both m and n are non-zero, however, additional spots are generated which are due specifically to multiple scattering. These reflect the coincidence mesh periodicity Pa_s and are indexed as $\{m + (P/Q)n\}$ with respect to the substrate (eqn (3.50)). The case $P = 3$, $Q = 4$ is illustrated in Fig. 3.34 and it is seen that all fourth-order spots are produced.

Having accepted the possibility of multiple scattering in diffraction by coincidence lattices it should also be considered in relation to simple overlayer structures. For these $P = 1$ and eqn (3.50) becomes

$$(m + n/Q)\lambda = a \sin \theta_n. \tag{3.53}$$

Once again multiple scattering (m and n non-zero) will generate all spots of order Q, but in this instance the complete set is also produced by single scattering from the overlayer alone ($m = 0$). Therefore multiple scattering may influence intensities but will not contribute any additional spots to the pattern.

When investigating diffraction by two-dimensional structures a convenient method for determining the possible positions of multiple-diffraction features in reciprocal space is based on the fact that the corresponding reciprocal mesh vectors $\boldsymbol{\sigma}$ may be generated by adding or subtracting integral multiples of the reciprocal mesh vectors of the separate lattices

$$\boldsymbol{\sigma} = m\mathbf{a}^* + n\mathbf{a}_s^* \tag{3.54}$$

$$= \frac{m}{\mathbf{a}} + \frac{n}{\mathbf{a}_s} \tag{3.55}$$

where $m, n = 0, \pm 1, \pm 2 \ldots$ For a diffraction maximum $\sin \theta/\lambda = \boldsymbol{\sigma}$; (this follows from Fig. 3.11, for example). Thus eqn (3.55) is equivalent to eqn (3.51).

As an illustration of a multiple-diffraction interpretation of a LEED pattern let us consider the case of silver deposition on a copper (111) surface (Bauer 1967). The silver layer would be expected to adopt a (111) configuration since this plane has the lowest specific free surface energy for f.c.c. metals. The individual reciprocal meshes of the Cu and Ag planes are shown in Fig. 3.35(a). The real-space mesh dimensions are $\mathbf{a}_{Cu} = a/\sqrt{2} = 0.2554$ nm, $\mathbf{a}_{Ag} = a/\sqrt{2} = 0.2886$ nm, and the ratio $\mathbf{a}_{Cu}/\mathbf{a}_{Ag}$ ($= Q/P$) is 0.884, or very nearly $\frac{8}{9}$. Thus the coincidence-lattice periodicity will be defined in both x- and y-directions by $8\mathbf{a}_{Ag} = 9\mathbf{a}_{Cu}$, and the vectors of the reciprocal mesh will be multiples of

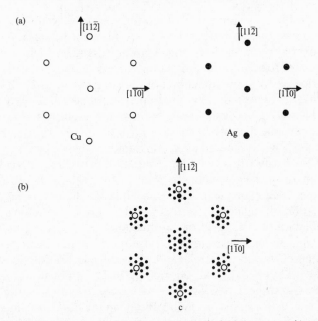

FIG. 3.35. Diffraction by six-fold symmetry on six-fold symmetry. (Ag on Cu {111} plane). Reciprocal lattice of (a) Cu, Ag and (b) superposition of Ag and Cu. (From Bauer 1967.)

$\frac{1}{8}\mathbf{a}_{Ag}^*$. The two-dimensional equivalent of eqn (3.54) becomes

$$\boldsymbol{\sigma} = (h\mathbf{a}_{Cu}^* + k\mathbf{b}_{Cu}^*) + (h'\mathbf{a}_{Ag}^* + k'\mathbf{b}_{Ag}^*)$$
$$= \mathbf{a}_{Ag}^*\{(h' + k') + \tfrac{9}{8}(h + k)\}$$

where h, h', k, k' take values 0, ±1, ±2, ..., etc. A drawing of the resulting mesh is shown in Fig. 3.35(b). It will be noticed that only those points are included for which h, h', k, k' are zero, or *small* integers. This is a general experimental observation, which may also be justified theoretically (Palmberg and Rhodin 1968); the only multiple-diffraction beams of significant intensity are those corresponding to combinations of low-order overlayer and substrate reciprocal-lattice vectors. Fig. 3.35(b) agrees well with the observed diffraction pattern.

3.6.3. *Parameter modulation*

While simple kinematic (or single scattering) theory cannot account for the diffraction patterns to be expected from coincidence lattices, there are a number of modifications to the theory which do produce the desired result. The requirement is that, for either the overlayer or the substrate considered independently, some parameter related to the diffraction process should modulate with a period equal to the dimension of the coincidence mesh. If again we take as an example the one-dimensional structure with $3\mathbf{a}_s =$ $4\mathbf{a}$ (Fig. 3.32) and examine a hard-sphere model, it is qualitatively obvious that, when the overlayer atoms are in equilibrium with respect to interactions ω_{AA} among themselves (i.e. are equally spaced), there will be a variation across the coincidence unit mesh in their positions normal to the surface. If substrate–adatom interactions ω_{AB} are also significant there may be some relaxation of adatoms parallel to the surface, resulting in a modulation of spacing in the x-direction across the unit mesh. It is assumed that substrate–substrate interactions ω_{BB} are strong and therefore substrate relaxation is negligible. Calculations carried out by Fedak and Gjostein (1967), although using the somewhat suspect pairwise interaction model (cf. Chapter 12), indicate that significant modulations in adatom spacings will occur. The result is that within the overlayer, considered in isolation, the period is now $4\mathbf{a}$ $(=3\mathbf{a}_s)$ rather than $\mathbf{a}_s$, and so kinematic scattering by this layer alone should produce $\frac{1}{4}$th-order spots. There still remains, however, the question of the intensities of these spots; will relaxations of a realistic magnitude give spots which are detectable?

This problem has been considered semiquantitatively by Palmberg and Rhodin (1968) who conclude that modulation of the atom co-ordinates normal to the surface, modulation of the x- and y-co-ordinates in the surface plane, or of the atomic scattering factors of both overlayer and substrate could all, individually or in combination, produce diffraction patterns characteristic of the coincidence mesh periodicity. Thus, while multiple diffraction provides a basically satisfactory interpretation of coincidence patterns, the effects of overlayer distortion are important in contributing to spot intensities, particularly for electron energies above about 50 eV.

3.7. Diffraction by imperfect structures

In all of the foregoing discussion of LEED it has been assumed that the surface region of the solid is structurally perfect, in which case the diffraction pattern will consist of sharp spots on a dark background. On the other hand, if imperfections occur they may generate additional features in the pattern and from an analysis of these one may obtain information on the surface forces which generated the imperfection. It is important to note, however, that the degree of disorder which must exist before the effects become observable is not trivial. For example, if we take a perfect array and remove scatterers at random the overall background intensity in the pattern will increase, but optical simulation indicates that even when 20–30 per cent have been removed the increase is not particularly noticeable. The sensitivity of both patterns and I–V curves to surface perfection has been discussed by various authors, including Park (1966), Jona (1967), and Heckingbottom (1969).

The basic concept in relation to defective surface layers is that the diffracted radiation will not be concentrated entirely in sharp spots. These may still occur, at positions determined by the average ideal mesh, but their intensity will be reduced compared with that of spots in a 'perfect' pattern. In other words, intensity is removed from the nodes in the reciprocal lattice and is redistributed in other regions of the reciprocal unit mesh. Six general classes of pattern imperfection can be identified: overall background intensity, diffuse spots, rings, split spots, streaks, and features due to surface faceting. These are discussed below, with emphasis being placed on split spots and streaks to illustrate the procedure involved (see also McKee, Roberts, and Williams 1977).

The theory of X-ray diffraction by imperfect crystals has been well documented (Guinier 1963) and many aspects, particularly those relating to planar disorder in crystals, can be applied directly to the LEED case. A simple kinematic treatment can account, at least qualitatively, for the phenomena observed and almost all cases can be discussed in terms of the basic kinematic expression for the scattered intensity, eqn 3.24, or a modification of this.

One further point must be mentioned here. The theory developed in previous sections has involved the implicit assumption that the incident electron beam is a single monochromatic wave emanating from a point source, but in practice the beam is generated by a filament of finite size and has a finite spread of energies. This means that in any cross-section of the beam the radiation at different points will not be exactly in phase. If diffraction effects are to be observed, however, there is a limit to the degree to which radiation reaching different parts of the grating can be out of phase (Smith and Thompson 1971). This restriction leads to the notion of a *coherence width* which may be thought of as that area over which the incident beam acts like a single monochromatic wave (see Park, Houston, and Schreiner 1971). Experiment (e.g. Park 1969) suggests that in LEED this width is of the order of 10 nm or less. Since the beam has a physical width of the order of 1 mm it may be considered to consist of a large number of pencils of coherent radiation, each producing a diffraction pattern from a small area of the surface. The picture seen on the screen is formed by superimposition of all these patterns. As a consequence, in the particular case of a defective structure, the defects will exert an influence on the pattern only if they occur on a scale less than the coherence width.

3.7.1. *Background intensity, diffuse spots, and ring patterns*

As mentioned above, an overall increase in background intensity occurs when random disorder is introduced into the surface, for example as random vacant sites in an otherwise ordered adsorbed layer. The effect has been examined analytically for one-dimensional cases by Estrup and Anderson (1967) and Park and Houston (1969). Diffuse spots are the result of a 'particle-size-broadening' effect. For example, suppose the adsorbate occurs in roughly circular patches, or islands, each of diameter equal to N unit mesh spacings. If the coverage is sufficiently low so that N is small and only about one island falls within the coherence width of the incident beam, the pattern will be effectively produced by a single grating of limited extent. Since the spot width is inversely proportional to N (Section 3.4.1), the spots will be broadened. Often heating sharpens the pattern as it allows particles initially absorbed at isolated sites to diffuse to an island and increase its size.

Ring patterns can arise in two ways. In the first place imagine an adsorbate which can exist in all possible orientations in the plane parallel to the substrate and suppose that a large number of these possible orientations occur on different parts of the surface.

If we think of the Ewald construction for this system (cf. Fig. 3.11) then it may be seen that a particular lattice rod, e.g. 01, will occur not just once in reciprocal space, but as many times as there are different mesh orientations. Each rod will lie at a distance $1/d_{01}$ from the origin of reciprocal space but there will be a random distribution in their azimuthal angles. Taken together the rods therefore form a cylinder about the incident-beam direction and when this cylinder intersects the screen a ring will result. An example of this type of imperfection has been observed in the case of graphitic layers on platinum (Lyon and Somorjai 1967, May 1969).

Any rings formed in the above case will have the 00 spot as their centre, but rings around different individual spots can also occur if adsorbed particles sit in normal surface sites but are randomly separated from their neighbours; the standard deviation of the average distance between nearest neighbours, however, must be small. This effect is very clearly illustrated by Ellis (1972) who used it to interpret patterns from the UO_2 (111) surface. Similar rings have been observed during alkali–metal adsorption on nickel (Gerlach and Rhodin 1969, see Chapter 12).

3.7.2. Split spots and streaks

In general, spot splitting will occur whenever a surface layer corresponding to approximately monolayer coverage contains regions of displaced structure (i.e. different 'phases') separated by 'anti-phase boundaries'. We may consider (McKee, Perry, and Roberts 1973) two identical areas of a (2×1) overlayer, each with structure factor F, which are shifted relative to each other by one substrate atom spacing in the $\mathbf{a}$ direction (Fig. 3.36). In the kinematic approximation the scattered amplitude is given (eqn 3.23) by

$$A(\mathbf{K}) \propto F \sum_{m_1=0}^{1} \exp\left(-im_1\mathbf{K}.\mathbf{a}\right) \sum_{m_2=0}^{1} \exp(-im_2\mathbf{K}.\mathbf{b})$$

$$= F(1 + \exp[-i\mathbf{K}.\{\mathbf{a}+(N-1+S)\mathbf{b}\}]). \quad (3.56)$$

The vectors locating the origins of the two regions have values 0 and $\{\mathbf{a}+(N-1+S)\mathbf{b}\}$. Since we are considering first the case of full coverage, the regions are 'separated' only by a single spacing b, and so in the last equation $S = 1$. If each (2×1) region consists

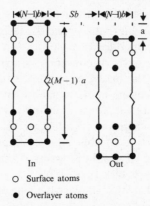

O   Surface atoms

●   Overlayer atoms

FIG. 3.36. Two out of phase, separated islands of adsorbate with a (2×1) structure. (From McKee, Perry and Roberts 1973.)

of $(M\times N)$ scatterers, F will be of the form

$$F \propto \frac{\sin(M\mathbf{K}.\mathbf{a})}{\sin(\mathbf{K}.\mathbf{a})} \frac{\sin(\frac{1}{2}N\mathbf{K}.\mathbf{b})}{\sin(\frac{1}{2}\mathbf{K}.\mathbf{b})}. \tag{3.57}$$

To consider the intensity profiles in the $\mathbf{b}^*$ direction in reciprocal space we set $\mathbf{K}.\mathbf{a}$ as follows:

$$\mathbf{K}.\mathbf{a} = 2\pi \times \text{integer (profile through integral-order spots)}$$

or

$$\mathbf{K}.\mathbf{a} = \pi \times \text{odd integer (profile through half-order spots)}$$

Then, with $S = 1$,

$$I(\mathbf{K})_{\mathbf{b}^*} \propto \frac{M^2 \sin^2(\frac{1}{2}N\mathbf{K}.\mathbf{b})}{\sin^2(\frac{1}{2}\mathbf{K}.\mathbf{b})} \{2 \pm 2 \cos(N-1+S)\mathbf{b}\} \tag{3.58}$$

The term in braces represents the interference between the two regions. The positive sign applies to the integral-order spot profile, and since an identical expression describes the case of a perfect (2×1) layer, without an anti-phase boundary, it follows that the *integral-order* spots are unaffected by the presence of the boundary. The negative sign applies to the half-order profile. If we examine the position which would correspond to the centre of a half-order spot in a perfect (2×1) pattern, by setting $\mathbf{K}.\mathbf{b} = \pi \times \text{odd}$ integer, it is seen (eqn (3.58)) that in the 'defective' pattern the intensity at this point is zero; the complete profile is shown in Fig. 3.37. The half-order spots are 'split' in a direction *normal* to that of the anti-phase boundary.

If boundaries occur on a substrate of four-fold symmetry they should lie in two mutually orthogonal directions so that spots will split into four rather than two parts. A simple method for determining which particular spots in a pattern will be affected by the presence of anti-phase domains has been outlined by Park (1969). The discussion leads to an interesting conclusion regarding the effect of domains in a layer of c(2×2) structure on a four-fold substrate. If the adsorbate is located in the sites of maximum co-ordination, in contact with four surface atoms, the fractional-order spots, but not the integral orders, will be split in the presence of anti-phase domains. On the other hand, if adsorption occurs at 'bridge' sites, between two surface atoms, a different set of beams will be split—those for which the sum of the indices is not an even integer.

The term 'split spot' is found in use in two additional contexts. The first is the 'continuous spot splitting' observed with systems such as Ni(110)/O_2 (see Chapter 11). The second is in relation to diffraction by stepped surfaces (see Ellis and Schwoebel 1968, Henzler 1970, 1974). In this context the term has on occasion been somewhat misused, since the observed spots are those expected from the surface periodicity and are not produced by an

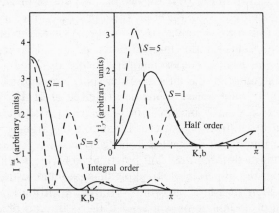

FIG. 3.37. Intensity profile for scattering by two narrow, out-of-phase islands ($N = 3$, eqn (3.56)). Solid curve, not separated ($S = 1$); broken curve, separated four spacings ($S = 5$).

intensity modulation due to any feature such as an anti-phase boundary.

If a number of anti-phase domains fall within the coherence area of the incident beam, streaks or diffuse spots tend to be observed rather than split spots. If the boundaries are linear and parallel, the streak length is inversely related to the average domain width, the effect again being analogous to particle-size broadening in X-ray diffraction. In practice it is more likely that the boundaries will be somewhat irregular and this will reduce the streak length and increase the width. In the case of irregular boundaries lying in more than one surface direction either crossed streaks or diffuse spots are produced.

The above considerations apply to layers at high coverage, in which case anti-phase boundaries, as we have seen, will generally affect certain spots only. At low coverage the situation may be different, since the adsorbate may be distributed over the surface in one of two ways. It may be concentrated in 'islands' with regions of clean surface between, or it may be more evenly spread with the vacant sites fairly randomly arranged. The first case can be easily considered by the extension of the argument used above to investigate the effect of an anti-phase boundary at full coverage. If we take two identical islands separated by a distance Sb (Fig. 3.36), the scattered intensity will be given by eqn (3.58), with $S > 1$ in this case. The intensity profiles with $S = 5$ are shown in Fig. 3.37 for two narrow, out-of-phase islands. When compared with the full coverage case, the half-order profile is found to be similar, the spots being split. The important difference occurs in the integral-order profile. The separation between the islands causes intensity to be displaced from the main maximum (i.e. the integral-order spot) into subsidiary maxima in the profile. Laser simulation (McKee *et al.* 1973) shows that these subsidiary maxima are of sufficient magnitude to be visible in the diffraction pattern. Extending the argument to the case of a number of elongated isolated islands we may conclude that all diffraction features will be streaked owing to the island separation, the direction of streaking being normal to the preferred direction of island growth. A further contribution to the streaking of certain spots will arise if there are anti-phase relationships between islands. If the islands tend to be circular then all spots will be diffuse.

3.7.3. *Patterns from faceted surfaces*

When the wavelength of the incident beam is varied, spots in the pattern converge to, or diverge from, the 00 spot which, using normal incidence, is at the centre of the screen. If a surface undergoes faceting in the course of an experiment (see Chapter 12) new surface planes, at various orientations to the original plane, are produced. Each new orientation will give rise to a separate 00 spot and these will no longer lie at the screen centre. If the wavelength is now varied, each set of spots produced by a particular orientation will move to a different point on the screen, providing an unmistakable indication of the presence of facets. From the spot movements the facet planes can be identified (Tucker 1967, Tracy and Blakely 1969, Kirby *et al.* 1979).

REFERENCES

ANDERSSON, S. (1969). *Surf. Sci.* **18**, 325.

ANDERSSON, S., and PENDRY, J. B. (1976). *J. Phys. C solid St. Phys.* **9**, 2721.

ANDERSSON, S., PENDRY, J. B., KASEMO, B., and VAN HOVE, M. (1973). *Phys. Rev. Lett.* **31**, 595.

BAUER, E. (1967). *Surf. Sci.* **7**, 351.

—— (1969*a*). In *Techniques of metals research* (ed. R. F. Bunshah), vol. 2, p. 502. John Wiley Interscience, New York.

—— (1969*b*). In *Techniques of metals research* (ed. R. F. Bunshah), vol. 2, p. 560. John Wiley Interscience, New York.

CREWE, A. V. (1971). *Sci. Am.* **224** (4), 26.

DEMUTH, J. E., JEPSEN, D. W., and MARCUS, P. M. (1973). *Phys. Rev. Lett.* **31**, 540.

DEMUTH, J. E., and RHODIN, T. N. (1974). *Surf. Sci.* **45**, 249.

DUKE, C. B., LIPARI, N. O., LARAMORE, G. E., and THEETEN, J. B. (1973). *Solid St. Commun.* **13**, 579.

DUKE, C. B., and SMITH, D. L. (1972). *Phys. Rev.* **B5**, 4730.

ELLIS, W. P. (1972). In *Optical transforms* (ed. H. S. Lipson), p. 229. Academic Press, New York.

ELLIS, W. P., and SCHWOEBEL, R. L. (1968). *Surf. Sci.* **19**, 159.

ERTL, G., and KÜPPERS, J. (1974). *Low energy electrons and surface chemistry.* Verlag Chemie, Weinheim.

ESTRUP, P. J. (1970). In *Modern diffraction and imaging techniques in material science* (eds S. Amelinckx, R. Gevers, G. Renault, and J. Van Landuyt), p. 377. North-Holland, Amsterdam.

ESTRUP, P. J., and ANDERSON, J. (1967). *Surf. Sci.* **8**, 101.

ESTRUP, P. J., and McRAE, E. G. (1971). *Surf. Sci.* **25**, 1.

FEDAK, D. G., and GJOSTEIN, N. A. (1967). *Surf. Sci.* **8**, 77.

FINK, M., MARTIN, M. R., and SOMORJAI, G. A. (1972). *Surf. Sci.* **29**, 303.

GERLACH, R. L., and RHODIN, T. N. (1969). *Surf. Sci.* **17,** 32.

GUINIER, A. (1963). *X-ray diffraction in crystals, imperfect crystals, and amorphous solids.* W. H. Freeman, San Francisco.

HECKINGBOTTOM, R. (1969). *Surf. Sci.* **17,** 394.

HENZLER, M. (1970). *Surf. Sci.* **19,** 159.

—— (1974). *Jap. J. appl. Phys.* Suppl. 2 (2), 389.

HIRABAYASHI, H. (1968). *J. phys. Soc. Japan* **24,** 846.

IGNATJEV, A., PENDRY, J. B., and RHODIN, T. N. (1971). *Phys. Rev. Lett.* **26,** 189.

JONA, F. (1967). *Surf. Sci.* **8,** 478.

JOYNER, R. W., and SOMORJAI, G. A. (1973). In *Specialist periodical reports: surface and defect properties of solids* (eds M. W. Roberts and J. M. Thomas), vol. 2, p. 1. The Chemical Society, London.

KAMBE, K. (1970). *Surf. Sci.* **20,** 213.

KIRBY, R. E., McKEE, C. S., RENNY, L. V., and ROBERTS, M. W. (1979). To be published.

LAGALLY, M. G., BUCHHOLZ, J. C., and WANG, G. C. (1975). *J. vac. Sci. Technol.* **12,** 213.

LAGALLY, M. G., NGOC, T. C., and WEBB, M. B. (1971). *Phys. Rev. Lett.* **26,** 1557.

—— (1972). *J. vac. Sci. Technol.* **9,** 645.

LANDER, J. J. (1965). In *Recent progress in solid-state chemistry* (ed. H. Reiss), p. 26. Pergamon Press, Oxford.

LANDER, J. J., and MORRISON, J. (1962). *J. chem. Phys.* **37,** 729.

—— (1963). *J. appl. Phys.* **34,** 3517.

LANDMAN, U., and ADAMS, D. (1974). *Phys. Rev. Lett.* **33,** 585.

—— (1975). *Surf. Sci.* **51,** 149.

LIPSON, H. S., and LEE, R. M. (1970). *Crystals and X-rays.* Wykeham, London.

LYON, H. B., and SOMORJAI, G. A. (1967). *J. chem. Phys.* **46,** 2539.

McDONNELL, L., WOODRUFF, D. P., and MITCHELL, K. A. R. (1975). *Surf. Sci.* **45,** 1.

McKEE, C. S., PERRY, D. L., and ROBERTS, M. W. (1973). *Surf. Sci.* **39,** 176.

McKEE, C. S., ROBERTS, M. W., and WILLIAMS, M. L. (1977). *Adv. Colloid and Interface Science* **8,** 29.

MARCUS, P. M., DEMUTH, J. E., and JEPSEN, D. W. (1975). *Surf. Sci.* **53,** 501.

MASON, R., and TEXTOR, M. (1976). In *Surface and defect properties of solids* (eds M. W. Roberts and J. M. Thomas), Vol. 5, p. 189. The Chemical Society, London.

MAY, J. W. (1969). *Surf. Sci.* **17,** 267.

—— (1970). *Adv. Catalysis* **21,** 151.

MAY, J. W, SZOSTAK, R. J., and GERMER, L. H. (1969). *Surf. Sci.* **15,** 37.

MITCHELL, K. A. R., WOODRUFF, D. P., and VERNON, G. W. (1974). *Surf. Sci.* **46,** 418.

NIELSEN, P. E. H. (1973). *Surf. Sci.* **35,** 194.

PALMBERG, P. W., and RHODIN, T. N. (1968). *J. chem. Phys.* **49,** 147.
PARK, R. L. (1966). *J. appl. Phys.* **37,** 295.
—— (1969). In *The structure and chemistry of solid surfaces* (ed. G. A. Somorjai), paper 28. John Wiley, New York.
PARK, R. L., and FARNSWORTH, H. E. (1964). *Rev. sci. Instrum.* **35,** 1592.
PARK, R. L., and HOUSTON, J. E. (1969). *Surf. Sci.* **18,** 213.
PARK, R. L., HOUSTON, J. E., and SCHREINER, D. G. (1971). *Rev. sci. Instrum.* **42,** 60.
PENDRY, J. B. (1971). *Phys. Rev. Lett.* **27,** 856.
—— (1974). *Low energy electron diffraction: the theory and its application to the determination of surface structure.* Academic Press, New York.
PRUTTON, M. (1975). *Surface physics.* Clarendon Press, Oxford.
RHODIN, T. N., and TONG, D. S. Y. (1975). *Phys. Today* **28** (Oct.), 23.
SMITH, F. G. and THOMPSON, J. H. (1971). *Optics.* John Wiley, New York.
SOMORJAI, G. A. (1972). *Principles of surface chemistry.* Prentice-Hall, Englewood Cliffs, N.J.
SOMORJAI, G. A., and SZALKOWSKI, F. J. (1971). *J. chem. Phys.* **54,** 389.
STONE, J. M. (1963). *Radiation and optics: an introduction to the classical theory.* McGraw-Hill, New York.
STROZIER, J. A., JEPSEN, D. W., and JONA, F. (1975). In *Surface physics of crystalline materials* (ed. J. M. Blakely). Academic Press, New York.
TAYLOR, N. J. (1966). *Surf. Sci.* **4,** 161
TOWNE, D. H. (1967). *Wave phenomena.* Addison-Wesley, Reading, Mass.
TRACY, J. C., and BLAKELY, J. M. (1969). *Surf. Sci.* **13,** 313.
TUCKER, C. W. (1966). *J. appl. Phys.* **37,** 3013.
—— (1967). *J. appl. Phys.* **38,** 1988.
TUCKER, C. W., and DUKE, C. B. (1972). *Surf. Sci.* **29,** 237.
VAN HOVE, M., and TONG, D. S. Y. (1975). *J. vac. Sci. Technol.* **12,** 230.
WEBB, M. B., and LAGALLY, M. G. (1973). *Solid St. Phys.* **28,** 301.
WOOD, E. A. (1964). *J. appl. Phys.* **35,** 1306.

4

ELECTRON SPECTROSCOPY

4.1. Introduction

IF WE consider Fig. 4.1(a), which has been referred to as 'the Propst diagram', then we have some idea of the number of possibilities available for spectroscopic studies of surfaces. Arrows pointing away from the solid represent secondary particles from which information concerning the solid may be extracted. Each combination of incident and outgoing particles constitutes in principle a technique for studying the solid surface. There are 36 such combinations, but there are many more possible techniques than this since a single combination of an 'in arrow' with an 'out arrow' may lead to several quite distinct spectroscopies depending on what properties of the probe particle and the emitted particle are being monitored. We shall not consider all possible combinations. An excellent review of the emerging spectroscopies was recently written by Park (1975a,c) and the current state of development with respect to surface phenomena is summed up by him as follows:

'An attempt to understand one spectroscopy thus frequently succeeds in spawning another. Each new spectroscopy provides a different view of the surface, and in untangling the complexities of surface phenomena we will need them all.'

The application of electron spectroscopy to study solid surfaces is a relatively recent development, although for gaseous molecules it has been very actively pursued over the last 15 years. There have been three main approaches. Firstly, the application of X-rays as used by Siegbahn for electron excitation, and sometimes referred to as core-level spectroscopy, secondly the use of somewhat gentler u.v. radiation where only the valence levels of molecules are potentially available for investigation, and thirdly the study of Auger electrons. The use of photons of increasing energy to probe the electronic structure of atoms and molecules is illustrated in Fig. 4.1(b). In this chapter we lay stress on the kind of fundamental information that has come to light from

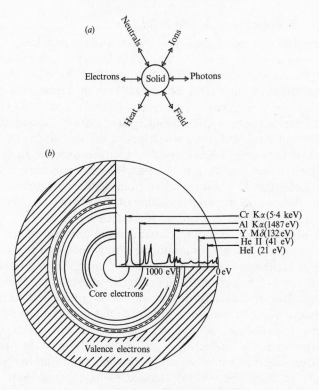

FIG. 4.1. (a) The 'Propst diagram'. (b) Excitation of electrons from various regions in a molecule using different photon energies. (From Siegbahn 1972.)

photoelectron spectroscopy of gases, the particular difficulties and interesting twists that arise from studies of solids, and the experimental techniques that had to be developed to ensure that worthwhile and meaningful surface data could be obtained. Clearly in the field of adsorption and catalysis we shall be interested to learn whether electron spectroscopy can provide the analytical sensitivity for studying submonolayer adsorption, if different states of adsorption can be observed, and whether the nature of the surface bond formed can be explored. However, before considering these it is apposite to make some brief comments on the approach that has been adopted in studies of the solid state *per se*.

4.1.1. *Photoemission and the solid state*

In the 1960's extensive use was made of photoemission to probe the band structure of solids; this arose from the recognition that the process was volume dependent and that the photoelectrons can originate from depths of several tens of nanometers below the surface. It was largely the work of Gobeli and Allen (1962), Van Laar and Scheer (1962), and Spicer (1963) on elemental and compound semiconductors which paved the way to the development of the subject. Their results, together with the theoretical work of Kane (1962), are of much wider significance than the understanding of photoemission from semiconductors and impinge directly on the understanding of the physics and chemistry of surfaces and the mechanism of surface reactions.

A phenomenological and classical representation of the three processes associated with photoemission is shown in Fig. 4.2: (1) optical excitation from an initial state of energy ε_i to a final state of energy ε_f (2) electron transport to the surface, and (3) electron escape. It is of course from the process of optical excitation that spectroscopic information is obtained, but factors that influence electron scattering and escape, such as electron–electron and electron–phonon scattering, are very relevant. In general electron–electron scattering can lead (when the mean free path is small) to losses of several electron volts in magnitude. It is therefore relatively easy to distinguish between those electrons that escape without energy loss and those which have undergone

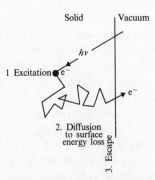

FIG. 4.2. An illustration of the three processes associated with photoemission: (1) optical excitation; (2) movement of the excited electron to the surface; (3) escape over the potential barrier at the surface. (From Spicer 1968.)

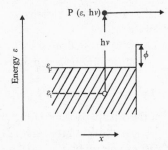

FIG. 4.3. Energy diagram for a metal. Distance normal to the surface is x, $P(\varepsilon, h\nu)$ is the probability of a photon of energy $h\nu$ exciting an electron to a final energy ε_f, ϕ is the work function, ε_i the initial energy of the electron and ε_F the Fermi energy. (After Spicer 1971.)

electron–electron scattering; those electrons that are unscattered provide the useful spectroscopic information. When the electron mean free path is such that many electron–phonon (i.e. electron–lattice) events can occur before electron escape the average energy loss is considerably smaller, but the energy distribution curves of the photoemitted electrons are distorted at all energies.

Two basic types of photoemission experiment may be carried out, the first involving measurement of the total number of emitted electrons as a function of photon energy $h\nu$ to give the spectral distribution, or yield curve. For frequencies below some threshold value ν_0 no emission will be observed; in the case of a metal at 0K, $h\nu_0$ is equal to the work function ϕ (see Chapter 1). In the second experiment a fixed frequency is used and the number of electrons $N(\varepsilon)$ emitted with a final state energy in the range ε_f to $(\varepsilon_f + d\varepsilon)$ is determined, giving an energy distribution curve (EDC); this distribution proves more useful than the yield for extracting band-structure information. In fact, such information would be obtained most directly from a knowledge of $P(\varepsilon, \nu)$, the normalized probability of a photon exciting an electron to a final state between ε_f and $(\varepsilon_f + d\varepsilon)$, since $P(\varepsilon, \nu)$ depends directly on the numbers of available empty states at ε_f and filled states at the initial energy $(\varepsilon_f - h\nu)$, and on the optical matrix elements coupling these states (Fig. 4.3). To relate $P(\varepsilon, \nu)$ to the experimental quantity $N(\varepsilon, \nu)$ two further terms must be introduced, one being the optical absorption coefficient $\alpha(\nu)$, which will determine the number of electrons excited at a particular depth x below the surface. The second term is the probability $B(\varepsilon, x)$ that an electron excited at x will eventually escape from the solid; to a first approximation it may be represented by the expression

$$B(\varepsilon, x) = B_0(\varepsilon) \exp\{-x/L(\varepsilon)\}$$

where $L(\varepsilon)$, known as the escape depth, is the depth at which B has been reduced by $1/e$ compared to its value, B_0, at the surface. The factor B_0 relates to escape over the surface barrier and will be zero for radiation frequencies below the threshold.

It is now possible to approximate $N(\varepsilon, \nu)$ by a relationship of the form (Spicer 1968)

$$N(\varepsilon, \nu) = \frac{B_0(\varepsilon)}{1 + \{1/\alpha(\nu)L(\varepsilon)\}} P(\varepsilon, \nu). \qquad (4.1)$$

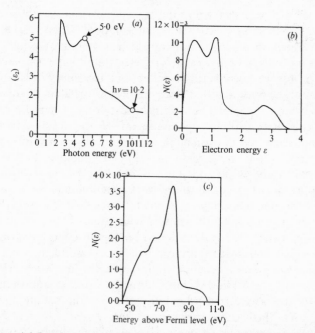

FIG. 4.4. (a) Imaginary part of dielectric constant ε_2 for Cu. (b) EDC obtained from Cu with Cs on the surface for $h\nu = 5$ eV. Note that this curve has several pieces of structure in it, whereas the ε_2 curve has only one peak at 5 eV. (c) EDC for clean Cu, $h\nu = 10 \cdot 2$ eV. Note that several pieces of structure occur in the EDC, whereas there is no strong structure near $10 \cdot 2$ eV in the ε_2 curve. (After Spicer 1971.)

It is the ability to determine $P(\varepsilon, \nu)$ that makes photoemission such a powerful tool for band-structure investigation. An alternative approach is to determine the optical constants of the solid; in this case ε_2, the imaginary part of the frequency dependent dielectric constant is related, not to $P(\varepsilon, \nu)$, but to the integral of $P(\varepsilon, \nu)$ over all possible final states. Clearly, the information obtained from EDC's will be much more valuable than that derived from the dielectric constant. This point is emphasized by comparison of the two methods for copper (Fig. 4.4); much more structure is observed in the EDC.

From the energy at which the structure appears the initial and final states involved in the transition can be identified. By examining the way in which EDC structure varies with photon energy $h\nu$ the relative importance of initial and final states can be ascertained. In order to explore fully the band structure it is sometimes advantageous to reduce the threshold for photoemission by adsorbing caesium on the surface of the solid being studied. This enables energy levels to be explored which otherwise would not be accessible using relatively low-energy photon sources (e.g. up to $h\nu = 5$ eV). That the caesium merely reduces the work function of the surface and does not perturb it in any other way must of course be established in order that the results have credence.

4.1.2. Quantum mechanical description and Koopmans'theorem

The problem of the interaction of light with a solid wherein light is absorbed and hole–electron pairs are produced may also be expressed in terms of Hartree–Fock time-dependent perturbation theory, the principal result of which is that

$$W_{i-f} = \frac{2\pi}{h} |\langle f | H' | i \rangle|^2 \rho_f(\varepsilon) \delta(\varepsilon_f - \varepsilon_i - h\nu)$$

where δ is the Kronecker delta symbol. This equation simply states that the transition rate W_{i-f} from some initial state i to a continuum of final states f is (a) zero unless the difference between the final-state energy and the initial-state energy is $h\nu$ (the photon energy) and (b) proportional to the product of the square of the matrix element of the interaction Hamiltonian coupling the initial and final states $\langle f | H' | i \rangle$ and the density of final states $\rho_f(\varepsilon)$.

The most important term is $\delta(\varepsilon_f - \varepsilon_i - h\nu)$ and the requirement of first-order time-dependent perturbation theory is that $\varepsilon_f - \varepsilon_i = h\nu$. Strictly speaking ε_i and ε_f refer to energies for the whole system before and after an absorption process. Now Koopmans' theorem associates the energy parameters in the Hartree–Fock equation (i.e. the electron energy levels) with the energy required to remove those particular electrons from the solid. This assumes that the absence of a particular electron from the valence band or the presence of an additional electron in the conduction band will have no significant influence on the other states. This will be the case if the hole and electron wave functions do not interact. Therefore from Koopmans' theorem we conclude that the energy required to excite an electron-hole pair in the crystal is equal to the difference between the electron energy and the hole energy, ε_i and ε_f.

The matrix element term in the expression for W_{i-f} depends on the details of the initial and final state wave functions; the model calculation of matrix elements most frequently referred to is based on the use of Bloch wavefunctions for the initial and final states and a classical plane wave for the light. The average matrix element is the arithmetic average of the matrix elements coupling each of the final states at energy ε to each of the initial states at energy $\varepsilon - h\nu$ and hence may be a function of both ε and $h\nu$. In the case of matrix elements which obey the strict **k** conservation condition the average element depends mainly on whether there happen to be very many final states at ε that have the same **k** vector as initial states at $\varepsilon - h\nu$. Consequently the average matrix elements in such a case would be strongly dependent on both ε and $h\nu$.

In order to predict qualitatively from the calculated electronic band structure of a material where strong transitions will occur, assuming that **k** conservation is favoured, one should look for areas where a conduction band is more or less parallel to a valence band over a reasonable range of **k** space. Of course, one should keep in mind the three-dimensional nature of **k** space since bands may be parallel in one direction and not in another.

We have introduced the two approaches to photoemission, the classical picture and that arising from a quantum-mechanical treatment. No attempt, however, has been made to analyse how the more physical and classical model can be reconciled with the quantum-mechanical approach. Fiebelman and Eastman (1974) have discussed this and also more general aspects of photoemission theory.

4.2. Photoelectron spectroscopy

Various abbreviations have been used to describe the different forms of photoelectron spectroscopy, ESCA (electron spectroscopy for chemical analysis and later extended to electron spectroscopy for chemical applications) was first introduced by Siegbahn. A more useful terminology which makes a distinction between the exciting radiation is XPS (X-ray photoelectron spectroscopy) and UPS (ultra-violet photoelectron spectroscopy). The electron events involved in electron spectroscopy are illustrated in Fig. 4.5; UPS represents emission from the valence levels and XPS emission from the core levels. Subsequent to emission, electron reorganization can lead to either the ejection of a photon (X-ray fluorescence) or internal electronic reorganization leading to the ejection of a second electron. This latter event is referred to as the Auger process and will be discussed in detail later.

The kinetic energy ε_{kin} of electrons emitted from a molecule by radiation of energy $h\nu$ is given by eqn (4.2) where ε_{bind} is the binding energy of the bound electron:

$$\varepsilon_{kin} = h\nu - \varepsilon_{bind}. \tag{4.2}$$

Thus from a knowledge of the magnitude of $h\nu$ and the determination of ε_{kin} the binding energy can be determined. A schematic

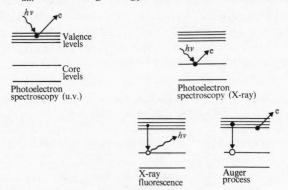

FIG. 4.5. Photoelectron processes.

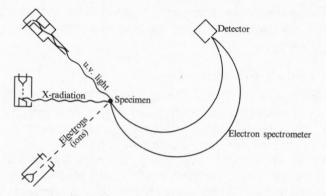

FIG. 4.6. Principle of electron spectroscopy showing its main components: specimen, radiation source, electron energy analyser, and detector.

representation of the apparatus is shown in Fig. 4.6; a monochromatic beam of photons (X-rays or u.v.) generates the photoelectrons.

Electron binding energies for three elements Si, Sr, and Os are shown in Table 4.1. Four sources of exciting radiation have in general been used: Mg $K\alpha = 1253$ eV; Al $K\alpha = 1486$ eV; He I = 21·4 eV; He II = 40·8 eV. Clearly with increasing atomic number (Table 4.1) only the outer levels can be probed. There are distinct advantages, however, of probing the electronic structure by more than one source of radiation, and some examples will be considered later.

In view of the fact that the electrons occupy discrete energy levels the resulting electron spectrum shows distinct and separate peaks corresponding to these levels. Fig. 4.7 shows the X-ray induced photoelectron spectra (XPS) of the inert gases He, Ne, and Ar, induced by Mg $K\alpha$ radiation. The different peaks correspond to electrons being ejected from the different occupied orbitals of the

TABLE 4.1
Electron binding energies (eV)

	K	L_1	L_2	L_3	M_1	M_2	M_3	M_4	M_5	N_1	N_2
Si	1839	149	100	99	8	3					
Sr	16105	2216	2007	1940	358	280	269	135	133	38	20
Os	73871	12968	12385	10871	3049	2792	2458	2031	1961	655	547

molecule. The width of the observed peak depends on (a) the inherent width of the exciting radiation, (b) the inherent width of the level being probed, and (c) the resolution of the analyser ($\Delta E/E \simeq 5 \times 10^{-4}$). With X-rays it is (a) which is usually most important.

The different electronic levels, from the valence to the core electrons, can be probed by using X-ray and u.v. sources of radiation (Fig. 4.1(b)). The higher the energy the larger is the linewidth, and so one suffers a loss in resolution. Using Mg Kα linewidths of ~1·0 eV for the core levels of gases are usually obtained.

The separation of 'core' and 'valence' electrons is well demonstrated by the electron spectrum of CO (Fig. 4.8); at low kinetic energy (i.e. high binding energy) we can observe two peaks

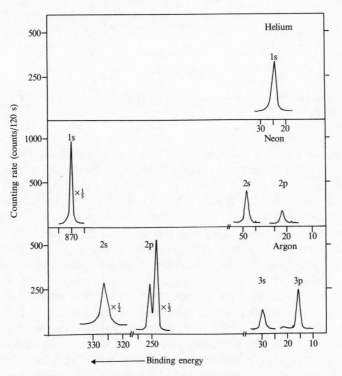

FIG. 4.7. ESCA spectra from the noble gases He, Ne, Ar, excited by Mg Kα radiation. (After Siegbahn *et al.* 1967.)

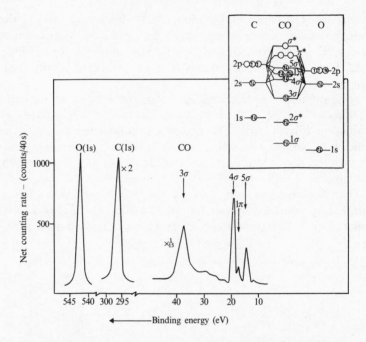

FIG. 4.8. Core- and valence-electron spectrum in CO excited by Mg Kα.

corresponding to ionization from the C(1s) and O(1s) core orbitals. There is then a wide region which is free of lines, and then in the range of ionization energies from 10 to 40 eV peaks due to valence-electron ionization are observed. By making use of Koopmans' theorem it is possible to relate (equate) the observed ionization energy with SCF calculations of orbital energies. This provides confidence regarding the assignment of peaks in the valence orbital region of the electron spectrum. If we look closely at the CO spectrum the intensities of the various peaks (i.e. the areas under the respective curves) vary from peak to peak. Thus the peak intensity does not reflect directly the number of electrons in a particular orbital. The same is true of the inert gases (Fig. 4.7); the Ne(2s) line is approximately three times as intense as the 2p peak even though there are three times as many 2p electrons as 2s electrons. This therefore means that the photoionization cross-section for a Ne(2s) electron is ~9 times greater than for a Ne(2p) electron.

Although the core levels are essentially inert as far as chemical bonding is concerned they are influenced by chemical environment. For example the C(1s) electron is more tightly bound in CO_2 than in CO, but the O(1s) binding energy is larger for CO than for CO_2. We have here an example of a *chemical shift*. The shifts can be qualitatively explained in terms of electronegativity. The O atom is more electronegative than C and therefore reduces the electron charge density on the C atom. Consequently the binding energy of the C core electrons is increased and that of the O decreased. Since CO_2 contains twice as many oxygens as CO, the increase in the C(1s) binding energy of CO_2 is correspondingly greater than that of CO. By analogy therefore we would anticipate the core-binding energy of a given type of atom to differ even within a given molecule if the immediate environment is different. This is illustrated by $(CH_3)_2CO$ where the C(1s) binding energy of the carbonyl carbon is greater than that of the CH_3 'carbons' because of the greater electronegativity of the oxygen compared with hydrogen.

A particularly instructive set of data (Fig. 4.9) illustrating how variation in electron density has an important influence on the

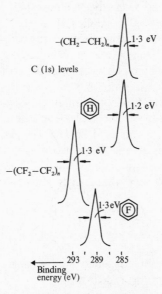

Fig. 4.9. Comparison of C(1s) peaks for polymers and monomeric compounds. (After Clark *et al.* 1973.)

observed binding energy is provided by various fluorocarbons (Clark, *et al.* 1973). Fluorine, being an excellent electron-withdrawing ligand, decreases the electron density around the carbon atom leading to a higher binding energy than observed with 'neutral' carbon.

A simple method of estimating chemical shifts is to consider the valence-electron orbitals as defining a 'valence' shell of electric charge with the core electrons residing at the centre of that shell of, say, radius r. If for example q electronic charges are removed from the valence shell and taken in infinity, the potential energy of the inner electrons is lowered by an amount $\Delta\varepsilon$ given by $\Delta\varepsilon = q/r$. Thus the binding energy of a core electron (the energy necessary to take it to infinity) is increased by the same amount. Clearly in order to calculate the value of $\Delta\varepsilon$ we must know the value of r. Ionic radii are available but these represent the distances at which the wavefunctions of neighbouring atoms begin to overlap. Siegbahn prefers the mean radius for the valence-electron orbitals according to the wavefunctions of atomic electrons. The mean radius is in all cases $\sim 0\cdot 1$ nm so that the corresponding shift in binding energy of inner electrons per unit charge removed from the valence shell is $\Delta\varepsilon \simeq 14\,\text{eV}$. Observed shifts are usually smaller than this owing to a charge transfer of less than one electron in the valence shell. Also no account has been taken above of changes in the Madelung potential arising from charges on the other atoms in the molecule; in many cases this has an overriding influence on the shift observed. The free-ion model can be improved to take account of the fact that valence electrons are not transferred to or from infinity when a chemical bond is formed. In an ionic bond between two atoms A and B electrons are transferred from the valence shell of A to the valence shell of B, and if the internuclear distance is R then the energy shift of the core electrons becomes

$$\Delta\varepsilon = q\left(\frac{1}{r} - \frac{1}{R}\right)$$

where q is the number of electrons transferred. The sign of the shift is different for the two atoms.

In atoms and molecules with closed-shell configurations the atomic 1s levels are spin degenerate, i.e. the binding energies of electrons with spin up are the same as those of spin down electrons. Thus no energy splitting is observed for the 1s lines in the XPS spectra. However, for many electron systems with open-shell configurations the core levels should be split by exchange interaction with the unfilled shell. The magnitude of the splitting has been estimated to be about 3 eV for the 2s and 2p subshells in free iron atoms with an unfilled d shell.

Those molecules that are paramagnetic (NO, O_2), i.e. with unpaired electrons in their outer orbitals, also exhibit core-level splitting, whereas in the 1s spectrum of diamagnetic N_2 no splitting occurs. In the case of O_2 the $O(1s)$ line is split into two with an energy separation of $1 \cdot 0$ eV and the relative intensities $2:1$. The molecule has two unpaired electrons so that the total spin of the oxygen molecule after emission of a 1s electron is either $\frac{3}{2}$ or $\frac{1}{2}$. This gives statistical weights of 4 and 2 respectively for the two states which agrees with the observed intensity ratio $2:1$. With only one unpaired electron the NO molecule should have an intensity ratio of $3:1$ as found experimentally for the $N(1s)$ line in NO.

Splittings of atomic core $p_{\frac{3}{2}}$ levels have also been observed in compounds of Th, U, and Pu; the magnitude of the splitting is apparently dependent on the chemistry. In the case of uranium ($^5p_{\frac{3}{2}}$) the largest splitting (10 eV) was observed with uranyl acetate and the smallest (3 eV) with UO_2. It is particularly interesting from a surface-chemistry viewpoint that Th, ostensibly metal but in fact undoubtedly covered by a thin oxide layer, exhibited splittings. The application of an external field enhances the splitting owing to the fact that it provides an additional contribution to the internal-field gradient.

4.2.1. *Applications in surface chemistry*

There are clearly three facts that emerge from studies of the photoelectron spectroscopy of gaseous molecules: (*a*) the orbital sensitivity of XPS and UPS (e.g. Figs. 4.7 and 4.9): (*b*) the sensitivity to the chemical environment in that the binding energy obtained by XPS is very sensitive to the charge on the atom in question (e.g. Fig. 4.9); (*c*) the high analytical sensitivity of XPS, i.e. its ability to pick out one atom amongst a large number of others, e.g. the cobalt atom in the vitamin B12 molecule (Siegbahn *et al.* 1967). There were therefore strong grounds for

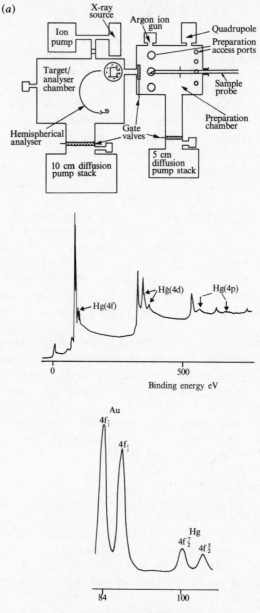

FIG. 4.10. (a) UHV electron spectrometer. (b) XPS of gold foil exposed to Hg; Au(4f) peaks and Hg(4f) peaks shown on an expanded scale. (After Brundle *et al.* 1974).

anticipating that photoelectron spectroscopy would be a powerful experimental technique in the study of solid surfaces and adsorption at solid surfaces. The results of Bordass and Linnett (1969) for the adsorption of CH_3OH on tungsten gave a clear lead as to the likely potential application of UPS in surface studies; that the experimental conditions were not well defined is of no particular consequence since they were clearly 'signposting experiments'. It was, however, only after the availability of an ultra-high-vacuum electron spectrometer that the true potential of electron spectroscopy in surface chemistry could be explored. The essential features of the ESCA-3 spectrometer are shown in Fig. 4.10(a); a fuller description can be obtained elsewhere (Brundle *et al.* 1974). Fig. 4.10(b) shows the X-ray spectrum of a gold film with some adsorbed Hg present (Brundle and Roberts 1972). The simplicity of this spectrum should be contrasted with the electron-impact-induced Auger spectrum of gold where some 80 transitions are possible and mercury can be observed only with some difficulty (Joyner and Roberts 1973). Less than 0·1 per cent of a monolayer of Hg present on a gold surface can be detected by XPS.

If a sample is not a good conductor charging of the sample can occur; this has the effect of shifting the binding energies of core levels. Relative binding energies for the same sample are unaffected, however. The most generally reliable methods for circumventing sample charging are as follows.

(a) Studying the sample as a very thin film on a conducting backing so that the sample is in electrical contact with the spectrometer.

(b) Deposition of a thin film of a reference sample on the surface of the material being investigated. Observation of a suitable core level of the reference then allows the energy scale to be corrected for sample charging provided the reference is in electrical contact with the sample.

4.2.2. *Surface sensitivity and escape depth of photoelectrons*

The first suggestion of the surface applicability of XPS had come from Siegbahn's experiments with iodostearic acid (Siegbahn *et al.* 1967); these led to the suggestion that the escape depth of X-ray-induced photoelectrons was between 5 and 10 nm. Whether this figure would apply generally for all solids was an open question. There is now quite clear experimental evidence that the escape depth is a function of electron energy and that it varies from about 0·5 to 3 nm in the electron-energy range 10–1000 eV (Fig. 4.11). Clark and Thomas (1977) have critically reviewed previous data and also reported new results for escape depths from various solids.

The escape depth has been in general determined by depositing, under controlled conditions, known amounts of adsorbate. In the case of metal overlayers the thickness of the deposit has been determined by a quartz-crystal microbalance. The approach used by Brundle and Roberts (1972) was to make use of the fact that at 77 K the sticking probability S of such molecules as CO_2 and H_2O was close to unity. Therefore from known exposures and assuming $S \simeq 1·0$ the surface concentration of adsorbate could be estimated. For the case of CO_2 (Fig. 4.12) it was concluded that at least 10 per cent of a monolayer could be detected, while for Hg on Au, a particularly favourable case in view of mercury's high cross-section for ionization, about 0·2 per cent of a monolayer could be observed.

There are two possible advantages that ultra-violet photoelectron spectroscopy (He I and He II) has over XPS. First is the obvious one that the valence electrons are being probed and this should give more direct information on surface bonding. The second is that the energy of the ejected electrons is likely to fall in the shortest escape depth region (Fig. 4.11). As to analytical surface sensitivity experimental data suggest that UPS is at a disadvantage compared with XPS, the spectra being frequently

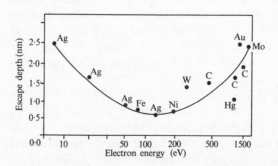

FIG. 4.11. Escape depth as a function of kinetic energy of photoelectron.

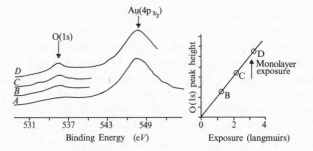

FIG. 4.12. O(1s) spectra for CO_2 condensed on a Au substrate at 77 K as a function of exposure. Curve A is for the clean Au surface.

broad and superimposed on the substrate band structure. However, distinct advantages accrue when both XPS and UPS spectra are determined and examples will be considered later.

In addition to the X-ray-induced core-level spectra there are frequently present satellite spectra which are closely associated with them. Of particular interest has been the suggestion that distinction can, for example, be made between the valence states Cu^+ and Cu^{2+}. Advantage of this has been taken by a number of investigators of the oxidation of copper, as discussed in Chapter 11.

4.3. Auger electron spectroscopy (AES)

4.3.1. *Introduction*

Some 50 years ago Auger (1923, 1925) investigated the trajectories of electrons emitted from argon by X-rays using a Wilson cloud chamber; he observed that the emission of an electron was often accompanied by the emission of a second electron. This electron was considerably slower than the first and the two electrons followed unrelated directions. He suggested that the first electron leaves a vacant level in the electron cloud of the excited atom, and an electron from a higher level can fall into this level liberating a characteristic quantum of energy. This quantum of energy can be absorbed by the atom itself resulting in the emission of a second electron. This mode of de-excitation is the Auger process which leads to secondary-electron emission; a second leads to the emission of electromagnetic radiation, that is

X-ray fluorescence (Fig. 4.5). It is clear that in principle similar information concerning the transitions involved should be obtainable from X-ray fluorescence and Auger spectroscopy. The relative importance of the two processes is illustrated in Fig. 4.13. It is clear that the Auger process involving K-level ionization is dominant until $Z \simeq 35$ after which X-ray fluorescence takes over.

The suggestion that the Auger process should be exploited in surface studies was made by Lander (1953) of the Bell Laboratories (some 30 years after its discovery). His comment went virtually unnoticed until Harris (1968) of the General Electric Company in Schenectady, N.Y., reported his extensive studies of the Auger spectra of metals. One of the inherent problems in Auger spectroscopy is that the Auger electrons are superimposed on the large secondary-electron background, and Harris's contribution was extremely significant in that he realized that these Auger electrons could be easily recognized by electronic differentiation of the measured energy distribution. The potential advantage of this approach is clear in Fig. 4.14.

The curve in Fig. 4.14(a) is the collector current as a function of the retarding voltage, i.e. I represents the number of electrons with energy greater than the retarding potential at each point on the curve. The curve in Fig. 4.14(b) is the derivative of the retarding-potential curve, which is the electron energy distribution $N(\varepsilon)$. The curve in Fig. 4.14(c) is the second derivative of the curve in Fig. 4.14(a) and leads to sharply defined peaks on a smaller background. It is this curve of $dN(\varepsilon)/d\varepsilon$ which is largely used in AES.

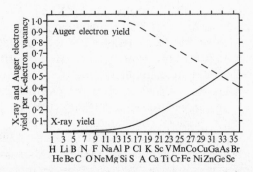

FIG. 4.13. Auger electron emission and X-ray fluorescence yields for K-shell electron vacancies as a function of atomic number. (After Somorjai 1972).

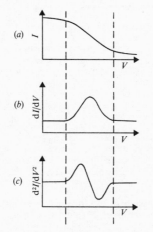

FIG. 4.14. Idealized shapes of detected Auger signals as functions of the retarding potential V.

Another important historical point to note is that Weber and Peria (1967) indicated that differentiated secondary-electron energy distributions could easily be obtained using retarding-field analysers as used in LEED studies. This came about through discussions between Harris and Peria (personal communication from R. E. Weber to M. W. Roberts 1975) and marked a significant acceleration in the use of Auger spectroscopy since LEED had been in vogue for a number of years and equipment was therefore available.

Auger electrons will of course be produced with both X-rays and electrons. Electron bombardment has nevertheless been the most widely used initiator, since it is comparatively easy to produce an electron beam of suitable energy. Three distinct energy regions can be distinguished in the secondary electrons emitted from a surface bombarded by electrons of primary energy E_p (Fig. 4.15). Region A consists mainly of primaries which have been elastically scattered; those that are coherently scattered produce the diffraction features in LEED. Region B of the electron distribution curve is known as the characteristic loss region, since it contains peaks separated from the elastic peak A by energies which are characteristic of the material and independent of the primary energy. Region C is a large 'slow' (low-energy) peak made up of true secondary electrons; superimposed

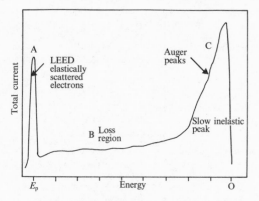

FIG. 4.15. Energy distribution of electrons back-reflected from a surface bombarded by E_p V primary electrons.

on its high-energy side are small maxima due to Auger transitions.

4.3.2. *Nomenclature in AES*

The notation adopted is that usually used in atomic spectroscopy:

$$K \rightarrow 1s$$
$$L_1 \rightarrow 2s$$
$$L_2L_3 \rightarrow 2p$$
$$M_1 \rightarrow 3s$$
$$M_2M_3 \rightarrow 3p$$
$$M_4M_5 \rightarrow 3d \quad \text{etc.}$$

where K, L, M etc. refer to the principal quantum numbers 1, 2, 3, ... The subscripts L_1, L_2, L_3, etc. indicate the multiplicity j which is a vector sum of the angular momentum l and the spin quantum number s: $J = l + s$. For example for Na $(1s^2 2s^2 2p^6 3s^1)$ we have $K(1s_{\frac{1}{2}})$ $L_1(2s_{\frac{1}{2}})$ $L_2(2p_{\frac{1}{2}})$ $L_3(2p_{\frac{3}{2}})$ and $M_1(3s_{\frac{1}{2}})$, each of which are distinct electronic levels with their own characteristic binding energies. If an electron vacancy in a K shell is filled by an electron from the L_1 shell and the energy is transferred to an electron in the L_2 shell the resulting Auger electron is designated KL_1L_2.

In considering Auger processes involving L and M holes consideration must be given to a special type of *radiationless* process called a Coster–Kronig transition. This is of the type $L_i L_j X$ or $M_i M_j X$ and involves transitions between subshells of a particular shell. These transitions are very strong when energetically allowed and cause a rapid redistrubution of core levels which strongly influences the relative intensities of the observed lines in the L and M groups.

4.3.3. Energies of Auger electrons

The kinetic energy of Auger electrons was given by Burhop (1952) as

$$\varepsilon_{vxy(Z)} = \varepsilon_{v(Z)} - \varepsilon_{x(Z)} - \varepsilon'_{y(Z)} \tag{4.3}$$

where $\varepsilon'_{y(Z)}$ is the ionization energy appropriate to level y in an atom of atomic number Z already singly ionized in an inner level. Further, $\varepsilon'_{y(Z)}$ was approximated as the ionisation energy of the y electron in the atom of next-highest atomic number $\varepsilon_{y(Z+\Delta)}$ where $\Delta = 1$ to account for the extra positive charge. Experimental values of Δ have been found to be generally between $\frac{1}{2}$ and $\frac{3}{4}$. Thus for any Auger transition involving a set of levels VXY with an element of atomic number Z a peak should appear at

$$\varepsilon_{(Z)} = \varepsilon_v - \varepsilon_x - \varepsilon_{y(Z+\Delta)} - \phi_A \tag{4.4}$$

where ϕ_A is the effective work function of the analyser material (e.g. in the LEED–Auger system it is the work function of the grid material). *It should be noted that the Auger energies are independent of the energy of the incident radiation.* When the observed and calculated values of ε are plotted against Z it can be seen that most elements should be identifiable (Chang 1971), even when several coexist on the surface.

Although X-ray energy-level data give a good approximation of Auger energies, they are strictly speaking inadequate in that the final (doubly ionized state) of the atom should be included in some way. This additional specification is needed because the energies of Auger processes are governed by the coupling schemes among electronic wavefunctions in the initial (singly ionized) and final (doubly ionized) electronic configurations. These coupling schemes not only govern the energies of the transitions but also determine the *number* of possible transitions.

A difficulty that does arise, however, is when x and y are different; eqns (4.3) and (4.4) do not yield values in agreement with experimental data. The following is the reason. When an inner level v is ionized, x is trapped by the vacancy v and the excess energy released causes the ejection of electron y. The kinetic energy of the electron y after leaving the ion is then given by eqn (4.5) (omitting ϕ_A) where β is an adjustable parameter close to unity.

$$\varepsilon_{v(Z)} - \varepsilon_{x(Z)} - \varepsilon_{y(Z)} - \beta\{\varepsilon_{y(Z+1)} - \varepsilon_{y(Z)}\}. \tag{4.5}$$

On the other hand if y is trapped by the vacancy and x ejected the kinetic energy of x after leaving the ion is

$$\varepsilon_{v(Z)} - \varepsilon_{y(Z)} - \varepsilon_{x(Z)} - \beta\{\varepsilon_{x(Z+1)} - \varepsilon_{x(Z)}\}. \tag{4.6}$$

Quantum mechanically these two cases are indistinguishable, since the initial and final stages of the transitions are identical in both cases. The kinetic energy of the electrons must be the same in both cases. Now eqns (4.5) and (4.6) can only be the same if β is altered drastically. It is clearly not desirable to alter β drastically.

Chung and Jenkins (1970) suggested that this difficulty can be overcome by taking the average of the expressions (4.5) and (4.6) so that

$$\varepsilon_{vxy(Z)} = \varepsilon_{v(Z)} - \varepsilon_{x(Z)} - \varepsilon_{y(Z)} - \beta'\{\varepsilon_{x(Z+1)} - \varepsilon_{x(Z)}\}$$
$$- \beta'\{\varepsilon_{y(Z+1)} - \varepsilon_{y(Z)}\} \tag{4.7}$$

or

$$\varepsilon_{vxy(Z)} = \varepsilon_{v(Z)} - \varepsilon'_{x(Z)} - \varepsilon'_{y(Z)}.$$

Thus $\varepsilon'_{x(Z)}$ and $\varepsilon'_{y(Z)}$ now become the effective ionization potentials of the electrons x and y respectively and have values between $\varepsilon_{(Z)}$ and $\varepsilon_{(Z+1)}$. It is usual to take

$$\varepsilon'_x = \frac{\varepsilon_{x(Z)} + \varepsilon_{x(Z+1)}}{2} \quad \text{and} \quad \varepsilon'_y = \frac{\varepsilon_{y(Z)} + \varepsilon_{y(Z+1)}}{2}.$$

Of course picking the energy half-way between the value it would have for an atom having atomic number Z and an atom having atomic number $(Z+1)$ is arbitrary, but it gives good results when compared with the experimental data.

Not only should it be important to locate from first principles the energy of a given Auger transition but it would also be advantageous to be able to know the intensity of each Auger line.

The rate of emission of Auger electrons is determined in the main by the rate of ionization of the X electron shell since the only other mode of de-excitation, that of X-ray emission, is relatively inefficient. The experimentally measured Auger current from a pure solid is typically $10^{-5} I_p$ where I_p is the primary beam current. In principle the Auger current can be calculated by determining the ionization cross-section. These calculations are complex and in any case to date have not yielded good agreement with experiment.

Let us consider, however, what parameters may be changed to maximize the Auger current. Firstly regarding the ratio E_p/ε_x where E_p is the energy of the incident electron and ε_x the ionization energy of the initial level X, maximum ionization is found to occur near $E_p \simeq 3\varepsilon_x$. The second factor is secondary ionization. Secondary electrons with any energy greater than ε_w can also cause ionization of a level w. This in effect increases the effective primary beam and can greatly enhance ionization. Although at first sight we might conclude that secondary ionization is helpful because it increases the ionization rate, its effect is difficult to calculate and presents problems in quantitative analyses. Some recent studies by Smith and Gallon (1974) are described later.

Clearly an important consideration in the application of AES is the width of a peak. Let us consider a transition such as the $L_{2,3}V_1V_1$ transition in Si; the energy ε_1 is given by

$$\varepsilon_1 = \varepsilon_{L_{2,3}} - 2\varepsilon_{V_1}.$$

We are ignoring for simplicity β and ϕ_A. Similarly for the $L_{2,3}V_2V_2$ transition ε_2 is given by

$$\varepsilon_2 = \varepsilon_{L_{2,3}} - 2\varepsilon_{V_2} = \varepsilon_1 - 2\delta$$

where $\varepsilon_{V_2} - \varepsilon_{V_1} = \delta$ and δ is the width of the valence band. Therefore the width of the Auger 'line' will be 2δ. In practice the situation is not as simple as this because of variations in density of states within a band and also because other transitions, e.g. $L_{2,3}V_1V_2$ etc., are possible.

Since escaping electrons frequently suffer small energy losses, broadening of the Auger peak will occur. Energy-loss processes impart to most peaks on the $N(\varepsilon)$ curve a characteristic tailing structure on the low-energy side. It is now accepted practice to record an Auger peak at the energy corresponding to the position of the sharp minimum in the $dN(\varepsilon)/d\varepsilon$ curve (Fig. 4.14(c)).

4.3.4. Quantitative analysis

A simple method of analysis can be achieved by comparing the Auger signal from the element X present in the sample with that from a pure elemental standard. In this case the concentration of X is given by

$$C_x = \frac{I_x}{I_x \text{ (standard)}}$$

where I_x and $I_{x(\text{standard})}$ are the Auger currents from the 'impurity' X and the standard (pure) X. The peak-to-peak height from each sample can therefore be compared directly under identical conditions. A number of inherent errors are nevertheless involved in this method: (a) chemical effects on peak shapes, (b) surface topography, and (c) variation of escape depth with the Auger electron energy.

Chemical effects can change the peak shape and thus lead to error when using 'peak-to-peak' heights in the differential spectrum for a measure of the Auger signal. It is therefore preferable to use $N(\varepsilon)$ rather than $N'(\varepsilon)$ data for quantitative work. Gallon (1969) developed simple and useful expressions for the Auger current when a second material is condensed on a substrate. For the condensate:

$$I_n^c/I_\infty^c = 1 - \exp(n/n_c) \tag{4.8}$$

where n_c is a constant often taken to be the inelastic mean free path (i.m.f.p.) of the condensate Auger electrons in the condensate and n is the condensate thickness. From the substrate the Auger current is given by

$$I_n^s/I_0^s = \exp(-n/n_c) \tag{4.9}$$

where the superscripts s and c refer to the substrate and condensate respectively and the subscripts 0, n, and ∞ to zero condensate, n-layer-thick condensate, and bulk material respectively.

Equations deduced by Seah (1972) take into account how the thickness and scattering properties of the condensate layer influence the rate of generation of both the substrate and condensate Auger electrons. For example, the back-scattered electron current varies with thickness and the atomic number of the condensate, and Smith and Gallon (1974) have recently shown that it increases with atomic number and the energy of the primary beam (Fig. 4.16).

The influence of electron trajectories on eqns (4.8) and (4.9) is also considered by Seah (1972) who derives eqn (4.10):

$$I_n^s(\theta)\, d\Omega = I_p A_s n_s^s \cos\theta \exp\left(\frac{-n}{n_c^s \cos\theta}\right) d\Omega \qquad (4.10)$$

where I_p is the primary electron beam, A_s is the Auger current per incident electron for the substrate atoms emitted into solid angle Ω at an angle θ to the surface normal, s is the thickness of each substrate atom layer, and n_s^s and n_c^s are the i.m.f.p.'s of the substrate Auger electrons in the substrate and condensate. It should be noted that the characteristic layer thickness n_c is replaced by $n_c^s \cos\theta$ which reduces to the same value for a detector along the surface normal.

What are the roles of n_c and n_s in the quantitative use of AES? The analogous equation to (4.10) for the condensate is equation (4.11):

$$I_n^c(\theta)\, d\Omega = I_p A_c n_c^c \cos\theta\left\{1 - \exp\left(\frac{-n}{n_c^c \cos\theta}\right)\right\} d\Omega \qquad (4.11)$$

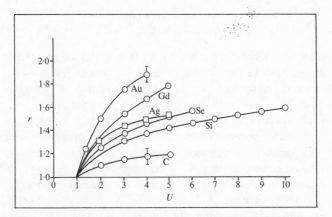

FIG. 4.16. Back-scattering factors r of C, Si, Se, Ag, Gd and Au; $U = E_p/\varepsilon_x$ where E_p is the primary energy and ε_x is the energy of the ionized level. (After Smith and Gallon 1974.)

where n_c^c is the i.m.f.p. of the condensate Auger electrons in the condensate. In both eqns (4.10) and (4.11) the Auger currents are proportional to the factors $n_s^s \cos \theta$ and $n_c^c \cos \theta$ respectively. These proportionalities are usually put equal to unity in calibration experiments which is clearly unjustified.

4.3.5. *Experimental techniques used in AES*

A number of different methods for analysing the kinetic-energy distribution of emitted electrons are feasible. However, in the context of AES there are just two, the retarding-field and the cylindrical-mirror analysers. Both have certain features in common: they are ultra-high-vacuum compatible, provide easy sample access, are electrostatic and compact.

Retarding-field analyser. The most successful method that has been adopted is to modulate the analysing energy and synchronously detect the output current. Thus a perturbing voltage $\Delta V = k \sin \omega t$ is superimposed on the analyser energy so that the collected electron current $I(\varepsilon)$ is energy modulated. The retarding-field analyser functions by repelling electrons with energy less than $E = eV$, where V is the voltage applied to the analysing grids. The collector therefore receives an electron current

$$I(\varepsilon) \propto \int\limits_E^\infty N(\varepsilon)\, d\varepsilon.$$

We therefore have $I'(\varepsilon) \propto N(\varepsilon)$ with this device so that $N(\varepsilon)$ can be obtained by tuning to the modulating frequency ω.

However, detection at 2ω for obtaining $I'' \propto N'(\varepsilon)$, i.e. $dN(\varepsilon)/d\varepsilon$ provides a number of distinct advantages.

(a) The capacitively coupled current which is larger than the electron current has no 2ω component and will therefore be ignored by the phase-sensitive detector.

(b) Most peaks on the $N(\varepsilon)$ curve have a low-energy tail so that the position of the peak is more readily observed on an $N'(\varepsilon)$ curve.

It was with this step that the interest in Auger spectroscopy escalated. The optics of the large number of LEED machines in existence in the late 1960's were ideally suited for determining

$N'(\varepsilon)$. Almost overnight chemical analysis of surfaces became possible with a sensitivity of ~5 per cent of a monolayer.

Cylindrical-mirror or hemicyclindrical velocity analyser (CMA and HCVA). There is little doubt that the CMA is the best analyser for Auger electron spectroscopy. It was first used by Palmberg, Bohn, and Tracy (1969) and a diagramatic representation of it is shown in Fig. 4.17(a). The main feature of the

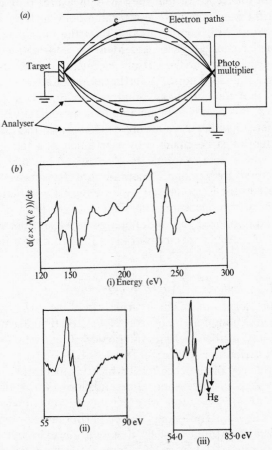

FIG. 4.17. (a) Schematic diagram of the coaxial cylindrical analyser used in the Vacuum Generators Auger spectrometer. (b) Auger electron spectra of a polycrystalline gold foil, $E_p = 3$ keV, $I_p = 0.5$ μA: (i) 120–300 eV, modulation voltage = 2·0 eV peak to peak; (ii) 55–90 eV, modulation voltage = 0·20 eV peak to peak; (iii) After 0.4×10^{-6} torr exposure to mercury. Compare with XPS (Fig. 4.10(b)).

analyser is that it accepts electrons around the full 360° of the cone of half-angle whose apex is at the target. This results in the collection and analysis of as many as 10 per cent of all the electrons leaving the surface. The ability of the CMA to focus correctly electrons of such widely varying angles is the result of second-order focusing. Now the collection efficiency of 10 per cent is similar to that in the LEED/Auger system, but there is an important difference in that the CMA measures only those electrons at the pass energy. The retarding-field analyser collects all the electrons with energy greater than the cut-off energy, and these higher-energy electrons contain no information about the energy distribution at the cut-off energy. Moreover they contribute a great deal of noise. A further point of advantage with the CMA is that the current is monitored by means of an electron multiplier; this results in being able to acquire data using a single gain whereas with the retarding-field analyser the gain for the grid system must be changed over as much as a factor of ~300 for the whole energy range. It is worth noting that the CMA manufactured by Vacuum Generators Ltd. has a novel feature in that the current is monitored by an externally mounted photo-multiplier which measures the light emitted from a phosphor screen and induced by the impinging Auger electrons (Fig. 4.17(a)). An Auger spectrum of Hg adsorbed on Au is shown in Fig. 4.17(b).

4.3.6. *Loss features in Auger spectroscopy*

Ionization loss peaks. These arise when the primary electron causes ionization of an inner (core) level and then leaves the solid without any further inelastic interaction. Clearly features in the energy distribution will appear at $\varepsilon = E_p - \varepsilon_B$ where ε_B is the binding energy of the core level which has been ionized.

When $E_p \gg \varepsilon_B$ the system after ionization will contain an excited ion of energy ε_B and two electrons the sum of whose kinetic energy is $E_p - \varepsilon_B$. This energy will be partitioned between the two electrons; an extreme case is that one electron will be at the Fermi energy (ε_F taken as zero) and the second electron leaves the solid with kinetic energy ($E_p - \varepsilon_B$). In general the loss energy ε_L for such an ionization process is given by

$$\varepsilon_L = E_p - E_p' = \varepsilon_B + \varepsilon_S$$

where E_p' is the energy of the scattered primary electron and ε_S is the energy above the Fermi level to which the secondary electron is scattered. This analysis neglects many body effects. It is clear that loss features should 'move' as the energy of the primary beam is varied and therefore can be distinguished from true Auger electrons.

One interesting example of the advantage of 'loss features' over 'Auger peaks' is the case of C contamination on Pd. The $C(KL_2L_2)$ and $Pd(M_4N_2N_2)$ transitions overlap, and since no other carbon Auger feature exists detection of small concentrations of C on Pd is difficult. However, the $C(K)$ and $Pd(M_5)$ ionization features are well separated making ionization spectroscopy (in this case) a more effective technique than Auger spectroscopy.

Plasmon loss peaks. When a primary electron moves through a solid it will perturb the electron potential, generating plasmons, or collective oscillations of the valence electrons. For free-electron metals the bulk plasmon frequency ω_p is given by

$$\omega_p = \left(\frac{4\pi n e^2}{m}\right)^{\frac{1}{2}}$$

where n is the electron density, m the electron mass, and e the electronic charge. At the solid surface the electric field is weakened by depolarization effects and so the restoring force for the plasma motion is decreased. This leads to a distinct surface oscillation of frequency ω_S which is lower than ω_p and given by $\omega_S = \omega_p/\sqrt{2}$. In the secondary-electron spectrum plasmon peaks generally appear as a series with energies separated from the primary energy by multiples of $h\omega$. They may therefore also appear as 'low-energy satellites' to Auger peaks. If the surface is covered by an adsorbate with dielectric constant η greater than unity the surface plasmon frequency is $\omega_p/(1+\eta)^{\frac{1}{2}}$.

4.3.7. *Applications of Auger spectroscopy*

One of the undoubted advantages of the availability of Auger spectroscopy has been in relation to its applicability to the definition on an atomic scale of solid surfaces. This point is highlighted by the viewpoint expressed some 10 years ago that provided a single crystal displayed a LEED pattern that was

expected of its surface geometry then the surface was clean. This
was certainly not the case for Cu(100). Although the pattern was
that expected of clean Cu(100), its chemical reactivity was not
(Joyner, McKee, and Roberts 1971). In fact its inactivity was
shown by AES to be due to an appreciable surface concentration
(20 per cent of a monolayer) of S which was present but did not
in any way influence the LEED pattern. Clearly the correct
surface structure is a prerequisite but not sufficient condition of
characterization. The preparation of atomically clean surfaces has
therefore been put very much on a quantitative basis with the
availability of AES. The case of tungsten is particularly interesting (Joyner, Rickman, and Roberts 1973) since this metal has
played a central role in surface chemistry (Fig. 4.18).

Although the X-ray-induced spectrum of an atom has been
shown to be sensitive to its chemical environment, the potential
of AES in being able to discriminate between different bonding
situations has not been at all clear. There are two points to look
for.

(a) Shifts in energy of an Auger electron arising from the
occurrence of charge transfer. If for example electrons are
transferred from a less to a more electronegative element,
then the binding energies of the remaining electrons in the
positively charged atom are increased and the energies at

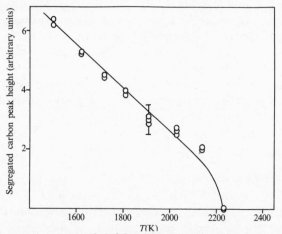

FIG. 4.18. Carbon Auger peak height observed when tungsten was heated to
2200 K *in vacuo*. The process of carbon segregation was reversible with temperature. (After Joyner, Rickman and Roberts 1973.)

which the Auger electrons appear will be decreased. The quantitative interpretation of observed shifts is not easy as we are dealing with a 'difference quantity' involving three electron levels.

(b) If the electrons involved in the Auger process include valence electrons, there is a possibility that a change in the transition probability occurs. This will lead to a change in the shape of a complex Auger spectrum.

We would therefore expect to see changes in the shape of an Auger spectrum when the chemical bonding of an adatom changes. Let us consider the case of carbon where the Auger spectrum is due to the KLL transition, the L electrons being the 2s and 2p electrons. Thus changes in the hybridization of these orbitals will change the shape of the KLL spectrum owing to changes in the transition probabilities for the various configurations of the final state of the doubly ionized carbon atoms after the Auger transition. Fig. 4.19 shows the results of Haas, Grant, and Dooley (1972). Each form of carbon appears to give quite distinct spectra, thus giving a fingerprint for the particular state. The general utility of this particular aspect of AES, although as yet unproven, is anticipated to be limited.

Concentration–depth profiles of oxidised stainless steel have been reported (Stoddart and Hondros 1972); these are obtained by ion-etching successive layers from the surface. In particular there is a strong enrichment of the outer layers of the oxide with Cr and a correspondingly low iron content which, however, rises rapidly as the metal–oxide interface is approached (Fig. 4.20). Another most interesting feature is the enrichment of the surface with such metallic elements as Mn (1·1 per cent total); V (0·05 per cent total). All these high surface enrichments from low bulk alloy concentrations decay rapidly with distance from the oxide surface until they become undetectable. On the other hand there is a depletion of Ni at the metal–metal oxide interface and in fact no Ni was detected within the sensitivity of the experiment. This at least is consistent with observations that the oxidation resistance of Fe–Cr–Ni alloys is chiefly related to the preferential oxidation of Cr. Recently Smith (1976) has carried out a detailed AES–ion-sputter profile study of oxides of aluminium. He concludes that the quantification of the relationship between peak-to-peak heights of Auger spectra and concentration profiles is

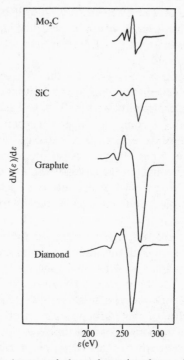

FIG. 4.19. Auger peak shapes for various forms of carbon.

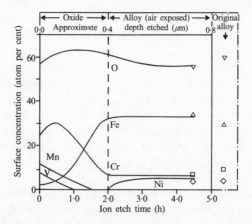

FIG. 4.20. Element concentration profiles in oxide layer formed on stainless steel after 2 h at 1200 K.

only semiquantitative. Even such stable oxides as Al_2O_3 and TiO_2 are influenced by the electron beam.

One area of surface chemistry where the Auger CMA spectrometer could be of unique advantage is in the direct study of desorption. A system which lends itself to this approach is the $W + N_2$ one and illustrates the possible potential of this particular spectrometer (Joyner, Rickman, and Roberts 1974) for kinetic studies. More detailed discussions of AES and its applications in surface chemistry and catalysis are to be found elsewhere (Joyner and Roberts 1975).

4.4. Appearance-potential spectroscopy (APS)

Park, Houston, and Schreiner (1970) devised this technique for determining the energy levels of adsorbed species at metal surfaces. The subject has been reviewed recently by Bradshaw (1974) and by Park (1975a,c). The essence of the method is that when the total electron-excited fluorescence of a solid is plotted as a function of its potential with respect to the electron source, sharp changes in slope occur at the appearance potentials of characteristic X-rays. By taking the derivative of these curves electronically these changes in slope can be easily detected (cf. Auger spectroscopy).

The appearance potentials correspond to the threshold energies at which core electrons are excited to unoccupied states. The main feature of the spectrometer (Fig. 4.21) is a vacuum diode

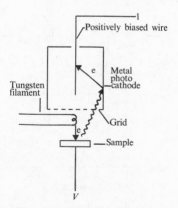

FIG. 4.21. Schematic arrangement of an appearance potential spectrometer.

and photomultiplier; electrons emitted thermionically from a tungsten filament cathode strike the surface under investigation. The soft X-ray photons produced at the surface strike a photocathode negatively biased so as to reject electrons. The photoelectron current is amplified by an electron multiplier. The potential of the surface being studied is programmed with a linear ramp in the range from 0 to 2 kV. A small a.c. signal is superimposed on the voltage applied to the surface; this allows the derivative of the fluorescence to be determined. The resulting variation in multiplier current is synchronously detected with a phase lock-in amplifier. The spectrometer is ultra-high-vacuum compatible and has been used up to background pressures of about 10^{-7} torr. No difficulty was encountered from ions produced either by electron-impact desorption or by gas-phase ionization.

More recently Park and Houston (1973a) have shown that it is not necessary to use an electron multiplier and have described a

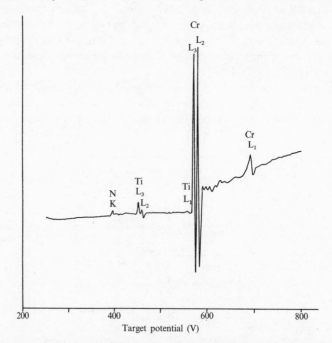

FIG. 4.22. Appearance potential spectrum of outgassed chromium sample. (After Park and Houston 1973(a).)

simpler spectrometer which is also claimed to have a performance superior to their original one. The spectrum of an outgassed polycrystalline chromium sample is shown in Fig. 4.22; the original surface contaminants were C and O which were removed by heating. These contaminants were replaced by nitrogen and titanium peaks presumably by diffusion from the bulk. Redhead and Richardson (1972), using a slightly modified spectrometer, have reported data for thorium on tungsten. Oxidation of the thorium-covered surface did not produce significant changes in shape and position of the thorium O_5, O_4, N_7, and N_6 peaks.

Recently Park and Houston (1973b) have used soft X-ray appearance-potential spectroscopy to determine the binding energies of the N_4 to N_7, O_1 to O_5, and P_1 to P_3 levels of 'clean' uranium surfaces. The measured levels are consistently a few electron volts lower than tabulated X-ray binding energies based on uranium which clearly had a surface layer of oxide present. It is interesting that in experiments where the 'clean' uranium was exposed to oxygen the appearance potentials did not shift. Park and Houston postulate that the discrepancy is in part due to a 'surface chemical shift'.

Suggestions by Tracy (1972) that appearance-potential spectroscopy would not be useful with semiconductors and insulators have recently been refuted by Park and Houston (1973a). These authors report appearance-potential spectra for Si and SiC and also refer to their previously published data for Cr_2O_3 and SiO_2.

A novel method based on the 'disappearance' of the primary electrons that produce the excitation has been described recently by Kirschner and Staib (1973). They show how, by using phase-sensitive techniques, the first and second derivatives of the elastically back-scattered electron current *versus* the energy of the incident electrons may be obtained. The authors claim that disappearance-potential spectroscopy has some advantages over APS.

4.5. Ion–neutralization spectroscopy (INS)

This particular form of electron spectroscopy is due to Hagstrum of Bell Laboratories who, some 30 years ago, was interested in phenomena occurring in a glow discharge but with particular concern as to what happened at the cold cathode. This led to an interest in the interaction of 'slow' atomic particles with solid surfaces, when these particles carried appreciable amounts of potential energy. INS therefore is an emission electron spectroscopy where the kinetic-energy distribution of electrons ejected from a solid by such positive ions as He^+ is monitored. Close to the surface a non-radiative, Auger-type electronic-transition process occurs in which an electron from a surface atom neutralizes the incident ion to the ground state of the parent atom. The energy liberated in this process simultaneously excites a second electron in the surface atom which may be ejected into vacuum. The kinetic-energy distribution of these ejected electrons provides information on the electronic structure in the surface region of the solid. Hagstrum (1975) has given an excellent account of how INS came into being, its particular advantages and problems, and its present status. We give here only a brief outline of the subject.

Hagstrum rightly emphasizes the difference between 'one-electron spectroscopy' such as UPS or XPS and 'two-electron spectroscopy' such as AES and INS. For an emission-electron spectroscopy to be viable it must yield what Hagstrum calls the

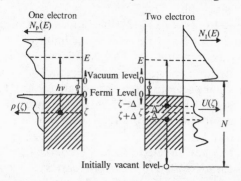

FIG. 4.23. Electron energy level diagrams indicating transitions and distribution functions appropriate to a one-electron spectroscopy, UPS, and a two-electron spectroscopy, INS. Note the two energy scales, E and ζ, having their respective zeros at the vacuum and Fermi levels. In the one-electron spectroscopy the final state at E always lies above the initial state at ζ by $h\nu$, the energy of the absorbed photon, and a one-to-one correspondence exists between $N_p(E)$ the measured distribution and one-electron transition density and the local density of states $\rho(\zeta)$. In INS on the other hand the final state at E may be achieved from an infinite set of paired initial states symmetrically disposed relative to the energy ζ halfway between the final state in question and the initially vacant level in the incident ion outside the solid. The measured $N_I(E)$ is the final-state or two-electron transition density and yields the fold function $F(\zeta)$ in eqn 4.12. N is the neutralization energy of the ion near the solid surface. (After Hagstrum 1975.)

one-electron transition density of the process. This is the relative probability that an electron, resident in the volume near the surface of the solid to which the spectroscopy is sensitive, becomes involved in the process; it is the initial-state transition density. However, experiment measures the final-state transition density, i.e. the relative probability that electrons appear at a specified final-state energy. In the case of one-electron spectroscopy the final-state transition density is equal to the initial-state transition density. For a two-electron spectroscopy (INS) the difference between these two transition densities is fundamental. The reason for this is that electrons at a given final-state energy can come from an infinite set of paired initial states (Fig. 4.23). Thus, the final-state or two-electron transition density $F(\zeta)$ which is obtained from the experimental kinetic-energy distribution $N_{I}(E)$ is, in fact, the self-convolution or fold of the initial-state or one-electron transition density $U(\zeta)$, i.e.

$$F(\zeta) = \int_{-\zeta}^{\zeta} U(\zeta + \Delta) U(\zeta - \Delta) \, d\Delta. \qquad (4.12)$$

The major problem in developing INS as a useful spectroscopy was to extract $U(\zeta)$ since it is the one-electron transition density that is most sensitive to the local density of initial states. The strategy used to solve this formidable problem is described by Hagstrum (1975).

Hagstrum and his colleagues have already obtained interesting results, and perhaps the outstanding fact that emerges is the need for complementary spectroscopic studies. For example, there are unresolved problems relating to UPS and INS studies of ordered multilayers of Te on Ni(100). Two orbitals are detected by UPS but one of these is absent in the INS spectrum. Hagstrum suggests that this is either due to the location of the orbital relative to the sensitive regions of each spectroscopy or to the difference in matrix elements of the two spectroscopies.

4.6. Summarizing comments

The advance in surface science arising from the development of electron spectroscopy has been the ability for the first time to analyse directly the surface and subsurface regions of solids. At the most elementary level this may just be analytical or compositional data, but at the more sophisticated level direct information on the nature of the chemical bonding of surface species is possible. In analysing the composition of the surface and subsurface regions, where XPS and AES are the most helpful, consideration has to be given to the following.

(1) The energy of the excited electron being monitored since this has an important bearing on the depth from which it can escape.

(2) The cross-section for ionization and the shape of the peak envelope since this may well depend on the chemical environment of the atom. For example the peak shapes of Fe in metallic and oxidized iron are distinctly different.

(3) The distribution of the element being analysed within the surface/subsurface region. When confined entirely to the surface, analysis of the data is straightforward, but with an unknown distribution function the data are ambiguous. It is usual to assume either (a) uniform distribution with depth or (b) a linearly decreasing concentration with depth.

(4) The intensity of the signal, which will depend on the photon and/or electron flux.

There are still problems to be solved and the possible damaging role of the electron beam in AES is one of them.

Many of the systems studied so far by XPS and UPS have been chosen for strategic reasons, namely that the surface chemistry was reasonably well established, an important methodology when new experimental techniques are being developed. Two approaches have in the main been used to date. The first is based on the observation that was made at an early stage that different adsorption states exhibit different core-level spectra. For example, the C(1s) value for atomic carbon is different from carbon present in molecular carbon monoxide. Similarly the O(1s) value for chemisorbed oxygen is different from the O(1s) value for molecularly adsorbed water. These are but two examples (Atkinson, Brundle, and Roberts 1974a,b) and are considered later. This may be referred to as the chemical-shift approach. No serious attempt has, however, been made to take into account relaxation effects and their influence on binding-energy values. In this sense the chemical-shift approach is not quantitative. The second approach, largely in UPS, is to compare difference

TABLE 4.2

	Experimental convenience (5)	Chemical information on inner shells (5)	Chemical information on valence electrons (5)	Ease of interpretation of chemical information (5)	Lack of surface damage (5)	Ease of quantification (5)	Sensitivity and speed of analysis (5)	Total Max. possible = 35
Auger electron spectroscopy (AES)	4·3	1·9	1·5	1·8	2·1	2·9	4·8	19·3
Appearance potential spectroscopy (APS)	3·7	2·3	1·0	1·5	0·9	1·4	2·5	13·3
Electron energy loss spectroscopy (EELS)	3·7	1·6	1·7	1·7	2·2	1·3	2·4	14·6
Ion-neutralization spectroscopy (INS)	1·2	0·0	3·5	1·9	3·7	1·4	2·1	13·8
Ultra-violet photo-emission spectroscopy (UPS)	3·1	0·2	4·1	2·7	4·4	1·8	3·1	19·4
X-Ray photoemission spectroscopy (XPS or ESCA)	3·2	4·3	2·7	3·4	4·3	3·4	3·1	24·4

We are grateful to the following for help in compiling this table: A. M. Bradshaw, C. R. Brundle, T. Edmonds, S. Evans, J. C. Fuggle, T. Gallon, R. Heckingbottom, K. Kishi, R. Mason, D. Menzel, F. Meyer, M. Prutton, M. L. Sims, G. A. Somorjai and C. J. Todd. From Joyner and Roberts 1975.

spectra, i.e. the UPS of the adsorbent plus adsorbate compared with the UPS of the adsorbent alone. This has also been success-ful, although recent angular UPS studies (Quinn 1975) have emphasized the possible interpretive limitations of difference spectra. Angular photoemission studies have not been discussed here, but are likely to become one of the more important techniques in studying surface bonding over the next five years. A recent review is that by Feuerbacher and Willis (1976), while excellent accounts of vacuum ultra-violet photoelectron spectros-copy are those of Price (1974) and Turner (1970).

There has been much debate recently on the reference level to be used in photoelectron spectroscopy. What is now clear is that the electron binding-energy values should be referred to the Fermi level; what is not so clear is what has to be added to convert this value to the vacuum level. It is not a question of simply adding the work function of the solid, as was first thought, but a value of about 6 eV seems to be the most appropriate. The theoretical reason is, however, as yet not obvious.

Which particular electron spectroscopic technique is best for surface studies? Table 4.2, taken from a recent review (Joyner and Roberts 1975), is an attempt to answer that question. Out of a possible maximum of 35 points, XPS scores the highest with 24·4, followed by AES and UPS with 19·3 and 19·4 points respectively.

REFERENCES

ATKINSON, S. J., BRUNDLE, C. R., and ROBERTS, M. W. (1974a). *Chem. Phys. Lett.* **24** (2), 175.

—— (1974b). *Faraday Disc. chem. Soc.* **58,** 62.

AUGER, P. (1923). *C. R. Acad. Sci. Paris* **177,** 169.

—— (1925). *C. R. Acad. Sci. Paris* **180,** 65.

BORDASS, W. T., and LINNETT, J. W. (1969). *Nature, Lond.* **222,** 660.

BRADSHAW, A. M. (1974). In *Specialist periodical reports: surface and defect properties of solids*, (eds. M. W. Roberts and J. M. Thomas), vol. 3. Chemical Society, London.

BRUNDLE, C. R., and ROBERTS, M. W. (1972). *Proc. R. Soc. A*, **331,** 383.

BRUNDLE, C. R., ROBERTS, M. W., LATHAM, D., and YATES, K. (1974). *J. Electron. Spectrosc.* **3,** 241.

BURHOP, E. H. S. (1952). *The Auger effect and other radiationless transitions.* Cambridge University Press.

CHANG, C. C. (1971). *Surf. Sci.* **25,** 53.

CHUNG, M. F., and JENKINS, L. H. (1970). **22,** 479.

CLARK, D. T., FEAST, W. J., KILCAST, D., and MUSGRAVE, W. K. R. (1973). *J. Polymer Sci.* **11**, 389.

CLARK, D. T. and THOMAS, H. R. (1977). *J. Polymer Sci.* **15**, 2843.

FEUERBACHER, B. and WILLIS, R. F. (1976). *J. Phys. C.* **9**, 169.

FIEBELMAN, P. J., and EASTMAN, D. E. (1974). *Phys. Rev. B* **10**, 4933.

GALLON, T. E. (1969). *Surf. Sci.* **17**, 486.

GOBELI, G. W., and ALLEN, F. G. (1962). *Phys. Rev.* **127**, 141, 150.

HAAS, T. W., GRANT, J. T., and DOOLEY, G. J. (1972). *Proc. Int. Symp. on Adsorption–Desorption Phenomena.* Academic Press, London, New York.

HAGSTRUM, H. D. (1975). *J. vac. Sci. Technol.* **12** (1), 7.

HARRIS, L. A. (1968). *J. appl. Phys.* **39** (3), 1419.

JOYNER, R. W., McKEE, C. S., and ROBERTS, M. W. (1971). *Surf. Sci.* **26** (1), 303.

JOYNER, R. W., RICKMAN, J. and ROBERTS, M. W. (1973). *Surf. Sci.* **39** (2), 445.

—— (1974). *J. Chem. Soc. Faraday Trans.* 1 **70**, 1825.

JOYNER, R. W., and ROBERTS, M. W. (1973). *J. Chem. Soc. Faraday Trans.* 1 **69**, 1242.

—— (1975). In *Specialist periodical reports: surface and defect properties of solids*, (Senior Reporters: M. W. Roberts and J. M. Thomas), vol 3. Chemical Society, London.

KANE, E. O. (1962). *Phys. Rev.* **127**, 131.

KIRSCHNER, J. and STAIB, P. (1973). *Phys. Lett.* **42A**, 335.

LANDER, J. J. (1953). *Phys. Rev.* **91**, 1382.

PALMBERG, P. W., BOHN, G. K., and TRACY, J. C. (1969). *J. appl. Phys.* **40**, 1740.

PARK, R. L. (1975a). *Chemical analysis of surfaces, surface physics of materials*, vol. 11. Academic Press, London, New York.

—— (1975b). *Surf. Sci.* **48**, 80.

—— (1975c). In *Experimental methods in catalysis*, vol. 2 (eds. R. B. Anderson and P. T. Dawson). Academic Press, London, New York.

PARK, R. L., and HOUSTON, J. E. (1973a). *J. appl. Phys.* **44**, 3810.

—— (1973b). *Phys. Rev. A* **7**, 1447.

PARK, R. L., HOUSTON, J. E., and SCHREINER, D. G. (1970). *Rev. sci. Instrum.* **41**, 1810.

PRICE, W. C. (1974). In *Chemical spectroscopy and photochemistry in the vacuum ultraviolet* (eds. C. Sandorfy, P. J. Ansloos, and M. B. Robin). Reidel, New York.

QUINN, C. M. (1975). *Faraday Disc. chem. Soc.* **60**, 139.

REDHEAD, P. A., and RICHARDSON, G. W. (1972). *J. appl. Phys.* **43**, 2970.

SEAH, M. P. (1972). *Surf. Sci.* **32**, (3), 703.

SIEGBAHN, K. (1972). In *Electron spectroscopy (Proc. Int. Conf., Asilomar, Calif.)* (ed. D. A. Shirley). North-Holland, Amsterdam.

SIEGBAHN, K., NORDLING, C., FAHLMAN, A., NORDBERG, R., HAMRIN, K., HEDMAN, J., JOHANSSON, G., BERGMARK, T., KARLSSON, S. E., LINDGREN, I., and LINDBERG, B. (1967). Atomic molecular and solid state structure studied by means of electron spectroscopy. *Nova Acta Regiae Soc. Sci. Upsal.* Ser IV, **20**.

SMITH, D. M. and GALLON, T. E. (1974). *J. Phys. D., app. Phys.* **74**, 151.

SMITH, T. (1976). *Surf. Sci.* **55**, 601.

SOMORJAI, G. A. (1972). *Principles of surface Chemistry.* Prentice Hall, Englewood Cliffs, N.J.

SPICER, W. E. (1963). *Phys. Rev. Lett.* **11**, 243.

—— (1968). In *A survey of phenomena in ionized gases.* Int. Atomic Energy Agency, Vienna p. 271.

—— (1971). *Nat. Bur. Stand. Spec. Publ.* **323**, 139.

STODDART, C. T. H., and HONDROS. E. D. (1972). *Nature, Lond.* **237** (75), 90.

TRACY, J. C. (1972). *J. appl. Phys.* **43**, (10), 4164.

TURNER, D. W. (1970). Molecular photoelectron spectroscopy. *Phil. Trans. R. Soc. A* **268**, 7.

VAN LAAR, J., and SCHEER, J. J. (1962). *Philips res. Rep.* **17**, 101.

WEBER, R. E., and PERIA, W. T. (1967). *J. appl. Phys.* **38** (11), 4355.

GENERAL REFERENCES

BRUNDLE, C. R. (1975). Elucidation of surface structure and bonding by photoelectron spectroscopy. *Surf. Sci* **48**, 99.

CARLSON, T. A. (1975). Photoelectron spectroscopy. *Ann. Rev. Phys. Chem.* **26**, 211.

DELGASS, W. N., HUGHES, T. R. and FADLEY, C. S. (1970). X-ray photoelectron spectroscopy: a tool for research in catalysis. *Catalysis Rev.* **4**(2), 179.

EASTMAN, D. E. and NATHAN, M. I. (1975). *Physics Today,* **28**, (4), 44.

FADLEY, C. S. (1976). Solid state and surface analysis by means of angular dependent X-ray photoelectron spectroscopy. In *Progress in Solid State Chemistry.* (Eds G. A. Somorjai and J. McCaldin). Vol. 11, p. 265. Pergamon Press, Oxford.

HERCULES, D. M. (1976). Electron spectroscopy: X-ray and electron excitation. *Anal. Chem.* **48**, (5), 294.

JOYNER, R. W. (1977). *Surf. Sci.* **63**, 291.

MASON, R. and TEXTOR, M. (1976). In *Specialist periodical reports: surface and defect properties of solids.* (Senior Reporters: M. W. Roberts and J. M. Thomas), vol. 6. Chemical Society, London.

MENZEL, D. (1975). Investigations of adsorption on metal surfaces by photoelectron spectroscopy, combined with other methods. *J. vac. Sci. Technol.* **12**, (1), 1975.

PARK, R. L. (1975). Inner shell spectroscopy. *Physics Today* **28**, (4), 52.

PRUTTON, M. (1975). *Surface Physics.* Clarendon Press, Oxford.

QUINN, C. M. (1973). *An introduction to the quantum chemistry of solids.* Clarendon Press, Oxford.

Rivière, J. C. (1973). *Auger electron spectroscopy*. Contemp. Phys. **14,** (6), 513.

Roberts, M. W. (1977). New perspectives in surface chemistry and catalysis. *Chem. Revs.* **6,** 373. Chemical Society, London.

Simmons, G. W. (1970). Auger electron spectroscopy and inelastic scattering in the study of metal surfaces. *J. Colloid and Interfac. Sci.* **34,** 343.

Somorjai, G. A. (1972). *Principles of surface chemistry*. Prentice Hall, Englewood Cliffs, N.J.

Spicer, W. E., Yu, K. Y., Lindau, I., Pianetta, P. and Collins, D. M. (1976). In *specialist periodical reports: surface and defect properties of solids*. (Senior Reporters: M. W. Roberts and J. M. Thomas), vol. 6. Chemical Society, London.

Thomas, J. M. (1974). The impact of electron spectroscopy and cognate techniques on the study of solid surfaces. *Prog. in surf. and membrane Sci.* **8,** 49.

5

INFRARED, RAMAN, AND PARTICLE IMPACT SPECTROSCOPIES

5.1. Introduction

IN this chapter we consider briefly how infrared spectroscopy, Raman spectroscopy, secondary-ion mass spectrometry, ion scattering, and electron-impact spectroscopy have contributed to the understanding of surfaces and their chemical reactivity. Each experimental technique is illustrated by a few examples of application in surface chemistry.

Our knowledge of molecular structure has relied heavily on infrared data, particularly in the early developments of the subject. The essence of the method is that we can 'see' particular vibrations of bonds within a molecule as a result of the energy they absorb. It turns out that the vibrational energy of molecules lies in the infrared part of the spectrum, i.e. in the wavelength range 10^{-2}–5×10^{-4} cm. The frequencies of the various vibrations are determined by the mechanical motion of the molecule and therefore depend on the force constants of the bonds and the masses of the relevant atoms. On the other hand the intensities of the infrared absorptions are determined by electrical factors such as dipole moments and polarizabilities. One of the important points concerning infrared spectroscopy is that certain structural groups give rise to vibrational bands in the same region of the spectrum irrespective of the complexity of the molecule in which the group is situated. The problem of applying infrared methods to surface studies arises from the difficulty of obtaining sufficient intensity from a monolayer of surface species. In view of the strong infrared absorption of the carbonyl group it is not surprising that the adsorption of carbon monoxide has received most attention in surface chemistry. Raman spectroscopy has been less successful than infrared, the main problem arising from the fact that very intense sources are required. Furthermore these must be as perfectly monochromatic as possible and of such a wavelength that they do not give rise to intense fluorescence. Recent laser sources have made the field of Raman spectroscopy more attractive in that the radiation is both more intense and monochromatic.

A new development, which will not be discussed in detail, is the technique of inelastic electron-tunnelling spectroscopy (IETS), introduced by Jaklevic and Lambe (1966). The specimen is a 'sandwich' of two metal films separated by an insulator layer $\sim$2–3 nm thick, for example Al–Al_2O_3–Pb. When a potential V (0–500 mV) is applied across the sandwich electrons tunnel quantum mechanically through the insulator resulting in a current I. If an impurity molecule (e.g. CH_3COOH) has been introduced at the oxide–metal interface (Al_2O_3/Pb) this can be vibrationally excited by the tunnelling electrons, which thereby lose energy. A peak is then produced in the electron energy distribution $d^2 I/d V^2$ *versus* V, which thus corresponds to a vibrational spectrum of the impurity species. The technique offers advantages in sensitivity and resolution compared with conventional i.r. spectroscopy of adsorbates and has been applied to a wide range of molecules from formic acid (Lewis, Mosesman, and Weinberg 1974) to DNA (Simonsen, Coleman, and Hansma 1974). (For a recent review see Keil, Graham, and Roenker 1976).

5.2. Infrared and Raman spectroscopy

5.2.1. *Transmission infrared*

Infrared spectroscopy has played one of the most significant roles in experimental studies of adsorbed species. This is in part due to its comparatively long-wavelength radiation being only weakly scattered at the surface so that transmission measurements are possible. In order to obtain high sensitivity to adsorbed molecules high-surface-area adsorbents are necessary, and these were available in the form of fine metal powders supported on silica in the pioneering work of Eischens and his colleagues (Eischens and Pliskin 1958, Eischens, Francis, and Pliskin 1956). The sample is frequently used as pressed porous discs and the accessible region of the spectrum is then determined by the absorption of radiation by the adsorbent. Quite clearly the fraction of the incident light transmitted will often be very small, so that in general the use of low- or medium-cost instruments will result in poor band resolution. Given a good instrument, however, the precision and sensitivity of the technique will depend on the sample. Radiation scattering by the solid particles can be responsible for a loss of intensity over the whole of the spectrum, but this scattering decreases as the particle size is reduced below the wavelength of the incident radiation (~700 nm). Scattering is also diminished by decreasing the interparticle distance or void volume, and this explains the extensive utilisation of pressed discs made from metals supported on high-area silica. Fig. 5.1 shows the spectra of carbon monoxide chemisorbed on silica-supported Pt (curve A) and alumina-supported Pt (curve B). The peak at 2070 cm^{-1} (curve A), which is also observed with platinum films, was interpreted as the linear structure Pt—C≡O. The interpretation is based on strong similarities with the carbon–oxygen bands in carbonyls such as $Fe(CO)_5$ and $Ni(CO)_4$. As to the type of bonding present on the alumina-supported platinum, a broad band at about 1820 cm^{-1} is assigned to the bridged structure

while the band at 2050 cm^{-1} is attributed to a shifted linear structure. Again support for the 1820 cm^{-1} designation comes from metal carbonyl chemistry. For example in $Fe_2(CO)_9$ the carbon monoxide bridges the two metal atoms and bands in

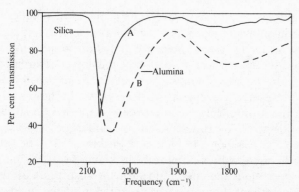

FIG. 5.1. Spectra of carbon monoxide chemisorbed on silica-supported Pt (curve A) and alumina-supported Pt (curve B). (From Eischens 1958.)

the same region as in the adsorption experiments (1800 cm^{-1}) are observed. On the basis of a comparison between the extinction coefficients of bridged and linear carbon monoxide Eischens (1958) estimated that about half of the carbon monoxide adsorbed on alumina-supported platinum (curve B) was in the bridge form. One of the earliest reported studies of hydrocarbon reactions at metal surfaces was that of Little, Sheppard, and Yates (1960). They showed that with palladium supported in porous silica glass ethylene when adsorbed gave evidence of acetylenic surface species which was interpreted as evidence of dissociative chemisorption: Pd—CH=CH—Pd. Also, some bands due to saturated hydrocarbons were detected which indicated either associative adsorption or partial self-hydrogenation from the excess ethylene present. A good review of this and similar early work is to be found elsewhere (Little 1966). It is likely that considerable progress will be made in the application of infrared spectroscopy with the recent availability of Fourier transform spectrometers. In fact recent work by Prentice, Lesiunas, and Sheppard (1976) has shown, using the exceptionally high sensitivity of a Digilab FTS-14 infrared interferometer, the presence of π-bonded species from ethylene chemisorbed on silica-supported Pd and Pt catalysts. The species apparently coexist with σ-bonded M—CH_2CH_2—M species but are more readily removed than the latter by hydrogen. The assignment of bands at $1510-1520 \text{ cm}^{-1}$ to π-complexed ethylene species on palladium is reinforced by the detection of absorption bands in this wavenumber region when palladium atoms are allowed to react with ethylene to give what was thought to be $Pd(C_2H_4)_3$ (Atkins *et al.* 1975).

Infrared studies with evaporated metal films are somewhat more restrictive than with supported metal in view of the smaller surface area involved, and therefore the relatively small number of molecules in the beam path. However, in view of the comparatively large extinction coefficient of CO infrared studies of its chemisorption on metal films have been feasible (Hayward 1971, Ford 1970).

5.2.2. *Reflection infrared*

This approach to the study of infrared spectra of adsorbed molecules is largely the result firstly of the theoretical work of

Greenler (Greenler 1966, 1969, Greenler, Rahan, and Schwartz 1971) on reflection at a metal surface covered by a thin absorbing film and secondly of the experimental fulfilment of Greenler's predictions by Pritchard and Sims (1970). The system investigated was the Cu+CO system. The type of reflection cell used is shown in Fig. 5.2; its main feature was that it contained two glass plates P clipped to a molybdenum hinge M. In the open position shown in Fig. 5.2 copper films could be deposited on the plates by evaporation from three bead sources B. The plates were then drawn together magnetically by the iron slug enclosed in glass (S) until they were parallel and 0·9 mm apart to provide a multi-reflecting light guide with high angles of incidence. The cell was evacuated to pressures in the ultra-high vacuum range ($\sim 1 \times 10^{-9}$ torr).

Monochromatic radiation from a Grubb–Parsons M2 grating monochromator was focused on the entrance of the parallel-plate system and the transmitted radiation monitored by a Golay cell. By placing a Perkin–Elmer gold grid polarizer at the monochromator exit plane polarized radiation was obtained. Spectrum averaging was carried out on a computer after digitally recording the signals on punched tape.

The spectrum of the clean film was first recorded, scanning at least five times for averaging purposes, before introducing CO at $\sim 10^{-2}$ torr and repeating the scan. A resolution of 5 cm^{-1} was

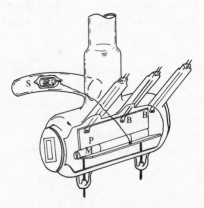

Fig. 5.2. Diagram of reflection cell showing hinged glass plates in the open position. (From Pritchard and Sims 1970.)

possible and the noise level was ±0·1 per cent of the transmitted intensity.

5.2.3. *Theoretical background to reflection method*

The analysis by Greenler (Greenler 1966, 1969, Greenler *et al.* 1971) indicated that the absorption per reflection for the parallel component of the incident radiation should be very sensitive to the angle of incidence with a sharp maximum at 88° to the surface normal. Optical constants for the bulk metal and the adsorbed material were assumed. Pritchard emphasizes that, although proper analysis of the reflection differences between the clean and gas-covered surface is complex, the general form of the variation of the absorption factor with angle can be understood qualitatively in terms of (*a*) the amplitude of the electric field of the radiation at the surface, (*b*) the orientation of the electric field with respect to adsorbed dipoles, and (*c*) the proportionality of the amount of adsorbed material in the radiation to the secant of the angle of incidence.

The experimental evidence for Greenler's theory is convincing in that Pritchard and Sims (1970) showed that a maximum in absorption occurs close to 88° which is the calculated value. Another aspect of Greenler's theory is that there should exist an optimum number of reflections for a maximum signal-to-noise ratio for a given absorption band. Tompkins and Greenler (1971) have recently investigated this experimentally by optimizing the spacing between the samples. By changing the spacing both the number of reflections and the amount of energy traversing the sample will change. They demonstrated that the optimum number of reflections is close to unity; too many reflections result in the absorption band being lost in the noise. It is concluded that it is imperative to have the experimental facility to vary the spacing and hence the number of reflections with an adsorbent not previously examined by the reflection–absorption technique. Recently Greenler (1975) has reported calculations relating to the design requirements for the study of 19 different metals by the reflection–absorption method.

5.2.4. *Infrared emission spectroscopy*

This topic has received much less attention than infrared absorption spectroscopy and it is often regarded as a 'complicating side issue' in high-temperature infrared absorption studies. Just as an absorbance measurement as a function of wavelength describes the absorption spectrum, the emittance as a function of wavelength describes the same spectrum in the emission mode. The emittance measurement will depend on the ratio of the emission radiation at any wavelength to that emitted by a perfect black body at the same temperature and at the same wavelength. Since the black-body distribution is complex and temperature dependent this results in a complex dependence of the energy emitted on wavelength.

Dewing (1970) has described an infrared emission spectrometer based on a gold filament on which a thin layer of the substrate under investigation is spread. The gold, which is highly reflective in the infrared and therefore is also a poor emitter, acted as a heater for the sample. The sample was coated on to the filament from a dispersion in water, the water being removed by heating the filament. Dewing has drawn attention to possible advantages of emission spectrometry: (a) the presence of gas-phase molecules will superimpose an absorption spectrum on the emission which may be more easily recognized than a combination of two absorption spectra in conventional absorption spectroscopy of surface species; (b) the absence of scattering effects at the short-wavelength end of the spectrum which is frequently a problem in absorption studies; (c) that the technique is applicable to very thin coatings of non-compacted material, whereas absorption spectrometry usually requires pressed disc samples.

Low and Coleman (1966) suggested that the energy limitation imposed by cold, weak emitters could be overcome by interferometry. They explored this possibility using a multiple-scan interference spectrometer based on a Michelson interferometer. The apparatus is shown schematically in Fig. 5.3. Radiation from a source S passes through the window W which constitutes the aperture of the spectrometer and is reflected and transmitted by a semi-transparent film on the beam splitter plate B. The dispersion introduced by B is compensated by plate C. The reflected ray is again reflected at the stationary mirror M_1 to pass through C and B to a bolometer detector D. The transmitted ray is reflected at mirror M_2, returned to the semi-reflecting metal film on B, and is reflected to the detector. If the two rays are in phase on arriving at D there will be enhancement; if they are out of

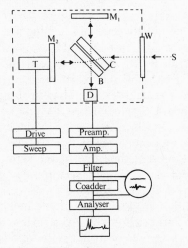

FIG. 5.3. Interference spectrometer.

phase there will be destructive interference. The phasing is adjustable by changing the geometry through the motion of M_2 to give interference fringes. The phasing is dependent on frequency; thus if multiple-component radiation enters the interferometer each component undergoes the interference phenomenon and a signal corresponding to the sum of interferences appears at the detector. The mirror M_2 mounted on a transducer T is fed a sawtooth signal by the sweep and drive electronics which results in it moving linearly forward and returning abruptly to its original position. This motion causes the path lengths of reflected and transmitted rays to vary so that the rays are alternately in and out of phase. This constitutes a single scan and produces a spectrum.

An interesting feature of the technique is that by suitable choice of mirror velocity the fringes can be produced at audiofrequencies. Instrumentation is also provided to increase the signal-to-noise ratio since one scan of a weakly emitting surface is barely detectable from noise. The signal of one scan is digitized and stored, and scanning repeated until a good signal-to-noise ratio is obtained.

5.2.5. *Raman spectroscopy*

Hendra (1970) has described fully the experimental arrangements usually used in recording laser Raman spectra of surface species.

The radiation intensity may be as high as $10^6\,\mathrm{W\,cm^{-2}}$, and if the sample absorbs the radiation serious heating will occur leading to chemical changes at the surface and also possibly in the bulk. With a specimen of polytetrafluoroethylene using the Ar^+ laser system (operating at 488 and 514·5 nm at ~1 W power) the temperature rise was small (<5°). Thus in the case of white or pale surfaces Hendra concluded that heating is unlikely to be a serious problem. On the other hand with a blackened thermocouple in an evacuated cell a temperature of ~573 K was achieved. At the present time one of the serious restrictions to the application of laser Raman spectroscopy to the study of adsorbed molecules is the associated fluorescence; the origin of this fluorescence has aroused considerable debate.

Buechler and Turkevich (1972) have reviewed the application of laser Raman spectroscopy to the study of adsorption on, and the nature of, solid surfaces. As in the early application of infrared spectroscopy in surface chemistry porous Vycor glass has also figured in the development of Raman spectroscopy and again the spectrum of water was of interest. The results of Lippincott *et al.* (1969) showed no bands at $3440\,\mathrm{cm^{-1}}$ characteristic of the OH bond in ordinary water. A strong band at $620\,\mathrm{cm^{-1}}$ was ascribed to a new type of water existing in the pores of Vycor glass which they called 'polywater'. This led to intense activity throughout the world to establish or otherwise the existence of 'polywater'. At the present time the concensus of opinion is that it has not been established that a new form of water had been discovered by Lippincott and his colleagues.

Buechler and Turkevich (1972) give examples of other systems they investigated including propylene adsorption on Vycor glass and also propylene on 'molybdenum-covered' Vycor. Many of the problems inherent in laser Raman studies of surfaces were apparent. They included surface heating generated by the laser beam leading to surface decomposition and also the occurrence of fluorescence. The results were therefore of limited value.

Recently Greenler and Slager (1973) have given attention to the possibility of using their reflection method to study the Raman excitation intensity of thin layers on silver and nickel. Since a beam of light reflected from a metal surface at normal incidence produces an electric standing wave field which has a node at the surface then molecules located at the surface cannot interact with

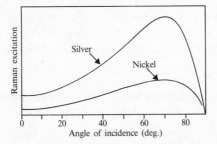

FIG. 5.4. The Raman excitation intensity of thin films on silver and nickel resulting from the standing-wave field of 488 nm laser light. The results are calculated for the laser light, polarized parallel to the plane of incidence, as a function of the angle of incidence.

the field and therefore cannot absorb energy. However, for an appropriate high angle of incidence the incident and reflected electric-field vectors combine to give a resultant standing-wave field with a significant amplitude at the surface largely normal to the surface. Fig. 5.4 shows the dependence of the Raman excitation on the angle of incidence for both Ag and Ni surfaces calculated for the blue (488 nm) line of an argon–krypton laser. At an angle of incidence of about 70° there is a maximum calculated for the Raman excitation. Greenler and Slager (1973) also considered the intensity of the Raman-shifted light at some distance from the surface. They conclude that the Raman-scattered light should be collected at an angle which is about 60° to the surface normal. A schematic diagram of a possible reflection-Raman sample system suggested by Greenler and Slager is shown in Fig. 5.5. From studies of thin

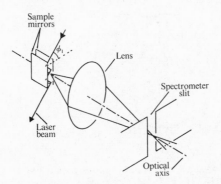

FIG. 5.5. Schematic view of a possible reflection Raman system.

layers of benzoic acid (up to 5 nm) the authors conclude that the technique is sensitive enough for layers a few nanometres in thickness. Whether it can be used for submonolayer quantities remains to be seen.

5.3. Electron-impact excitation of the vibrations of adsorbed species

Even before the wave nature of the electron had been established Farnsworth (1925) noticed that at energies of a few electron volts most electrons are reflected from a metal surface without measurable loss in energy. However, if sufficiently high resolution facilities had been available to Farnsworth he would have observed loss peaks some 0·5 eV lower than the incident electron energy. These losses which are due to the creation of surface vibrational modes are of the order of 0·1 eV and the experimental difficulties involved in their detection are appreciable, particularly with relation to the precision required in the energy of the primary electron beam. At somewhat higher energies the loss structure can be related to transitions between valence and conduction states (Rowe, Ibach, and Froitzheim 1975, Küppers 1973, Ohtani, Terada, and Murata 1974).

It is very much to the credit of Propst and Piper (1967) that they were able to overcome these difficulties; they reported energy-loss spectra for hydrogen adsorbed on W(100) with maxima at 139 and 69 meV. There had previously been a number of

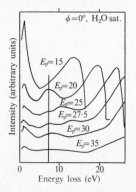

FIG. 5.6. The secondary electron energy distribution of W(100) as a function of energy loss for a H$_2$O-covered surface for various primary energies. In this figure a 7·5 eV loss is clearly seen.

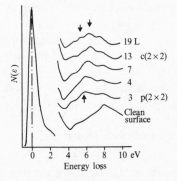

FIG. 5.7. Energy-loss spectra of 100 eV primary electrons reflected from a Ni (100) plane. Energy-loss peaks associated with the chemisorption bond of Ni—O are shown by arrows. LEED patterns corresponding to these spectra are indicated by p(2×2) and c(2×2). The ordinate represents distribution of electron energy.

electron energy-loss spectra for tungsten but none for chemisorbed species. More recently Edwards and Propst (1971) have observed energy-loss spectra for H_2, N_2, H_2O, O_2, and CO-covered tungsten surfaces; Fig. 5.6 shows a loss peak at 7·5 eV for a H_2O-covered tungsten surface. The position and width of this peak at $E_p = 20$ eV is quite similar to a prominent peak seen in the loss spectrum of H_2O in the gas phase measured by Lassettre et al. (1968). Other features of the gas-phase spectrum were, however, not observed. The instrument used was a high-resolution (35 meV), ultra-high-vacuum electron spectrometer incorporating two 127° electrostatic analysers. The tungsten surface was cleaned by heating to ~2000 K for 10 s.

More recently Ohtani et al. (1974) have shown how electron energy-loss spectroscopy can provide information on the electronic energy levels associated with chemisorbed oxygen on the Ni(100) surface. Energy-loss spectra were measured by means of a retarding-field analyser, and Fig. 5.7 shows the loss spectra of 100 eV primary electrons reflected from the (100) surface after various oxygen exposures. Clean surface peaks at 1·8, 4·3, and 8·0 eV and can be assigned to the transition between the bands near the W and K symmetry points, the transition from the L_2' to the upper L_1 state, and surface-plasmon excitation respectively. A loss peak not observed with the clean surface occurred at 5·8 eV and was associated with the p(2×2) oxygen overlayer, i.e.

at formally 0·25 surface coverage. With increasing oxygen exposure (and coverage) this peak splits into two peaks at 5·4 and 6·4 eV; the surface structure at this stage is the c(2×2), i.e. a coverage of 0·5. The peaks are caused by excitation of electrons in chemisorption nickel–oxygen bonds and the splitting shows that a change in the nature of the bond occurs as the oxygen coverage increases.

5.4. Electron-stimulated desorption (ESD) or electron-impact desorption (EID)

In addition to elastic scattering of electrons incident on solid surfaces and the inelastic excitation of adsorbate vibrations discussed above, there are a number of other processes, leading to energy loss, that can occur. For example, there may be both physical and chemical changes in the adsorbed layer, desorption of ionic and neutral atomic and molecular species, perturbation of the bonding of surface species, and polymerization. The term electron-stimulated desorption (ESD) is used in a general way to describe such processes that occur when a solid is bombarded by low-energy electrons (<500 eV). ESD, as well as being of intrinsic interest, has provided useful information on the nature of adsorbed species, particularly on tungsten, and has been reviewed by Madey and Yates (1971a).

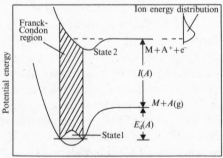

FIG. 5.8. Schematic potential curves for interaction between a surface M and an atom A, and between M and the ion A^+. A possible electronic transition resulting in electron-stimulated desorption of A^+ is indicated by the Franck–Condon region. $E_d(A)$ is the binding energy of A to M and $I(A)$ is the ionization potential of A.

A theoretical model for the electron-induced desorption process was put forward by Redhead (1964) and Menzel and Gomer (1964); the essential features of their model may be discussed in terms of the potential-energy diagram (Fig. 5.8). This shows potential energy as a function of distance from the solid surface for two particular electronic states which may exist between an adatom (or adsorbed molecule) and a metal surface. State 1 is the normal attractive state in which the adatom vibrates at a given mean distance from the surface and state 2 is a repulsive state arising from the adatom being in the ionized form A^+. ESD occurs when an electron incident on the surface with an energy greater than $E_0 = E_d + I$ excites the adatom from state 1 to state 2. Since the ionization potential I is known the activation energy E_d for desorption of neutral A may be determined.

The positive ion formed at the surface by electron bombardment is therefore desorbed from the surface with a range of kinetic energies, the assumption being made that the Franck–Condon principle is applicable, in other words that during an electronic transition in a molecule (an adatom in the present case) the nuclear separation is essentially unchanged. The range of possible internuclear separations in the ground state is represented by the width of the shaded area. Of course not all ions formed by this Franck–Condon-type process will desorb; neutralization of the ion at the surface is possible in view of the almost infinite supply of electrons in the metal capable of tunnelling through the surface to the positive ion. Thus if neutralization, e.g. by an Auger process, occurs, the atom will either be ejected as a neutral with the kinetic energy already gained or be readsorbed. Not only must transitions of the Franck–Condon type from ground to ionic states be considered but also transitions from ground into anti-bonding states are possible resulting in desorption of neutrals.

It is clear that the model implies that the two electrons involved in the reaction finally remain in the gas phase. There is no reason why this should be the case, and if both electrons come to rest at the Fermi level in the metal then a term 2ϕ eV must also be considered where ϕ is the work function of the metal. Until detailed information is available concerning bonding of ground-state adatoms it will not be possible to proceed very far. Nevertheless data concerning cross-sections for desorption, threshold energies for desorption, and ion–energy distributions are already providing important information on the nature of bonding, particularly with relation to the question of the role of different adsorbed states for any one metal–gas system.

5.4.1. Experimental methods in ESD studies

Mass and energy analysis. The most common approach has been to use a quadrupole mass spectrometer of the type used by Sandstrom, Leck, and Donaldson (1968). The source of the mass spectrometer is placed in line of sight of the target so that a fraction of the desorbed ions travel directly into the source; the spectrometer electron beam is of course off. It is important to make sure that no gas-phase ions enter the spectrometer. The quadrupole has some advantages over the magnetic-sector instrument being easier to operate and also in that it is less likely to be selective with regard to the energy spread of the ion flux. Neutrals will be more difficult to detect, while metastable or excited species can be detected with a photoelectron multiplier since they can trigger off an avalanche of secondary electrons. Menzel (1970), using a magnetic-sector analyser investigated neutrals produced in ESD by adjusting the voltages in the target region so that surface ions could not enter the mass spectrometer; neutrals were detected by observing an increase of a particular gas-phase peak on turning on the electron beam to the surface. One of the major problems, however, is that the ion-transmission probabilities, sensitivities, and angular distribution of the ejected ions are usually not known. Therefore absolute determination of the ionic-desorption cross-sections can not be achieved.

Although mass spectrometric energy analysis of the desorbed ions has been reported by Coburn (1968) for CO on W, the non-ideal ion optics introduce broadening effects. It is preferable to couple a mass analyser to a hemispherical analyser as has been done by Madey and Yates (1971b).

5.4.2. Cross-sections of ion desorption

The effective desorption cross-section Q^+ can be measured by observing the rate at which the surface adlayer is depleted by the electron beam. Moore's (1961) data suggested that the ion current produced was directly proportional to the electron bombarding current. Thus

$$i^+ = i^- Q^+ \sigma \qquad (5.1)$$

where σ is the coverage of adsorbed species involved in the desorption process and i^+ is the ion current desorbed from the surface bombarded by an electron current i^-. The ion currents are easily measurable and estimates of σ may be made by a variety of different approaches (e.g. flash desorption, work function, etc.).

For a first-order desorption, which Moore's data indicated,

$$-\frac{d\sigma}{dt} = nQ\sigma \qquad (5.2)$$

where n is the electron flux (electrons $cm^{-2} s^{-1}$) or i^-/Ae, A being the specimen area (cm^2) and e the electronic charge. Therefore, differentiating eqn (5.1) and using eqn (5.2) gives

$$-\frac{di^+}{dt} = nQi^+ = \left(\frac{J}{e}\right)Qi^+$$

where J is the current flux ($A\,cm^{-2}\,s^{-1}$). On integration we have

$$\frac{i^+}{i_0^+} = \exp\left(-\frac{JQ}{e}t\right). \qquad (5.3)$$

Therefore Q may be determined from the time constant of the exponential decay of i^+ with time. It should be noted that Q is in this case the total desorption cross-section, i.e. it includes ions, neutrals, and any other process that results in depopulation of the adsorbed state being investigated.

Total cross-section values Q for electron-stimulated desorptions are $\leqslant 10^{-17}\,cm^2$; for Q^+ the values are usually $\leqslant 10^{-20}\,cm^2$. These contrast with cross-section values of $\sim 10^{-16}\,cm^2$ for electron + gas-phase molecular reactions of the type

$$e + M_2 \rightarrow M^+ + M + 2e.$$

5.4.3. Threshold values

The minimum bombarding electron energy for a reaction of the kind $e + M_2$ (see above) is referred to as a threshold energy. As mentioned earlier it is important to know the fate of both the bombarding and ionized electrons. Some authors have assumed that both electrons have zero kinetic energy after reaction (Redhead 1964); others have assumed that the ionization electron is promoted to the Fermi level of the substrate while the bombarding electron and the ion are released with zero kinetic energy (Menzel and Gomer 1964), while Nishijima and Propst (1970) have discussed the case where both the impinging electron and the ionization electron come to rest at the Fermi level of the substrate when the ion is desorbed with zero kinetic energy. In general the ion kinetic energies observed in ESD show that virtually no zero-kinetic-energy ions are released during the process. In measuring threshold values cognizance must be taken of values being too high owing to lack of instrumental sensitivity. Nishijima and Propst showed that when the gain of the detector was increased by a factor of 10^3 the apparent threshold of O^+ ions from chemisorbed oxygen on W decreased by some 2 or 3 V.

5.5. Ion scattering and secondary-ion mass spectroscopy

By analysing the energies of scattered noble-gas ions the nature of solid surfaces may be investigated. The basis of the technique is a single elastic binary collision of primary ions and surface atoms. If penetration of the ion beam into the subsurface occurs, then it has a much smaller probability of escaping from the solid without undergoing multiple collisions. Ions which have undergone multiple collisions will not conform to the elastic binary collision model on which the surface analysis is based. It is therefore a technique which is 'monolayer sensitive'.

For high- and medium-energy 'particles' (0.5 keV—2 MeV) the back-scattering is of the Rutherford type in which the bare nuclei of ion and target atom interact. This makes the quantitative analysis of the scattered yield reasonably tractable. Since there is conservation of kinetic energy and momentum, it can be shown (Buck and Poate 1974) that

$$\frac{E_1}{E_0} = \frac{M_1^2}{(M_1 + M_2)^2} \left\{ \cos \theta + \left(\frac{M_2^2}{M_1^2} - \sin^2 \theta \right)^{\frac{1}{2}} \right\}^2 \qquad (5.4)$$

where θ is the scattering angle, M_1 and M_2 are the masses of the projectile ions and the target atoms respectively, and E_0 and E_1

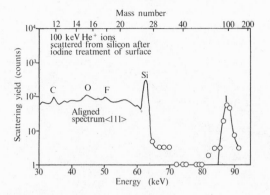

Fig. 5.9. Aligned ⟨111⟩ spectrum for 100 keV ^{4}He$^+$ ions scattered from Si target after iodine–methanol treatment ($\theta = 120°$). (From Buck and Poate 1974.)

the energies of the incident and scattered particles. It is interesting to note that for $\theta = 90°$ eqn (5.4) reduces to

$$\frac{E_1}{E_0} = \frac{M_2 - M_1}{M_2 + M_1}.$$
(5.5)

Thus the energy scale becomes a mass scale for target atoms at the surface, higher energy indicating a larger mass.

High-energy scattering is the most quantitative ion-scattering analytical technique since the physics of the scattering and energy-loss mechanisms are understood. The fundamental basis of the process is Rutherford scattering from which eqn (5.4) follows. Fig. 5.9 is a spectrum of the scattering yield of ^{4}He$^+$ ions from a silicon target after it had been treated with iodine in methanol. Peaks due to carbon, oxygen, fluorine, and iodine are present; the technique is clearly surface sensitive.

Medium energy (0.5—3 keV) noble–gas ions striking a surface cause sputtering of surface atoms, i.e. they transfer their energy to the substrate atoms leading to the emission of secondary ions. When these ions are mass analysed the technique is referred to as secondary-ion mass spectrometry (SIMS). Intuitively we would regard the process as too destructive for surface studies but this is not necessarily the case, as discussed recently by Barber and Vickerman (1976). The essential feature of a SIMS apparatus is a combination of an ion gun for producing the primary ion beam

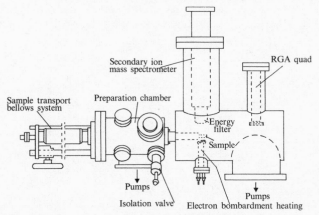

FIG. 5.10. Schematic representation of a SIMS apparatus. (After Barber and Vickerman 1976.)

and a mass spectrometer to monitor the sputtered charged particles. During early work Fogel (1967) used magnetic mass spectrometers, but more recently quadrupole mass spectrometers have been used. It is essential for quantitative and meaningful surface work to use ultra-high-vacuum conditions.

Fig. 5.10 shows the arrangement of the instrument described by Barber and Vickerman. A SIMS spectrum of a nickel surface before and after interaction with carbon monoxide at 77 K is shown in Fig. 5.11. It is significant to note that, although the predominant CO-containing ion is $NiCO^+$, a peak from Ni_2CO^+ is also present. The ratio of these two species is approximately constant between 77 K and 295 K, but at 380 K the $NiCO^+$ was not present while the Ni_2CO^+ remained. It is concluded that there are two distinct surface species and that one does not derive from the other. The authors (Barber, Vickerman, and Wolstenholme 1975) not unnaturally suggest a bidentate mode of CO chemisorption:

In the case of copper after saturation at room temperature Cu_2CO^+ species were observed, while at 390 K and a CO pressure of 10^{-6} torr various carbide species CuC_3^+, CuC_4^+, Cu_2C^+,

$Cu_2C_3^+$, and $Cu_3C_2^+$ were observed together with small peaks which were assigned to $Cu_2CO_2^+$. The authors suggested that the presence of $Cu_2CO_2^+$ indicated that the carbon is formed by the disproportionation of the carbon monoxide followed by the desorption of carbon dioxide.

Clearly the technique of SIMS is at an early stage of development and its potential for chemisorption studies is as yet not fully assessed. Barber and Vickerman (1976) claim that it is

'able to observe species which have not been reported previously in the literature (e.g. a bridged structure for adsorbed carbon monoxide on copper)'.

Adsorption of CO on copper is itself an area of some controversy (Joyner, McKee, and Roberts 1971, Chesters and Pritchard 1971, Tracy 1972) and adsorption at 290 K has been attributed (Tracy 1972) to surface impurities. Recent studies using combined LEED and electron spectroscopy (X-ray and uv) have shown, however, that Cu(100) does adsorb carbon monoxide at 290 K. The adlayer is somewhat disordered but the He spectra indicate clearly the presence of peaks characteristic of molecular carbon monoxide (Isa, Joyner and Roberts, 1977). At the present time the main difficulty with SIMS is to relate 'the spectra' to different surface structures. Benninghoven (1975), Niehus and Bauer (1975) and Werner (1975) have recently also reviewed the

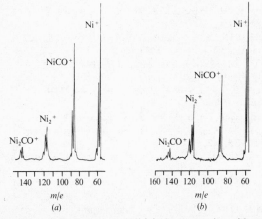

FIG. 5.11. Comparison of SIMS spectra of Ni (a) after saturation with carbon monoxide at 77 K and (b) after warming the saturated surface to room temperature. Note that (b) was run at 10 times greater sensitivity. (From Barber et al. 1975.)

applications and potential of SIMS in surface studies and Morabito (1974) has compared the quantitative aspects of surface analysis by AES and SIMS, while the determination of surface structure by ion scattering has been discussed by Brongersma (1974). An interesting comparison of ion scattering and secondary-ion emission as tools for the study of metal surfaces has been reported recently (Grunder, Heiland, and Taglaner 1974).

REFERENCES

ATKINS, R. M., MACKENZIE, R., TIMMS, P. L., and TURNEY, T. W. (1975). *J. chem. Soc. chem. Commun.* 764.

BARBER, M., and VICKERMAN, J. C. (1976). In *Specialist periodical reports: surface and defect properties of solids* (eds. M. W. Roberts and J. M. Thomas), vol. 5. Chemical Society, London.

BARBER, M., VICKERMAN, J. C., and WOLSTENHOLME, J. (1975). *J. chem. Soc. Faraday Trans. 1* **72**, 40.

BENNINGHOVEN, A. (1975). *Surf. Sci.* **53**, 601.

BRONGERSMA, H. H. (1974). *J. vac. Sci. Technol.* **11**, 231.

BUCK, T. M., and POATE, J. M. (1974). *J. vac. Sci. Technol.* **11** (1), 289.

BUECHLER, E., and TURKEVICH, J. (1972). *J. phys. Chem.* **76**, 2325.

CHESTERS, M. A., and PRITCHARD, J. (1971). *Surf. Sci.* **28**, 460.

COBURN, J. W. (1968). *Surf. Sci.* **11**, 61.

DEWING, J. (1970). In *Chemisorption and catalysis* (ed. P. Hepple). Institute of Petroleum, London.

EDWARDS, D., and PROPST, F. M. (1971). *J. chem. Phys.* **55**, 5175.

EISCHENS, R. P. (1958). Address to American Chemical Society, Division of Petroleum Chemistry.

EISCHENS, R. P., FRANCIS, S. A., and PLISKIN, W. A. (1956). *J. phys. Chem.* **60**, 194.

EISCHENS, R. P., and PLISKIN, W. A. (1958). *Adv. Catalysis* **10**, 1.

FARNSWORTH, H. E. (1925). *Phys. Rev.* **25**, 41.

FOGEL, Ya. M. (1967). *Sov. Phys. Uspekhi* **10**, 17.

FORD, R. R. (1970). *Adv. Catalysis* **21**, 51.

GREENLER, R. G. *J. chem. Phys.* (1966). **44**, 310.

—— *J. chem. Phys.* (1969). **50**, 1963.

—— *J. vac. Sci. Technol.* (1975). **12** (6), 1410.

GREENLER, R. G., RAHAN, R. R., and SCHWARTZ, J. P. (1971). *J. Catalysis* **23**, 42.

GREENLER, R. G., and SLAGER, T. L. (1973). *Spectrochim. Acta* **29A**, 193.

GRUNDER, M., HEILAND, W., and TAGLANER, E. (1974). *Appl. Phys.* **4**, 243.

HAYWARD, D. O. (1971). In *Chemisorption and reactions on metallic films* (ed. J. R. Anderson). Academic Press, New York.

HENDRA, P. J. (1970). In *Chemisorption and catalysis* (ed. P. Hepple). Institute of Petroleum, London.

ISA, S., JOYNER, R. W. and ROBERTS, M. W. (1977). *J. Chem. Soc. Chem. Commun.* 377.

JAKLEVIC, R. C., and LAMBE, J. (1966). *Phys. Rev. Lett.* **17**, 1139.

JOYNER, R. W., McKEE, C. S., and ROBERTS, M. W. (1971). *Surf. Sci.* **26**, 303.

KEIL, R. G., GRAHAM, T. P., and ROENKER, K. P. (1976). *Appl. Spectrosc.* **30**, 1.

KÜPPERS, J. (1973). *Surf. Sci.* **36**, 53.

LASSETTRE, E. N., SKERBLE, A., DILLON, M., and ROSS, K. (1968). *J. chem. Phys.* **48**, 5066.

LEWIS, B. F., MOSESMAN, M., and WEINBERG, W. H. (1974). *Surf. Sci.* **41**, 142.

LIPPINCOTT, E. R., STRONBERG, R. R., GRANT, W. H., and CESSAC, G. L. (1969). *Science* **164**, 1482.

LITTLE, L. H. (1966). *Infrared spectra of adsorbed species.* Academic Press, London, New York.

LITTLE, L. H., SHEPPARD, N., and YATES, D. J. C. (1960). *Proc. R. Soc. A* **259**, 242.

LOW, M. J. D., and COLEMAN, I. (1966). *Spectrochim. Acta* **22**, 369.

MADEY, T. E., and YATES, J. T. (1971*a*). *J. vac. Sci. Technol.* **8** (4), 525.

—— (1971*b*). *J. vac. Sci. Technol.* **8**, 39.

MENZEL, D. (1970). *Angew. Chem.* **9**, 255 (int. edn.).

MENZEL, D., and GOMER, R. (1964). *J. chem. Phys.* **41**, 3311.

MOORE, G. E. (1961). *J. appl. Phys.* **32**, 1241.

MORABITO, J. M. (1974). *Analyt. Chem.* **46**, (2), 189.

NIEHUS, H., and BAUER, E. (1975). *Surf. Sci.* **47**, 222.

NISHIJIMA, M., and PROPST, F. M. (1970). *Phys. Rev. B* **2**, 2368.

OHTANI, S., TERADA, K., and MURATA, Y. (1974). *Phys. Rev. Lett.* **32**, 415.

PRENTICE, J. D., LESIUNAS, A., and SHEPPARD, N. (1976). *J. chem. Soc. chem. Commun.* 76.

PRITCHARD, J., and SIMS, M. L. (1970). *Trans. Faraday Soc.* **66**, 427.

PROPST, F. M., and PIPER, T. C. (1967). *J. vac. Sci. Technol.* **4**, 53.

REDHEAD, P. A. (1964). *Can. J. Phys.* **42**, 886.

ROWE, J. E., IBACH, H., and FROITZHEIM, H. (1975). *Surf. Sci.* **48**, 44.

SANDSTROM, D. R., LECK, J. H., and DONALDSON, E. E. (1968). *J. chem. Phys.* **48**, 5683.

SIMONSEN, M. G., COLEMAN, R. B., and HANSMA, P. K. (1974). *J. chem. Phys.* **61**, 3789.

TOMPKINS, H. G., and GREENLER, R. G. (1971). *Surf. Sci.* **28**, 194.

TRACY, J. C. (1972). *J. chem. Phys.* **56**, 2748.

WERNER, H. W. (1975). *Surf. Sci.* **47**, 301.

EXPERIMENTAL ASPECTS OF
SURFACE KINETICS

6.1. Introduction

In 1953 Becker and Hartman (1953) carried out an experiment which was to initiate one of the main areas of development in surface chemistry over the following 15 years, and the technique still plays an important part in many investigations. They were applying the then recently introduced Bayard–Alpert (BA) ionization gauge to study an aspect of surface reactions about which a large amount of information can be obtained from pressure measurements alone, namely the kinetics of the adsorption and desorption processes. The particular significance of the BA gauge was that it enabled one to measure pressures below $\sim 1 \times 10^{-7}$ torr accurately for the first time, thus making it possible to study adsorption on atomically clean surfaces at coverages as low as a few per cent of one monolayer.

The relative experimental simplicity of pressure measurement lends considerable appeal to kinetic investigations, and they were taken up and rapidly developed by Ehrlich (1961 a,b), Redhead (1961), and other workers to yield extensive and accurate data. This led to the identification of different states of adsorption of a given molecule on a given surface and to detailed models for the adsorption and desorption processes. In general these models involve a number of conceptually simple steps. Initially there will be a collision between a gas-phase molecule and a surface, which may involve an exchange of energy and may result in the molecule being trapped at the surface. Eventually a trapped species will transfer to a strongly bound chemisorbed state or it will be desorbed; in either case its ultimate fate may be preceded by a period of diffusion over the surface.

To provide data upon which the constructional details of this type of model can be based four different experiments must be carried out relating to the initial molecule–surface collision, to surface diffusion, and to the overall kinetics of both adsorption and desorption. In this chapter we shall consider the more practical aspects of these experiments, while the results and their consequences will be discussed in Chapter 8.

6.2. Gas–solid collisions

For the purposes of subsequent surface processes the critical factor in the initial collision of a gas molecule with the surface is the probability of trapping as opposed to reflection back into the gas phase. This in turn will depend on the efficiency of energy transfer from the molecule to the solid which must be controlled by the form of the gas–solid interaction potential. A beam of neutral molecules provides an appropriate means for investigating this potential, while measurement of thermal accommodation coefficients gives an estimate of energy-transfer efficiency.

6.2.1. *Molecular-beam scattering*

This approach is based on the assumption that, if a particle is elastically scattered from a surface, it suffers no change in kinetic energy but only a change in momentum in the direction normal to the surface. Hence it is *specularly reflected,* that is the angle of reflection is equal to the angle of incidence. At the other extreme, if the particle is trapped for a 'finite' time (implying at least several vibration periods within the potential well at the surface) then its re-emission will be diffuse, the direction of emission being given by the cosine distribution law (see Loeb 1961). The nature of the scattering is investigated experimentally by directing a molecular beam onto the surface and scanning over all scattering angles using a movable detector. Low incident energies ($<20\,\text{eV}$) are used to avoid penetration of the solid. Ideally the energy distributions of the particles before and after collision would also be measured, but this is not often done because of experimental complexity. The results are generally presented in the form of plots of trajectory distributions in the plane containing the incident beam and the surface normal, and some examples are shown in Fig. 6.1. For a review of the subject see Weinberg (1975).

6.2.2. *Thermal accommodation coefficient measurements*

The thermal accommodation coefficient α attempts to represent directly the efficiency of energy exchange between a gas and

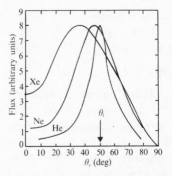

FIG. 6.1. Experimental scattering distributions for various gases scattered from Ag(111): angle of incidence $\theta_i = 50°$; angle of reflection θ_r; surface temperature $T_s = 560$ K; gas temperature $T_g = 300$ K. (From Saltsburg and Smith 1966.)

surface and is defined by the relationship

$$\alpha = \frac{E_r - E_i}{E_s - E_i} \tag{6.1}$$

where E_i and E_r are the total energies of the particles incident on and reflected from the surface and E_s is the total energy of the reflected particles in the limiting case in which the particles come to complete thermal equilibrium with the surface. If attention is restricted to translational energies then E is equal to $\nu(2kT)$ where $2kT$ is the mean translational energy per molecule issued from a body of gas at temperature T and ν is the number of molecules striking the surface per unit area per second. Thus we can rewrite eqn (6.1)

$$\alpha = \frac{T_r - T_i}{T_s - T_i}. \tag{6.2}$$

If there is zero energy exchange between particle and solid, $T_r = T_i$ and $\alpha = 0$; this would correspond to the case of specular reflection of a molecular beam. In the other limit, if energy exchange is complete, $T_r = T_s$ and $\alpha = 1$, corresponding to diffuse beam reflection.

The experimental system consists of a fine wire (radius r_1) mounted axially in a cylindrical tube (radius r_2). The filament is heated electrically to a constant temperature T_s, and the cell is immersed in a thermostat to give a constant wall temperature T_w.

The gas to be studied is introduced at a pressure sufficiently low that free molecular flow (see Dushman 1962) prevails. Gas molecules transport energy from the heated filament to the walls, but because the mean free path is larger than the cell dimensions there is no transfer of energy to other gas molecules on the way. Also, because $r_1 \ll r_2$ a molecule makes many wall collisions for every collision with the filament, and so the gas temperature is effectively $T_g = T_w$. The filament temperature T_s is calculated from its measured resistance, and the power input W_f required to maintain this temperature is also determined. Now

$$W_f = E_r - E_i \tag{6.3}$$

since $(E_r - E_i)$ is just the energy removed from the filament by the gas. Also $E = 2\nu k T$ and so

$$E_s - E_i = 2\nu k (T_s - T_i) = 2\nu k (T_s - T_w). \tag{6.4}$$

However, from the Hertz–Knudsen equation (see eqn 6.15)

$$\nu = P(2\pi m k T_w)^{-\frac{1}{2}} = PZ. \tag{6.5}$$

Thus, from eqns (6.1), and (6.3)–(6.5) we obtain

$$\alpha = \frac{W_f}{2PZk(T_s - T_w)}. \tag{6.6}$$

6.3. Surface diffusion

The diffusion of metal atoms on the surfaces of their own lattices is of considerable interest, since it represents one possible mechanism for bulk rearrangements, such as recrystallization and grain growth, and since it is involved in the nucleation and growth of thin metal films. Surface self-diffusion has therefore been extensively studied (Gjostein 1963) using three types of technique.

(i) Mullins (1957) showed that the surface self-diffusion coefficient D_s could be derived from macroscopic measurement of mass-transfer processes such as thermal grooving and the flattening of corrugated surfaces, and many investigations have adopted this approach.

(ii) Müller (1937) devised the field emission microscope with which metal surfaces can be observed with a resolution of about 2 nm. The instrument basically distinguishes areas of

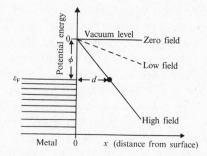

Fig. 6.2. Energy diagram for field electron emission from a metal at 0 K; ε_F represents the Fermi level.

different work function on the surface and so can be used to follow any processes which lead to work-function changes (Gomer 1961; Swanson and Bell 1973).

(iii) Müller (1951) introduced a second novel microscope, the field ion microscope, which is capable of resolution down to about 0·2 nm. With this technique, therefore, individual metal surface atoms, and their movements, can be directly observed (Müller and Tsong 1969; Bowkett and Smith 1970).

Since in the present discussion we are interested in events on the atomic scale we shall consider only the two types of microscope, but it is worth remembering that extensive data has been accumulated using the macroscopic techniques and that it may be of relevance to surface processes (see, for example, the discussion of faceting in Chapter 12).

6.3.1. *Field electron emission*

If we take the simple Sommerfeld picture of a free-electron metal at 0 K (Chapter 1) we find that the valence band is filled to the Fermi level, while all levels above this are empty. The electrons are constrained to remain in the solid by a potential barrier at the surface which may first be taken as a simple step function (Fig. 6.2); the potential-energy zero is set at the vacuum level. Normally electron emission from the solid involves excitation of electrons *over* this barrier, as in thermionic or photoelectric emission, but in the presence of an electric field this need not necessarily be the case. If the field strength is F (V cm^{-1}) and the metal is negative, the potential energy of an electron at a distance x cm from the surface is $-eFx$ (Fig. 6.2). As F is increased the slope of the potential-energy plot outside the surface increases, and if F eventually reaches a value in the range $3-6 \times 10^7$ V cm^{-1} the width d of the barrier in the region of the Fermi level (Fig. 6.2) will fall to $\leqslant 1$ nm. Fowler and Nordheim (1928) predicted that under these conditions quantum-mechanical tunnelling of electrons *through* the barrier should become possible, leading to the process known as field electron emission. As the field is increased further the possibility of tunnelling becomes available to electrons in the valence band below the Fermi level, and the total emission current density J increases. Straightforward quantum-mechanical arguments may be used to calculate J (see e.g. Gomer 1961); an image potential term $-e^2/4x$ is included which modifies the shape of the barrier (Fig. 1.7) but has relatively little effect on the emission. The result is known as the Fowler–Nordheim equation, and is of the form

$$J = BF^2 \exp\left(\frac{-b\phi^{\frac{3}{2}}}{F}\right). \tag{6.7}$$

The terms B and b involve universal constants and slowly varying elliptical functions of the quantity $(e^3 F)^{\frac{1}{2}}/\phi$ which represents the ratio of the Schottky lowering of the barrier (see e.g. Blakemore 1974) to the work function of the surface ϕ. Tabulated values of these functions are available, but for many purposes B and b may be treated as constants independent of F. The experimentally accessible quantities are the total emission current I and the applied voltage V:

$$I = JA \quad \text{and} \quad F = \beta V$$

where A is the emitter area and β a term dependent on emitter geometry. Eqn (6.7) may therefore be rewritten in the form

$$\frac{I}{V^2} = AB\beta^2 \exp\left(\frac{-b\phi^{\frac{3}{2}}}{\beta V}\right). \tag{6.8}$$

This relationship applies strictly at 0 K only, but the temperature dependence of the field emission process is weak and the error involved in using it at finite temperatures is small, about 3 per cent at 300 K and about 50 per cent at 1000 K. Observed current densities are in the range 10^2-10^3 A cm^{-2}.

In order to enhance the field strength at the surface the specimen in a field emission experiment is in the form of a fine needle tip, with a radius of curvature of the order of $r \approx 100$–200 nm. Since

$$F \approx V/5r$$

it is seen that for $F = 3$–6×10^7 V cm^{-1} the applied voltage must be in the range $V \approx 1$–6 kV. In the field emission microscope the tip is situated opposite a conducting fluorescent screen (Fig. 6.3), which is held at a few kilovolts positive with respect to the tip. Emitted electrons travel along the lines of force which diverge almost radially from the tip and cause fluorescence on striking the screen. Because of its small size the tip is generally monocrystalline, and because of its shape different crystal planes will be exposed on its surface. These planes in turn will have different work functions, and from eqn (6.8) it can be seen that the emission from planes of low work function will be high, and *vice versa*. The electrons striking the screen therefore produce an image which is a highly magnified 'work-function map' of the surface in the form of a stereographic projection (cf. Chapter 2). Planes of high ϕ, such as (110) on tungsten, appear as dark patches, while those of low ϕ (e.g. W(111)) appear as bright patches (Plate 3).

Two outstanding features of this very simple instrument are its magnification M and resolution δ. The former is given approximately by the relationship

$$M \approx \frac{\text{tip-screen distance (10 cm)}}{\text{radius of tip } (\sim 10^{-5} \text{ cm})} \approx 10^6. \tag{6.9}$$

In practice, however, this may be reduced slightly owing to the presence of the shank of the tip. Resolution is limited by the statistical distribution of electron momenta transverse to the emission direction to ~2 nm. The exact value depends on the square root of the tip radius but is independent of tip–screen distance and of the applied field F. The field emission microscope thus offers a method for directly viewing individual surface planes of the order of a few nanometres in diameter. Further, any process such as adsorption or surface diffusion which may alter the work function of various surface regions can be followed on this scale by visual observation of changes in the pattern. The

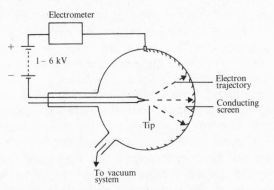

FIG. 6.3. Field emission microscope.

effects of adsorption of CO on rhenium, and of O_2 on rhodium, are shown in Plate 4.

Since the field required for emission is high, the stress on the emitter tip is very large, of the order of 10^4–10^6 N cm^{-2} over linear dimensions of 10^{-5}–10^{-4} cm (Gomer 1961). Owing to the presence of lattice imperfections such stress can be withstood by a metal only if its melting point is above about 1300 K. Thus tips can generally be made from ordinary wires only for the transition metals, but if metal whiskers, which are essentially defect free, can be grown, lower-melting-point metals can also be investigated.

In order to study surface diffusion the field emission tube is equipped with a source for evaporation of the adsorbate which is exposed to one side of the tip only. The tube walls are then cooled to a temperature such that no adsorbate molecules striking them will be reflected, and the tip is cooled to a temperature at which surface diffusion is negligible (this is conveniently achieved by immersing the complete tube in liquid hydrogen or liquid helium). The evaporation source is next activated and an immobile adsorbed layer deposited on one side of the tip. Finally, if the tip temperature is gradually increased the diffusion of adsorbate onto the initially clean regions can be followed from the accompanying changes in the emission pattern. In areas where emission from the clean surface is low, such as the (110) plane of W (Plate 3), it may not be possible to detect changes visually, but diffusion in these regions can still be followed by measuring work-function changes.

Although field electron emission was introduced in this section because of its use in diffusion studies, it is important to emphasize that its applications far exceed the mere visual observation of surface processes. For completeness the two most significant of these additional applications are listed below.

(i) Quantitative determination of work function changes during adsorption, for individual crystal planes on the tip. The tip is mounted on a moveable support so that the image of a particular plane can be positioned over a probe hole in an internal fluorescent screen. By measuring the emission current passing through the hole the variation of work function on the plane can be determined using the Fowler-Nordheim equation (van Oostrom 1966). Absolute work functions can also be obtained, although difficulties are introduced by uncertainties regarding the geometric factor β in eqn (6.8) (see Swanson and Bell 1973).

(ii) Measurement of the total energy distribution of the field emitted electrons. This provides a powerful means of studying the effect of adsorption on electronic structure and for the identification of clean metal surface states (Gadzuk and Plummer 1973; Plummer 1975).

It is also interesting to note that the field emission process need not necessarily occur into a vacuum, but can take place into liquids or films in contact with the emitting surface (Gomer 1972). This can be used to give information on the valence bands in these media, and on electron mobilities and breakdown phenomena in liquids.

6.3.2. *Field ion emission*

The resolution of the field emission microscope is not sufficient to reveal individual surface atoms, but using an instrument of essentially the same construction, but different in its mode of operation, this major goal can be attained. This is of course the field ion microscope (FIM). The metal specimen tip is in this case held at a high positive potential (field strength $F \approx 1\text{--}3 \times 10^8$ V cm^{-1}) with the screen negative, and an 'imaging' gas, usually helium, is introduced at low pressure ($\sim 1\text{--}4 \times 10^{-3}$ torr).

In the presence of the high field gas atoms are polarized and attracted to the tip, so that the impingement rate is much higher than that given by the Hertz–Knudsen equation (eqn (6.15)).

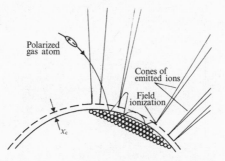

FIG. 6.4. The image-formation process in the field ion microscope. Ionization of the image gas occurs over the high field regions of the tip; x_c is the critical distance for ionization. (Bowkett and Smith 1970.)

Also, the trapping probability at the surface is much greater than normal since, if not accommodated at the first collision, the polarized atom tends to be held in the surface region by the field and makes a series of hops (Fig. 6.4), losing energy to the solid in the process. Eventually the atom is accommodated and then ionized by loss of an electron to the tip. The mechanism involved is illustrated in Fig. 6.5. In the absence of any external field the outermost electron in a free gas atom lies at a potential I below the vacuum level (Fig. 6.5(a)); in the presence of a field the potential-energy diagram changes to that shown in Fig. 6.5(b) so that the electron is now confronted by a barrier of finite width d. When the atom approaches a metal surface (Fig. 6.5(c)) there is an attraction between the electron and the image charge induced in the metal and this further reduces the barrier width. Provided the field is sufficiently high, tunnelling of the electron from the atom into the metal is then possible; the probability p of this event is given (Gomer 1961) by a relationship of the approximate form

$$p = \exp\left\{-0.68\,\frac{I^{\frac{3}{2}}}{F}\left(1 - \frac{7.6F^{\frac{1}{2}}}{I}\right)^{\frac{1}{2}}\right\}. \qquad (6.10)$$

As would be expected from Fig. 6.5 ease of ionization by tunnelling increases as I decreases, at a given applied field. If, however, the atom approaches the surface *closer* than a critical distance $x_c \approx (I - \phi)/F$ cm, ionization becomes very unlikely since the electron level in the atom now lies below the Fermi level in the metal and there are no vacant levels in the conduction band

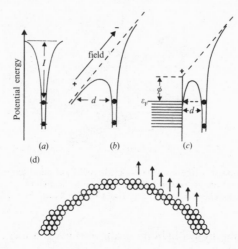

FIG. 6.5. Field ionization of a helium atom at a metal surface: (a) free He atom; (b) atom in high field; (c) atom in high field close to metal surface; (d) schematic cross-section through field emission tip (arrows indicate positions of maximum field strength). (After Gomer 1961.)

into which the electron can tunnel. For the image gases generally used $x_c \approx 0.5$ nm.

When an ion is formed it is repelled from the positively charged tip to the screen where it causes fluorescence. The ionization probability will be greatest at those points in the ionization zone above the tip surface where the field is highest, and to a first approximation these points will be above atoms which 'protrude' from the surface, i.e. which have the lowest number of nearest neighbours. If we consider a ball model of a hemispherical tip (Fig. 6.5(d)) it is seen that such atoms lie at the 'step' edges of low-index planes, or in higher-index planes, and therefore these are the surface features which we expect to show up most clearly in a field ion micrograph. Comparison of the ball model with the micrograph of an (001) oriented platinum tip (Plate 5) confirms this expectation. Micrographs in general correspond approximately to a stereographic projection of the tip structure, but the exact form of the projection is rather complex (Bowkett and Smith 1970).

Ions have a low phosphor efficiency, so the image in the FIM is much weaker than that in the FEM and photography may involve exposure times of up to an hour. This problem can be overcome by using some form of image intensifier, such as a channel-plate converter, which consists of a large number of fine hollow fibres coated with an electron-emitting material. A potential is applied across the plate, and an ion striking one end of a fibre generates an electron which moves down the tube generating further electrons as it collides with the walls. The emerging electrons then fall on a phosphor screen to produce a much intensified image.

When we consider a micrograph such as that in Plate 5 the fundamental question arises as to what exactly the bright spots in the image represent. With this instrument resolution will be limited by the transverse velocity of the ions (an effect equivalent to that limiting resolution in the field emission microscope) and also by the limits within which the field distribution at the critical ionization distance x_c accurately reflects the contours of the surface atoms; overlap of the equipotentials centred on individual atoms will lead to some loss of structure. In addition, it has been recently realized (Tsong and Müller 1970) that all brightly imaged surface atoms are capped by an adsorbed image gas atom which, in the high applied field, is polarized and held at the surface by electrostatic attraction. It is believed that these atoms are fortunately located in registry with the surface atoms and so generally will not affect the basic interpretation of the micrographs, but the possibility of imperfect registry causing image distortions must always be borne in mind (Bassett 1973).

The potential resolution of the FIM is given by (Müller 1965)

$$\delta \approx \left(\frac{6 \times 10^{-4} Tr}{F}\right)^{\frac{1}{2}} \text{nm} \qquad (6.11)$$

where T (K), r (nm) and F (V nm^{-1}) are the tip temperature, tip radius, and field strength respectively. The temperature term appears principally because of the transverse-velocity effect; obviously it is desirable to work at liquid helium temperature (4·2 K), but for economic reasons liquid nitrogen (77 K) is often used instead. While resolution improves with increasing field F this must be balanced against the phenomenon of field evaporation, discussed below, which limits the range of metals which can be used as specimens. Resolution also depends on the ionization potential I of the image gas through a relationship of the type

(Gomer 1961)

$$\delta \propto \frac{I - \phi}{I^{\frac{3}{2}}}.$$ (6.12)

The best resolution is obtained using helium (cf. Table 6.1). In practice the potential resolutions calculated from eqn (6.11) and listed in Table 6.1 are not achieved, but atom spacings down to ~0·224 nm have been observed (Müller and Tsong 1969) and spacings of 0·3–0·4 nm are readily detected; in Plate 6 for example the atoms of W(111) are separated by 0·446 nm.

Although each spot is associated with a surface atom, as pointed out above it is not, however, a direct image of the atom but rather of the potential surface above an image gas atom sitting on the substrate atom. This is to some extent a philosophical point, but at least it still leaves some room for discussion of a remark made by Ostwald in 1900 in relation to the 'billiard-ball' concept of the atom (see Toulmin and Goodfield 1962):

'One does not require, therefore, to give up the advantage of the atomic hypothesis if one bears in mind that it is an illustration of the actual relations in the form of a suitable and easily manipulated picture, but which may, on no account, be substituted for the actual relations. One must always be prepared for the fact that sooner or later the reality will be different from that which the picture leads one to expect.'

In spite of whether or not spots in the image finally confirm the existence of 'atoms' one cannot but be excited by the beauty of micrographs such as those of Graham and Ehrlich (1974) shown in Plate 6 which reveal each individual 'atom' in the central (111)

TABLE 6.1

Data on field-ion microscopy with various gases

Gas	Ionization potential (eV)	Image field (V nm^{-1})	Potential resolution for 50 nm radius tip (from eqn (6.11)) (nm)
He	24·6	45·0	0·12 (20 K)
Ne	21·6	37·0	0·13 (20 K)
H$_2$	15·6	22·8	0·16 (20 K)
A	15·7	23·0	0·32 (80 K)

From Müller 1965.

plane of a tungsten tip and also demonstrate the crystalline perfection of the solid.

Eqn (6.11) implies that the resolution of the FIM will be improved by increasing the applied voltage, but this suffers two limitations.

(a) There is a general blurring of the image as emission becomes significant at an increasing number of points on the surface; for each image gas there will thus be a 'best image voltage' (BIV).

(b) Field evaporation of the specimen will occur. This is the process by which, above some critical value of the applied field, parent metal atoms are ionized and removed from the tip by the action of the field. The field at which it occurs is very critical, at 0 K being in the range 40–50 V nm^{-1} for most refractory metals and in the range 30–40 V nm^{-1} for the other transition metals. For helium the best image voltage is 45 V nm^{-1}, making metals such as tungsten the most obvious choice of tip material. Other metals can, however, be examined by using a different image gas with lower ionization potential I since, if F is the viewing field, $I^{\frac{3}{2}}/F$ is approximately constant. As shown in Table 6.1 there will be some loss of resolution. Neon, hydrogen, and argon are the alternative gases most commonly used.

Tip stability can also be increased in two further ways: (a) by working at a lower temperature, since for a given gas this reduces the minimum field required for ionization (Southon and Brandon 1963), and (b) by using image intensification. Thus for example it is possible to obtain images of metals such as nickel (m.p. 1726 K) using a converter with neon imaging at 63 K (see Bowkett and Smith 1970).

While it limits the choice of tip material, field evaporation can be used to advantage in three ways. Firstly, it enables specimen tips to be cleaned, since at high fields it may be assumed that impurity atoms will be rapidly removed. Micrographs taken after this treatment show no evidence of contamination, although this does not guarantee that small impurity levels do not persist. Secondly, field evaporation allows one to examine the detailed structure of the specimen in depth, since it is possible to remove atom layers one by one from the surface, examining the image between each operation. This controlled evaporation is extremely important for the investigation of defect structures in metals, which is one of major applications of field ion microscopy.

The third application of field evaporation, the atom probe, is, for the surface chemist at least, the most spectacular. Yet again, it was introduced by Müller (Müller, Panitz, and McClane 1968) and involves the combination of a field ion microscope and a time-of-flight mass spectrometer (Müller 1974). The emission tip is mounted on a universal joint and can be moved to locate any chosen spot in the image over a small hole in the screen. A field evaporation pulse is then applied so that the imaged atom is desorbed as an ion and passes through the hole and down the flight tube of the mass spectrometer. The time-of-flight t, i.e. the interval between the application of the evaporation pulse and the arrival of the ion at the spectrometer detector, is given by

$$t = \frac{lm^{\frac{1}{2}}}{(2Vne)^{\frac{1}{2}}} \tag{6.13}$$

where l is the tip-to-detector distance, V is the desorption potential, and m and ne are the mass and charge of the ion. Thus single atoms on the tip can be identified in terms of their mass/charge ratio.

FIM observation of adsorption. Field ion images of surfaces after gas exposure often show new bright spots, and it was at one time thought that these represented adsorbate species (Ehrlich 1966). They are now recognized, however, as being due to a field-induced rearrangement of surface metal atoms in the presence of the adsorbate (Van Oostrom 1970). For example, Brenner and McKinney (1970) used an atom probe to investigate bright spots observed after adsorption of CO and N_2 on tungsten and found that in almost all cases the corresponding desorbed ions were tungsten species, generally single ions, and in some instances clusters of two or three ions. After CO adsorption some spots were associated with carbon atoms.

At present it appears that 'electronegative' adsorbates such as N_2, CO, O_2, and probably many others are not directly visible in FIM images. The exact reasons for this are not entirely clear, but will include the following (Lewis and Gomer 1971).

(i) In the presence of the image gas assisted, or promoted,

field desorption occurs at fields considerably lower than those required in vaccum, possibly through formation of compound ions involving adsorbate, substrate, and image gas atoms. In addition to removing the adsorbate the process creates considerable surface disorder.

(ii) The 'visibility' of an adsorbed gas species, even if it remains on the surface in the presence of the field, may be much lower than that of substrate atoms. Again, the exact cause is not known, but consider an image gas atom in the neighbourhood of an adsorbed atom. For ease of ionization we shall require, among other factors, that the adsorbed atom should have a high density of empty states near the Fermi level and that the local field normal to the surface should be as high as possible (see the discussion of the field ionization process earlier in this section). Neither of these criteria is likely to be fulfilled by an electronegative adsorbate; it may tend, for example, to act as a dielectric, which would not lead to field enhancement. However, in spite, of these problems associated with observation of adsorbed gas molecules, the FIM has produced useful information on adsorption-induced processes such as surface reconstruction (or its absence) (Cranstoun and Pyke 1971) and faceting (Rendulic 1970, Bassett 1968).

The situation regarding adsorption of metals is quite different. Field desorption is more difficult than with adsorbed gases because of the higher heats of adsorption, and visibility is good since there will be vacant states just above the Fermi level; there will also be local field enhancement. A considerable amount of work has therefore been carried out on thin-metal-film growth and on the surface diffusion and interaction of metal adatoms.

6.4. Kinetics of adsorption

When we consider adsorption on the molecular level the kinetic parameter of fundamental interest is the sticking probability S, defined as the probability that on striking the surface a gaseous molecule will become adsorbed for a finite time rather than be directly reflected back into the gas phase. (The term 'finite time' is of course rather ambiguous and it is discussed further in Section 6.5). In its operational definition S is equated with the ratio of the rate of adsorption R_a to the rate of impingement ν of gas-phase molecules with the surface:

$$S(\theta) = \frac{R_a(\theta)}{\nu}. \tag{6.14}$$

The general experimental arrangement (Fig. 6.6) consists of a cell which contains the adsorbent and which is equipped with a gas inlet, an outlet to a pumping system, and a gauge for pressure measurement. Some examples of sticking probability curves are shown in Fig. 6.7; note that in general S is quite strongly dependent on fractional surface coverage θ.

In principle determination of the two factors ν and R_a is straight-forward, but considerable difficulties may be encountered in practice. The techniques involved are discussed in some detail below, both because of these difficulties, and because they are instructive in the application of basic concepts relating to the behaviour of molecules in the gas phase at low pressures, and on surfaces.

6.4.1. *The rate of impingement of gas molecules on a surface*

The rate of impingement ν is related to the pressure P above the surface by the Hertz–Knudsen equation (see Dushman 1962; or Loeb 1961)

$$\nu(\text{molecules cm}^{-2}\ \text{s}^{-1}) = \frac{P}{(2\pi mkT)^{\frac{1}{2}}} = ZP \tag{6.15}$$

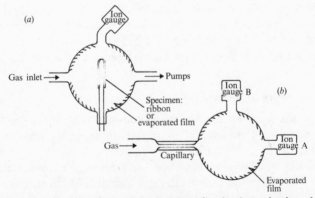

FIG. 6.6. Alternative experimental arrangements for the determination of sticking probabilities.

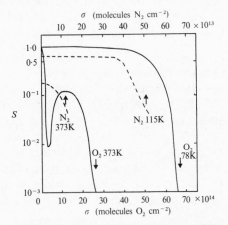

Fig. 6.7. Variation of sticking probability with coverage: solid curves, oxygen on nickel films at 78 K and 373 K (Horgan and King 1969); broken curves, nitrogen on polycrystalline tungsten wire (Ehrlich 1961a.)

where $Z = (2\pi mkT)^{-\frac{1}{2}}$, m is the mass of a single molecule, and k is the Boltzmann constant.

Although determination of ν involves only determination of the pressure P, two types of problem arise, the first with the pressure-sensing devices and the second with the motion of gas molecules within the system (McKee and Roberts 1967, Hayward and Taylor 1967).

A. Pressure-gauge operation. To avoid contamination experiments of any kind which involve submonolayer surface coverages must generally be carried out at pressures below, say, 10^{-6} torr, and so in general ionization gauges or mass spectrometers must be used for pressure measurement. The majority of sticking-probability determinations have involved Bayard–Alpert ionization gauges but these instruments present a number of problems, a fact which often appears to be ignored. (For a general discussion of gauges and associated problems see Redhead, Hobson and Kornelsen 1968). At least an approximate allowance can usually be made for gauge pumping effects, but calibration uncertainties and interactions with adsorbate molecules to produce both new gas phase species and activated parent species are so widespread that they have limited satisfactory kinetic measurements to relatively few gases such as N_2, H_2, O_2, CO, and the inert gases. Even in these cases care must be taken in the interpretation of the gauge readings.

B. *Non-Maxwellian effects.* The Hertz–Knudsen eqn (6.15) is derived on the assumption that the gas behaves in a 'Maxwellian' fashion, i.e. that the molecules are in completely random motion. Since, however, sticking-probability experiments are carried out in the free-molecular-flow regime, molecules move in straight lines until they strike a solid surface and their motion cannot be randomized by gas-phase collisions. This means that it is relatively easy for directional effects to be introduced. (It means also that it is more appropriate to think in terms of the flux of molecules across a surface rather than in terms of the pressure at the surface).

Directional effects are most serious in the case of adsorbents with a high external surface area, such as evaporated films, and when the sticking probability is high. The fact that these effects were overlooked by many of the earlier workers in this field invalidates the results obtained by them for initial sticking probabilities on films. When S has fallen to less than $0.1–0.01$, however, the exact form of the gaseous motion becomes unimportant. Under these conditions molecules make many collisions with surfaces within the system before they are adsorbed and so there is a high probability that departure from Maxwellian behaviour introduced at any point will be removed in subsequent collisions.

The ion-gauge pressure reading P_G, when substituted in the Hertz–Knudsen equation, gives the rate of impact $\nu_{G,out}$ of molecules *within the gauge* on the opening of the tubing connecting the gauge to the reaction cell. In the absence of gauge pumping there will be no *net* flow of molecules from the cell into the gauge, and so the rate of impact $\nu_{G,in}$ of molecules *within the cell* on the opening of the connecting tubing will be equal to $\nu_{G,out}$. So, if P_G substituted in eqn (6.15) is to lead to the correct value for the rate of impact ν_s on the adsorbent surface we require

$$\nu_{G,in} = \nu_s.$$

If this condition is to hold, the opening to the ion gauge, and the high area adsorbent surface, should both form parts of a spherical surface (of area A). If we consider first n molecules reflected from any point on this surface, then provided they are reflected according to the cosine law (i.e. diffusely rather than specularly; see Loeb 1961) all other parts of the surface will receive a uniform flux, n/A (Knudsen 1934). We are now left with the motion of the molecules between their point of entry into the reaction vessel and their first encounter with any surface within it. Again an ideal system should preserve spherical symmetry, the gas emanating from a point source in the centre of a spherical vessel. A good approximation to this arrangement has been used by Hayward, King, and Tompkins (1965) and Hayward, Taylor, and Tompkins (1966) in which the source takes the form of a small, thin-walled glass sphere with a large number of holes in its surface (Fig. 6.8(a)). An alternative system used by McKee and Roberts (1967) involves effusion of molecules from a large gas reservoir through a hole in a very thin platinum sheet set into the spherical wall of the reaction cell (Fig. 6.8(b)). Provided motion in the reservoir is Maxwellian the molecules passing through the hole will emerge with a cosine distribution of velocities and, as with diffuse reflection from the surface of a sphere mentioned above, will provide all points on the reaction-cell surface with an equal flux.

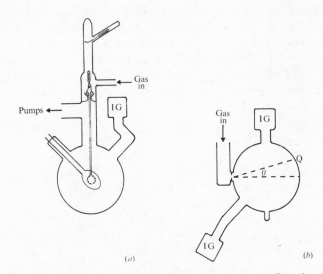

FIG. 6.8. Experimental arrangements which avoid directional effects in sticking probability determinations using evaporated films.

The erroneous initial sticking probabilities obtained by many earlier workers were mainly due to the fact that capillary tubes leading directly into the reaction vessel were used for gas introduction. The incoming molecules were thus collimated, to varying degrees, into a molecular beam (Dayton 1956) and their initial distribution over the adsorbent surface was non-uniform. Further, the pressure measured by the ion gauge was not necessarily related even to some average rate of impingement over the surface; depending on the exact location of the gauge the calculated value of ν could be either unrealistically high (for example gauge A, Fig. 6.6(b)) or unrealistically low (gauge B, Fig. 6.6(b)).

C. Surface roughness. When the impingement rate is related to the pressure it is implicitly assumed that a molecule approaches the surface, collides, and if not adsorbed is reflected back into the main volume of the reaction cell. This will always occur with a plane surface (Fig. 6.9(a)), but if the surface is rough, or porous, as is the case with many evaporated films for example, a molecule reflected on its first impact may make one or more further collisions with the surface before it is finally adsorbed or once again returns to the cell (Fig. 6.9(b)). When this occurs the Hertz–Knudsen equation will give a value for ν which is too low and hence lead to a calculated sticking probability (S_{cal}) which is higher than the probability of adsorption per single collision with the surface (Takaishi 1957, McKee and Roberts 1967).

A molecule striking the surface for the first time will be adsorbed with probability S or reflected with probability $(1-S)$. Let C_1 be the probability that if reflected the molecule will strike

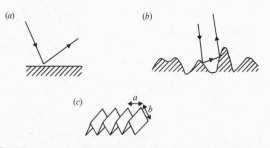

FIG. 6.9. Collision of a gaseous molecule with a surface: (a) plane surface; (b) rough surface; (c) surface of hypothetical geometry.

some other point on the surface. The total probability of a second collision is thus $C_1(1-S)$, of adsorption on this collision $SC_1(1-S)$, and of reflection $SC_1(1-S)^2$. So, the probability of a third collision leading to adsorption will be $SC_1C_2(1-S)^2$ where C_2 is the probability of a surface impact following the second reflection, and so on. The overall probability of adsorption S_{cal} will thus be

$$S_{cal} = S + SC_1(1-S) + SC_1C_2(1-S)^2$$
$$+ \ldots SC_1C_2 \ldots C_j(1-S)^j + \ldots . \quad (6.16)$$

Owing to differences in surface topography from point to point in general $C_i \neq C_j$, but if $\langle c \rangle$ is some average over all C_j then

$$S_{cal} = S \sum_{j=0}^{\infty} \{\langle c \rangle (1-S)\}^j = \frac{S}{1 - \langle c \rangle (1-S)}. \quad (6.17)$$

(This follows since $(1-x)^{-1} = 1 + x + x^2 + \ldots$ provided $x^2 < 1$.) Thus

$$\frac{S_{cal}}{S} = \{1 - \langle c \rangle (1-S)\}^{-1}. \quad (6.18)$$

It is seen that $S_{cal}/S = 1$ when $S = 1$, and increases to a limiting value of $(1 - \langle c \rangle)^{-1}$ as S decreases. Values of $\langle c \rangle$ can be calculated exactly for regular structures such as that shown in Fig. 6.9(c) (Takaishi 1957). For this $\langle c \rangle = (1 - a/2b)$, while the surface roughness factor is $\rho = 2b/a$; ρ is defined by the relationship

$$\rho = \frac{\text{total film surface area}}{\text{area of substrate (assumed to be smooth)}}.$$

A typical value of ρ for evaporated metal films might be 5, and so $\langle c \rangle = 0 \cdot 8$ and

$$\frac{S_{cal}}{S} = \{1 - 0 \cdot 8(1-S)\}^{-1}. \quad (6.19)$$

If the actual sticking probability was $S = 0 \cdot 5$, then $S_{cal} = 0 \cdot 83$, about 66 per cent too large; as S decreases the difference $(S_{cal} - S)$ tends to a maximum of 400 per cent of S. Obviously this effect must be borne in mind when comparing sticking probabilities determined on films with those on bulk samples of the same material.

6.4.2. *The rate of adsorption*

The rate of adsorption can be derived from the variation of any physical quantity which is sensitive to surface coverage, and a wide variety of methods has been used. The degree of accuracy achieved, however, is not high in some cases, and only the more reliable methods will be discussed.

A. *Gravimetry.* The most direct approach is to weigh the adsorbate as a function of time during the experiment. This can be done either by means of a vacuum microbalance (Czanderna 1971) or a quartz-crystal oscillator (Levenson 1967, Bryson, Cazcarra, and Levenson 1974). In the latter case thin, suitably cut quartz crystals are made to resonate at their fundamental resonance frequency (~ 5 MHz) by a radiofrequency oscillator. If their mass is increased by adsorption of n molecules (molar mass M g) then the frequency will change by an amount Δf where

$$n = (1 \cdot 079 \times 10^{16}) M^{-1} \Delta f. \qquad (6.20)$$

The accuracy of the technique depends on the accuracy with which Δf can be determined, and is such that rates of adsorption down to 1×10^{12} molecules cm^{-2} s^{-1} have been measured (for xenon condensing on xenon, Levenson 1967).

B. *Thermal desorption.* If at some time t during an adsorption experiment the adsorbent can be heated to a temperature sufficiently high to cause complete desorption of all adsorbed species, then by measuring the resulting pressure increase ΔP the surface population at time t may be calculated, knowing the system volume and the gas temperature T_g. The *average* rate of adsorption up to time t will be given by

$$\langle R_a \rangle = \left(\frac{V \Delta P}{RT} \right) \frac{1}{t} \text{ molecules } s^{-1} \qquad (6.21)$$

where R is the gas constant. Obviously this method is most readily applied to refractory metals in the form of ribbons or wires since these can generally be 'flashed' electrically to temperatures high enough to ensure desorption of even the most strongly held species. Lower-melting-point bulk metals, oxides, and also metal films, have been used successfully, however, in a number of cases to investigate the desorption of at least the more weakly bound species. Complications may arise if a compound of the metal and adsorbate is formed. In certain cases this may be desorbed rather than the adsorbate alone; for example with the tungsten–oxygen system WO_3 evaporates at ~ 1200 K.

C. *Flow methods.* These are the most widely used since they give a continuous measure of adsorption rate. Gas enters the reaction vessel containing the adsorbent through a leak at a controlled rate L (torr $l\,s^{-1}$) and is continuously removed to the pumps at a rate $S_E P$ (torr $l\,s^{-1}$) where $S_E(l\,s^{-1})$ is the pumping speed of the pumps and P is the pressure in the reaction vessel. The gas input and exit routes may be separate (Fig. 6.6(a)) or they may be combined in the form of a uniform bore capillary tube of known dimensions (Fig. 6.6(b)).

During adsorption, when the sample is taking up gas at a rate R_a (molecules s^{-1}), the rate of pressure rise in the cell will be given by

$$V\frac{dP}{dt} = L - kTR_a + kTR_d - S_E P \qquad (6.22)$$

where R_d (molecules s^{-1}) is the rate of desorption from the sample. Generally VdP/dt is negligible compared with the other terms and this is also the case with R_d, at least at low coverages. (The determination of R_d is discussed in Section 6.4.3.) If there is any significant pumping by the ion gauge or by surfaces in the system other than that of the adsorbent, this will introduce additional terms su·:h as $(-S_G P)$ and $(-S_{wall} P)$ on the right-hand side of eqn (6.22). Since some estimate of their magnitude can usually be obtained from blank experiments, we shall for simplicity omit them from the present discussion. Given the above approximations eqn (6.22) can be rewritten as

$$R_a = \frac{(L - S_E P)}{kT}. \qquad (6.23)$$

When the adsorbent is in the form of a metal ribbon or wire and adsorption measurements are to be combined with investigation of desorption kinetics using the thermal desorption technique (see Section 6.5), the arrangement (a) in Fig. 6.6 is preferred over arrangement (b) since in thermal desorption it is important to be able to vary the pumping speed S_E of the system. Whereas this is not possible with the fixed capillary in system (b) it can be done easily in (a) by using an adjustable valve at the exit from the cell.

Using arrangement (a) the pumping constant for the system S_E/V is determined by introducing a single dose of gas into the reaction vessel with the adsorbent held at a temperature sufficiently high to prevent adsorption. The rate of pressure decrease is then measured and S_E/V calculated using

$$V\frac{dP}{dt} = -S_E P$$

or

$$\ln\left(\frac{P}{P_0}\right) = \left(\frac{S_E}{V}\right)t \tag{6.24}$$

where P_0 is the initial pressure and P the pressure at time t.

A continuous leak is next established, still in the absence of adsorption, and a steady-state pressure P' will be obtained when the input and exit rates balance each other, i.e.

$$\frac{dP'}{dt} = 0 \quad \text{and} \quad L = S_E P'.$$

Hence L may be determined. The gas is stored in a reservoir at a pressure sufficiently high so as to be unaffected by the flow into the reaction system, and thus L remains constant throughout the experiment. Having determined S_E and L, adsorption is started by cooling the sample to the desired temperature, causing an abrupt decrease in pressure from P' to P_0. The subsequent pressure behaviour reflects the variation of S with coverage. Normally the pressure rises continuously (Fig. 6.10) since S is a monotonically decreasing function of coverage but in some cases the reverse may be true (see e.g. Fig. 6.7).

Using eqn (6.23) for the rate of adsorption and eqn (6.15) for the rate of gas impingement gives the following expression for the sticking probability on a surface of area A:

$$S = \frac{(L - S_E P)/kT}{ZAP} = \left(\frac{L}{P} - S_E\right)C \tag{6.25}$$

where L and C $(=1/kTZA)$ are constants for the system. Hence

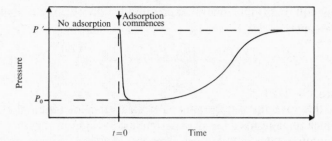

FIG. 6.10. Variation of the pressure in the reaction cell during a sticking probability determination on a metal wire or ribbon.

determination of S reduces to measurement of the pressure in
the system.

With the fixed-capillary arrangement (b) (Fig. 6.6) the rate of
gas input L is equal to the rate of flow through the capillary from
the reservoir. Provided conditions are such that the gas flow is in
the free molecular regime this will be (see e.g. Dushman 1962)

$$L = FP_R \tag{6.26}$$

where P_R (torr) is the reservoir pressure and $F(l\,s^{-1})$ is the
capillary conductance, which may be calculated from the capillary
dimensions (Lund and Berman 1966). Similarly, the rate of
removal of gas from the cell will be the rate of flow out through
the capillary, i.e.

$$S_E P = FP \tag{6.27}$$

where P is the pressure in the cell. Inserting (6.26) and (6.27)
into (6.23) gives for the rate of adsorption

$$R_a = \frac{F(P_R - P)}{kT}. \tag{6.28}$$

Note that when the sticking probability is high the experimental
conditions are always such that

$$P_R \gg P$$

and so R_a then becomes

$$R_a = \frac{FP_R}{kT}. \tag{6.29}$$

Using eqn (6.28) the general expression for the sticking probability becomes

$$S = \frac{(F/kT)(P_R - P)}{ZAP} = FC\left(\frac{P_R}{P} - 1\right) \qquad (6.30)$$

where $C = 1/kTZA$.

In this case the only preliminary measurement required is of F, which remains fixed for all experiments. Sticking-probability determination then entails measurement of two pressures, in the reservoir and in the reaction vessel. This procedure is thus rather more convenient than that involved with arrangement (a) (Fig. 6.6), and has been preferred in measurements on evaporated metal films since quantitative desorption experiments are of only limited practicability with such adsorbents.

With both flow-method arrangements the total surface coverage σ (molecules cm^{-2}) at any time t is obtained by taking the integral of $R_a(\sigma)$ between $t = 0$ and t:

$$\sigma = \frac{1}{A} \int_0^t R_a(\sigma) \, dt. \qquad (6.31)$$

D. *The reflection-detector technique.* This is a modification of the flow technique particularly suited to the accurate determination of high sticking-probability values on evaporated films or on single-crystal surfaces. In the capillary-flow method, when S is high (say $\geqslant 0.1$), it is necessary in practice to operate the system with the reservoir pressure P_R several orders of magnitude greater than the cell pressure P—typical values might be $P_R \sim 10^{-5}$–10^{-4} torr, $P \sim 10^{-9}$–10^{-8} torr. For accurate calculation of S the two ionization gauges measuring P_R and P should be accurately calibrated relative to each other, but since they are operating in very different pressure ranges this is extremely difficult. Further, because of the form of eqn (6.30) any errors in pressure determination are reflected directly in the value of S.

Accuracy is increased in any arrangement in which gas molecules entering the cell must strike the adsorbent surface *before* they can reach the gauge measuring the cell pressure. One possible system for use with films (Hayward and Taylor 1967) incorporates a retractable baffle to shield the gauge opening from the point of gas entry. If P_1 is the pressure measured with the baffle in position, then ZAP_1 will be the rate of collision with the film surface of molecules which have been reflected at least once from the film. The *total* rate of collision will then be $(ZAP_1 + R_a)$ since the rate of entry of gas into the reaction cell will be equal to the rate of adsorption R_a when S is high (cf. eqns (6.26) and (6.29)). Hence in this case

$$S = \frac{R_a}{ZAP_1 + R_a}. \tag{6.32}$$

If S approaches unity ZAP_1 must be small compared with R_a and so errors in measurement of P_1 will introduce a relatively small error into S.

This experimental arrangement also offers an extremely sensitive method for determining whether S is actually equal to 1, or is very slightly less than unity (King 1966). In the former case the pressure P_1 would not rise above the background value on admission of gas to the cell.

The problems associated with determination of high sticking probabilities are further reduced if pressure measurement can be made in the cell by *two* gauges, one which can, and the other which cannot, receive molecules directly from the gas source (Fig. 6.8(b)). If the pressures measured by these gauges are respectively P and P_1 we may combine eqns (6.30) and (6.32), assuming $P_R \gg P$. Elimination of R_a then gives

$$S = 1 - \frac{P_1}{P}. \tag{6.33}$$

In this case the conductance of the gas inlet need not be known (although it is required for calculation of coverage using eqn 6.31). Also, the two gauges concerned now operate in very similar pressure ranges so that calibration problems are greatly reduced. They can in fact be completely eliminated using the system with the retractable baffle by making a measurement of pressure P_1 with the baffle in place and then immediately retracting it to give P.

An elegant variation of this technique has been used by King and Wells (1972) to measure sticking probabilities on single-crystal surfaces. The gas enters the cell in the form of a molecular beam of ~ 0.04 cm^2 cross-sectional area (Fig. 6.11). Two successive experiments are run under identical conditions, in the first of

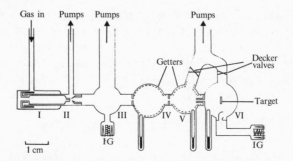

FIG. 6.11. Diagrammatic representation of the molecular-beam apparatus for sticking probability determination, approximately to scale. The molecular beam is generated in sections I to V. (From King and Wells 1972.)

which the incoming beam falls directly on the adsorbent surface; only molecules which have been reflected at least once from the surface can reach the gauge, which therefore measures a pressure equivalent to P_1 (eqn (6.32)). In the second experiment the crystal is turned out of the incoming beam which therefore strikes the cell wall, from which it is diffusely reflected. The gauge and adsorbent experience the same molecular flux in this case and the measured pressure is equivalent to P in eqn (6.30). So, from the pressure–time traces for the two experiments a sticking-probability coverage curve may be constructed using eqns (6.31) and (6.33). The method is only suitable for cases in which $S \geqslant 0.01$ but it is considerably more accurate than other procedures which can be applied to single crystals. Another very sensitive variation of the technique involving the use of a field emission tip as a detector for CO reflected from a tungsten substrate has been developed by Bell and Gomer (1966).

The principle of the 'two-gauge' technique has been employed by Levenson (1967; Bryson, Cazcarra and Levenson 1974) in what is probably the most direct determination of adsorption probability. Two quartz-crystal oscillators, one (A) directly exposed to the gas source and the other (B) capable of receiving only molecules reflected from the target, were used in measurements of the condensation coefficients c of inert gases, and of CO_2, H_2O, and N_2O. The oscillators were held at 50 K, at which temperature all molecules striking them were adsorbed. If df_A/dt and df_B/dt are their respective rates of frequency change, then it may be deduced (cf. eqns (6.20) and (6.33)) that

$$c = 1 - K \, df_B/df_A \qquad (6.34)$$

where K is an experimentally accessible constant related to the reaction cell geometry.

Analysis of the system was simplified and sensitivity was increased by immersing the cell in liquid nitrogen or liquid helium so that any molecules reaching the walls were adsorbed with a probability greater than 0·9999.

E. *Other methods for measurement of the rate of adsorption.* The rate of adsorption may also be determined by direct measurement of the rate of change of surface coverage by XPS or AES. Since these techniques are sensitive to coverages of 1 per cent of a monolayer or less, the accuracy of this method is potentially high, and in addition it should be applicable to a very wide range of adsorbate molecules. It does, however, require absolute calibrations of the spectroscopic peak heights in terms of coverage. At present these are not available theoretically with any degree of certainty, and have been determined accurately experimentally only for systems for which the sticking probability is already known! This then is a promising method for the future.

Various other physical phenomena, such as work-function changes and the development of particular LEED patterns, have been used to estimate surface coverages, but in general these methods are very imprecise and lead to sticking probabilities which at best are average values over a considerable fraction of a monolayer.

6.4.3. *Rate of desorption in relation to sticking probability*

The methods described above measure the net rate of adsorption. If, however, the residence time of a molecule in the adsorbed state is not effectively infinite on the time scale of the adsorption experiment, there will be a finite rate of desorption R_d, and so we must write

$$\text{net (measured) rate of adsorption } R_{net} = (R_a - R_d) \quad (6.35)$$

where R_a is the absolute rate of adsorption, which is the quantity required for sticking-probability calculation. The rate of desorption may be determined at any time by isolating the reaction vessel from both the source of gas and the pumps. Suppose that just before this is done the pressure and sticking probability are P and S. At that instant

$$SZAP = R_{net} + R_d. \tag{6.36}$$

After isolation the pressure will decrease as gas within the cell continues to be taken up by the surface; the rate of adsorption must simultaneously decrease until an equilibrium is attained, at a pressure P_{eq}, when the absolute rate of adsorption $R_{a,eq}$ just balances the rate of desorption $R_{d,eq}$. Then

$$R_{a,eq} = SZAP_{eq} = R_{d,eq}. \tag{6.37}$$

(Note that, since S is identical in eqns (6.36) and (6.37), we have made the assumption that the sticking probability remains unchanged by the uptake of gas during the period of isolation. This is justified since the increase in coverage involved will generally be negligible.) From eqns (6.36) and (6.37) it follows that

$$S = \frac{R_{net}}{ZA(P - P_{eq})}. \tag{6.38}$$

Hence allowance can be made for desorption by periodically isolating the reaction cell and measuring P_{eq}.

6.5. Kinetics of desorption

The principal method for investigation of the kinetics of desorption involves thermal displacement of the adsorbate. The pressure variation in the reaction vessel as a function of time, or more appropriately of temperature, during the process is known as a desorption spectrum. For reviews of the subject see Ehrlich (1963) and King (1975).

If the reaction vessel is sealed the pressure is often found to increase in a number of steps as temperature increases; the classic example of nitrogen desorption from polycrystalline tungsten, following adsorption at 115 K, is shown in Fig. 6.12 (Ehrlich 1961(a)). On the other hand, if the reaction vessel is continuously pumped during desorption the spectrum will consist of a series of separate peaks; the way in which the appearance of a single desorption feature is influenced by the pumping speed is illustrated by some calculated curves in Fig. 6.13 (Redhead 1962).

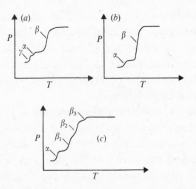

FIG. 6.12. Flash desorption spectra: (a) N_2/W, 115 K; (b) N_2/W, 298 K; (c) CO/W, 298 K. (After Ehrlich 1961(a) and (b).)

The simplest interpretation of such multiple-feature spectra is that there exist on the surface distinct states of adsorption, each with a characteristic activation energy for desorption E_d which is constant for all adsorbed particles in the given state. As the temperature of the surface is increased the most weakly bound adsorbed phase, i.e. the one with the lowest value of E_d, will be desorbed first, followed by the one with the next-lowest E_d, and so on. This reasoning suggests that each separate feature in the spectrum, step or peak, indicates the existence of a distinct adsorption state. We shall first consider an analysis of the experimental data based on this assumption and shall find that it appears to be adequate in many cases. What has not been

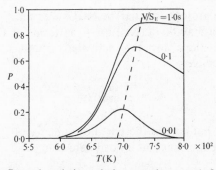

FIG. 6.13. The effect of variation of the pumping speed S_E on a thermal desorption peak; V is the reaction cell volume and P the normalized pressure. (After Redhead 1962.)

considered, however, is the possible influence of adsorbate–adsorbate interactions and later in this section we shall show that in certain cases the operation of such interactions within a *single* adsorbed state can give rise to multiple peaks in the desorption spectrum. Caution must therefore be exercised in assigning a single peak to a single state.

By analysing the spectrum more closely it becomes possible in principle to characterize the adsorbed states quantitatively, the main parameters of interest being the activation energy for desorption E_d and the order of the desorption process. The first of these, E_d, is often estimated from the desorption temperature by applying an equation first proposed by Frenkel which gives an expression for the mean residence time τ of a molecule on the surface.

$$\tau = \tau_0 \exp(E_d/RT) \tag{6.39}$$

where τ_0 is a constant.

The way in which this equation is derived may be illustrated by considering a first-order desorption process. If the surface concentration is σ molecules cm^{-2} and the rate of desorption is $-d\sigma/dt$ molecules $cm^{-2} s^{-1}$ then the average probably of desorption per second is $-(1/\sigma)(d\sigma/dt)$.

The reciprocal of this quantity will be the average residence time τ. For a first-order process

$$-\frac{1}{\sigma}\frac{d\sigma}{dt} = k_1 = A \exp\left(\frac{-E_d}{RT}\right) \tag{6.40}$$

where A is a pre-exponential factor. Using transition-state theory an expression may be derived for k_1 in terms of Q^* and Q^a, the partition functions for the activated complex and adsorbed species respectively:

$$k_1 = \frac{kT}{h}\frac{Q^{\ddagger}}{Q^a}\exp\left(\frac{-E_d}{RT}\right) \tag{6.41}$$

where $Q^{\ddagger}$ is that part of the partition function Q^* not relating to motion along the reaction co-ordinate. Thus

$$\tau = \frac{1}{k_1} = \frac{h}{kT}\frac{Q^a}{Q^{\ddagger}}\exp\left(\frac{E_d}{RT}\right) \tag{6.42}$$

and

$$\tau_0 = \frac{h}{kT} \frac{Q^a}{Q^\ddagger}. \tag{6.43}$$

If we make the assumption (not necessarily valid in all cases) that the adsorbed molecule and activated complex differ only in the nature of the vibration of their bond to the surface, and also possibly in their degree of translational freedom over the surface, then the rotational, and the major part of the vibrational, contributions to Q^a and $Q^\ddagger$ cancel, so that the expression for τ_0 becomes

$$\tau_0 \approx \frac{h}{kT} \frac{Q_t^a \, Q_{v,s}^a}{Q_t^\ddagger} \tag{6.44}$$

where $Q_{v,s}^a$ is the partition function for vibration of the adsorbate–surface bond. If ν_s is the vibration frequency of this bond then

$$Q_{v,s}^a = \left\{ 1 - \exp\left(\frac{-h\nu_s}{kT}\right) \right\}^{-1} = 1 + \exp\left(\frac{-h\nu_s}{kT}\right) + \exp\left(\frac{-2h\nu_s}{kT}\right) + \dots$$

If the bond is strong ν_s will be high so that

$$Q_{v,s}^a \approx 1.$$

If

$$Q_t^a = Q_t^\ddagger$$

this would give

$$\left.\begin{aligned} \tau_0 = h/kT &= 1 \cdot 6 \times 10^{-13} \text{ s at } 300 \text{ K} \\ &= 4 \cdot 8 \times 10^{-14} \text{ s at } 1000 \text{ K} \end{aligned}\right\}. \tag{6.45}$$

In this case, however, the complex, since it is less strongly bound than the adsorbed molecule, may be expected to have the greater degree of translational freedom. Thus $Q_t^\ddagger > Q_t^a$ and τ_0 would be rather less than h/kT. In contrast, if the adsorbate–surface bond is weak we may expect $Q_t^a \approx Q_t^\ddagger$ and also $Q_{v,s}^a \approx h\nu_s/kT$. Then $Q^a/Q^\ddagger \approx h\nu_s/kT$ and $\tau \approx 1/\nu_s$. Vibration frequencies of chemical bonds are typically of the order of 10^{13} s^{-1}.

The cases just considered represent just two from many possibilities, but they indicate that, although τ_0 may vary by several orders of magnitude, as a rough guide a value of 10^{-13} s is not unreasonable.

Fig. 6.14 shows E_d as a function of temperature for various values of τ/τ_0 (eqn (6.39)). The exact significance of the absolute value of τ depends, among other factors, on the experimental conditions in use. If the experiment is such that the adsorbed species are to be observed only after the removal of gas-phase molecules, then obviously desorption must be negligible on the time scale of the experiment, and we would require τ to be greater than say 10^2–10^3 s.

If on the other hand we can conduct the experiment in the presence of gas, lower values of τ may be tolerated. The surface coverage σ will be related to τ by

$$\sigma = S\nu\tau$$

where S is the sticking probability and ν is the rate of impingement of molecules from the gas phase. If we assume for example that Auger spectroscopy enables us to detect 1 per cent of a monolayer of an adsorbate, i.e. $\sim 5 \times 10^{12}$ adsorbed species per

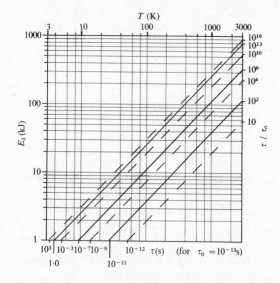

FIG. 6.14. Activation energy for desorption E_d as a function of temperature T, for various values the average surface residence time τ (eqn (6.39)).

cm^2, then we require

$$\tau \gtrsim \frac{5 \times 10^{12}}{S\nu} \text{ s.} \qquad (6.46)$$

Consider a gas of molecular weight 28 at 298 K, for which $\nu = 3 \cdot 85 \times 10^{20} P$ (torr) molecules cm^{-2} s^{-1}. Assume an average value of $0 \cdot 1$ for S; then relationship (6.46) becomes

$$\tau \gtrsim \frac{1 \times 10^{-9}}{P}.$$

So, if $P = 1 \times 10^{-9}$ torr adsorption would be detectable only if $\tau \gtrsim 1$ s, whereas if $P \approx 1 \times 10^{-4}$ torr (an upper limit for AES operation) τ could be as low as 10^{-5} s. The situation would be rather different in the case of a thermal desorption experiment; desorption of 5×10^{12} molecules into a volume of, say, 1 l would give a pressure of $\sim 6 \times 10^{-8}$ torr at 298 K, which could easily be detected above a background pressure of 1×10^{-9} torr, but obviously not against 1×10^{-4} torr.

It is important to remember of course that a significant role in the adsorption process may be played by species which exist on the surface for a time much too small to be directly detectable. This point is further discussed in Chapter 8.

For a more rigorous determination of E_d and the other parameters involved in desorption we must consider a comprehensive expression for the rate of variation of pressure in the reaction vessel (Ehrlich 1961c, Redhead 1962). This will allow for a number of processes which may occur simultaneously, including for example the following.

(a) A controlled leak of molecules into the reaction vessel from the gas source at a rate L (torr l s^{-1}).

(b) Loss of molecules in various ways; to the pumps, by ion-gauge pumping, by adsorption on the walls of the system, etc. These processes may be accounted for in terms of pumping speeds such as S_{pump}, S_{wall}, S_{gauge} (l s^{-1}), etc.

(c) Loss of molecules by re-adsorption on the sample, pumping at a speed S(l s^{-1}).

(d) Desorption from the sample at a rate $-AkT d\sigma/dt$ (torr l s^{-1}), where A (cm^2) is the sample area and σ (molecules cm^{-2}) is the concentration of adsorbed species on the surface.

If P is the pressure and n (molecules cm^{-3}) the concentration of gas-phase molecules in the reaction vessel at any instant, we can then write

$$V\frac{dP}{dt} = kT\frac{dn}{dt} = L - (S_{pump} + S_{wall} + S_{gauge} - S)P - AkT\frac{d\sigma}{dt}. \quad (6.47)$$

To simplify discussion we shall make two assumptions.

(1) Re-adsorption on the sample may be neglected ($S = 0$). This is generally valid except when the rate of removal of gas from the reaction vessel (S_{pump}) is low. Further, if different chemisorbed states exist on the surface it is to be expected that the most strongly held state will fill first and the most weakly held last. In desorption the reverse is true and so when the weakest state has been removed from the surface the higher heat states will still be fully populated and will not offer sites for re-adsorption (Redhead 1962).

(2) We also assume that adsorption by the walls, by gauges, etc. is negligible ($S_{wall} = S_{gauge} = 0$). This may not be true in practice, but these effects can generally be minimized and their contribution estimated with reasonable accuracy.

Before the heating cycle is started the system will be at equilibrium under a pressure P_{eq}. Adsorption on the sample has stopped and the rate of gas entry L is exactly balanced by the rate of removal to the pumps, i.e.

$$L = S_{pump}P_{eq} \text{ (torr } l \, s^{-1}). \quad (6.48)$$

By combining eqns (6.47) and (6.48) and incorporating the simplifying assumptions made above, we obtain for the rate of pressure change during desorption

$$\frac{dP}{dt} = \frac{S_{pump}}{V}P_{eq} - \frac{S_{pump}}{V}P - A\frac{kT}{V}\frac{d\sigma}{dt}. \quad (6.49)$$

Two extreme cases can be distinguished.

(i) If the pumping speed is negligible ($S_{pump} \approx 0$), because the system is deliberately isolated from the pump and/or because the rate of temperature increase is very high, then the desorption rate is measured by the first derivative of the pressure. The pressure–time trace will consist of a number of plateaus, one for each state desorbed (Fig. 6.12). These curves may be analysed directly, or the desorption rates may be determined by electronic differentiation. This condition is the more suitable for use with inert gases.

(ii) If the pumping speed is very high then the desorption rate is proportional to $(P - P_{eq})$:

$$\frac{-d\sigma}{dt} = \frac{S_{pump}(P - P_{eq})}{AkT}. \tag{6.50}$$

The pressure–time curve now consists of a number of peaks, one for each adsorbed state (Fig. 6.15(a)). This condition is more suitable for operation with chemically active gases. For high pumping speeds S_{pump} is generally difficult to determine accurately, so that absolute desorption-peak heights are rather uncertain. They can, however, be compared on a relative basis; peak temperatures and shapes are not affected.

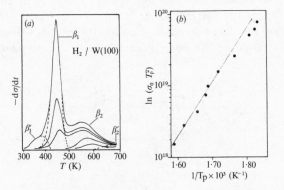

FIG. 6.15. (a) Flash desorption sequence for successively larger amounts of H_2 adsorbed on W(100) at 300 K. The broken curves for β_1 and β_2 states represent computed flash desorption traces; the broken curves β_1' and β_2' represent additional states observed, probably on edges of the crystal, etc. (b) Plot to determine rate parameters for the β_2 state. The coverage varies from 3.9×10^{12} molecules cm^{-2} to saturation, 2.5×10^{14} molecules cm^{-2}. The straight line represents $E_d = 135$ kJ mol^{-1} and $\nu_2 = 4.2 \times 10^{-2}$ cm^2 molecule^{-1} s^{-1}. (After Tamm and Schmidt 1969.)

Experimental conditions intermediate between (i) and (ii) above are of course possible but data are most easily analysed if either (i) or (ii) is satisfied.

Having experimentally determined the desorption rate we may write a rate equation in the usual form:

$$-\frac{d\sigma}{dt} = \nu_n \sigma^n \exp\left(-\frac{E_d}{RT}\right) \tag{6.51}$$

where ν_n is the frequency factor for the desorption reaction of order n. If this expression is to be combined with eqn (6.49), the relationship between time and temperature must be known. Various heating schedules may be experimentally convenient, but analysis of the results is particularly simple when there is a linear variation of either $1/T$ or T with time; we shall restrict this discussion to the latter case, i.e.

$$T = T_0 + \beta t. \tag{6.52}$$

Then

$$\frac{d\sigma}{dt} = \beta \frac{d\sigma}{dT}. \tag{6.53}$$

The parameter which can be measured most accurately from the experimental curves is the temperature T_p at which the desorption rate is a maximum. Combining eqns (6.51) and (6.53) and differentiating the desorption rate with respect to temperature gives

$$-\frac{d}{dT}\left(\frac{d\sigma}{dT}\right) = \frac{d}{dT}\left\{\frac{\nu_n}{\beta}\sigma^n \exp\left(-\frac{E_d}{RT}\right)\right\}$$

$$= \frac{n\nu_n}{\beta}\sigma^{n-1}\exp\left(-\frac{E_d}{RT}\right)\frac{d\sigma}{dT}$$

$$+\frac{\nu_n}{\beta}\sigma^n\left(\frac{E_d}{RT^2}\right)\exp\left(\frac{-E_d}{RT}\right). \tag{6.54}$$

Note that in this procedure we have made the assumptions that the activation energy E_d and the frequency factor ν_n are independent of coverage.

For a maximum in the desorption rate the right-hand side of eqn (6.54) will be zero, and substituting for $d\sigma/dT$ from eqns

(6.51) and (6.53) we can write

$$\frac{E_d}{RT_p^2} = \frac{n\nu_n}{\beta}\sigma_p^{n-1}\exp\left(\frac{-E_d}{RT_p}\right) \tag{6.55}$$

$$= \frac{\nu_1}{\beta}\exp\left(\frac{-E_d}{RT_p}\right) \qquad \text{for} \quad n = 1 \tag{6.55a}$$

$$= 2\sigma_p\frac{\nu_2}{\beta}\exp\left(\frac{-E_d}{RT_p}\right) \qquad \text{for} \quad n = 2. \tag{6.55b}$$

A very important conclusion from these equations is that for a first-order process with constant activation energy the temperature at which the corresponding peak in the desorption-rate curve occurs is independent of coverage (eqn 6.55a). In contrast, for a second-order process with fixed activation energy (or for one which is first order with an activation energy which is coverage dependent), the peak shifts to lower temperatures as the coverage increases; this is readily seen by rearranging eqn (6.55b) as follows:

$$\ln\left(\frac{\beta E_d}{2\nu_2 R\sigma_p}\right) = 2\ln T_p - \frac{E_d}{RT_p}. \tag{6.56}$$

It is also found that in favourable cases, where there is no overlap of desorption peaks arising from different states, the peak shape may be used to distinguish the order of the process. For a first-order desorption the desorption peak is assymetric about its maximum, while in the second-order case the peak is symmetrical.

From eqn (6.55a) it follows that the first-order activation energy can be obtained by assuming a value for ν_1. A first-order desorption process involves simply the breaking of the adsorbate–surface bond during one of its vibrations normal to the surface and as a first approximation the frequency factor may be taken as $\nu_1 \approx 10^{13}\,\text{s}^{-1}$. Any assumption regarding ν_i may be avoided, however, by varying the heating rate (i.e. β). From eqn (6.55a)

$$\beta E_d/\nu_1 RT_p^2 = \exp(-E_d/RT_p). \tag{6.57}$$

Hence

$$\ln(T_p^2/\beta) = (E_d/RT_p) + \ln(E_d/\nu_1 R). \tag{6.58}$$

Thus the value of E_d can be found from the slope of a plot of $\ln(T_p^2/\beta)$ against $(1/T_p)$ and ν_1 can then be determined by substitution in eqn (6.55a). The method is not accurate unless β is varied over one or two orders of magnitude; the validity of the procedure for second order desorption processes has been demonstrated by Lord and Kittelberger (1974).

For a second-order process, since the peak in the desorption-rate–time curve is approximately symmetrical, it follows that $\sigma_p \approx \frac{1}{2}\sigma_0$ where σ_0 is the initial surface concentration of the species considered. This may be obtained by measuring the area under the desorption rate curve i.e.

$$\int_{t=0}^{\infty} \left(\frac{d\sigma}{dt}\right) dt = \sigma_0. \tag{6.59}$$

Eqn (6.55b) may then be written in the form

$$\ln(\sigma_0 T_p^2) = \frac{E_d}{RT_p} + \ln\frac{\nu_2 R}{\beta E_d} \tag{6.60}$$

so that a plot of $\log(\sigma_0 T_p^2)$ against $1/T_p$ gives a straight line of slope E_d/R. If a linear plot is not obtained, and also if the reaction is known not to be first order with constant activation energy, then a coverage dependence of the activation energy, or of ν_2 (see Section 10.3.2) is indicated.

This type of analysis has been applied to multiple-peak desorption spectra in a number of cases. Results for H_2 adsorbed on W(100) (Tamm and Schmidt 1969) at 300 K are shown in Fig. 6.15(b). The β_2 peak for example shifts to lower temperatures as coverage increases, indicating a second-order desorption process. The rate parameters are then obtained using eqn (6.60), as shown in Fig. 6.15(b). The quantities E_d and ν_2 appear to be independent of coverage, at least to a first approximation but it is important to emphasize that the success of the analysis may be illusory. The methods currently used to resolve spectra such as that in Fig. 6.15(a) into their component peaks have been shown (Pisani, Rabino, and Ricca 1974) to be insensitive to the model chosen for the adsorbed layer, i.e. the number of adsorbed states i, the values of the orders n_i, the coverage dependence of ν_{ni} and E_{di} and so on. Nor is this the only complication, as further consideration reveals.

It has been so far implicitly assumed that the thermal desorption spectrum faithfully reflects the equilibrium surface population of the adsorbate—that each peak represents a separate surface species and that the number of molecules desorbed in the peak corresponds to the surface population of that species at the adsorption temperature. There are two ways in which this assumption may be invalidated. In the first place there may be interconversion between different states *on the surface* as the temperature is raised but before the desorption temperature is reached. This situation is discussed in some detail in Chapter 10 for the case of nitrogen adsorption on polycrystalline tungsten.

The second problem, which reaches far beyond the confines of the flash desorption technique, is that our analysis has ignored the possibility of interactions between adsorbed species on the surface. There is considerable evidence, however, both experimental and theoretical (see Chapters 10 and 12) that such interactions are important. If we assume that they do operate then it is possible to show in principle that multiple desorption peaks may arise from a single adsorbed state (Goymour and King 1973; Adams 1974). Some considerable time ago J. K. Roberts investigated the effect of adsorbate–adsorbate interactions on the heat of adsorption (see Roberts and Miller 1939; Miller 1949). Let us consider non-dissociative adsorption; (the dissociative case is discussed by Goymour and King 1973). If the only interaction takes the form of a pairwise repulsion ω between molecules on nearest-neighbour sites, then for coverages less than half a monolayer the minimum energy configuration will have nearest-neighbour sites unoccupied and the differential heat of adsorption H will remain constant at its initial value H_0 up to this coverage. Further adsorption will lead to a stepwise decrease in H to $(H_0 - z\omega)$, where z is the number of nearest-neighbours, owing to the operation of the repulsive interactions (curve A, Fig. 6.16). When the configurational entropy of the system is taken into consideration the picture is modified in that at low coverages some nearest-neighbour sites will in fact be occupied and at high coverages some nearest-neighbour pairs will be vacant. This effect tends to smooth out the $H-\theta$ curve and at high temperature H would decrease linearly with θ (curve C, Fig. 6.16). In cases where adsorption is non-activated the differential heat of adsorption is equal to the activation energy for desorption, at least to a first approximation, and so one should investigate more closely the effect of variation in E_d on the desorption spectrum. This may be done in an approximate way by using the values of H illustrated in Fig. 6.16 for E_d in eqn (6.55) but the results are not accurate. The more rigorous approach uses eqn (6.41) or its second order equivalent and attempts to evaluate the partition functions $Q^{\ddagger}$ and Q_a on the basis of a model including interactions, such as that used to produce Fig. 6.16. In eqn. (6.41) E_d is identified as the activation energy for desorption at zero coverage and is taken as an adjustable parameter the value of which is determined approximately using eqn (6.55a) (or (6.55b)). The nearest-neighbour adsorbate–adsorbate interaction ω, and adsorbate–activated complex interaction ω^*, act as the remaining adjustable parameters in the model.

On this basis one arrives at the very significant result that, if there is a sigmoidal variation in H with θ for a single adsorbed state (i.e. curve B, Fig. 6.16), desorption from this *one state* can give rise to *two* peaks in the desorption spectrum even when the repulsive energy ω is as low as 5 per cent of the zero-coverage desorption energy (Adams 1974); this is illustrated in Fig. 6.17. For $\omega = 0$ a single peak is obtained and this is also the case for initial coverages less than half a monolayer. At higher coverages however a second peak appears; its separation in temperature from the first increases as ω is increased.

When these results are compared with experimental data, for example $H_2/W(100)$ and $CO/W(100)$, the peak temperatures and half-widths can be brought into agreement by parameter adjustment but there are difficulties with relative peak heights. Also it

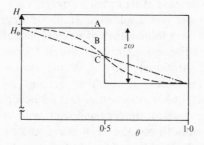

Fig. 6.16. Change in differential heat of adsorption with coverage: (A) for perfect order; (B) for imperfect order; (C) for random order. (From Adams 1974.)

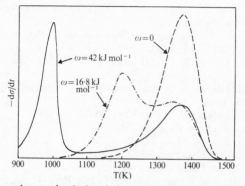

FIG. 6.17. Dependence of calculated desorption spectra on nearest-neighbour interaction energy ω, for first order desorption from a one-dimensional lattice with $E_d = 318$ kJ mol^{-1}. (From Adams 1974.)

is unfortunately found that the calculations are insensitive to the order of the desorption process. The problems may be partly due to the assumptions involved in the derivation of the absolute rate theory expressions but also they indicate that the model used for the surface interactions may not be sufficiently sophisticated.

The above discussion has emphasized the need for caution in interpretation of desorption spectra, both in the case of a single peak when E_d or ν_n, or both, may be coverage dependent and, *a fortiori*, in the case of multiple peak spectra. A proper analysis must rely heavily on other experimental information, especially with regard to surface structure (King 1975).

REFERENCES

ADAMS, D. L. (1974). *Surf. Sci.* **42,** 12.

BASSETT, D. W. (1968). Trans. Faraday Soc. **64,** 489.

—— (1973). In *Surface and defect properties of solids* (ed. M. W. Roberts and J. M. Thomas), vol. 2, p 34. The Chemical Society, London

BECKER. J. A., and HARTMAN, C. D. (1953). *J. phys. Chem.* **57,** 157.

BELL, A. A., and GOMER, R. (1966). *J. chem. Phys.* **44,** 1065.

BLAKEMORE, J. S. (1974). *Solid state physics.* Saunders, Philadelphia.

BOWKETT, K. M., and SMITH, D. A. (1970). *Field-ion microscopy.* North-Holland, Amsterdam.

BRENNER, S. S., and McKINNEY, J. T. (1970). *Surf. Sci.* **23,** 88.

BRYSON, C. E., CAZCARRA, V., and LEVENSON, L. L. (1974). *J. vac. Sci. Technol.* **11,** 411.

CRANSTOUN, G. K. L., and PYKE, D. R. (1971). *Appl. Phys. Lett.* **18,** 341.

CZANDERNA, A. W., (ed.) (1971). *Vacuum microbalance techniques.* Plenum Press, New York.

DAYTON, B. B. (1956). *Trans. 3rd Vacuum Symp.*, p 5. American Vacuum Society, New York.

DUELL, M. J., DAVIS, B. J., and MOSS, R. L. (1966). *Disc. Faraday Soc.* **41**, 43.

DUSHMAN, S. (1962). *Scientific foundations of vacuum technique.* John Wiley, New York.

EHRLICH, G. (1961a). *J. chem. Phys.* **34**, 29.

—— (1961b). *J. chem. Phys.* **34**, 39.

—— (1961c). *J. appl. Phys.* **32**, 4.

—— (1963). *Adv. Catalysis.* **14**, 256.

—— (1966). *Disc. Faraday Soc.* **41**, 7.

FOWLER, R. H., and NORDHEIM, L. (1928). *Proc. R. Soc.* **A119**, 173.

GADZUK, J. W., and PLUMMER, E. W. (1973). *Rev. mod. Phys.* **45**, 487.

GJOSTEIN, N. A. (1963). In *Metal surfaces: structure, energetics and kinetics*, p 99. American Soc. Metals, Ohio.

GOMER, R. (1961). *Field emission and field ionisation.* Harvard University Press, Cambridge, Mass.

—— (1972). *Accts chem. Res.* **5**, 41.

GOYMOUR, C. G., and KING, D. A. (1973). *J. chem. Soc. Faraday I* **69**, 749.

GRAHAM, W. R., and EHRLICH, G. (1974). *Surf. Sci.* **45**, 530.

HAYWARD, D. O., KING, D. A., and TOMPKINS, F. C. (1965). *Chem. Commun.* 178.

HAYWARD, D. O., and TAYLOR, N. (1967). *J. sci. Instrum.* **44**, 327.

HAYWARD, D. O., TAYLOR, N., and TOMPKINS, F. C. (1966). *Disc. Faraday Soc.* **41**, 75.

HORGAN, A. M., and KING, D. A. (1969). In *The structure and chemistry of solid surfaces.* (ed. G. A. Somorjai), paper 57. John Wiley, New York.

KING, D. A. (1966). *Disc. Faraday Soc.* **41**, 63.

—— (1975). *Surf. Sci.* **47**, 384.

KING, D. A., and WELLS, M. G. (1972). *Surf. Sci.* **29**, 454.

KLEIN, R. (1970). *Surf. Sci.* **20**, 1.

KNOR, Z. and MÜLLER, E. W. (1968). *Surf. Sci.* **10**, 21.

KNUDSEN, M. (1934). *Kinetic theory of gases: some modern aspects.* Methuen, London.

LEVENSON, L. L. (1967). *Nuovo Cim. Suppl.* **5**, 321.

LEWIS, R. T., and GOMER, R. (1971). *Surf. Sci.* **26**, 197.

LOEB, L. B. (1961). *The kinetic theory of gases*, 3rd edn. Dover Publications, New York.

LORD, F. M., and KITTELBERGER, J. S. (1974). *Surf. Sci.* **43**, 173.

LUND, L. M., and BERMAN, A. S. (1966). *J. appl. Phys.* **37**, 2489 and 2496.

McKEE, C. S., and ROBERTS, M. W. (1967). *Trans. Faraday Soc.* **63**, 1418.

MILLER, A. R. (1949). *The adsorption of gases on solids.* Cambridge University Press.

MÜLLER, E. W. (1937). *Z. Phys.* **106,** 541.

—— (1951). *Z. Phys.* **131,** 136.

—— (1965). *Science.* **149,** 591.

—— (1974). *Japan. J. appl. Phys. Suppl.* **2,** 1.

MÜLLER, E. W., PANITZ, J. A., and McLANE, S. B. (1968). *Rev. sci. Instrum.* **39,** 83.

MÜLLER, E. W., and TSONG, T. T. (1969). *Field-ion microscopy.* Elsevier, Amsterdam.

MULLINS, W. W. (1957). *J. appl. Phys.* **28,** 335.

PISANI, C., RABINO, G., and RICCA, F. (1974). *Surf. Sci.* **41,** 277.

PLUMMER, E. W. (1975). In *Topics in applied physics.* (ed. R. Gomer). Springer-Verlag, Berlin.

REDHEAD, P. A. (1961). *Trans. Faraday Soc.* **57,** 641.

—— (1962). *Vacuum.* **12,** 203.

REDHEAD, P. A., HOBSON, J. P., and KORNELSEN, E. V. (1968). *The physical basis of ultra-high vacuum.* Chapman-Hall, London.

RENDULIC, K. D. (1970). *Surf. Sci.* **21,** 401.

ROBERTS, J. K., and MILLER, A. R. (1939). *Proc. Camb. Phil. Soc.* **35,** 293.

SALTSBURG, H., and SMITH, J. N. (1966). *J. chem. Phys.* **45,** 2175.

SOUTHON, M. J., and BRANDON, D. G. (1963). *Phil. Mag.* **8,** 579.

SWANSON, L. W., and BELL, A. E. (1973). *Adv. electronics and electron phys.* **32,** 193.

TAKAISHI, T. (1957). *J. phys. Chem.* **61,** 1450.

TAMM, P. W., and SCHMIDT, L. D. (1969). *J. chem. Phys.* **51,** 5352.

TOULMIN, S., and GOODFIELD, J. (1962). *The architecture of matter.* Pelican Books, Harmondsworth.

TSONG, T. T., and MÜLLER, E. W. (1970). *Phys. Rev. Lett.* **25,** 911.

VAN OOSTROM, A. (1966). *Philips. res. Rep. Suppl.* **11,** 102.

—— (1970). *Appl. Phys. Lett.* **17,** 206.

WEINBERG (1975). *Adv. colloid and interface sci.* **4,** 301.

7

PHYSICAL ADSORPTION

7.1. Introduction

As mentioned in Chapter 1 criteria for distinguishing between physical adsorption and chemisorption are pragmatically useful but fundamentally unsatisfactory since any one criterion that holds for a given system breaks down for another. For example, specificity supposedly distinguishes between physical adsorption and chemisorption, the latter being regarded as specific and the former unspecific. However, there is now clear evidence for structural specificity in the adsorption of xenon on various crystal planes of metals with a variation of heat from plane to plane. Plate 7 shows field emission microscope (FEM) results (Ehrlich 1963) for Xe on Mo. Flash filament work by Ehrlich (1963) on Xe adsorption on W also revealed Xe adatoms in at least two states of adsorption at 80 K (Fig. 7.1). Leaving aside the question of the possibility of the adsorbed state being perturbed during thermal desorption the results taken together with the FEM data are convincing evidence for structural specificity. Furthermore, there is a problem in that it is now well established that chemisorbed species have a significant influence on the form of a physical-adsorption isotherm. This can lead to different estimates

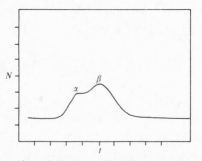

Fig. 7.1. Flash desorption of Xe from W after adsorption at 80 K. Amount evolved proportional to gas density N. Time scale, 1 division $\equiv$ 50 ms; surface concentration, $\sim 6 \times 10^{12}$ molecules cm^{-2}.

TABLE 7.1

System	C (constant in BET eqn)	V_{m_1}/V_{m_2}
Kr–Mo	2300	
Kr–Mo + CO (ads)	1170	1·4
Kr–Ni	1200	
Kr–Ni + C$_2$H$_4$ (ads)	400	1·8
Kr–Fe	420	
Kr–Fe + O$_2$ (ads)	1040	3·0

for the monolayer capacity V_m and therefore estimates of the surface area. Very frequently this is reflected in a variation in the C constant of the BET relationship (Little, Quinn, and Roberts 1964)

$$\frac{P}{V(P_0 - P)} = \frac{1}{V_m C} + \frac{C-1}{V_m C} \frac{P}{P_0} \qquad (7.1)$$

where V is the volume of gas adsorbed at pressure P, P_0 the saturation vapour pressure at the adsorbent temperature and V_m the monolayer value. Since if certain assumptions are made C can be shown to be related to the heat of adsorption by the relationship $C = \exp\{(q_1 - q_L)/RT\}$, where $q_1 - q_L$ is the difference between the heat of adsorption in the first layer and subsequent layers, then such a variation in C may be interpreted as being due to a change in the interaction energy in the first layer. Some examples are given in Table 7.1; V_{m_1}/V_{m_2} is the ratio of the monolayer before and after chemisorption (Little, Quinn, and Roberts 1964).

We shall not be concerned here with the applications of physical adsorption for the characterization of solid surfaces such as the determination of pore radii, pore volume, or surface areas, but will concentrate on the fundamental information now available which throws light on the forces involved and the nature of physical adsorption on metal surfaces. The former topics are covered in detail elsewhere (see general references).

7.2. Forces in physical adsorption

The attractive forces involved in physical adsorption can in a classical sense be divided into three distinct categories.

(a) If the adatom (or admolecule) possesses no permanent dipole then attraction to the solid is due to non-polar dispersion forces.

(b) If an electric field is associated with the solid surface then an induced dipole may exist in the adatom.

(c) If the adatom is polar in nature then there will be additional interactions with the adsorbent.

The potential energy of interaction between two bodies was shown by London to fall off with the sixth power of distance, but at very close distances repulsion dominates. This is reflected in eqn (7.2) which is usually referred to as the Lennard–Jones (6:12) potential, although the author was careful not to attach any significance to the power in the r^{12} term other than it was a convenience in computation (Lennard-Jones 1924):

$$E = \frac{-K_L}{r^6} + \frac{K_r}{r^{12}}. \tag{7.2}$$

The usual approach is to obtain some theoretical estimate of the long-range London constant K_L and to adjust the constant K_r to give a 'correct' value of the separation of the two molecules at the potential minimum or equilibrium distance. The potential-energy–distance curve has the correct qualitative shape but there are distinct quantitative limitations to the relationship.

For any one molecule interacting with a solid surface the total energy of interaction is an infinite series composed of terms such as those in eqn (7.2) with the appropriate values of r. Over large distances the repulsive r^{12} term can be neglected and we have

$$E = \frac{\pi N_0}{6} K_L r^{-3}$$

where N_0 is the number of atoms per cm^3 in the solid and K_L the London constant (eqn (7.2)).

To take this approach further it is necessary to consider the various possibilities for the London constant K_L. London derived the expression

$$K_L = \tfrac{3}{2}\alpha_1\alpha_2 \frac{I_1 I_2}{I_1 + I_2} \tag{7.3}$$

where $I_{1,2}$ and $\alpha_{1,2}$ are the ionization potentials and the polarizabilities of the respective atoms (gas and solid). On the other hand Müller (1936) modified the Slater–Kirkwood expression to give

$$K_L = \frac{6mc^2\alpha_1\alpha_2}{\alpha_1/x_1 + \alpha_2/x_2} \qquad (7.4)$$

where $x_{1,2}$ are the diamagnetic susceptibilities of the atoms and c the velocity of light. Now the constants required in the Kirkwood–Müller expression (eqn (7.4)) are available in tabular form so that predictions concerning interaction energies are possible.

Pierrotti and Halsey (1959) have compared interaction energies determined from krypton adsorption data on evaporated films of iron, copper, sodium, and tungsten with those predicted by several dispersion-force theories. The most satisfactory relationship was found to be the Kirkwood–Müller equation; it yields semi-quantitative agreement with experiment and it is possible to predict the general form of isotherms of the rare gases using the Kirkwood–Müller energies in an isotherm equation developed by Singleton and Halsey. However, success in ascribing heats of adsorption to dispersion energies alone may have possibly involved a somewhat wide latitude in selection of values for the variables that make up the dispersion constant or the parameters of the force law. Nearly 20 years ago Graham (1960) drew attention to the fact that dispersion energies failed to reproduce heats of nitrogen adsorption on carbon, and he suggested 'the possible importance of other forms of electronic interaction' (see also Barrer 1966).

There are also problems with the Lennard-Jones formulation which neglects the finite response of the metal electrons to charge fluctuations; this substantially decreases the interaction energy. Mavroyannis (1963) derived the following expression for the adsorption energy U of an isolated atom on a structureless metal surface:

$$U = \frac{\alpha^{\frac{1}{2}}n_e^{\frac{1}{2}}}{8r^3} \frac{h(\omega_p/\sqrt{2})}{(n_e/\alpha)^{\frac{1}{2}} + h(\omega_p/\sqrt{2})} \qquad (7.5)$$

where α is the polarizability of the adatom, n_e the number of its electrons, r the effective distance from the surface, and ω_p the plasmon frequency of the adsorbent. Engel and Gomer (1970) report that this equation gives reasonable order of magnitude values for the adsorption energy. For r these authors used the atomic radius plus a screening length of $0 \cdot 07$ nm to give values of U of 13, $18 \cdot 8$, and $23 \cdot 8$ kJ for Ar, Kr, and Xe on W. We cannot, however, dismiss on this basis alone the significance of dispersion forces in adsorption.

In all the theoretical approaches the surface has been idealized as a mathematical plane and the detailed distribution of electrons and ion cores ignored for simplicity. Structural specificity of inert-gas interaction with metal surfaces is now well accepted, and in general the strongest binding appears to occur over regions that are on the atomic scale rough. Ehrlich argued that for metals the relative magnitudes of binding energy as a function of substrate structure should be given by carrying out appropriate sums over pairwise interactions. On this basis Ehrlich and Hudda (1959) predicted that heats of adsorption should decrease in the order

$$(116) > (130) \simeq (120) > (100) > (110).$$

The more recent data of Engel and Gomer (1970) do not support this; in particular the heat of adsorption of Ar, Kr, and Xe is greater on W(110) than any of the planes (120), (100), (211), and (111). Thus either we accept that dispersion forces do not play as dominant a role in inert-gas adsorption on metals as previously envisaged or that pairwise summations are invalid because of the non-localized nature of the electron gas. Engel and Gomer (1970) are inclined to. reject the latter and so we are left with considering the role played by dispersion forces in physical adsorption on metals. Relevant to such a discussion are (*a*) information on structural specificity obtained from field emission studies, (*b*) changes in work function on adsorption, (*c*) the influence of pre-adsorbed species on inert-gas adsorption (accommodation, dipole moment, heat of adsorption), and (*d*) structural information from LEED.

We shall consider first the important experimental fact that the inert gases exhibit surprisingly high surface potentials (~ 1 eV) or dipole moments when adsorbed on transition-metal surfaces. Table 7.2 summarizes some surface-potential data for various inert-gas molecules on metal surfaces; these values have been obtained by capacitor, diode, and photoelectric techniques. A review of some of the earlier work is to be found elsewhere (Culver and Tompkins 1959).

TABLE 7.2
Surface potentials (SP) of inert gases on metals

System	S.P (eV)	Method	Author
Ni + Xe	0·85	Capacitor	Mignolet 1950
W + Xe	1·38	Field emission	Ehrlich and Hudda 1959
Ti + Xe	0·84	Capacitor	Mignolet 1950
K + Xe	0·0	Photoelectric	Van Oirschot and Sachtler 1970
Fe + Xe	0·66	Photoelectric	Bouwman (see Van Oirschot and Sachtler 1970)
W(211) + Kr	0·6	Field emission	Engel and Gomer 1970
W(110) + Kr	2·0	Field emission	Engel and Gomer 1970
Cu(111) + Xe	0·48	Diode	Chesters *et al.* 1973

In addition to dispersion forces electrostatic polarization forces have been invoked; these arise from the external field at the metal surface so that the heat of adsorption q_a is made up of two terms $q_p + q_d$, where q_p is the polarization interaction energy and q_d the dispersion term. Now

$$q_p = \tfrac{1}{2}\alpha F^2$$

and

$$\alpha F = \frac{-\Delta\phi}{2\pi\sigma} = \mu$$

where α is the polarizability of the adatom, F the surface field $\Delta\phi$ the surface potential of the adsorbate (i.e. the negative of the work function change) σ the surface concentration of adatoms, and μ the dipole moment (see Chapter 12). Since the field F is dependent on surface structure, being largest for the most close-packed planes, we have a possible explanation for the observed dipole moments or surface potentials of adsorbed inert gas atoms and their variation with surface structure. There have been a number of attempts to argue a case for or against reconciling the surface potentials of inert gases with the degree of polarization of the adatom by the surface electric field. Gundry and Tompkins (1960) were of the opinion that it was difficult to decide which of these two approaches was more appropriate. Further analysis by Hall (1966) strengthened the polarization viewpoint; his rationale was as follows. The surface potential (SP) of the majority of adsorbed gases conforms roughly to the relationship $SP \sim k\alpha$ where k is a constant for a given metal. Hence k may be obtained from a plot of SP against α for a series of adsorbates. Now if k is related to the field strength then there is a *prima facie* case for correlating SP with field strength. A parameter such as heat of sublimation of the metal ΔH_s is also related to field strength. Fig. 7.2 is a plot of k against ΔH_s of the metal for the adsorbate xenon which establishes that $k \simeq 0\cdot53\Delta H_s \times 10^{21}\,\mathrm{V\,cm^{-3}}$ with ΔH_s in $\mathrm{kJ\,mol^{-1}}$ so that $SP = 0\cdot53\Delta H_s \times 10^{21}\alpha$. Müller (1970) has more recently reconsidered this relationship and believes it to be more accurately represented by $SP = 1\cdot6\Delta H_s \times 10^{21}\,\alpha$; he points out that Hall's relationship is based on data obtained with adsorption systems other than inert-gas atoms (e.g. CO, C_2H_4, C_2H_6). Klemperer and Snaith (1971) have also drawn attention to the relationship existing between surface potential and latent heat of sublimation (in fact they plotted SP *versus* the melting point of the metal). Although they are proponents of the polarization theory they recognise the possible role of surface roughness in determining the SP value in that the higher-melting-point metals will tend to produce surfaces which on the atomic scale will be appreciably rougher than those

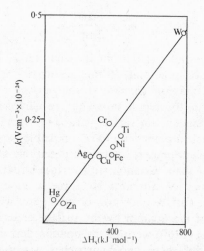

FIG. 7.2. Surface potential correlation constant k as a function of the heat of sublimation ΔH_s of various metals. (From Hall 1966.)

of low-melting-point metals. The surface field would be expected to be enhanced by atomic roughness. Although useful, the approach adopted by Hall, Müller, and Klemperer is essentially macroscopic and does not consider the finer details of the effect of atomic structure on the surface-potential values. This can be studied in detail by field emission and is discussed later. Engel and Gomer (1970), however, are of the opinion that, since the surface field decays within a distance of 0·05 to 0·1 nm, polarization cannot be sufficient to account for the observed dipole moments.

A very different approach is that based on a quantum-mechanical model analogous to the Mulliken no-bond charge-transfer complex. This was first put forward by Mignolet (1953) and later discussed by Matsen, Makrides, and Hackerman (1954), Tuck (1958), Gundry and Tompkins (1960), and more recently by Van Oirschot and Sachtler (1970), and Engel and Gomer (1970). The essential feature in the Mulliken concept is the suggestion that an otherwise endothermic electron transfer $A +$ $M \rightarrow A^+ + M^-$ can be stabilized by resonance between A^+ and the metal. The energy decrease $\Delta \varepsilon$ is given by

$$\Delta \varepsilon = \frac{(H_{01} - S\varepsilon_0)^2}{\varepsilon_1 - \varepsilon_2}$$

where ε_0 is the energy of the system in the neutral state, ε_1 the energy of the $M^- + A^+$ state, S the overlap integral, and H_{01} the Hamiltonian. Following Engel and Gomer $\Delta \varepsilon$ can be written in a simplified manner as

$$\Delta \varepsilon = \frac{S\{I - \phi - e^2/4r - f(\psi)\}^2}{I - \phi - e^2/4r} \tag{7.6}$$

where r is the adatom–surface distance, and $f(\psi)$ incorporates terms representing the first-order perturbation energy of the ion core by the metal potential and the exchange integral for the electron transferred from A to M. The actual value of $\Delta \varepsilon$ depends on small differences between rather large terms and is therefore not easy to evaluate, but it is at least possible to conclude that maximum bonding and maximum charge transfer will be favoured by a high work function ϕ because this minimizes the denominator and maximizes the numerator for a given S (eqn (7.6)). It is relevant to point out that, since S

incorporates the wavefunction at the surface, the condition for achieving resonance between a rare-gas ion core and the surface is the availability of electron levels of comparable energy in the surface. Whether such levels are inherently present in some metal surfaces, absent in others, or are introduced by chemisorption is therefore crucial to probing further the relevance of the charge-transfer-no-bond (CTNB) theory in physical adsorption. The central theme in such discussions has been the interpretation of changes in work function observed during inert-gas adsorption and it is logical to consider this aspect first.

7.3. Work-function studies

Values of the surface potential of a number of inert gases are summarized in Table 7.2. For transition metals the surface potential is close to 1 eV, but for any one adsorbent (e.g. W) the usual trend is for the observed surface potential to decrease in the order $Xe > Kr > Ar > Ne$. Also of significance is that alkali metals exhibit very little change in work function when exposed to Xe or Kr at 77 K; this is not surprising in view of the earlier adsorption studies with Na of Pierrotti and Halsey (1959) who observed little uptake at relative pressures less than 0·9. This fact, these authors concluded, was compatible with interaction energies calculated using the Kirkwood–Müller theory. Van Oirschot and Sachtler (1970) reported that even at the vapour pressure of solid xenon at 77 K no appreciable adsorption took place.

The variation of the observed surface potential (*SP*) with uptake (σ) was suggested by Pritchard (1962) to provide a very satisfactory method for determining the monolayer capacity of metal surfaces. The monolayer point was taken to correspond to where the change in the slope of the $SP-\sigma$ plot occurred. This approach has a particular advantage over the more conventional adsorption method in that it does not require particularly precise measurement of pressure or an inconveniently small-volume apparatus. Both these are requirements essential for determining the monolayer value of solids of very small surface area. The obvious disadvantage is that we are assuming that a sudden change in the $SP-\sigma$ plot corresponds to the completion of the monolayer.

In order to discuss some of the general points raised above we shall consider two recent studies, the first by Engel and Gomer

(1970) and the second by Nieuwenhuys and Sachtler (1974). Both groups of workers have made use of the field emission probe-hole technique which enables adsorption to be investigated on a number of different crystal planes under identical circumstances. Nieuwenhuys, van Aardenne, and Sachtler (1974) also considered data obtained with metal films.

7.3.1. *Adsorption of inert gases on single crystal planes*

Engel and Gomer (1970), using a tungsten field emitter, have determined changes in work function as a function of coverage and adsorption energies (or to be more precise energies relative to that observed with the W(110) plane) under both non-equilibrium and equilibrium conditions. The authors have assumed that for any particular plane $\theta = 1$ corresponds to the point for each gas where the '$\Delta\phi$–dose added' curve levels off; this may not always be justified. We should recall that, since the temperatures used are usually below 20 K, multilayer formation would be expected with all the gases.

Fig. 7.3 shows $\Delta\phi$–dose plots for Ar, Kr, and Xe for a number of crystal planes and Table 7.3 shows dipole moments and adsorbate charges q at $\theta = 0$ for monolayer densities (atoms cm^{-2}) of $7 \cdot 9 \times 10^{14}$ for Ar, $7 \cdot 2 \times 10^{14}$ for Kr, and $6 \cdot 0 \times 10^{14}$ for Xe. The most striking result is the relatively high value for q with both Kr and Xe on the (110) plane, the most closely packed plane in the b.c.c. system. Fig. 7.3(d) shows the variation of $\Delta\phi$ and also the pre-exponential term A' of the Fowler–Nordheim equation (eqn (7.7); cf. Section 6.3.1) during adsorption on W(211).

$$\ln \frac{I}{V^2} = \ln A' - \frac{6 \cdot 8 \times 10^7 \phi^{\frac{3}{2}}}{\beta V}. \qquad (7.7)$$

The pre-exponential is plotted in the form $B' = \ln\{A'(\theta = 0)/A'(\theta)\}$. Changes in B' are observed for all planes and can be accounted for in terms of the polarization of the adsorbate by the applied field as suggested by Menzel and Gomer (1964):

$$B' = 1 \cdot 28 \times 10^{-15} \alpha \frac{\sigma \phi^{\frac{1}{2}}}{K} \qquad (7.8)$$

where α is the adcomplex polarizability and K an effective dielectric constant given by $(1 + 4\pi\alpha\sigma f(\sigma)/d)$ where $f(\sigma)$ is the fraction of the induced dipole-layer potential active at the dipole site. In fact the second term is close to zero so that K may be taken to be unity. The observed values of B' (Fig. 7.3) can be explained by eqn (7.8) except for the W(110) plane where the variation is much too large. If as Duke and Alferieff (1967) have suggested changes in B' may be due to tunnel-resonance effects then a possible interpretation invoking such a process is feasible. Whether this is the case remains to be seen, but if correct then the $\Delta\phi$ values for W(110) are in doubt.

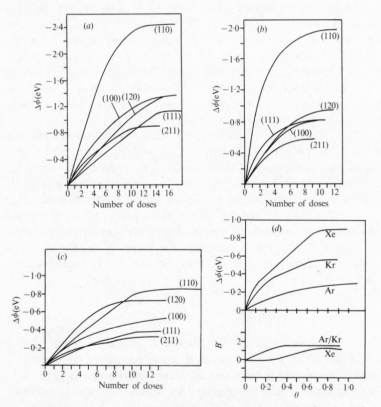

FIG. 7.3. Work-function changes $\Delta\phi$ *versus* number of doses impinged on the (211), (100), (120), (111), and (110) regions of a tungsten field emitter. Dose sizes have been normalized to correct for variations in source-tip normal angle: (*a*) xenon doses at 20 K; (*b*) krypton doses at 15 K; (*c*) argon doses at 4·2 K. (d) Work-function change $\Delta\phi$ and Fowler–Nordheim pre-exponential change B' *versus* θ for the (211) region. (From Engel and Gomer 1970.)

TABLE 7.3

Dipole moments and adsorbate charges at θ = 0 for monolayer densities of inert gases adsorbed on a tungsten field emitter

Plane	Argon		Krypton		Xenon	
	μ (debye)	q/e	μ (debye)	q/e	μ (debye)	q/e
(110)	0·29	0·023	1·93	0·15	1·67	0·12
(120)	0·40	0·031	0·43	0·033	0·60	0·043
(100)	0·31	0·025	0·38	0·029	0·97	0·070
(111)	0·16	0·013	0·50	0·038	0·41	0·029
(211)	0·29	0·023	0·44	0·034	0·81	0·058

From Engel and Gomer 1970.
See text for details.

The influence of pre-adsorbing oxygen was to decrease the interaction energy with Kr and Xe to such an extent that the characteristics of the inert-gas–clean-metal interaction were completely absent. Now oxygen chemisorption increases ϕ so that if charge transfer is the predominant mechanism on the clean surface then the increase in ϕ is more than compensated by the decrease in overlap which results from the presence of chemisorbed oxygen adatoms which physically keep the inert-gas atoms away from the metal surface. On the other hand, the influence of chemisorbed oxygen can also be explained on the basis of dispersion forces, which will be altered by the fact that O(ads) is less polarizable than the metal and also by the increase in distance between the Xe atom and the metal atoms. The role of presorbed oxygen in influencing inert-gas interaction has been known for some 20 years (Table 7.1) and in general it always results in a decrease of the extent of inert-gas adsorption compared with the clean metal. Accompanying the decrease in monolayer adsorption is a decrease in the heat of adsorption as reflected in changes in the C constant in the BET equation. There is, however, some ambiguity in the interpretation of such data since they were usually obtained with evaporated metal films which are inherently thermally unstable. Recent data of Dresser, Madey, and Yates (1974) obtained with W(111) are free of such possible problems, and their results, obtained by a combination of flash desorption and work-function measurements, are very instructive. The energy of Xe adsorption increased from

39 kJ mol^{-1} with the clean (111) surface to between 50 and 54 kJ mol^{-1} for a surface partially covered by oxygen. On the other hand, Engel and Gomer (1970) reported that pre-adsorbed oxygen reduced the heat of xenon adsorption on tungsten. There is need for further studies, and the interesting analogies drawn by Dresser *et al.* (1974) with various oxides of xenon (e.g. XèO$_3$) must remain speculative at this stage.

Nieuwenhuys and Sachtler (1974) and Nieuwenhuys, Bouwman, and Sachtler (1974) have described two separate approaches to the problem of physical adsorption. The first is the field emission probe-hole technique for studying the adsorption of Xe on single-crystal planes of the f.c.c. metals Ir and Pt and the second is adsorption on evaporated films. With regard to the field emission data the highest values of surface potential are, as with tungsten, on the closest-packed planes, (111) and (100) in this case. On the (110) region the value is only about half that on the (111) or (100) regions while on the atomically rough plane (210) the surface potential is between those on the (110) and (111) or (100) regions. Table 7.4 summarizes the results for iridium and includes information on the heat of adsorption which also varies with the crystal face. The initial and maximum heat of adsorption decreases in the order (321); (111) and (100); (531), (731), and (210); (110).

We recall here first the large surface potential of Xe and its high heat of adsorption on W(110). Both are considered to be generally incompatible with dispersion-force interaction. However, we cannot *a priori* rule out dispersion forces and it would be reasonable to suggest that the most likely situation for the participation of a significant dispersion force contribution would be when the adsorbate atoms are in close contact with a large number of substrate atoms. Nieuwenhuys and Sachtler have examined the packing of a Xe adatom in the available sites on different f.c.c. crystal planes.

Table 7.5 (taken from Nieuwenhuys and Sachtler 1974) gives the co-ordination of Xe atoms adsorbed above sites B_x for crystal faces (hkl). The index x refers to the number of lattice atoms with which one adatom with the same diameter as the metal atom is in contact. For Xe, which is 1·6 times larger than an Ir atom, the number of direct-contacting metal atoms, a, is in many cases much smaller than x (see Table 7.5). The surface-metal atoms

TABLE 7.4

Work function (ϕ), maximum surface potential of Xe (SP), change in Fowler–Nordheim pre-exponential ($\Delta \log A$) by Xe adsorption, and heats of adsorption (Q) calculated from the desorption times at temperatures T on different regions of an iridium field emitter.

Tip region	ϕ (eV)	SP (V)	$\Delta \log A'$	T_{min}	T_n	T_{max}	q_{min} (kJ mol^{-1})	q_n (kJ mol^{-1})	q_{max} (kJ mol^{-1})
Total emission	5·00	1·18 ±0·03	−0·42 ±0·05						
(111)	5·79 ±0·05	1·8	−1·9	103	111	118	27·3	29·4	31·5
(100)	5·67 ±0·05	1·6	−1·8	103	109	118	27·3	29	31·5
around									
(110)	5·0	0·8	−0·0	99	103	110	26·5	27·3	29·4
(210)	5·0	1·3	−0·5	105	110	115	28	29·4	30·2
(321)	5·4	1·0	−0·0	110	119	124	29·4	32	32·8
(511)				103	107	115	27·3	28·5	30·2
(531)–(731)	4·9			106	110	115	28	29·4	30·2

TABLE 7.5

The co-ordination of Xe atoms adsorbed on sites above the centres B_n at an f.c.c. crystal plane (hkl)

Crystal face	Site	a	b	c
111	B_3	$3(C_6)$	$3(C_9)$	$3(C_9)$
100	B_4	$4(C_8)$		—
110	B_5	$4(C_7)$		$1(C_{11})$
210	B_6	$3(C_7)$		$1(C_{11})$
	B_3	$3(2C_6+1C_8)$		
321	B_6	$3(2C_6+1C_9)$	$6(2C_6+2C_8+1C_9+1C_{10})$	$2(1C_6+1C_{11})$
	B_4	$3(1C_6+1C_8+1C_9)$	$4(2C_6+1C_8+1C_9)$	$1(C_{10})$
	B_3 type 1	$3(1C_6+1C_8+1C_9)$	$4(1C_6+1C_8+1C_9)$	$3(2C_6+1C_8)$
	B_3 type 2	$3(1C_6+1C_8+1C_9)$	$4(2C_6+1C_8+1C_9)$	$2(1C_8+1C_{10})$
	B_3 type 2	$3(C_7)$	$3(1C_6+1C_8+1C_9)$	$2(1C_8+1C_{10})$
311	B_5	$3(C_7)$	$3(C_7)$	$4(2C_7+2C_{10})$
	B_3	$3(2C_6+1C_9)$		$2(C_7)$
211	B_5	$3(1C_7+2C_9)$	$4(3C_7+1C_{10})$	$2(C_9)$
	B_3 type 1	$3(1C_6+2C_9)$	$5(2C_7+1C_9+2C_{10})$	$3(C_7)$
	B_3 type 2	$3(2C_7+1C_9)$	$4(1C_7+2C_9)$	$3(2C_7+1C_{10})$
	B_3 type 3	$3(2C_6+1C_8)$	$3(2C_7+1C_9)$	$2(C_9)$
531	B_6	$3(2C_6+1C_8)$	$6(3C_6+2C_8+1C_{10})$	$1(C_{11})$
	B_4	$4(2C_6+2C_9)$	$4(2C_6+1C_8+1C_9)$	
	B_3	$3(1C_7+2C_9)$	$3(1C_6+1C_8+1C_9)$	$2(1C_6+1C_8)$
331	B_5	$3(2C_7+1C_9)$	$5(3C_6+2C_9)$	$1(C_{11})$
	B_3 type 1	$3(2C_6+1C_8)$	$3(1C_7+2C_9)$	$5(4C_7+1C_{11})$
	B_3 type 2	$3(2C_7+1C_9)$	$3(2C_7+1C_9)$	$2(C_9)$
310	B_6	$4(1C_6+2C_8+1C_9)$	$5(2C_6+1C_8+2C_9)$	$1(C_{11})$
	B_4	$3(2C_6+1C_8)$	$5(2C_6+2C_8+1C_9)$	
	B_3	$3(1C_6+2C_7)$	$3(2C_6+1C_8)$	—
320	B_6	$4(2C_6+2C_7)$	$5(1C_6+2C_7+2C_9)$	—
	B_5	$3(1C_6+2C_7)$	$4(2C_6+2C_7)$	$1(C_{11})$
	B_3	$3(2C_6+1C_9)$	$4(2C_6+1C_7+1C_9)$	$1(C_{11})$

(a) Number of direct contacting metal atoms.

(b) Total number of neighbour atoms at a distance <0·04 nm from the adatom.

(c) Number of neighbour atoms at a distance between 0·04 and 0·09 nm from the adatom.

are distinguished by the symbol C_m in order to examine the influence of surface co-ordination on the heat of adsorption. The symbol m refers to the number of first-nearest-neighbour atoms, so that the number of missing nearest neighbours is $12 - m$. Further data (Table 7.5) refer to numbers of atoms within various distances of an adatom, for example c represents the number of neighbours at a distance between 0·04 and 0·09 nm from the hard sphere representing the Xe adatom. According to an r^{-6} Lennard-Jones potential their contribution to the bonding is significant but less than that anticipated from atoms of type b (Table 7.5).

If close packing with the metal substrate lattice is the only necessary condition for strong interaction, then the heat of adsorption of Xe on f.c.c. planes should decrease in the order (321); (531); $331 \sim 210 \sim 320 \sim 310$; (110); (100); (111). This sequence agrees with the heats of adsorption (Table 7.4) except for (111) and (100). It is logical to argue that the high heats observed on the close-packed (111) and (100) regions are indicative of forces other than dispersion being involved in the interaction between Xe and the metal, and as suggested by Nieuwenhuys and Sachtler this unknown factor may also account for the large surface potentials observed in general, and in particular on the (111) and (100) planes. Whether the unknown factor arises from polarization of the adsorbate by the electric field at the metal surface or from CTNB interaction is as yet not resolved.

7.4. Structural aspects of the adsorbed layer

One of the important recent problems in physical adsorption has been the question of deciding whether the adsorbed layer conforms to a site-type adsorption or is close packed (Brennan, Graham, and Hayes 1963, Anderson and Baker 1962). This is in itself of considerable intrinsic mechanistic interest, but it also has a bearing on the question of the value to be used for the cross-sectional area of a physically adsorbed adatom (Cannon 1963). The latter has generally been obtained from comparisons of monolayer values for a given adsorbent using nitrogen as a standard with a cross-sectional area of 0·162 nm². The problem at issue is highlighted by the range of values reported by Singleton and Halsey (1955) for the cross-sectional area of adsorbed xenon. With carbon it was 0·202 nm², with silver iodide

$0 \cdot 182$ nm^2, and with anatase $0 \cdot 273$ nm^2. Pritchard (1962), using his surface-potential method, quotes $0 \cdot 24$ nm^2 for Cu films while Cannon obtained a value of $0 \cdot 25$ nm^2 with nickel powder.

The work of Ponec and Knor (1962) and Anderson and Baker (1962) suggested that the above approach should be viewed with caution, and in particular Anderson and Baker emphasized the possible importance of lattice parameters in determining the availability of sites for adsorption. This point is also obvious when we consider the known influence of chemisorbed molecules (Table 7.1) on inert-gas interaction.

Brennan *et al.* (1963) investigated very carefully this particular aspect of physical adsorption and concluded that in the case of Kr and Xe there was a marked similarity between the *number* of krypton and xenon atoms in the monolayer. They concluded from comparisons of Kr and Xe isotherms that in general the effective areas of krypton and xenon atoms on the metals Co, Ni, Pt, Mo, Ta, W, Ti, and Zr were virtually the same. This was interpreted to mean that both adatoms occupy the same surface site and that their effective area is determined by the site area, and the authors rejected the close-packed configuration where each adatom exerted its own characteristic area related to its hard-sphere diameter. Brennan and Graham (1965) explored this point further by calculating interatomic energies using the Lennard-Jones (6:9) and (6:12) potentials and the Buckingham exp:6 potential for various configurations of adatoms with the object of finding the densest occupation of available sites. Their calculations in general support their adsorption experiments.

More recently more definitive information on the structural nature of the adlayer has come from LEED studies. In particular Chesters and Pritchard (1971), and Chesters, Husain, and Pritchard (1973), have investigated the adsorption of Xe on Cu(100) at 77 K. Up to an exposure of 45 L, at a xenon pressure of 10^{-6} torr, there was only a faint diffuse circle around the 00 position but this became more intense and sharper reaching a maximum after 100 L. At this stage 12 equally spaced diffuse spots could be discerned in the ring, but with varying electron energy the intensity of each spot varied with the intensity of the beams associated with the substrate Cu atoms. This the authors suggested was due to double diffraction and on this basis interpreted the ring patterns in terms of two hexagonal domains of

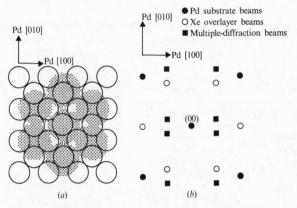

FIG. 7.4. (a) Atomic arrangement of ordered xenon layer on Pd (100); (b) schematic representation of observed LEED pattern from hexagonal overlayer structure. (From Palmberg 1971.)

xenon adatoms, one domain rotated at 30° from the other, and the domains having preferred orientations with respect to the substrate with Xe[01] parallel to Cu[110] or Cu[1$\bar{1}$0].

For Xe on Pd(100) at 77 K Palmberg (1971) observed ordering only as maximum coverage was approached and his LEED data suggested the formation of a close-packed hexagonal structure (Fig. 7.4). Simultaneous Auger and work-function studies were carried out and two further relevant pieces of information obtained: (a) the isosteric heat was initially about 34 kJ mol^{-1} and fell to about 29 kJ mol^{-1} with increasing coverage, the sticking probability being close to unity at low coverage, and (b) by making use of the Helmholtz formula for an array of point dipoles in the form (eqn 12.15))

$$\Delta\phi = 2\pi\mu_0\sigma(1 + 9\alpha\sigma^{\frac{3}{2}})^{-1} \tag{7.9}$$

and the observed change in work function $\Delta\phi$, the dipole moment and polarizability of an 'isolated' adsorbed xenon atom were calculated. It is interesting that the polarizability α of a xenon adatom ($8\cdot2\times10^{-24}$ cm^3) is much higher than an isolated xenon atom ($\sim4\times10^{-24}$ cm^3). The dipole moment at zero coverage (μ_0) was calculated to be 0·95 debye. With increasing surface coverage dipole–dipole repulsive interactions occur with a decrease in the heat of adsorption. Table 7.6 summarizes surface structures of adsorbed xenon on metal surfaces. Somorjai, Morabito, and

TABLE 7.6

The structures of xenon monolayers on metal single crystals

Substrate	Monolayer structure	Orientation (substrate direction parallel to Xe(01) rows)	Surface potential (eV)
Graphite (0001)	hcp ($\sqrt{3} \times \sqrt{3}$) R 30° $a = 0.426$ nm	[$1\bar{1}00$]	—
Pd(100)	hcp, $a = 0.448$ nm	[001] or [010]	0.94
Cu(100)	hcp, $a = 0.45$ nm	[011] or [$01\bar{1}1$]	0.47
Cu(111)	hcp ($\sqrt{3} \times \sqrt{3}$) R 30° $a = 0.442$ nm	[$11\bar{2}$]	0.48
Cu(110)	centred rectangular $a = 0.46$ nm, $b = 0.72$ nm	[$1\bar{1}0$]	0.61
Ag(111)	hcp, $a = 0.45$ nm	[$10\bar{1}$]	0.44
Ag(110)	centred rectangular $a = 0.44$ nm, $b = 0.813$ nm	[$1\bar{1}0$]	0.45
Ag(211)	hcp, $a = 0.44$ nm	[$01\bar{1}$]	0.45

From Chesters *et al.* 1973.

Müller (1969) have used a multi-technique approach to the study of physical adsorption, in particular of Kr and Xe on Ag(110). Ellipsometry indicated clearly that appreciable adsorption occurred at pressures below 10^{-8} torr and temperatures below 273 K but only at higher pressures ($\geq 10^{-7}$ torr) did the intensity in the background of the diffraction pattern increase. Both the intensity of background and the 00 spot returned to their original values when the gas was removed. No ordering of the physically adsorbed layer was revealed in the LEED pattern, and the estimated heats of adsorption at low coverage were 25·6 kJ mol^{-1} (Xe) and 16 kJ mol^{-1} (Kr). What is not clear in these LEED studies is whether the electron beam is inhibiting the formation of ordered structures. These results differ from those reported by Chesters *et al.* (1973). More recently McElhiney and Pritchard (1976) have combined LEED with AES and electron energy loss spectroscopy to investigate the adsorption of Xe and CO on Au(100).

It could be argued that the interaction of inert gases with clean metal surfaces is a special case in the field of physical adsorption and that other non-metallic substrates are more amenable to theoretical analysis. Two substrates that have received particularly detailed investigation are graphite and boron nitride. Barrer

(1937), Crowell and Young (1953), and Pace (1957) calculated the heat of adsorption of an isolated argon atom over various positions on the basal plane of graphite and concluded that the graphite surface was energetically homogeneous with respect to the adsorption of argon. Boron nitride was first chosen by Pierotti and Petricciani (1960) on the grounds that its hexagonal modification has a crystal structure which is very similar to graphite. The nearest-neighbour distance in boron nitride is 0·144 nm and in graphite it is 0·142 nm; the interlaminar spacing in boron nitride is 0·333 nm and in graphite it is 0·335 nm. These early calculations have been followed up by Thomas, Ramsey, and Pierotti (1973) who have obtained high-precision isotherm data at very low surface coverage (~3 per cent) of argon with both BN and graphitized carbon black P33 (2700), the latter having been used extensively by Halsey and his colleagues in their virial-theory analysis of high-temperature physical adsorption (Constabaris, Sams, and Halsey 1961). The BN data were fitted to a two-surface model corresponding to the basal plane and edge plane of the BN structure. Adsorption was assumed to obey Henry's law on one surface and a submonolayer isotherm on the other. The contribution of the edge plane was estimated to be about 5 per cent which is consistent with data from electron micrographs. Comparison of the interaction energies with the basal plane of BN and graphite indicate that the graphite interaction is between 2 and 4 per cent stronger which is in qualitative accord with dispersion-energy calculations.

7.5. Analysis of physical-adsorption isotherm data

It is pertinent to comment briefly on the various approaches that have been made to the analysis of experimental isotherms. The BET equation is the most frequently used relationship for obtaining surface areas from adsorption isotherm data. It is worth noting, however, that for the adsorption of Kr and Xe on metal surfaces the 'knee' in the isotherm frequently occurs at values of the pressure P such that $P_0 - P \simeq P_0$. Since $(C-1)/C \approx 1$ (see Table 7.1), eqn (7.1) then becomes the Langmuir equation

$$\frac{1}{V} = \frac{1}{aV_m P} + \frac{1}{V_m}. \tag{7.10}$$

A plot of $1/V$ *versus* $1/P$ is, however, a somewhat insensitive way
of ascertaining whether a set of adsorption results conforms to the
Langmuir model; Eley (1966) has emphasized this in connection
with ascertaining whether adsorption is molecular or dissociative
when, in the latter case, a $P^{\frac{1}{2}}$ term should replace P (eqn (7.10)).

Ricca and Bellardo (1967) have advocated the Dubinin–
Radushkevich (DR) relationship

$$\ln V = \ln V_m - \beta \left(RT \ln \frac{P}{P_0} \right)^2 \qquad (7.11)$$

and shown it to apply to a large number of physical adsorption
systems. Fig. 7.5 shows its applicability over a 10^5-fold range of
pressure ($10^{-5} - 10^{-10}$ torr) for Kr and Xe on Mo films at 77·3 K
and 90·2 K respectively. The slope of the plot, β (eqn (7.11)), is
related to the adsorption potential for the particular metal–gas
system and is somewhat analogous to the equivalent term in
the Singleton–Halsey (1955) relationship.

Ricca and Bellardo (1967) also analyse their data in terms of
the BET equation and conclude that different plots, and therefore V_m values, are obtained in different relative pressure ranges.
They suggest that the non-applicability of the BET equation to
these particular metal + gas systems possibly arises from the
heterogeneity of the film surface and strongly advocate the adop-
tion of the DR relationship for determining V_m values. Two
points must, however, be made: (*a*) the BET equation, although

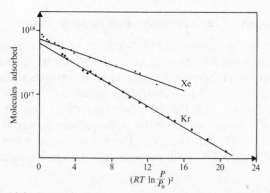

FIG. 7.5. Dubinin–Radushkevich plot for adsorption of Kr and Xe on Mo film.
(From Ricca and Bellardo 1967.)

based on a plausible physical model for an adsorbed layer, is in essence only a convenient method of determining mathematically a 'knee' in an isotherm, and (b) it was originally adopted for nitrogen isotherm data where the 'knee' usually occurred at P/P_0 values in the range 0·1–0·3. In the case of both Kr and Xe adsorption on metals the 'knee' occurs at much lower relative pressure ($P/P_0 \leqslant 0·01$) and in the data analysed by Ricca and Bellardo the knee had not been approached in that their P/P_0 values were in the range $10^{-6}–10^{-2}$. This is not to suggest that no problems exist with the application of the BET equation to extract meaningful monolayer values. There are in fact many situations discussed by Gregg and Sing (1967) where the limitations of the approach are clearly recognized. One example concerns the apparently trivial decision which has to be made regarding the value to use for the saturation vapour pressure of the gas (P_0). Haynes (1962) has shown that the precise value chosen influences markedly the slope and the linearity of the BET plot. Furthermore, choice of the best P_0 value to use apparently depends on the particular range in which the constant C in the BET equation lies. At 90·1 K for Kr adsorption on a series of minerals with C values in the range 30–100 the 'best' P_0 value is 30 torr.

In analysing adsorption data, where the pressure-measuring device and the sorbent are not at the same temperature, it is necessary to consider carefully whether the measured pressure is the same as that above the sorbent. The phenomenon of thermal transpiration, whereby a gradient of temperature gives rise to a gradient of pressure, will clearly be important in physical adsorption where there usually exists a temperature differential of 200 K between the measuring device and the sorbent. Bennett and Tompkins (1957) critically reviewed the data available before 1957 and also reported new data for CO, CO_2, Kr, and O_2. A similar approach was adopted for adsorption studies at high temperature (Hardy and Roberts 1971), where the measuring device was at a lower temperature than the sorbent. At pressures below 10^{-3} torr the ratio of the measured to the corrected pressure $\dfrac{P_1}{P_2}$ is often approximated to $\left(\dfrac{T_1}{T_2}\right)^{\frac{1}{2}}$. When very accurate results are required, however, a careful investigation by Edmonds and Hobson (1965) has shown that this relationship is not strictly applicable.

7.6. Conclusions

Emphasis has been given to the experimental information that has become available during the last decade regarding the microscopic aspects of physical adsorption of inert gases on metals. Although the forces in physical adsorption are intrinsically easier to define than in chemisorption, there is still a dearth of theoretical background. Work-function studies have provided unambiguous evidence for the role of charge transfer in physical adsorption, but our understanding is far from complete and attempts, for example, to relate core-level shifts in an X-ray photoelectron spectroscopic study of xenon adsorbed on tungsten with simple charge distributions have not been successful (Yates and Erickson 1974). This may, however, partly reflect the limitations of the experimental approach in that relaxation effects may play an overriding role. The subject of physical adsorption has been reviewed recently by Landman and Kleiman (1977), the authors emphasizing both the theoretical and experimental aspects.

REFERENCES

ANDERSON, J. R., and BAKER, B. G. (1962). *J. Phys. Chem.* **66,** 482.

BARRER, R. M. (1937). *Proc. R. Soc. A* **161,** 476.

—— (1966). *J. colloid interface Sci.* **21,** 415.

BENNETT, M. J., and TOMPKINS, F. C. (1957). *Trans. Faraday Soc.* **53,** 185.

BRENNAN, D., and GRAHAM, M. J. (1965). *Phil. Trans. R. Soc.* **258,** 41.

BRENNAN, D., GRAHAM, M. J., and HAYES, F. H. (1963). *Nature, Lond.* **199,** 1152.

CANNON, W. A. (1963). *Nature, Lond.* **197,** 1000.

CHESTERS, M. A., HUSAIN, M., and PRITCHARD, J. (1973). *Surf. Sci.* **35,** 161.

CHESTERS, M. A., and PRITCHARD, J. (1971). *Surf. Sci.* **28,** 460.

CONSTABARIS, G., SAMS, J. R., and HALSEY, G. D. (1961). *J. phys. Chem.* **65,** 367.

CROWELL, A. D., and YOUNG, D. M. (1953). *Trans. Faraday Soc.* **49,** 1080.

CULVER, R. V., and TOMPKINS, F. C. (1959). *Adv. Catalysis* **11,** 67.

DRESSER, M. J., MADEY, T. E., and YATES, J. T. (1974). *Surf. Sci.* **42,** 533.

DUKE, C. B., and ALFERIEFF, M. C. (1967). *J. chem. Phys.* **46,** 923.

EDMONDS, T., and HOBSON, J. P. (1965). *J. vac. Sci. Technol.* **2,** 182.

EHRLICH, G. (1963). *Ann. N.Y. Acad. Sci.* **101,** 722.

EHRLICH, G., and HUDDA, F. G. (1959). *J. chem. Phys.* **30,** 493.

ELEY, D. D. (1966). *Disc. Faraday Soc.* **41,** 185.

ENGEL, T., and GOMER, R. (1970). *J. chem. Phys.* **52** (11), 5572.

GRAHAM, D. (1960). *J. phys. Chem.* **64,** 1089.

GREGG, S. J., and SING, K. S. W. (1967). *Adsorption, surface area and porosity.* Academic Press, New York.

GUNDRY, P. M., and TOMPKINS, F. C. (1960). *Trans. Faraday Soc.* **56,** 846.

HALL, P. G. (1966). *Chem. Commun.* p. 877.

HARDY, J., and ROBERTS, M. W. (1971). *J. chem. Soc.* p. 1683.

HAYNES, J. M. (1962). *J. phys. Chem.* **66,** 182.

KLEMPERER, D. F., and SNAITH, D. (1971). *Surf. Sci.* **28,** 209.

LANDMAN, U., and KLEIMAN, G. G. (1977). *In specialist periodical reports: surface and defect properties of solids* (eds M. W. Roberts and J. M. Thomas). Vol. 6. Chemical Society, London.

LENNARD-JONES, J. E. (1924). *Proc. R. Soc.* A **106,** 463.

LITTLE, J. G., QUINN, C. M., and ROBERTS, M. W. (1964). *J. Catalysis* **3,** 57.

McELHINEY, G., and PRITCHARD, J. (1976). *Surf. Sci.* **60,** 397.

MATSEN, F. A., MAKRIDES, A. C., and HACKERMAN, N. (1954). *J. chem. Phys.* **22,** 1800.

MAVROYANNIS, C. (1963). *Mol. Phys.* **6,** 593.

MENZEL, D., and GOMER, R. (1964). *J. chem. Phys.* **41,** 3311.

MIGNOLET, J. C. P. (1950). *Disc. Faraday Soc.* **8,** 105.

—— (1953). *J. chem. Phys.* **21** (7), 1298.

MÜLLER, A. (1936). *Proc. R. Soc.* A **154,** 624.

MÜLLER, J. (1970). *Chem. Commun.* p. 1173.

NIEUWENHUYS, B. E., BOUWMAN, R., and SACHTLER, W.M.H. (1974). *Thin Solid Films* **21,** 55.

NIEUWENHUYS, B. E., and SACHTLER, W.M.H. (1974). *Surf. Sci.* **45,** 513.

NIEUWENHUYS, B. E., VAV AARDENNE, O. G., and SACHTLER, W.M.H. (1974). *J. chem. Phys.* **5,** 418.

PACE, E. L. (1957). *J. chem. Phys.* **27,** 1341.

PALMBERG, P. W. (1971). *Surf. Sci.* **25,** 598.

PIEROTTI, R. A., and HALSEY, G. D. (1959). *J. phys. Chem.* **63,** 680.

PIEROTTI, R. A., and PETRICCIANI, J. C. (1960). *J. phys. Chem.* **64,** 1596.

PONEC, V., and KNOR, Z. (1962). *Czech. chem. Commun.* **27,** 1091.

PRITCHARD, J. (1962). *Nature, Lond.* **194,** 38.

RICCA, F., and BELLARDO, A. (1967). *Z. phys. Chemie* **52,** 318.

SINGLETON, J. H., and HALSEY, G. D. (1955). *Can. J. Chem.* **33,** 184.

SOMORJAI, G. A., MORABITO, J. M., and MÜLLER, R. (1969). In *The structure and chemistry of solid surfaces* (ed. G. A. Somorjai). John Wiley, New York.

THOMAS, H. E., RAMSEY, R. N., and PIEROTTI, R. A. (1973). *J. chem. Phys.* **59,** 6163.

TUCK, D. G. (1958). *J. chem. Phys.* **29,** 724.

VAN OIRSCHOT, G. J., and SACHTLER, W. M. H. (1970). *Ned. Tijdschr. vac. Tech.* **8,** 96.

YATES, J. T., and ERICKSON, N. E. (1974). *Surf. Sci.* **44,** 489.

GENERAL REFERENCES

GREGG, S. J., and SING, K. S. W. (1967). *Adsorption, surface area, and porosity.* Academic Press, New York.

LANDMAN, U., and KLEIMAN, G. G. (1977). Microscopic approaches to physisorption. In *Specialist periodical reports: surface and defect properties of solids* (eds M. W. Roberts and J. M. Thomas). Vol. 6. Chemical Society, London.

STEELE, W. A. (1974). In *The interaction of gases with solid surfaces* (eds D. D. Eley and F. C. Tompkins). *The international encyclopaedia of physical chemistry and chemical physics.* Pergamon Press, Oxford.

YOUNG, D. M., and CROWELL, A. D. (1963). *Physical adsorption of gases.* Butterworth, London.

8

MODELS FOR THE ADSORPTION AND DESORPTION PROCESSES

IN CHAPTER 6 we discussed various experimental techniques used to investigate the possible steps involved in transfer of a particle between the gas phase and an adsorbed state on a solid surface. In this chapter we first outline a possible generalized model for the adsorption–desorption process and then present some of the relevant experimental data. Finally, certain aspects of the model will be reconsidered in the light of these results.

8.1. Generalized model for the interaction of a gas-phase particle with a solid

When a gas-phase molecule collides with a solid surface it is possible to envisage a number of processes, some or all of which may occur. These are illustrated schematically in Fig. 8.1 and are listed below. (In this chapter we shall refer to particles in the general case as 'molecules', this term being taken to include atoms.)

  (i) The molecule may be elastically scattered back into the gas phase (step 1, Fig. 8.1). As with elastic scattering of electrons (Chapter 3) this can give rise to diffraction effects.

 (ii) The molecule may lose to the solid part of the translational component of its gas-phase kinetic energy normal to the surface and become trapped in a weakly bound state (step 2, Fig. 8.1). This state, which we shall refer to as state A, corresponds to the shallowest of the potential energy wells shown in Fig. 1.8, Chapter 1, and in very general terms would correspond to physical adsorption. The process of energy exchange between the incoming molecule and the solid is known as *accommodation*.

(iii) When the molecule is initially trapped in state A the molecule–surface bond will be vibrationally excited, but there may be further exchange of energy with the solid so

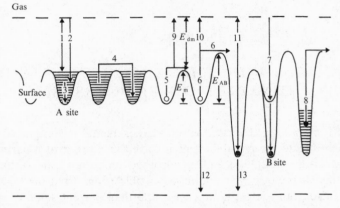

FIG. 8.1. Schematic representation of processes which may occur during the interaction of a gas-phase particle with a solid.

that the molecule moves to the ground state (step 3, Fig. 8.1).

(iv) As an alternative to (iii) the molecule may not reach the ground state at its initial point of encounter with the surface, but may hop to neighbouring A sites, becoming de-excited as it moves (step 4, Fig. 8.1).

(v) A molecule in the ground state at an A site also may diffuse to neighbouring A sites by taking up thermal energy from the lattice to overcome the activation energy barrier E_m for this type of movement (step 5, Fig. 8.1).

(vi) If there exists on the surface a second state B of higher binding energy, which would correspond to chemisorption, a molecule in an adjacent A site may obtain activation energy E_{AB} for diffusion into this site (step 6, Fig. 8.1). In this case state A acts as a mobile *precursor* for state B.

(vii) It is also possible that states A and B occur at effectively the same point on the surface, and that a molecule entering state A from the gas phase passes directly into state B without undergoing any lateral movement on the surface (step 7, Fig. 8.1). This is the type of process envisaged in the Lennard-Jones diagram (Fig. 1.8); in this case state A could be considered to be an immobile precursor for state B.

PLATES

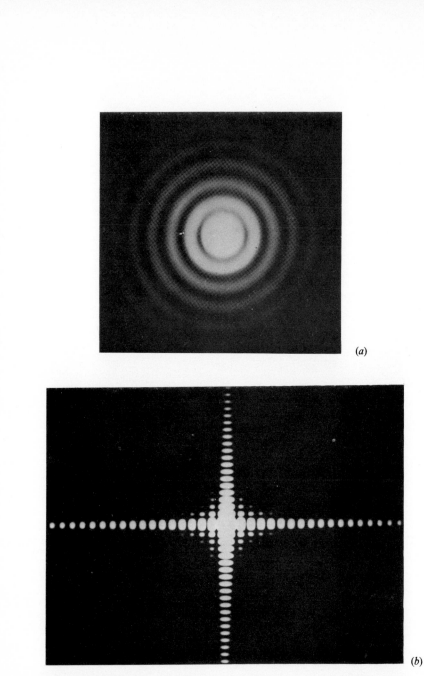

(a)

(b)

PLATE 1. Optical diffraction patterns produced by (a) a circular and (b) a rectangular aperture. (From Towne 1967.)

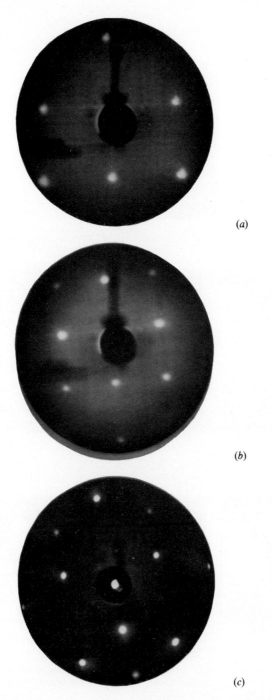

PLATE 2. Low energy electron diffraction (LEED) patterns produced by (a) Al (100) at 80 eV primary energy, (b) Al (100) at 120 eV, and (c) Cu (210) at 93 eV.

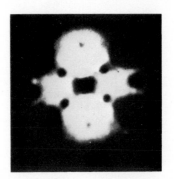

PLATE 3. Field electron emission microscope (FEM): micrograph of an [011] oriented clean tungsten tip. (From Duell, Davis and Moss 1966.)

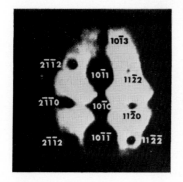 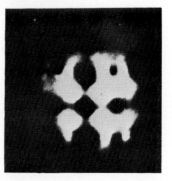

(a)   (b)

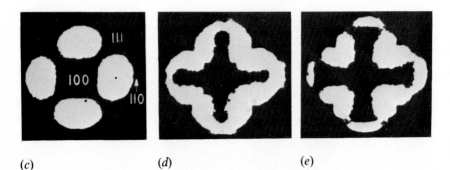

(c)   (d)   (e)

PLATE 4. FEM micrographs: (a) clean rhenium (b) CO covered rhenium (c) clean rhodium (d-e) rhodium exposed to oxygen for increasing time intervals. (a, b from Klein 1970; c-e from Ehrlich 1966.)

PLATE 5. Field ion emission microscope (FIM): micrograph of an [001] oriented platinum tip. (From Knor and Müller 1968.)

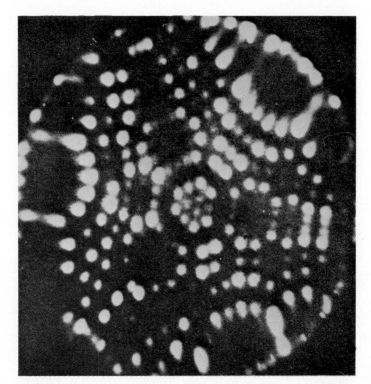

(a)

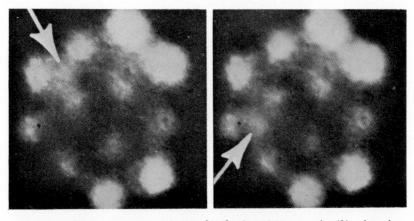

(b)

PLATE 6. FIM micrographs: (a) clean [111] oriented tungsten tip; (b) enlarged view of the (111) plane showing the displacement of a W adatom (indicated by arrow) on warming the tip to 600 K. (From Graham and Ehrlich 1974.)

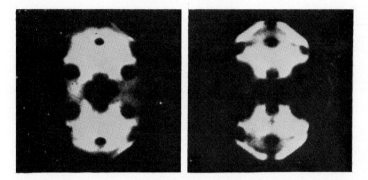

at 18 min   35 min

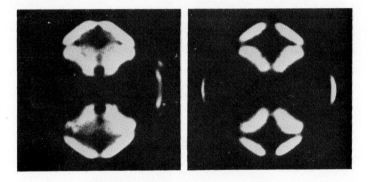

46 min   66 min

PLATE 7. FEM micrographs of the adsorption of xenon on a [110] oriented molybdenum tip. Emission from the region around the {100} planes (top and bottom of the image) is increased while that from the area around the {111} planes (left and right) fades away. Experiments at 20 K show that enhanced emission is due to an enhanced concentration of adsorbate around the {100} planes. (From Ehrlich 1963.)

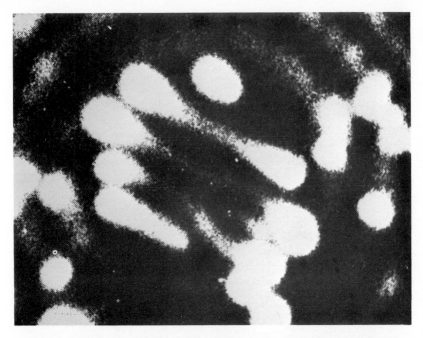

PLATE 8. FIM micrograph of parallel chains of adatoms, with a separation of ~1.5 nm, in a deposit of 80 ± 5 iridium atoms formed on a W(110) plane by alternately vapour depositing Ir atoms at 78 K and heating at 380 K. (From Bassett 1973.)

PLATE 9. LEED pattern (106 eV primary energy) observed from a polycrystalline gold surface after a number of cycles of ion bombardment and annealing at 600 K. (From Isa, Joyner and Roberts 1976).

(viii) The formation of strong adsorbate–surface bonds in the chemisorbed state B may release considerable amounts of energy and as with an A molecule in step 4 above, this energy may not be lost to the solid at the initial chemisorption site. The molecule will then diffuse in the B state until sufficient energy has been dissipated to the lattice for it to become localized at a particular site (step 8, Fig. 8.1).

(ix) At any time a molecule on the surface may undergo desorption. It is useful to distinguish desorption during diffusion (migration), with activation energy E_{dm} (step 9, Fig. 8.1) and desorption from the adsorbed states A and B with activation energies E_{dA} and E_{dB} respectively (steps 10 and 11, Fig. 8.1).

(x) Up to this point adsorbed species have been considered to act independently on the surface, but the possible influence of adsorbate–adsorbate interactions on the various steps outlined above must also be taken into account.

(xi) In general molecules in states A and B may also have the possibility of movement into the bulk of the solid to form a three-dimensional compound layer with the substrate (steps 12 and 13, Fig. 8.1), but for the purposes of the present chapter this possibility will not be considered.

8.2. The initial gas–surface collision

Experimental data on accommodation coefficients and gas–surface scattering obtained up to the early 1960's are in some confusion, undoubtedly caused at least in part by inadequate definition of surface conditions. As early as 1935, for example, J. K. Roberts (1935) observed that the accommodation coefficient α was sensitive to the mode of preparation of the surface and suggested that the value of α for, say, neon could be used as an indicator for adsorption of other gases. In spite of these problems, however, a number of generalizations emerged from this earlier work which have since been broadly confirmed.

Much of the more recent effort in this area has concentrated on the scattering of molecular beams from metal surfaces; for reviews of the subject see, for example, Logan 1973, Saltsburg 1973, Weinberg 1975. Three basic types of experiment may be envisaged, involving non-reactive, reactive and diffractive scattering. Diffraction of an inert gas beam provides information on the structure of the solid surface but little work has been done on this effect using metals (see Tendulkar and Stickney 1971; Stoll and Merrill 1973; Hayward and Walters 1974). In a reactive scattering experiment the surface catalyses a reaction between component gases in the incident beam and the desorbed products are detected using a movable mass spectrometer. The elimination of any influence of the vacuum chamber walls on the progress of the reaction is a major attraction of this technique but, as with diffractive scattering, it has been used in relatively few investigations. Palmer and Smith (1974) have studied the oxidation of CO over Pt(111), while Bernasek, Siekhaus and Somorjai (1973) have observed that H_2–D_2 exchange occurs readily on a stepped Pt(111) surface (actually Pt(997)) but not on the smooth (111) plane.

Non-reactive scattering of beams of inert gas atoms has been extensively studied. The distribution of particles scattered from a surface without having been trapped provides, in principle, a sensitive means for investigating the nature of the gas–solid interaction potential. Precisely defined experimental data are not easy to obtain, however, since the results are sensitive to a number of complex experimental parameters such as the spread in incident-beam energies and the detailed structure of the surface. On the other hand accommodation coefficients *can* be accurately measured, but because they involve an average over incident and reflected conditions of the gas particles they cannot provide such detailed information about the interaction. Nevertheless, they do offer a useful means of comparing theory and experiment.

As far as the theory of chemisorption is concerned a further parameter of interest is the probability that an incident molecule will become trapped at the surface for a time long enough to allow it to undergo further surface processes. This trapping probability f_t sets an upper limit to the sticking probability for the system, but again it is rather difficult to measure experimentally.

It is obviously very desirable to have a suitable theory which will enable interaction potential parameters, and f_t, to be extracted from accommodation-coefficient and scattering data. There has been a considerable number of attempts to establish such a theory, and these have adopted either a quantum- or a classical-mechanical approach. None has yet been entirely successful, although most are in qualitative agreement with experiment and in some cases reasonable quantitative agreement is also obtained. A drawback with the quantum-mechanical treatments has been that the simplifying assumptions involved restrict consideration to one-phonon transitions only, limiting the maximum energy transferred in a collision to $k\Theta$, where Θ is the surface Debye temperature. Thus these theories apply to weakly interacting systems only.

Three approaches which are illustrative of the classical-mechanical method are outlined below. They have been chosen because of their conceptual simplicity, not because they lead to comprehensive quantitative agreement with experiment. Indeed, no existing model does, although that due to McClure (1972) may offer the best prospects for development in this direction (see Weinberg 1975).

8.2.1. *Linear harmonic oscillator chain model*

This is a one-dimensional model (Fig. 8.2(a)) in which the lattice consists of a semi-infinite chain of particles of mass m bound to their nearest neighbours by harmonic forces, characterized by a force constant K (McCarroll and Ehrlich 1963). The interaction between the incident particle (mass m_0) and the first particle of

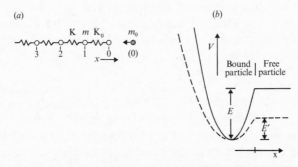

FIG. 8.2. (a) Gas atom (0) colliding with a one-dimensional linear chain. K is the harmonic force constant for the lattice. The incident particle (mass m_0) interacts with the lattice with a force constant K_0; its equilibrium position is 0. (b) Potential energy V as a function of relative separation x of particles 0 and 1. Plots for two different values of K_0 are shown. At a constant cut-off distance the binding energy E is proportional to K_0. (From McCarroll and Ehrlich 1963.)

the lattice is described by a modified harmonic potential (force constant K_0) which cuts off at some fixed separation of the two particles. Thus when the incident particle arrives it experiences no forces until it attains this separation (Fig. 8.2(b)).

The equations of motion for the system are set up and solved to give the maximum energy which the incoming particle may possess if it is to be captured (the critical energy for capture E_c), as a function of $\beta = K_0/K$. Two features become apparent.

(a) E_c, and hence the trapping efficiency, increases as the mass of the incident particle relative to that of the lattice particles increases and reaches its maximum value for a particle condensing on its parent lattice ($m_0 = m$).

(b) The trapping efficiency increases as β increases, up to $\beta \sim 0.75$. Since the cut-off distance was kept constant throughout, an increase in β ($= K_0/K$) corresponds to an increase in the depth of the potential well attracting the incident particle or (qualitatively at least) to an increase in the heat of adsorption.

A further interesting feature which emerges from the analysis is the relative displacements of adjacent pairs of atoms in the chain as a function of time after the initial collision. The oscillations of the link between the trapped atom and the first lattice atom continue for a considerable time, but the binding energy is rapidly transmitted to the lattice; for example, when the force constant for the incident atom is three-quarters that of the lattice ($\beta = 0.75$) 99·5 per cent of the binding energy is lost after two oscillations. Therefore in this particular model the limiting step for trapping is the initial exchange of translational energy. Once this is successful and the particle is bound, transfer of the binding energy to the lattice is very rapid.

This model suffers of course from several unrealistic features, including the following: (1) the lattice is initially at rest, i.e. at 0 K; (2) anharmonicity in particle binding is ignored; (3) the lattice is one dimensional. These deficiencies make quantitative prediction from the model impossible, but should not drastically affect its qualitative aspects.

Using the same concepts McCarroll (1963) has also investigated the effects of impurities on the trapping process. Consider first a *surface* impurity; it is found that increasing the strength of binding of the impurity to the main lattice allows the lattice to accept more energy and therefore increases the condensation efficiency. Also, if the surface impurity is identical to the colliding atom (analogous to a gas-phase atom striking a surface covered with an adsorbed monolayer) an increase in the mass of the gaseous species *decreases* the trapping probability.

Auger spectroscopy is at present one of the most widely used techniques for determining 'surface' purity, but it may sample only the first one to two atom layers into the solid. It is therefore extremely interesting that McCarroll (*op. cit.*) found that an impurity located in the third 'layer' of a one-dimensional lattice could have an effect on events following collisions at the surface. A decrease in the strength of bonding of this impurity decreases the condensation efficiency but, of more significance, it causes a decrease in the rate of energy transfer to the bulk since it reflects energy back to the surface. It is important to emphasize that the rate of dissipation of the energy of condensation may have a considerable influence on the eventual fate of the adsorbing particle.

8.2.2. *The hard- and soft-cube models*

Although only representing a one-dimensional lattice initially at 0 K, models of the linear-chain type are relatively complex and incorporation of more realistic features presents very considerable difficulties. Attempts have therefore been made to construct models which, if no more realistic, are at least more easily handled mathematically. One of these, known as the hard-cube model, involves the following assumptions (Logan and Stickney 1966).

   (i) Both gas particle and surface atom behave as rigid elastic spheres.

   (ii) The intermolecular potential between gas particle and surface is uniform in the plane of the surface.

   (iii) The surface atoms are represented by independent particles confined by square-well potentials, i.e. rigid boxes. A gas particle interacts with a surface atom by entering the box, colliding with the surface particle at an angle θ_i, and then leaving at an angle θ_r.

These assumptions lead to a picture of the surface atoms behaving as cubes with one face parallel to the surface and with motion only in the direction normal to the surface. Each gas particle interacts with just one of these cubes (Fig. 8.3(a)). Because of the simplicity of this system a further assumption, which represents an important advance on the linear lattice model is now possible.

(iv) A temperature-dependent velocity distribution (e.g. a one-dimensional Maxwellian distribution function) may be assigned to the surface atoms. Thus the lattice is no longer constrained to be at 0 K.

Using this model angular distributions of inert gases scattered from platinum can be derived which are for the most part in good qualitative agreement with experimental data (Hinchen and Foley 1965) (Fig. 8.3(b)). There are some instances of disagreement, however, and in an attempt to eliminate these one may move to a 'soft-cube' model (Logan and Keck 1968). In the hard-cube variant the gas–solid interaction is impulsive, i.e. instantaneous. In reality there will be a finite collision time and this 'soft' collision is introduced by allowing the gas to interact with the surface through a potential which is in two parts (Fig. 8.4(a)): a stationary or square-well attractive part (as in the hard-cube case) and an exponential repulsive part which is responsible for the finite interaction time. The surface atom is considered to be connected by a single spring to the remainder of the lattice, the latter being fixed; this represents the second major difference from the hard-cube model and allows for the introduction of the

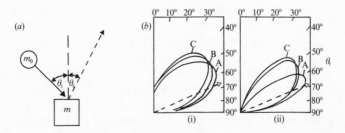

FIG. 8.3. (a) The 'hard-cube' model. (b) Mass dependence of scattering patterns. The magnitude of the radius vector at any angle θ_r represents the relative intensity of particles scattered at that angle. The broken line represents the specular direction. (i) Experimental results for He, Ne, and Ar on Pt (Hinchen and Foley 1965); (ii) analytical results. In both cases the angle of incidence $\theta_i = 67 \cdot 5°$ and the surface-gas temperature ratio is $T_s/T_g = 3 \cdot 60$. Curve A, He on Pt; curve B, Ne on Pt; curve C, Ar on Pt. (From Logan and Stickney 1966.)

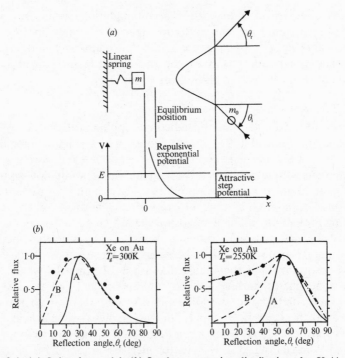

FIG. 8.4. (a) Soft-cube model. (b) In-plane scattering distributions for Xe/Au. Curve A, theory; curve B, experiment corrected for atoms adsorbed and then re-emitted. (From Logan and Keck 1968.)

force constant of this bond which, as we have seen from consideration of the linear-chain model, may play an important role in the interaction. There are three adjustable parameters: (i) the well depth E, which may be derived from experimental heats of adsorption; (ii) the range b of the repulsive potential; (iii) the oscillator frequency ω (a function of the force constant of the linear spring and the mass of the surface atom).

The model is compared with experimental data by choosing the above parameters to give the best fit for the angular position of the maximum in the scattering distribution. The forms of the scattering distributions are not *particularly* sensitive to these parameters but the agreement with experiment is reasonable in certain cases (Fig. 8.4(b)), although it is more suspect in others (see Weinberg 1975).

A comparison of measured accommodation coefficients with those derived from the soft-cube theory is shown in Fig. 8.5 (Logan 1969); trapping probabilities f_t are also included. Again, although agreement is good, the theoretical results are not very sensitive to the values chosen for b and E.

8.2.3. *The square-well model*

With the particular aim of predicting trapping probabilities Weinberg and Merrill (1971) have chosen a very simple model in which a gas particle approaches a smooth solid surface and interacts with it via an attractive square-well potential of depth E and an impulsive repulsive potential. Using a number of assumptions and approximations an expression is derived for the trapping probability which depends on the accommodation coefficient and the well depth. To estimate the latter they note that there is an approximate correlation between the square root of the heat of adsorption and the Lennard-Jones 6-12 potential-well depth for the corresponding gas–gas interaction and this is used to estimate the gas–solid well depth. Accommodation coefficients are taken either from experiment or, if these are not available, they are calculated using a semi-empirical equation of Goodman and Wachman (1967). Results for inert-gas trapping on tungsten are compared with experimental values (Weinberg and Merrill 1972) in Fig. 8.6. The latter were obtained by identifying the fraction of incident atoms initially trapped with the cosine contribution to the distributions of gases scattered from a W(110) surface. It is seen that the square-well model, in spite of its

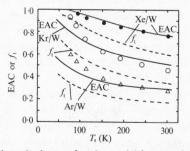

Fig. 8.5. Theoretical results from soft-cube model for experimental accommodation coefficient (EAC) and trapping probability f_t at various surface temperatures T_s. Corresponding experimental points for EAC for Ar/W ($\triangle$), Kr/W ($\bigcirc$), and Xe/W ($\bullet$). (From Logan 1969.)

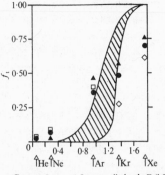

Gas-gas Lennard-Jones well depth E (kJ)

FIG. 8.6. Trapping probabilities f_t for inert gases on tungsten versus gas–gas Lennard-Jones 6-12 potential-energy parameter E. The gas temperature is 295 K and surface temperature T_s is a parameter. The cross-hatched region represents calculated values of f_t for $375 < T_s < 1300$ K. Experimental data are also shown. ▲ $T_s = 375$ K; □ $T_s = 575$ K; ● $T_s = 775$ K; ◇ $T_s = 1300$ K. (From Weinberg and Merrill 1972.)

simplicity, qualitatively predicts the trend in f_t, at least for these particular gas–metal combinations.

From these and other results one would expect that f_t would be near unity for the trapping of an inert gas on its own solid lattice. This has been demonstrated to be the case by Levenson (1967) using the quartz-crystal-oscillator technique described in Chapter 6. This experiment actually measured sticking probabilities but since these were found to be unity it follows that $f_t = 1$.

8.2.4. *Conclusions regarding the initial gas–surface collision*

Much work, both experimental and theoretical, remains to be done in this area, but for the purpose of applying existing results to our consideration of the chemisorption process some general conclusions can be drawn from the above discussion.

Both scattering distributions and accommodation coefficients appear to behave in the following ways.

    (*a*) For clean surfaces scattering tends to be specular and the accommodation coefficient is usually rather low (0·01–0·3).

    (*b*) Adsorption increases the diffusely scattered component and raises α.

    (*c*) As the attractive interaction between gas particle and surface increases scattering shifts towards the surface normal and becomes more diffuse.

(d) As the gas temperature T_g decreases, or as the surface temperature T_s increases, scattering again moves towards the surface normal. Under these conditions α tends to decrease slightly. In the case of gases which interact strongly with the surface, however, the effect of T_s is much reduced.

(e) The trapping probability f_t increases (i) with increasing mass of the gas particle, (ii) with decreasing gas temperature, and (iii) with decreasing surface temperature, although this effect is small.

8.3. Surface diffusion

There has been extensive study of surface diffusion using the field emission microscope, but this has largely been confined to the systems N_2, CO, O_2, H_2, and the inert gases on tungsten (Folman and Klein 1968, Ehrlich and Hudda 1959, 1961; Engel and Gomer 1970, Gomer 1961). Also, field ion microscopy has been used to investigate the diffusion of metals on metals (Bassett 1973). The experimental techniques involved have been outlined in Chapter 6 and the more important features of the surface-diffusion process to emerge are discussed below, with particular reference to overall models for chemisorption.

For chemically active gases adsorbed on tungsten it is found (Gomer 1961) that in general three types of diffusion can occur, although all three are not necessarily observed with a given gas–metal combination.

8.3.1. *Type 1 diffusion*

If adsorbate is deposited at 4·2 K to coverages in excess of a monolayer and the tip then carefully heated, a sharp boundary moving over the tip is observed at low temperatures, for example

Adsorbates on W	H_2	O_2	N_2
Spreading temperature (K)	$\ll 20$	~ 27	40.

This behaviour does not occur, however, above some critical temperature (~ 70 K for O_2/W). It is suggested that the initial deposit consists of a chemisorbed layer with a second, physically adsorbed layer resting on top of it. When the temperature is raised the physically adsorbed molecules can diffuse over the chemisorbed deposit and on reaching the edge of this they themselves become chemisorbed. Hence the chemisorbed layer advances over the region of the tip which was initially clean. The boundary will be sharp provided the distance a molecule can travel in a single diffusive hop (in this case of the order of a few tenths of 1 nm) is small compared with the resolution of the microscope (~ 2 nm). At the observed upper temperature limit physically adsorbed molecules desorb before they can reach the edge of the chemisorbed layer.

Boundary diffusion of this type is observed for H_2, N_2, O_2, CO, and CO_2 on top of a layer of the corresponding gas adsorbed on tungsten. The data can be analysed in a way analogous to that used for bulk diffusion (see Chapter 2) (Gomer 1961, Folman and Klein 1968). The theory of diffusion with precipitation or trapping is not simple, but can be solved accurately for the case of a semi-infinite plane solid (Fujita 1953) in which the boundary displacement x at time t is given by

$$x = (bt)^{\frac{1}{2}} \approx \left(D \frac{\sigma}{N_t} t \right)^{\frac{1}{2}} \tag{8.1}$$

where $b \approx D\sigma/N_t$, D is the diffusion coefficient, σ is the coverage in the deposit infinitely far from the boundary, and N_t is the trap-site density. This relationship is of the same type as that for the root-mean-square displacement of a particle diffusing in the bulk of a solid. It has been observed experimentally in a number of cases, for example hydrogen on H_2/W (Gomer, Wortman, and Lundy 1957). Gomer has assumed that $\sigma \approx N_t$ and therefore the diffusion coefficient may be obtained as the square of the slope of the x versus $t^{\frac{1}{2}}$ plot. Obviously if the spreading rate can be determined at various temperatures a value for the activation energy for migration E_m may be derived from the semi-logarithmic plot of b (eqn (8.1)) against $1/T$. Accurate measurements at different temperatures may prove difficult, however, in which case E_m may be estimated from x data at a single temperature by assuming (cf. eqn (2.48)) that D is of the form

$$D = \tfrac{1}{2} a^2 \nu_s \exp \left(\frac{-E_m}{kT} \right)$$

$$= D_0 \exp \left(\frac{-E_m}{kT} \right) \tag{8.2}$$

and inserting values for the jump distance a and the vibration frequency ν_s. This may not be such an accurate method as an Arrhenius-type plot of course, since there is some uncertainty as to the value of ν_s and since the exact form of the pre-exponential term D_0 in eqn (8.2) depends upon the particular model used for the diffusion process. For example, Folman and Klein (1968), following Kruyer (1953), have used the expressions $\frac{1}{4}a^2\nu_s$ for D_0 (cf. eqn (2.49)), and $6D$ for b in eqn (8.1). The order-of-magnitude values obtained for E_m lie in the range $0 \cdot 4 \text{ kJ mol}^{-1}$ (H_2/W) to $3 \cdot 7 \text{ kJ mol}^{-1}$ (O_2/W). These were derived using values of a between $0 \cdot 15$ and $0 \cdot 3 \text{ nm}$ and $\nu_s \sim 1 \times 10^{12} - 5 \times 10^{13} \text{ s}^{-1}$ (Gomer 1961, Folman and Klein 1968).

Up to this point we have not considered the possibility that the diffusing particle may desorb, but if this is included an estimate of the activation energy for desorption (E_d) may be obtained in the following way (Gomer et al. 1957, Gomer 1961). If the diffusing species is to traverse a mean distance $\bar{x}$ without evaporating its residence time τ on the surface must be given, from eqn (8.1), by

$$\bar{x} = (b\tau)^{\frac{1}{2}} \approx (D\tau)^{\frac{1}{2}}. \tag{8.3}$$

From the Frenkel equation (see Chapter 6)

$$\tau = \tau_0 \exp\left(\frac{E_d}{kT}\right) \tag{8.4}$$

where τ_0^{-1} is of the order of the vibration frequency of the adsorbate–surface bond. If it is assumed (Gomer 1961) that τ_0^{-1} is identical to ν_s in eqn (8.2), then, from eqns (8.2)–(8.4)

$$\bar{x} \approx \left\{ a^2 \exp\left(\frac{-E_m}{kT}\right) \exp\left(\frac{E_d}{kT}\right) \right\}^{\frac{1}{2}}$$

or

$$38 \cdot 28 \, T \log\left(\frac{\bar{x}}{a}\right) \approx E_d - E_m \tag{8.5}$$

where $38 \cdot 28 = 2 \times 2 \cdot 303 R \text{ J K}^{-1} \text{ mol}^{-1}$. Values of $\bar{x}$ may be determined by depositing identical doses of gas on the tip, heating to a particular temperature, and observing the maximum advance of the boundary. For the oxygen/tungsten system, for example, it is found (Gomer and Hulm 1957) that $\bar{x} \approx 30 \text{ nm}$ at 73 K and $\sim 120 \text{ nm}$ at 55 K. Using the previously determined value of $3 \cdot 7 \text{ kJ mol}^{-1}$ for E_m, these data lead to $E_d \approx 9 \cdot 3 \text{ kJ mol}^{-1}$.

For type 1 diffusion it was originally thought (Gomer 1961) that the boundary moved uniformly over the emitter, but more recent work (Folman and Klein 1968) has shown that it may spread preferentially in certain crystallographic directions. The case of H_2 on W is illustrated in Fig. 8.7. Spreading at first occurs radially from centres about halfway between the (011) and (010) planes (Fig. 8.7(a)). A cusp shape then developes (Fig. 8.7(b)) and finally the hydrogen flows in from both sides to eliminate the cusp (Fig. 8.7(c)). The significant point about this series of events is the indication that the (011) plane, which has the largest step edge in the b.c.c. system, is acting as a barrier to migration (Folman and Klein 1968). Other factors, such as a non-uniform distribution of trap sites on the bare surface, will contribute to the anisotropy in diffusion, but evidence for a potential barrier at surface steps also emerges from FIM studies of metal-adatom diffusion (Ehrlich and Hudda 1966, Bassett 1973) (see Chapter 12). Some caution must, however, be exercised when interpreting the FEM results; Engel and Gomer (1970), for example, have observed oxygen diffusion into (110) at 50 K but not at 25 K, which again might suggest the presence of a barrier to migration. These authors argue against this (but do not completely rule it out) on the grounds that at 25 K there is at least some diffusion over some of the vicinal regions of (110). Also the steps between

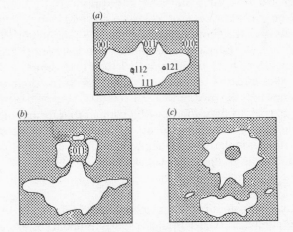

FIG. 8.7. Schematic field emission micrographs for H_2 spreading on W. (From Folman and Klein 1968.)

(110) planes along certain directions consist of (100) type configurations, and diffusion on (100) is rapid. It is suggested that lack
of adsorption on (110) may be due not to lack of diffusion but to
lack of binding, which implies a small but finite activation energy
for dissociative oxygen chemisorption (E_{diss}) on this plane. By
analogy with eqn (8.5) we can write

$$38 \cdot 28 T \log\left(\frac{\bar{x}}{a}\right) \approx E_{diss} - E_m \qquad (8.6)$$

where $\bar{x}$ is the mean diffusion distance before trapping. If lack of
adsorption at 25 K on the (110) region of the tip is taken to imply
that $\bar{x}$ is at least as large as the dimension of the (110) plane
($\sim$10 nm) then $E_{diss} - E_m \approx 2$ kJ mol^{-1}. Further, if it is assumed
that E_m for O_2 on bare W is similar to the value for second-layer
migration of O_2 on W, i.e. $\approx$2 kJ mol^{-1}, then $E_{diss} \approx 4$ kJ mol^{-1}.

Second-layer spreading of CO and N_2 on tungsten also exhibits
non-uniform boundary movement. Diffusion of CO initially occurs radially outwards from the central (011) plane, followed by
covering of the (121) planes and movement down the [111]
zones (Folman and Klein 1968). The behaviour of nitrogen
is similar, except for the interesting fact that the (001) planes
remain uncovered for a considerable time.

If nitrogen is deposited at 77 K on tungsten it is known (see
Chapter 10) that three states of adsorption occur, β, α, and γ,
with heats of adsorption of $\sim$350, $\sim$80, and $\sim$40 kJ mol^{-1} respectively. As the tip is heated above 77 K one might expect to
see diffusion of the γ-state, but this is not observed (Ehrlich and
Hudda 1961). The conclusion is that the mean diffusion distance
in this state must be small in comparison with the dimensions of
the planes on the tip surface ($\leqslant$10 nm). Thus the activation-
energy barrier to diffusion must be comparable to that either for
desorption or for conversion of γ species to a more strongly
bound state. Similarly at higher temperatures ($\sim$250–400 K) no
diffusion of the α state is detectable.

8.3.2. *Type 2 diffusion*

For initial deposits of 0·8–1·0 monolayers diffusion is observed
which is similar to type 1 in that a sharp boundary is present, but
because of the temperature ranges in which it occurs (180–240 K
with H$_2$/W (Gomer *et al.* 1957) and 400–530 K with O$_2$/W

(Engel and Gomer 1970)) it must involve diffusion within chemisorbed layers. The general observation is that migration takes place most easily on the most closely packed surface regions since binding energies on these should be minimal.

In the case of tungsten and other b.c.c. surfaces the highest density plane is the (110), which on a [110]-oriented field emission tip forms the central area. In all directions, and particularly along the zones {110}–{111}, it is surrounded by atomically 'rough' planes; the one exception is along zones such as {110}–{211}, which consist of terraces of {110} orientation. Diffusion on (110) and along these latter zones will thus be rapid, with trapping on the remainder of the surface. When trap sites in a given area are saturated diffusion may still be possible, however, if excess adsorbate is present, since *every* area of the surface consists, in some measure, of small sections of (110) planes.

8.3.3. *Type 3 diffusion*

If the coverage in the initial deposit is low, so that not all trap sites are saturated, boundary-free diffusion is observed owing to migration from trap to trap. As would be expected this has a higher activation energy than the boundary type just discussed. Thus Gomer *et al.* (1957) have found that for H_2 on tungsten $E_m = 67$ kJ mol^{-1} at a coverage of ~ 0.01 of a monolayer, falling to 40 kJ mol^{-1} in the range $\theta \sim 0.1$–0.3. When $\theta > 0.4$ boundary migration sets in, with $E_m = 25$ kJ mol^{-1}.

The experimental observations on behaviour of diffusion of types 2 and 3 may be illustrated by the systems N_2/W (Ehrlich and Hudda 1961) and CO/W and O_2/W (Engel and Gomer 1969, 1970); H_2 acts in a very similar way to O_2. In the studies of CO and O_2, adsorbate layers were deposited on the tip at low temperature (~ 60 K), with the gas source located so that the area to be investigated remained bare. Second-layer boundary migration was then initiated to bring the monolayer up to the edge of this area, the tip was rapidly heated to ~ 100 K to remove excess physisorbed material, and diffusion of chemisorbate onto the region of interest was then followed at higher temperatures. Work-function changes measured by the probe-hole technique were used to detect the arrival of diffusing species.

For CO initially in the virgin state on tungsten diffusion first occurs (at 275–300 K) in the {110}–{211} zones (see Fig. 8.8). It

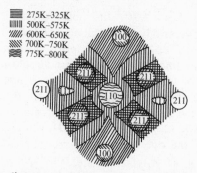

FIG. 8.8. Schematic diagram of field emission pattern showing the onset of diffusion temperatures as determined by emission changes for the probed region for virgin layers of CO deposited on W at 60 K. (From Engel and Gomer 1969.)

then proceeds outwards towards (111) and (120), reaching the latter, the vicinals of (100), and (100) itself, but appreciable work-function changes are observed on (110) only at 800 K. In view of our previous discussion this behaviour of (110) appears to be anomalous, but it may be rationalized by supposing that, although diffusion on (110) is indeed very rapid, there is no binding of CO on this plane below 800 K and hence no detectable emission changes. On the {110}–{211} zones diffusion is also rapid, but (110) terrace edges are present which act as trap sites and so lead to a work-function increase at low temperature.

Diffusion temperatures for O_2 into various planes are shown in Table 8.1. Here it is clear that diffusion on (110) occurs at lower temperature and with a lower activation energy ($\sim$63 kJ mol^{-1}) than on other parts of the surface. On these other areas there is little variation in the onset temperature, which contrasts with the behaviour of CO on W. It is suggested (Engel and Gomer 1970) that because of its smaller size an oxygen atom experiences a similar atomic environment on most planes, while the larger CO molecule does not. The possibility of surface reconstruction in the

TABLE 8.1.

Diffusion temperatures for O_2 on W (tip dosed at 25 K).

Plane	(110)	(111)	(120)	(110) vicinals	(211)	(100)	(211)–(110) zone
Temperature (K)	400	450	475	475	500	500	500

case of oxygen and the presence of multiple-binding states with Co may also contribute in part to the observed differences. (see Chapters 9 and 12).

Diffusion of nitrogen on tungsten has been studied in the higher temperature range by Ehrlich and Hudda (1961). Gas chemisorbed at 290 K is immobile up to 400 K, when boundary-free migration begins in the vicinity of the (111) planes. At 500 K movement over (110) takes place, the activation energy being estimated as 84 kJ mol^{-1}. Finally, at temperatures above 650 K bondary diffusion starts in the area around (100), with $E_m \sim$ 147 kJ mol^{-1}.

8.4. Kinetic models for chemisorption

When we examine sticking-probability curves such as those for the nitrogen-polycrystalline tungsten system (Fig. 8.9(a)) two rather unusual features are apparent.

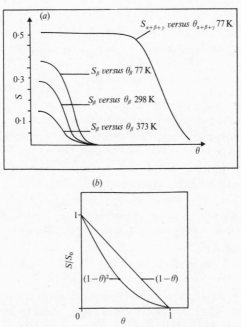

FIG. 8.9. (a) Sticking probabilities of N_2 on polycrystalline W wire. (From Ehrlich 1961.) (b) Form of eqns (8.7) and (8.8); S_0 (= c) is the initial sticking probability.

(i) At low coverage the sticking probability S *decreases* as the temperature of the surface is increased. This immediately suggests that chemisorption under these conditions cannot be a single-step process with a simple activation energy and points to an intermediate step adversely affected by temperature increase. At high coverages S becomes essentially temperature independent.

(ii) The sticking probability S is constant with coverage θ at low θ, then decreases sharply, and finally levels off again as saturation is approached. As the temperature is raised the fall off in S occurs at lower values of θ.

If non-dissociative adsorption followed a simple Langmuir model, in which any gaseous molecule striking an empty surface site is chemisorbed, while those striking occupied sites are reflected, the rate of adsorption at a given coverage would be proportional to the number of empty sites present at that time, i.e.

$$S = c(1 - \theta) \tag{8.7}$$

where c is the condensation coefficient. If dissociative chemisorption required the availability of a pair of adjacent vacant sites then

$$S = c(1 - \theta)^2. \tag{8.8}$$

The forms of eqns (8.7) and (8.8) are indicated in Fig. 8.9(b) and obviously do not describe the experimental curves.

From our previous discussion we know that a molecule striking a metal may become adsorbed in a weakly bound precursor state in which it can diffuse over the surface until it either is trapped into a more strongly held state or desorbs. This idea of a mobile precursor acting as a reservoir for subsequent chemisorption was put forward by Langmuir in 1929; the first experimental observation was made by Taylor and Langmuir (1933) when they found that the sticking probability of caesium on tungsten was unity up to coverages as high as 98 per cent of a monolayer. The model was also invoked by Becker and Hartman (1953) to explain their data for flash desorption of nitrogen from tungsten and was then taken up by Ehrlich (1955, 1956) in a slightly more refined version, which we shall consider first.

The events leading to chemisorption are illustrated schematically in Fig. 8.10. Gaseous molecules strike the surface at a rate ν

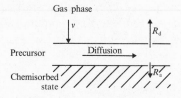

FIG. 8.10. Schematic representation of chemisorption via a mobile precursor state.

and enter a mobile precursor state with probability c, where c is the condensation coefficient. From the precursor a molecule may either be desorbed (at a rate R_d) or become chemisorbed (at a rate R_a). If σ_p is the surface concentration in the precursor and k_d, k_a are the rate constants for desorption and chemisorption respectively, then

$$\frac{d\sigma_p}{dt} = \nu c - k_d\sigma_p - k_a\sigma_p \qquad (8.9)$$

where, from the Hertz–Knudsen equation (Chapter 6),

$$\nu = \frac{P}{(2\pi mkT)^{\frac{1}{2}}} = ZP. \qquad (8.10)$$

Also, the rate of chemisorption will be, for unit area,

$$R_a = k_a\sigma_p \qquad (8.11)$$

while by definition the sticking probability is given by

$$S = \frac{R_a}{ZP}. \qquad (8.12)$$

Thus, combining eqns (8.9)–(8.12) we have

$$S = \frac{k_a\sigma_p}{ZP} = \frac{k_a}{ZP}\left(\frac{ZPc - d\sigma_p/dt}{k_d + k_a}\right). \qquad (8.13)$$

A number of assumptions are introduced at this point as follows.
   (i) Adsorption occurs on a perfect homogeneous plane on which all surface sites are identical.
   (ii) The condensation coefficient c is approximately unity.
   (iii) The binding energy of the precursor, and k_d, are the same above both filled and empty sites.
   (iv) The concentration of the precursor species is constant.

(v) Desorption from the chemisorbed state is negligible.

Assumptions (ii) and (iv) lead to a simplification of eqn (8.13), since $c = 1$ and $d\sigma_p/dt = 0$. Thus

$$S = \frac{k_a}{k_d + k_a} \tag{8.14}$$

Now, we can also write

$$k_a = f(\theta)\kappa_a \nu_a \exp\left(\frac{-E_a}{kT}\right) \tag{8.15}$$

and

$$k_d = \kappa_d \nu_d \exp\left(\frac{-E_d}{kT}\right). \tag{8.16}$$

Here κ is the probability that a vibration (frequency $\nu_{a(d)}$) of the precursor–surface complex in which the energy exceeds the activation energy E will actually lead to chemisorption (or desorption) and f is a factor which accounts for the availability of empty sites for chemisorption. For example, for dissociative adsorption of diatomic molecules, with the adsorbed atoms randomly distributed, $f = (1 - \theta)^2$. We shall consider this situation for illustrative purposes, remembering that it implies surface diffusion of chemisorbed species and negligible adatom–adatom interactions. Using eqns (8.15) and (8.16), eqn (8.14) becomes

$$S = \frac{1}{\{C/(1-\theta)^2\} \exp\{(E_a - E_d)/kT\} + 1} \tag{8.17}$$

where

$$C = \frac{\kappa_d \nu_d}{\kappa_a \nu_a}.$$

We are now in a position to examine the way in which this last equation accounts, at least qualitatively, for the main features of the experimental sticking-probability curves. The effect of temperature is best determined by rearranging eqn (8.17) and taking logarithms:

$$\ln\left(\frac{S}{1-S}\right) - \ln\left\{\frac{C}{(1-\theta)^2}\right\} = \left(\frac{E_d - E_a}{kT}\right). \tag{8.18}$$

This equation is now differentiated with respect to $(1/kT)$ at constant θ:

$$\left\{ \frac{\partial \ln S/(1-S)}{\partial(1/kT)} \right\}_\theta = E_\mathrm{d} - E_\mathrm{a}. \tag{8.19}$$

Thus the temperature dependence of the experimentally determined quantity $S/(1-S)$ depends on the relative magnitudes of E_d and E_a. Let us look at two cases.

(i) $E_d > E_a$. In this situation $S/(1-S)$ will increase as $1/T$ increases; in other words S will increase as T decreases (until $kT < E_\mathrm{a}$). In physical terms desorption from the precursor is the overall rate-controlling process, and this will decrease as the temperature is lowered. Precursor molecules then remain on the surface for longer periods, increasing the chance of encounters with vacant sites for chemisorption. When such a site is reached there will be a reasonable probability of chemisorption provided $kT > E_\mathrm{a}$. The fact that the precursor can 'seek out' vacant sites accounts for the observation that the sticking probability remains constant up to some critical coverage and then falls rather sharply. At this coverage the average distance between vacant sites will be of the order of the mean diffusion distance of precursor molecules before evaporation; at lower converges any molecule entering the precursor will have a high probability of finding an empty site and most will therefore be chemisorbed, while at higher coverages desorption from the precursor will become significant and S will fall. Furthermore, with increasing temperature the precursor diffusion distance will decrease and so the coverage at which S begins to fall will also decrease, as observed.

(ii) $E_a > E_d$. In this case S will increase as T *increases* (cf. eqn (8.19)) since the chemisorption step is now rate determining. Also, at a temperature corresponding to some given rate of chemisorption R_a the lifetime of diffusing molecules before desorption will be less than in case (i) for the same value of R_a, and so S will be a more sensitive function of θ in case (ii).

The model we have just discussed, with $E_\mathrm{d} > E_\mathrm{a}$ (case (i) above), is found to describe qualitatively the behaviour of the sticking coefficients of nitrogen on polycrystalline tungsten, but the agreement is not quantitative—hardly surprising in view of the assumptions made. Some improvement was achieved by Ehrlich (1956) by considering the surface to be heterogeneous, composed of active and inactive patches. The estimated values for E_d and E_a were ~19 kJ mol^{-1} and ~4–8 kJ mol^{-1} respectively.

The magnitude of E_d is consistent with the identification of the precursor as a physically adsorbed state, but at first sight it may seem rather unrealistic when we use this value in the Frenkel equation to calculate the mean residence time of a precursor molecule on the surface at, say, 300 K:

$$\tau = 10^{-13} \exp\left(\frac{19 \times 10^3}{8 \cdot 314 \times 300}\right) \approx 2 \times 10^{-10} \text{ s}.$$

The important factor, however, is not the absolute value of τ but the number of diffusive hops a molecule can make during this time. Taking the jump frequency as ~10^{13} s^{-1} we find this to be ~2000, and so a large number of potential chemisorption sites can be visited in spite of the seemingly low residence time. It is also interesting to calculate the concentration of precursor species σ_p. This will be of the order of $(\nu c \tau)$ where ν is the impingement rate and c the condensation coefficient. For N_2 at 300 K $\nu = 3 \cdot 85 \times 10^{20} P$ (torr); we shall assume for convenience that $c \approx 1$. Then taking $\tau = 2 \times 10^{-10}$ s

$$\sigma_p = 7 \cdot 7 \times 10^{10} P \text{ molecules cm}^{-2}. \qquad (8.20)$$

Since experiments of the types we have been examining are generally carried out at pressures lower than say 10^{-2} torr, it follows that $\sigma_p \lesssim 10^9$ molecules cm^{-2}. When this is compared with the typical substrate atom density of ~$5 \times 10^{+14}$ cm^{-2}, it is seen that interactions between precursor species will be negligible.

The quantitative description of sticking probability data was further investigated by Smith (1964), who specifically rejected the assumption that the precursor concentration is constant at all times, arguing that there must be a period at the beginning of the adsorption process during which the precursor accumulates. He also relaxed the restriction that the condensation coefficient should be unity and then proceeded along the lines outlined above to derive an expression for S which gave good agreement with the experimental curves for both N_2/W and CO/W. The main objection to this particular treatment is that the precursor was identified with the α-state of nitrogen, which has a heat of adsorption of $\sim 84\,\text{kJ mol}^{-1}$ (see Chapter 10). In FEM experiments diffusion of nitrogen deposited at 290 K is not detected up to 400 K, at which temperature α-N_2 is in fact desorbing rapidly (Ehrlich and Hudda 1961). This indicates that the binding energy and activation energy for diffusion must be comparable.

The model developed by Ehrlich, which we have discussed above, achieved limited quantitative success only by introducing the assumption of a heterogeneous or 'patchy' surface with regions of rapid chemisorption and other regions where adsorption was negligible. Using the same basic picture of the chemisorption process an alternative model was introduced by Kisliuk (1957, 1958); this has met with greater success because of its greater flexibility. The reasoning proceeds by considering the fate of a precursor species as follows (see e.g. Kohrt and Gomer 1970). For the first site visited in its migration over the surface let the probability of chemisorption be

$$P_{a_1} = P_a f(\theta) \tag{8.21}$$

where P_a is the probability of adsorption on an empty site (or pair of sites if adsorption is dissociative) and $f(\theta)$ is a function related to the adsorption mechanism which determines whether or not the site (or site pair) is empty. The probability of desorption from the first site will be

$$P_{d_1} = P_d f(\theta) + P_d''\{1 - \theta - f(\theta)\} + P_d'\theta \tag{8.22}$$

where the various probabilities relate to the following events: P_d, desorption from an empty site (or site pair) which is available for adsorption; P_d'', desorption from an empty site not available for adsorption (equal to P_d if adsorption is non-dissociative; in the formulation by Kisliuk (1958) of dissociative adsorption P_d'' was not distinguished from P_d'); P_d', desorption from a filled site. From (8.21) and (8.22) the probability of migration to another site will be

$$P_{m_1} = 1 - (P_{a_1} + P_{d_1}). \tag{8.23}$$

Now for the precursor species arriving at a second site the overall probabilities of adsorption, desorption, and further diffusion to a third site will be $P_{m_1}P_{a_2}$, $P_{m_1}P_{d_2}$, $P_{m_1}P_{m_2}$. Assuming that the probabilities of a particle experiencing a particular fate once it has reached a given site are identical for all sites these overall probabilities become $P_{m_1}P_{a_1}$, $P_{m_1}P_{d_1}$, and $P_{m_1}^2$. Similarly, for a species which reaches a third site these probabilities are $P_{m_1}^2 P_{a_1}$, $P_{m_1}^2 P_{d_1}$, $P_{m_1}^3$, and so on for succeeding sites visited. The overall probability S' that the precursor species will eventually be chemisorbed is

$$S' = \sum_{i=1}^{\infty} P_{a_i} = P_{a_1}[1 + P_{m_1} + P_{m_1}^2 + \ldots]$$

$$= \frac{P_{a_1}}{1 - P_{m_1}}$$

$$= \frac{P_{a_1}}{P_{a_1} + P_{d_1}}. \qquad (8.24)$$

Substituting for P_{a_1} and P_{d_1} then gives

$$S' = \frac{P_a f(\theta)}{(P_a + P_d)f(\theta) + P_d''\{1 - \theta - f(\theta)\} + P_d'\theta}. \qquad (8.25)$$

At zero coverage $f(\theta) = 1$, provided surface reconstruction does not occur, and so the initial sticking probability will be

$$S_0' = \frac{P_a}{P_a + P_d}. \qquad (8.26)$$

Eqn (8.25) may then be rewritten in the form

$$\frac{S'}{S_0'} = f(\theta)\left[f(\theta) + \frac{P_d''}{P_a + P_d}\{1 - \theta - f(\theta)\} + \frac{P_d'\theta}{P_a + P_d}\right]^{-1}. \qquad (8.27)$$

The next step is to insert expressions for $f(\theta)$ appropriate to various types of adsorption (Kohrt and Gomer 1970), recalling that this function is the probability that the required sites will be vacant; the first two cases were treated by Kisliuk (1957, 1958).

(a) Non-dissociative adsorption on a single site. In this situation $f(\theta) = (1 - \theta)$ and so

$$\frac{S'}{S_0'} = \left\{1 + \frac{P_d'}{P_a + P_d}\left(\frac{\theta}{1 - \theta}\right)\right\}^{-1} = \left\{1 + K_1\left(\frac{\theta}{1 - \theta}\right)\right\}^{-1} \qquad (8.28)$$

where

$$K_1 = \frac{P_d'}{P_a + P_d}. \qquad (8.29)$$

(b) Dissociative adsorption, requiring a pair of adjacent vacant sites for the dissociation step, but with eventual adsorption of the molecular fragments on a random distribution of sites. For many adsorbates the bond dissociation energy D will be considerably greater than twice the activation energy for surface diffusion of the chemisorbed fragments (E_m); taking N_2/W as an example $D = 945$ kJ mol^{-1}, while $E_m \sim 84$–147 kJ mol^{-1} for diffusion in the atomically bound β-state. If the dissociation energy is not immediately equilibrated with the lattice the fragments will therefore move across the surface losing energy as they go until equilibrium is finally attained. For N_2/W Ehrlich (1963) has estimated that this may require up to 50 diffusive hops. Provided occupied sites are randomly distributed, the probability that a given site is vacant is $(1 - \theta)$. There will be an identical chance that any one of its nearest neighbours is also vacant, and so $f(\theta) = (1 - \theta)^2$. Then eqn (8.27) becomes, after some rearrangement,

$$\frac{S'}{S_0'} = (1 - \theta)\left\{1 - \theta\left(S_0' - \frac{P_d'' - P_d}{P_a + P_d}\right) + K_1\left(\frac{\theta}{1 - \theta}\right)\right\}^{-1}$$

$$= (1 - \theta)\left\{1 - K_2\theta + K_1\left(\frac{\theta}{1 - \theta}\right)\right\}^{-1} \tag{8.30}$$

where

$$K_2 = S_0' - \frac{P_d'' - P_d}{P_a + P_d}. \tag{8.31}$$

If the probability of desorption is the same at *any* empty site then $P_d = P_d''$ and $K_2 = S_0'$. Eqn (8.30) then becomes

$$\frac{S'}{S_0'} = (1 - \theta)\left\{1 - K_2'\theta + K_1\left(\frac{\theta}{1 - \theta}\right)\right\}^{-1} \tag{8.32}$$

where $K_2' = S_0'$. This is the relationship derived by Kisliuk (1958) for dissociative adsorption.

(c) Dissociative adsorption, with only adjacent sites occupied. In this case it can be shown (Roberts and Miller 1939) that $f(\theta)$ is approximately equal to $(1 - \theta)^2(1 - \theta/z)^{-1}$ where z is the number of nearest-neighbour sites to a given site. If for example $z = 4$ then

$$\frac{S'}{S_0'} = (1-\theta)\left\{1 - K_2''\theta + K_1\left(\frac{\theta - \theta^2/4}{1-\theta}\right)\right\}^{-1} \quad (8.33)$$

where

$$K_2'' = S_0' - \frac{\frac{3}{4}P_d'' - P_d}{P_a + P_d}.$$

Up to this point we have examined only the probability S' that a molecule in the *precursor state* will be chemisorbed, whereas the experimentally accessible sticking probability S must include the condensation coefficient c. If c is independent of coverage

$$S = cS' \quad \text{and} \quad S/S_0 = S'/S_0'. \quad (8.34)$$

If, however, c has a different value (c') over a filled site compared with an empty site (c) then

$$S = S'[c(1-\theta) + c'\theta] \quad (8.35)$$

and

$$\frac{S}{S_0} = \left\{1 + \frac{\theta(c'-c)}{c}\right\}\frac{S'}{S_0''}. \quad (8.36)$$

Thus the value of the condensation coefficient, in addition to determining the absolute magnitude of S, can alter its functional form if c is coverage dependent.

Theoretical plots of S'/S_0' against θ according to eqn (8.28) (non-dissociative adsorption) for various values of K_1 are shown in Fig. 8.11(a) (Kisliuk 1957). Equivalent plots according to eqn (8.30) (dissociative adsorption) are shown in Fig. 8.11(b), and virtually the same curves are obtained if K_2 is replaced by K_2'' and eqn (8.33) is used instead (Kohrt and Gomer 1970). From these figures it may be concluded that S'/S_0' is relatively sensitive to K_2, i.e. to S_0', but relatively insensitive to K_1, i.e. to the probabilities of desorption over filled and empty sites.

Before considering the application of these relationships to analysis of experimental data it is interesting to compare the Kisliuk approach with the exponential rate theory of Ehrlich and others. There are two main points of difference. The exponential rate theory obviously will be discussed in terms of rate

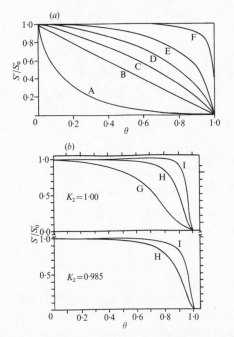

FIG. 8.11. (a) A plot of S'/S_0' (eqn (8.28)): A, $K_1 = 10$, B, $K_1 = 1\cdot0$; C, $K_1 = 0\cdot50$; D, $K_1 = 0\cdot25$; E, $K_1 = 0\cdot10$; F, $K_1 = 0\cdot01$. (After Kisliuk 1957.) (b) S'/S_0' (eqn 8.30)) for various values of K_1 and K_2: G, $K_1 = 0.001$; H, $K_1 = 0.01$; I, $K_1 = 0\cdot1$. (After Kohrt and Gomer 1970.)

expressions; also, it assumes that every gaseous molecule striking the surface is physisorbed. Kisliuk's theory and its derivatives, on the other hand, consider probabilities for diffusion etc. and allow for the possibility that the condensation coefficient may be less than unity and coverage dependent. The probability of diffusion (migration) P_m is not merely proportional to the inverse of the rate of diffusion since it also depends on the average time τ spent at a site; the same applies to adsorption and evaporation. Thus in terms of the rate constants for the various possible processes, which are given by equations such as

$$k_a = A_a \exp\left(-\frac{E_a}{kT}\right) \qquad (8.37)$$

$$k_d = A_d \exp\left(-\frac{E_d}{kT}\right) \qquad (8.38)$$

the average time spent above an empty site available for chemisorption will be

$$t_{av} = \frac{1}{k_m + k_a + k_d} \qquad (8.39)$$

and above a site not available for chemisorption

$$t''_{av} = \frac{1}{k''_m + k''_d} \qquad (8.40)$$

or

$$t'_{av} = \frac{1}{k'_m + k'_d}. \qquad (8.41)$$

The various probabilities are then given by

$$
\begin{aligned}
P_a &= k_a t_{av} & P_d &= k_d t_{av} \\
P''_d &= k''_d t''_{av} & P'_d &= k'_d t'_{av}
\end{aligned}
\qquad (8.42)
$$

In the Kisliuk scheme the case which corresponds to Ehrlich's model is that in which further migration is the most probable event at a given site and the probabilities of desorption above empty and unavailable sites are equal, i.e. $P_d = P'_d = P''_d$. Suppose this condition is imposed on eqn (8.30), along with the substitutions

$$K_1 = \frac{P_d}{P_a + P_d} = \frac{k_d}{k_a + k_d} \qquad (8.43)$$

and

$$K_2 = S'_0 = \frac{P_a}{P_a + P_d} = \frac{k_a}{k_a + k_d}. \qquad (8.44)$$

Then we find that eqn (8.30) transforms to

$$S' = \frac{1}{k_d/\{k_a(1-\theta)^2\} + 1} \qquad (8.45)$$

which is identical to eqn (8.17) of the exponential rate model. The lack of success of this model is due to the fact that this equation contains no adjustable parameters. The form of the entire sticking-probability curve is effectively controlled by the ratio k_a/k_d, in other words by the initial sticking probability S'_0 (cf. eqn (8.44)). As we saw above, in the Kisliuk model the form of the curve is still sensitive to the value of S_0', but does in addition depend on the relative probabilities of desorption over filled and empty sites through the parameter K_2 (eqn (8.31)).

The probability that a molecule will acquire thermal energy in excess of some value E is given by a Boltzmann term $\exp(-E/kT)$ only for systems at thermal equilibrium, and so it is only for systems of this type that rates may be expressed by equations such as (8.37) and (8.38). When a gaseous molecule is first trapped at a surface, however, it may take some time to equilibrate thermally with the lattice. A detailed description of its behaviour during this period would be difficult, but Kisliuk (1958) avoided the problem by way of a neat formal manoeuvre. He assumed that during the equilibration period a molecule may desorb or diffuse, but that the probability of chemisorption during this time is negligible. As far as the final rate of chemisorption is concerned, a molecule which desorbs during equilibration may equally well have been reflected on its initial collision with the surface and so all molecules which do not reach thermal equilibrium in the precursor state may be included in a reflection coefficient r. We may then write

$$S_0' = (1-r)S_0 \qquad (8.46)$$

where S_0' is the initial probability of chemisorption from the equilibrated precursor reservoir and S_0 is the observed initial sticking probability. The Kisliuk theory takes the precursor at thermal equilibrium as its starting point, and experimental data in the form S/S_0 are fitted to a theoretical expression for S'/S_0' such as eqn (8.30) by adjusting the parameters K_1 and K_2. If the value of S_0' corresponding to the chosen K_2 is less than the observed value of S_0, then the difference is formally ascribed to the term $(1-r)$; this is equivalent to setting $(1-r)$ equal to the condensation coefficient c. There is no discussion therefore as to what fraction of the incident molecules is reflected on first contact with the surface and what fraction is actually trapped for a finite time, that is of the distinction between a true reflection coefficient and the condensation coefficient.

The application of Kisliuk's theory as just discussed may be illustrated by reference to the work of Kohrt and Gomer on the systems W(110)/CO (Kohrt and Gomer 1968) and W(110)/O_2 (Kohrt and Gomer 1970); the most recent results are discussed by Wang and Gomer (1977). In these cases problems related to surface heterogeneity were of course greatly reduced by the use of single crystals. Sticking probabilities as functions of coverage at various temperatures are shown in Fig. 8.12. In Fig. 8.13 experimental values of S/S_0 *versus* θ for oxygen are compared with the best theoretical fits given by eqn (8.30). From these a number of conclusions may be drawn.

   (a) For O_2/W S_0' is close to unity even at temperatures where $S_0 < 1$, and so reflection is appreciable and increases with increasing temperature. Also, from eqns (8.26) and (8.42) it follows that $k_a \gg k_d$. For CO/W in contrast $S_0 \geqslant 0\cdot7$ for all $T \leqslant 700$ K and reflection must be small, at least on the clean surface. This is not unexpected since CO is known to adsorb, at least in part, without dissociation on W, and there will thus be a strong interaction potential for CO/W collisions, while O_2 molecules are only weakly bound and the potential is presumably much weaker.

   (b) For $O_2/W(110)$ $K_1 = k_d''/(k_a + k_d)$ falls between $0\cdot01$ and $0\cdot001$. The theoretical curves are fairly insensitive to the value of K_1 and so it cannot be determined accurately. Since, however, k_d' is the rate of desorption of molecular oxygen from above filled sites it is probable that E_d' is less than about 9 kJ mol^{-1}. Thus k_d' will be relatively high and the fact that K_1 is small must mean that k_a is even higher, indicating that the activation energy for chemisorption is very low. This is confirmed by the observation of unit

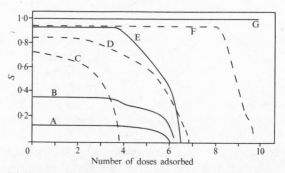

FIG. 8.12. Sticking probability data for CO and O_2 on W(110). Curve A, O_2, 400 K; B, O_2, 300 K; C, CO, 600 K; D, CO, 300 K; E, O_2, 63 K; F, CO, 100 K; G, O_2, 20 K. (Data from Kohrt and Gomer 1968, 1970.)

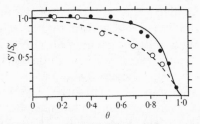

FIG. 8.13. Plots of S'/S_0' versus θ for W(110)-O$_2$. Points, experimental data; lines calculated from eqn (8.30); solid curve, 63 K, $K_1 = 0.01$, $K_2 = 0.99$; broken curve, 400 K, $K_1 = 0.001$, $K_2 = 0.80$. (From Kohrt and Gomer 1970.)

sticking probability at 20 K (Fig. 8.12). Note that if $E_d' \leqslant$ 9 kJ mol^{-1} the Frenkel equation indicates that the residence time τ for a precursor species moving over filled sites at 300 K will be $10^{-13} \exp(9/2.5) = 3.7 \times 10^{-12}$ s. If the jump frequency is of the order of 10^{13} s the species will make about 40 hops during this time. This does not mean that it will visit 40 distinct sites, since on any hop there will be a finite probability (e.g. 0.25 on a four-fold symmetric surface) of it returning to the site it left on the previous hop. Nevertheless, in spite of its very short residence time the precursor molecule will be able to move over a considerable area of the surface in its search for a vacant site.

(c) While eqn (8.30) provides a reasonable fit for the oxygen data, it fails to account for the constant values of S up to fairly high critical coverages θ_c ($0.5 \leqslant \theta_c \leqslant 0.7$, depending on temperature) which are evident in Fig. 8.12. Kohrt and Gomer (1970) suggest that at low coverage one may have to consider the possibility that a molecule trapped after its initial collision with the surface may become chemisorbed before reaching thermal equilibrium; this situation was specifically excluded in the derivation of eqn (8.30), which would not therefore accurately apply in this low-coverage region. At θ_c, however, the time required to migrate from the point of impact to an available site becomes greater than the equilibrium time and eqn (8.30) becomes valid.

8.5. The effect of adsorbate–adsorbate interactions

We have already considered the effect of adsorbate–adsorbate interactions on the desorption process and the Kisliuk approach may be easily modified to take account of the influence of such interactions on the kinetics of adsorption; this has been done for the case of $N_2/W(100)$ by King and Wells (1974) by the inclusion of a pairwise lateral repulsive interaction energy ω between nearest-neighbour (n.n.) chemisorbed adatoms. Thus short-range order is introduced into the chemisorbed layer, whereas in Kisliuk's model the layer was taken to be completely disordered. If, as assumed by King and Wells, a pair of adjacent empty sites is required for dissociation, $\theta = 0\cdot5$ represents maximum coverage if order is complete since all empty n.n. pairs have then been eliminated (Fig. 8.14(a)). Adsorption to any higher coverage requires some degree of disorder in the chemisorbed layer.

King and Wells did not distinguish the probability of desorption for a precursor molecule over a filled chemisorption site (P_d') from that for desorption over an empty site not available for chemisorption (P_d''). We therefore take as our starting point eqn (8.27) with $P_d'' = P_d'$, and determine $f(\theta)$ in the presence of the repulsive interaction. This may be done (King and Wells 1974) by using a method known as the 'quasi-chemical' approximation developed for discussion of mixing in liquids. The following is a non-rigorous outline of the procedure adopted (Rice 1967); a more detailed discussion is given by Fowler and Guggenheim (1960). We consider the process

$$AA + OO \rightarrow 2AO$$

which is shown diagrammatically in Fig. 8.14(b); that is, the conversion of a pair of empty sites (OO) and a pair of filled sites

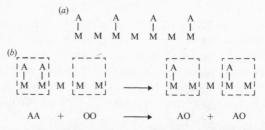

FIG. 8.14. (a) Maximum coverage in a completely ordered adlayer in the case in which a pair of adjacent empty sites is required for dissociation. (b) The conversion of a pair of filled sites (AA) and a pair of empty sites (OO) to two site pairs (AO), each of which is half-filled.

(AA) to two site pairs of type AO, in each of which one site is occupied (A) and the other empty (O).

Let N_{AA}, N_{AO} be the total numbers of AA and AO site pairs. Then the number of sites which are nearest neighbours of an A site will be $(2N_{AA} + N_{AO})$. If N_A is the total number of occupied single sites and if every surface site has z nearest neighbours, then

$$zN_A = (2N_{AA} + N_{AO}). \qquad (8.47)$$

In the same way we can write

$$zN_O = (2N_{OO} + N_{AO}) \qquad (8.48)$$

where N_O, N_{OO} are the total number of empty single sites and of empty site pairs respectively. Suppose now that the repulsive interaction energy is zero and that the distribution of chemisorbed species is random; for any occupied site the probability that any one of its n.n. sites will be empty is N_O/N, where N is the total number of surface sites. Thus the total number of AO pairs will be

$$N_{AO} = \frac{zN_A N_O}{N} = \frac{zN_A N_O}{(N_A + N_O)}. \qquad (8.49)$$

From eqns (8.47)–(8.49) we can now write

$$N_{AO} = \frac{(2N_{AA} + N_{AO})(2N_{OO} + N_{AO})}{2(N_{AA} + N_{AO} + N_{OO})} \qquad (8.50)$$

which immediately reduces to

$$\frac{N_{AO}^2}{N_{AA} N_{OO}} = 4. \qquad (8.51)$$

This relationship is simply an expression of the assumption that the particles are randomly distributed. The quasi-chemical approximation is invoked at this point by recognizing that the left-hand side of the last equation is analogous to an expression for the equilibrium constant of the 'reaction' under consideration; therefore we proceed by treating it as if this actually was the case. It is interesting to recall in passing that for a gaseous reaction such as oxygen isotopic exchange

$$^{16}O^{16}O + {}^{18}O^{18}O \rightarrow 2\,^{16}O^{18}O$$

which occurs without energy change the equilibrium constant is also 4 (neglecting differences in moments of inertia which appear in the rotational partition functions). This value results from the configurational entropy change involved, $\Delta S^{\ominus} = k \ln 4$. The behaviour of the 'reaction'

$$AA + OO \rightarrow 2AO \qquad (8.52)$$

is exactly analogous at a coverage of $\theta = 0 \cdot 5$. For any process occurring at constant volume the equilibrium constant is given by

$$\ln K = -\frac{\Delta U^{\ominus}}{kT} + \frac{\Delta S^{\ominus}}{k}$$

where $\Delta U^{\ominus}$ is the standard internal energy change. So for process (8.52), taking place without energy change, we can write

$$\ln K = \ln 4 = \Delta S^{\ominus}/k. \qquad (8.53)$$

We now assume that a repulsive pairwise interaction energy ω is involved. Then $\Delta U^{\ominus} = -\omega$ and the equilibrium constant will be given by

$$\ln K = +\frac{\omega}{kT_{\mathrm{s}}} + \ln 4$$

or

$$K = \frac{N_{\mathrm{AO}}^2}{N_{\mathrm{AA}} N_{\mathrm{OO}}} = 4 \exp\left(+\frac{\omega}{kT_{\mathrm{s}}}\right) \qquad (8.54)$$

where T_{s} is the surface temperature. It is now a straightforward matter to eliminate N_{AA} and N_{AO} from eqns (8.47), (8.48), and (8.54) to derive a quadratic equation for $f(\theta) = (N_{\mathrm{OO}}/\frac{1}{2}zN)$, the probability that a pair of n.n. sites will both be empty. The final solution is

$$f(\theta) = 1 - \theta - \frac{2\theta(1-\theta)}{\{1 - 4\theta(1-\theta)B\}^{\frac{1}{2}} + 1} \qquad (8.55)$$

where $\theta = N_{\mathrm{A}}/N$ and $B = 1 - \exp(-\omega/kT_{\mathrm{s}})$.

It was assumed by King and Wells that the condensation coefficient c is coverage independent and so, from eqn (8.27) with $P_{\mathrm{d}}'' = P_{\mathrm{d}}'$, the expression for the sticking probability becomes

$$\frac{S}{S_0} = \frac{S'}{S_0'} = \left\{1 + K_1\left(\frac{1}{f(\theta)} - 1\right)\right\}^{-1} \qquad (8.56)$$

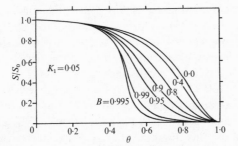

FIG. 8.15. The effect of the short-range order parameter B on computed sticking-probability curves, for a fixed value of K_1 (eqns (8.55) and (8.56)). (From King and Wells 1974.)

where

$$K_1 = \frac{P'_d}{P_a + P_d} \qquad (8.29)$$

and, from eqns (8.26) and (8.34)

$$S_0 = cS'_0 = \frac{cP_a}{P_a + P_d}. \qquad (8.57)$$

The effect of the short-range order parameter B is shown in Fig. 8.15, which represents computed sticking-probability curves for a fixed value of K_1 and various values of B; $B = 0$ corresponds to Kisliuk's model. Experimental data for the system $N_2/W(100)$ obtained (King and Wells 1972) using the molecular-beam technique (see Chapter 6) are shown in Fig. 8.16 for

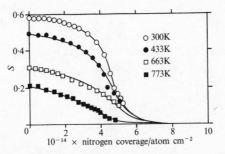

FIG. 8.16. Comparison of experimental sticking-probability data for $W(100)-N_2$ (symbols) with values calculated using eqns (8.55) and (8.56) (solid lines). (From King and Wells 1974.)

various adsorbent temperatures T_s. The solid curves represent computer-fitted curves from eqns (8.55) and 8.56); the agreement is seen to be good. The total density of chemisorption sites N was included as an adjustable parameter, and values for B, K_1, N, and S_0 are given in Table 8.2 (King and Wells 1974).

The mean value of B was 0.984 ± 0.005, indicating a high degree of short-range order, but no systematic variation of B with surface temperature T_s in the range 300–773 K is apparent from these data. This suggests that order results from limited migration of atoms following dissociation during the period in which the dissociation energy is being dissipated to the lattice. This process might be expected to be less sensitive to T_s than the alternative, namely continuous migration of chemisorbed species which have attained thermal equilibrium with the solid. This is in general agreement with the field emission microscope result (Ehrlich and Hudda 1961) that migration of $\beta - N$ on W is not appreciable below 650 K (cf. Section 8.3).

The best-fit values for the surface site density N derived from the data at various temperatures are very consistent, and the average value, $(9.8 \pm 0.5) \times 10^{14} \, \text{cm}^{-2}$, agrees well with the W(100) surface atom density, $10 \times 10^{14} \, \text{cm}^{-2}$. The saturation nitrogen coverages (at $S = 0.02$) correspond to ~60 per cent of a monolayer $(6 \times 10^{14} \, \text{atoms cm}^{-2})$ which correlates with the existence of considerable, but not total, short-range order and with the LEED observation (Adams and Germer 1971) of a c(2×2) pattern at saturation. They also confirm that the nitrogen is dissociatively adsorbed, since otherwise the coverage would have been of the order of 6×10^{14} *molecules* cm^{-2}.

TABLE 8.2

Computed best-fit values of kinetic parameters for $T_g = 300 \, K$.

$T_s(K)$	$N(10^{18} \times \text{m}^{-2})$	K_1	B	S_0
300	9·5	0·082	0·989	0·585
433	10·0	0·157	0·987	0·49
493	8·6	0·211	0·989	0·43
663	10·5	0·256	0·977	0·31
773	9·5	0·517	0·986	0·21
773	9·5	0·483	0·985	0.22

Examination of eqns (8.44) and (8.57) shows that if the rate of desorption from the precursor state (over empty sites) is less than the rate of chemisorption, i.e. $E_d > E_a$, then k_d/k_a tends to zero as surface temperature decreases and so there will be a low-temperature limit to S_0 such that

$$\lim_{T \to 0} S_0 = c. \tag{8.58}$$

Such a limit is observed for $N_2/W(100)$ (King and Wells 1974), S_0 being virtually temperature independent between 77 and 195 K. This yields $c = 0.60$ for this system. Further, from eqns (8.57), (8.44), and (8.37), (8.38) it follows that

$$\ln\left(\frac{c}{S_0} - 1\right) = -\frac{E_d - E_a}{kT_s} + \ln\left(\frac{A_d}{A_a}\right). \tag{8.59}$$

A plot of $\log(c/S_0 - 1)$ against $1/T_s$ gives $(E_d - E_a) = 18 \text{ kJ mol}^{-1}$. As for the case of $O_2/W(110)$ discussed previously it may be supposed that E_a for nitrogen on tungsten is small, and thus E_d, which is equal to the heat of adsorption of the precursor, must be of the order of 20 kJ mol^{-1}.

8.6. Conclusion

In this chapter we have seen how the flash desorption experiment of Becker and Hartman led to the construction of models for the chemisorption process which involve a mobile precursor state. It must be emphasized, however, that it is not to be expected that all chemisorption systems will be described by a precursor model—we have concentrated on this for the purpose of illustration only. Nevertheless, for a number of systems, including H_2, N_2, O_2, and CO, principally on W, a rather satisfying and detailed picture emerges from the accumulated data and fairly comprehensive sets of parameters characterizing the steps involved in the process are available. Much work remains to be done of course, particularly in relation to the initial gas–surface collision, to the effects of adsorbate–adsorbate interactions, to the behaviour of mixed adsorption systems, etc. In addition, the extension of investigations on a large scale to adsorbate gases other than those just mentioned awaits the introduction of a straightforward and accurate method for measurement of impingement rates and absolute surface coverages which will be applicable under a wide range of experimental conditions.

REFERENCES

ADAMS, D. L., and GERMER, L. H. (1971). *Surf. Sci.* **26,** 109.

BASSETT, D. W. (1973). In *Specialist periodical reports: surface and defect properties of solids* (eds. M. W. Roberts and J. M. Thomas), vol. 2, p. 34. Chemical Society, London.

BECKER, J. A., and HARTMAN, C. D. (1953). *J. phys. Chem.* **57,** 157.

BERNASEK, S. L., SIEKHAUS, W. J., and SOMORJAI, G. A. (1973). *Phys. Rev. Lett.* **30,** 1202.

EHRLICH, G. (1955). *J. phys. Chem.* **59,** 473.

—— (1956). *J. Phys. Chem. Solids.* **1,** 3.

—— (1961). *J. chem. Phys.* **34,** 29.

—— (1963). In *Metal surfaces: structure, energetics and kinetics*, p. 221. American Society of Metals, Ohio.

EHRLICH, G., and HUDDA, F. G. (1959). *J. chem. Phys.* **30,** 493.

—— (1961). *J. chem. Phys.* **35,** 1421.

—— (1966). *J. chem. Phys.* **44,** 1039.

ENGEL, T., and GOMER, R. (1969). *J. chem. Phys.* **50,** 2428.

—— (1970). *J. chem. Phys.* **52,** 1832.

FOLMAN, M., and KLEIN, R. (1968). *Surf. Sci.* **11,** 430.

FOWLER, R. H., and GUGGENHEIM, E. A. (1960). *Statistical thermodynamics.* Cambridge University Press, London.

FUJITA, H. (1953). *J. chem. Phys.* **21,** 700.

GOMER, R. (1961). *Field emission and field ionisation.* Harvard University Press, Cambridge, Mass.

GOMER, R., and HULM, J. K. (1957). *J. chem. Phys.* **27,** 1363.

GOMER, R., WORTMAN, R., and LUNDY, R. (1957). *J. chem. Phys.* **26,** 147.

GOODMAN, F. O., and WACHMAN, H. Y. (1967). *J. chem. Phys.* **46,** 2376.

HAYWARD, D. O., and WALTERS, M. R. (1974). *Japan. J. appl. Phys. Suppl.* **2,** 587.

HINCHEN, J. J., and FOLEY, W. M. (1965). In *Proc. 4th Int. Symp. on Rarefied Gas Dynamics. Adv. appl. Mech. Suppl.* **3** (2). Academic Press, New York.

KING, D. A., and WELLS, M. G. (1972). *Surf. Sci.* **29,** 454.

—— (1974). *Proc. R. Soc. A* **339,** 245.

KISLIUK, P. (1957). *J. Phys. Chem. Solids.* **3,** 95.

—— (1958). *J. Phys. Chem. Solids.* **5,** 78.

KOHRT, C., and GOMER, R. (1968). *J. chem. Phys.* **48,** 3337.

—— (1970). *J. chem. Phys.* **52,** 3283.

KRUYER, S. (1953). *Koninkl. Ned. Akad. Wetenschap. Proc.* **568,** 274.

LEVENSON, L. L. (1967). *Nuovo Cim. Suppl.* **5,** 321.

LOGAN, R. M. (1969). *Surf. Sci.* **15,** 387.

—— (1973). In *Solid State surface Science* (ed. M. Green) vol. 3, p. 1. Marcel Dekker, New York.

LOGAN, R. M., and KECK, J. C. (1968). *J. chem. Phys.* **49**, 860.

LOGAN, R. M., and STICKNEY, R. E. (1966). *J. chem. Phys.* **44**, 195.

McCARROLL, B. (1963). *J. chem. Phys.* **39**, 1317.

McCARROLL, B., and EHRLICH, G. (1963). *J. chem. Phys.* **38**, 523.

McCLURE, J. D. (1972). *J. chem. Phys.* **57**, 2823.

PALMER, R. L., and SMITH, J. N., Jr. (1974). *J. chem. Phys.* **60**, 1453.

RICE, O. K. (1967). *Statistical mechanics, thermodynamics and kinetics*, chap. 12. W. H. Freeman, San Francisco.

ROBERTS, J. K. (1935). *Proc. R. Soc. A* **152**, 464.

ROBERTS, J. K., and MILLER, A. R. (1939). *Proc. Camb. phil. Soc.* **35**, 293.

SALTSBURG, H. (1973). *Ann. Rev. phys. Chem.* **24**, 493.

SMITH, T. (1964). *J. chem. Phys.* **40**, 1805.

STOLL, A. G., and MERRILL, R. P. (1973). *Surf. Sci.* **40**, 405.

TAYLOR, J. B., and LANGMUIR, I. (1933). *Phys. Rev.* **33**, 423.

TENDULKAR, D. V., and STICKNEY, R. E. (1971). *Surf. Sci.* **27**, 516.

WANG, C. and GOMER, R. (1977). In *Proc. 7th Int. Vac. Cong. and 3rd Int. Conf. on Solid Surfs.*, p. 1155.

WEINBERG, W. H. (1975). In *Adv. colloid and interface science* (eds. J. T. G. Overbeek, W. Prins and A. C. Zettlemoyer) vol. 4, p. 301. Elsevier, Amsterdam.

WEINBERG, W. H., and MERRILL, R. P. (1971). *J. vac. Sci. Tech.* **8**, 718.

—— (1972). *J. chem. Phys.* **56**, 2881.

9

CARBON MONOXIDE

9.1. Introduction

CARBON monoxide chemisorption on metals ranks with nitrogen as the most extensively studied system, largely due to the fact that it is generally considered to be a problem that is reasonably tractable by a number of different experimental techniques and also since it has been considered to be the antithesis of N_2. Nitrogen is generally regarded as the archetype of dissociative chemisorption whereas CO has been considered to retain its molecular identity in chemisorption. The development of ideas regarding CO chemisorption has been based on kinetic, isotopic, infrared, electron-impact desorption, and electron-spectroscopic data. We shall see, however, that the case for CO being *in general* chemisorbed in an undissociated state is not unambiguous. The adsorption of CO on transition metals has recently been reviewed by Ford (1970) and Hayward (1971), and we shall merely highlight here some of the more important features and indicate how recent structural and electron-spectroscopic studies have had an important influence on current thinking.

9.1.1. *Molecular-orbital description of carbon monoxide*

Since we shall frequently refer to spectroscopic aspects of carbon monoxide adsorption we recall here the molecular-orbital diagram for CO (Fig. 9.1). It shows that the levels of the isolated oxygen atom are below those of the isolated carbon atom. The two hybrid orbitals on oxygen do not match the energies of those on carbon; the lower-energy sp hybrid on carbon matches approximately the higher-energy hybrid on oxygen with the result that the electrons in the non-bonding orbitals σ_{nb} (the lone pairs) are on the lower-energy hybrid of oxygen and the higher-energy hybrid of carbon. The 'oxygen' lone pair is extremely stable while the 'carbon' lone pair has a high degree of p character and is responsible for the donor properties of carbon monoxide. The bonding between carbon and oxygen results mainly from the sigma bond, mostly on the oxygen, and the set of π orbitals, π_x,

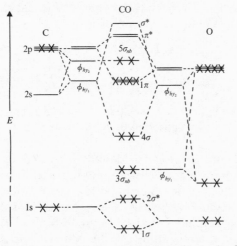

FIG. 9.1. The molecular-orbital energy diagram for CO.

π_y, also mainly on the oxygen atom. The anti-bonding π^* orbital is shown lying closer to the carbon p's than the oxygen p's. This is an important point in the interpretation of the acceptor properties of carbon monoxide in metal–carbonyl complexes. It is also relevant to the understanding of the spectroscopic data of chemisorbed CO on metals discussed later (Primet *et al.* 1973, Bradshaw and Pritchard 1970, Guerra 1969). Grimley (1969) has made calculations on chemisorbed carbon monoxide using a simple linear model Ni—C≡O. The 5σ orbital of CO and the $d\sigma$ orbital of the nickel atom are assumed to interact, forming the bonding and anti-bonding orbitals σ and σ^* of the 'surface compound'. Similarly the π^* orbitals of the CO and the $d\pi$ orbitals of the nickel atom form the bonding and anti-bonding orbitals π and π^*. The final result of the quantum-mechanical calculation is that there is a net shift of electrons from the nickel atom to the CO molecule, and the energies of the 5σ and π^* orbitals of the isolated CO molecule are raised and those of the $d\sigma$ and $d\pi$ orbitals of the nickel atom lowered.

9.2. Kinetic and adsorption studies

9.2.1. *Polycrystalline tungsten + CO*

Three different approaches have been used: (*a*) flash desorption, (*b*) enthalpy studies, and (*c*) adsorption from the gas phase from which sticking-probability data can be obtained directly. One fact that emerges quite unequivocally is that there is a multiplicity of adsorption states:

$$
\begin{array}{lll}
CO_{(g)} \to CO_{(ads)} & \gamma\dagger & \text{77 K and below} \\
\quad\; \to CO_{(ads)} & \alpha & \text{293 K and below} \\
\quad\; \to CO_{(ads)} & \beta \text{ (includes} & \text{293 K and above.} \\
& \text{substates } \beta_1\; \beta_2\; \beta_3) &
\end{array}
$$

The nomenclature α, β, and γ describes the various states observed, but it must be emphasized that this does not imply that, for example, the γ state observed at low temperature is a single state. On the contrary there is clear evidence that there is for example a range of different high-energy adsorption states β_1, β_2, β_3. Also at low temperature in the Mo+CO system (Little, Quinn, and Roberts 1964) a range of heats of adsorption, the magnitudes of which decrease linearly with θ as θ approaches unity, have been reported. There is therefore no reason to suggest at the present time that the γ state for CO on W does not also incorporate various substates, which may overlap with the α states. We also have to be cautious with our definition of an 'adsorption state', since states that by one criterion (e.g. E_d) are distinct may, by another experimental criterion, be indistinguishable.

The first desorption studies were reported by Ehrlich (1961) and Redhead (1961) using polycrystalline ribbons, and in both cases β-CO desorption was shown to be complex and various substates were designated β_1, β_2, β_3. These correspond to increasing temperatures of desorption and hence heats of chemisorption. The kinetics were first order from which it was concluded that the CO entity remained undissociated in the chemisorbed state.

Adsorption into the α state is generally thought to be significant only after population of the β states has been completed. Moreover, the concentration in the α state is both pressure and temperature dependent, and as already mentioned whether any clear distinction is possible between the α and γ states is doubtful. However, an estimate of the fractional coverage in the α- and

† Also referred to as the virgin state by Bell and Gomer (1966).

β-CO states can be obtained by integrating the respective desorption peaks (Redhead 1961).

Sticking-probability data have in general conformed to a pattern of (a) being close to unity at 100 K and less, (b) decreasing with increasing substrate temperature, and (c) being more invariant with coverage the lower the substrate temperature (Gomer 1966).

King (1975) has recently reviewed the kinetic studies with tungsten thought to be (100) oriented. Goymour and King (1973a) conclude that the kinetic evidence suggests that the β state is probably dissociated. Adams (1974), applying the interaction model to thermal desorption data (see Chapter 8), concluded that *non-dissociative adsorption* in the β state, but involving a coverage-dependent repulsive interaction energy of $(10\cdot5 + 29\theta)$ kJ mol^{-1}, gave a reasonable description of the data. On the other hand, Goymour and King (1973b) applied the interaction model to the β-state desorption traces assuming *dissociative chemisorption* and derived a repulsive interaction energy of 20 kJ mol^{-1}. Clearly kinetic data alone are not sufficient to make a firm decision and other evidence is discussed later.

Brennan and Hayes (1965) have compared the heats of chemisorption of CO on tungsten obtained by different experimental methods (Fig. 9.2). It should be emphasized that the coverage was estimated by assuming $0\cdot185$ nm^2 for the effective cross-sectional area of krypton on tungsten, derived from the assumption that the W films used were composed of equal areas of (100), (110), and (211) planes and that certain configurations of krypton adatoms argued by Brennan and Graham (1965) were the most probable. Excepting the high initial heats obtained calorimetrically by Brennan and Hayes with films, there is reasonable and remarkable agreement between the various data obtained under quite different circumstances. For example a calorimetric measurement at 300 K, where mobility of the adsorbed CO is restricted, need not give an enthalpy of adsorption which bears any relationship to an activation enthalpy of desorption in a flash desorption (high-temperature) experiment (see Chapter 6).

The high initial heats, some 160 kJ mol^{-1} higher than the plateau value of $\sim$320 kJ mol^{-1}, are undoubtedly genuine. Artefacts such as calorimeter design, etc., are quite clearly ruled

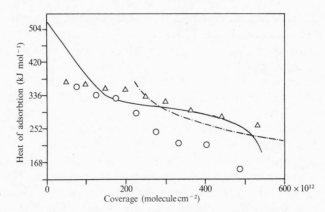

Fig. 9.2. Comparison of the heats of adsorption of carbon monoxide on tungsten as a function of coverage derived from different experimental methods: —— calorimetric data (Brennan and Hayes 1965); △○ the integral and differential heats from flash filament desorption studies (Ehrlich 1961); −·− the differential heats from field emission microscopy.

out, since such behaviour is absent with other metals and in particular absent with molybdenum. Why tungsten differs from such a close congener as molybdenum is as yet not clear.

9.2.2. CO adsorption on W(110)

Kohrt and Gomer (1971) have modified the step-desorption technique of Bell and Gomer (1966) for investigating sticking coefficients, desorption spectra, and isothermal desorption rates for CO on W (110). The method is ingenious and makes use of a field emitter placed in front of the crystal as a detector of the molecules reflected or desorbed from the (110) surface being investigated. To ensure that only molecules leaving the surface are detected by the emitter the entire walls of the apparatus are cooled in liquid H_2 which effectively makes them an ideal 'sink' for irrelevant molecules.

The desorption spectrum of a full virgin layer is shown in Fig. 9.3; the spectrum is qualitatively similar to that of polycrystalline tungsten in that there are clearly defined virgin (200–450 K) and β (850–1225 K) peaks. Re-exposure to CO after thermal cycling to ~500 K results in population of the α state which desorbs in the range 100–400 K. This α state is formed in the presence of the β adlayer and is not apparent during the subsequent desorption of the virgin layer formed at 77 K (Fig. 9.3).

Isothermal desorption of the β_2 state was shown to be first order with an activation energy of $292 \pm 12 \text{ kJ mol}^{-1}$ which is apparently invariant with coverage. The β_1 state also desorbed with first-order kinetics and an activation energy of 231 kJ.

The initial sticking coefficients S_0 increase with decreasing substrate temperature from a value of 0·73 at 600 K to 0·98 at 100 K. If the data are normalized by defining $\theta = \sigma/\sigma_{max}$, where σ is the quantity adsorbed and σ_{max} is the maximum adsorbed, plots of S/S_0 versus θ for $300 \leqslant T \leqslant 600$ K can be fitted to a Kisliuk-type isotherm (eqn (9.1) cf. eqn (8.28))

$$\frac{S}{S_0} = \left(1 + \frac{K\theta}{1-\theta}\right)^{-1} \qquad (9.1)$$

where $K = 0·10$. It is, however, clear that normalization with a single K value is strictly acceptable only when one state (say β) exists on the surface. At lower temperatures different K values are needed for normalization.

9.2.3. Molybdenum and other transition metals

There have been fewer kinetic studies of CO adsorption on Mo than W, but those that have been reported show a marked similarity to W. Degras (1967a,b) has reported, in a flash-filament study, three similar adsorption states, one α and two β. Both β states have a lower energy of desorption than the

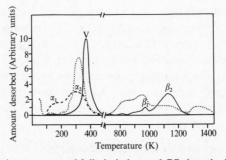

FIG. 9.3. Desorption spectrum of full virgin layer of CO deposited on W(110) at 77 K, heating time per step 5 s (full curve); α layer adsorbed at 77 K after heating full virgin layer to 600 K (broken curve). For comparison a spectrum from polycrystalline tungsten is shown as a dotted curve. (From Kohrt and Gomer 1971.)

three β states observed by the same author with tungsten

	W (kJ mol^{-1})	Mo (kJ mol^{-1})
α	77	85
β	216	127
	282	275
	324	

Relevant to the question of distinguishability of separate adsorption states and also of substates within any one designated state are the data reported by Little *et al.* (1964). In this work CO isotherms were determined on Mo films in the temperature range 250–316 K; the isotherms were shown to be reversible and by assigning $\theta = 1.0$ at a CO pressure of $\sim 4 \times 10^{-2}$ torr at 273 K the heat of adsorption was shown to decrease linearly with coverage in the θ range 0·75–0·95:

$$\Delta H_{\text{a}} = 50\theta - 50.5. \tag{9.2}$$

The isotherms were determined at CO pressures of between $\sim 10^{-3}$ and 5×10^{-2} torr and would presumably reflect adsorption states observed at both lower temperatures and CO pressures (e.g. the γ state). There is therefore good evidence for a continuous variation of adsorption states at high coverage based on the enthalpy of adsorption.

Using a combined radio-tracer–work-function method Mathews (1971) has claimed that five states of CO exist on Mo (100). This may be compared with the recent studies of Lecante, Riwan, and Guillot (1973) who have carried out LEED, AES, flash desorption, and work-function studies to follow the kinetics of CO adsorption on Mo (100). These authors distinguish three states at room temperature ($\beta_1, \beta_2, \beta_3$); β_1 and β_2 correspond to an increase in work function while β_3 results in a decrease. At the maximum coverage of the ($\beta_2 + \beta_3$) states the LEED pattern indicates a c(2×2) structure, corresponding to about half a monolayer; at the completion of the monolayer a p(1×1) structure forms.

CO interaction with Ni. Horgan and King (1969) have reported that the sticking probability S of CO on nickel films was very close to unity at both 78 K and 290 K ($\geqslant 0.99$). These are higher values than reported previously by Oda (1955, 1956) (0·6), Wagener (1957) (0·4), and Degras (1967a,b) ($\sim 5 \times 10^{-4}$). There are clearly two stages to the adsorption process at 78, 195, and 290 K (Fig. 9.4) in that about 1×10^{15} molecules cm^{-2} are adsorbed with close to unit sticking probability at 78 and 195 K. Subsequently S decreases and the form of the $S-\theta$ curves is dependent on temperature. At $S \simeq 5 \times 10^{-3}$ there is an abrupt change in the curve at all temperatures (Fig. 9.4). A film saturated with CO at 298 K was capable of adsorbing further CO, to the extent of about 10 per cent of the total, on cooling to 78 K.

The second stage of CO adsorption, when S decreases to below unity at 290 K, involves a strongly bound β state. Horgan and King attribute the decrease in S to below unity at 290 K to simultaneous desorption from a low heat state which they designate the α state. Coverage-dependent adsorption heats for the α state were obtained which varied from 100 kJ mol^{-1} to 71 kJ mol^{-1}. These values compare with heats decreasing steadily with coverage from $\sim$147 to 20 kJ mol^{-1} observed by Beeck (1940) and McBaker and Rideal (1955).

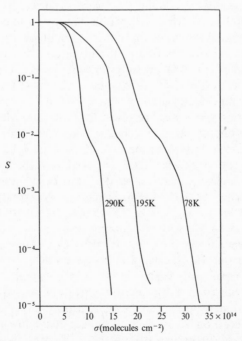

FIG. 9.4. Variation of sticking probability with coverage for carbon monoxide on nickel films at 78 K, 195 K, and 290 K. (From Horgan and King 1969.)

Although no CO_2 was observed at 290 K in the gas phase when CO interacted with Ni films deposited on smooth pyrex glass, at 373 K a small amount of disproportionation was detected at high coverage. On the other hand using nickel films deposited on HF-etched pyrex disproportionation was observed at 290 K. Disproportionation of CO on Ni had previously been reported by McBaker and Rideal (1955), Gregg and Leach (1966), and Leidheiser and Gwathmey (1948), and also invoked by Pitkethly and co-workers in their LEED studies. The possible role of the electron beam in inducing CO dissociation must, however, not be overlooked in LEED work (Pitkethly, 1970).

A nickel film saturated with O_2 at 290 K is capable of adsorbing small quantities of CO but no $CO_2(g)$ is desorbed. On the other hand pre-adsorbed CO is quantitatively displaced by oxygen with the formation of $CO(g)$ and $CO_2(g)$.

CO interaction with rhenium. The sticking coefficients of CO adsorption on polycrystalline Re were reported by Yates and Madey (1969), and from desorption studies various adsorption states designated. Some of the impetus for this work arose from the range of results reported in the literature for CO on refractory metals; S_0 for CO on tungsten was generally accepted as close to unity at 300 K whereas Gasser, Thwaites, and Wilkinson (1967) had reported a value of about 4×10^{-2} for CO on Re.

Five binding states were resolved by Yates and Madey (1969), the highest-energy state (β_3) being filled first and the lowest-energy (α_1) state last. With the exception of the splitting of the α state into two separate states $(\alpha_1$ and $\alpha_2)$ the behaviour is phenomenologically identical to the flash desorption states observed with the $W + CO$ system. First-order desorption behaviour is observed for β_1, α_1, and α_2 CO states; information on the β_2 and β_3 states on Re was not obtainable owing to overlapping.

No influence of the ion gauge on the CO kinetics was detected which contrasts with the Re and N_2 system where evidence was obtained (Yates and Madey 1969, McKee *et al.* 1971) for a surface species formed by electron activation of $N_2(g)$ in the ion gauge. These new binding states are reminiscent of the λ nitrogen states produced on W by electron impact of the low-temperature γ state (see also Chapter 10).

Qualitative studies indicated (Yates and Madey 1969) that

isotopic mixing occurred in all the β states but no clear evidence was available for mixing in the α states. The experiments were carried out by flashing the Re in a flow system containing a mixture of the isotopic CO molecules. The $^{12}C^{18}O + ^{13}C^{16}O$ mixture contained about 1·4 per cent of $^{13}C^{18}O$, and the small α peak in the temperature range 300–500 K would correspond to just about this amount. The isotopic data therefore resemble quite closely the behaviour observed with the W+CO system.

CO interaction with Mn, Zr, and Ta. Bickley, Roberts, and Storey (1971) have studied the kinetics of the interaction of CO with Mn and obtained unequivocal evidence for reaction well beyond the monolayer. At a CO pressure of ~10^{-2} torr and a temperature of 453 K a limiting uptake of the equivalent of four CO monolayers had occurred. The rate of CO uptake dθ/dt was shown to conform to

$$\frac{d\theta}{dt} = \frac{A}{\theta} - B \qquad (9.3)$$

where both A and B are temperature dependent. The form of eqn (9.3) suggests that the rate of uptake (incorporation into the lattice) is a balance between a diffusion process obeying Fick's law and desorption of weakly held chemisorbed CO. The fate of the CO within the lattice is not clear, but it should be noted that the most likely process, dissociation to give manganese carbide and oxide, is thermodynamically favourable. Similar observations have been made with zirconium and tantalum (Quinn and Roberts 1963).

9.3. Work-function studies on single-crystal planes of tungsten

Madey and Yates (1967), using a thermionic retarding-potential method, reported changes in work function for CO on (100)- and (110)-oriented tungsten surfaces. The crystals were heated to ~2200 K in 4×10^{-6} torr of oxygen to remove the carbon impurity found in tungsten and subsequently outgassed at temperatures up to 2500 K.

On exposure of W(100) to CO at 4×10^{-8} torr the work function increased (Fig. 9.5) by about 0·45 eV. Changes in work function on thermal desorption of the CO were also reported. The only evidence, albeit slight, for more than one CO state on

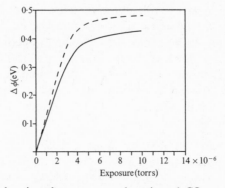

FIG. 9.5. Work-function changes upon adsorption of CO on (100) tungsten *versus* exposure to CO (torr s): (100) cooled before adsorption (broken curve); (100) flashed in flowing gas (full curve). (From Madey and Yates 1967.)

(100) is recovery in $\Delta\phi$ at high temperature. On the other hand Anderson and Estrup (1967) reported more drastic changes in work function on heating CO adsorbed on W (100), $\Delta\phi$ going from $+0.43$ eV to -0.16 eV. The high-temperature electropositive state (decrease in $\Delta\phi$) they designated β_H.

The variation of the work function of W (110) on adsorption and desorption of CO provides convincing evidence for at least two states on this plane (Fig. 9.6). One state is electropositive and this is formed at low exposures, but after an exposure of $\sim2\times10^{-6}$ torr s the work function begins to increase which is clear

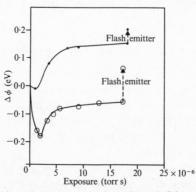

FIG. 9.6. Work-function changes upon adsorption of CO on (110) tungsten *versus* exposure to CO (torr s): ○ (110) flashed in flowing CO; ● (110) cooled before CO admitted. (From Madey and Yates 1967.)

evidence for an electronegative adsorbate. Madey and Yates draw attention to the problem associated with the reference emitter surface which is a W (100) crystal maintained at 1800 K in flowing CO. There is a gradual change in emitter work function which clearly influenced the $\Delta\phi$ values reported in Fig. 9.6. This, however, in no way affects the trends observed in $\Delta\phi$ or the conclusion that an electropositive state populates rapidly on the warm crystal (the β state) followed by an electronegative state as the crystal cools. Adsorption on the initially cool surface is largely of the electronegative species, the β state.

9.4. Isotopic studies

The first isotopic-exchange studies of CO on metals were reported by Webb and Eischens (1955) some 20 years ago using iron surfaces. They showed that the exchange

$$^{13}CO + C^{18}O \rightleftharpoons {}^{13}C^{18}O + CO \qquad (9.4)$$

at 195 K was 20 per cent complete, the extent of exchange increasing with temperature. They concluded that

'the nature of chemisorbed carbon monoxide is more complicated than expected from the simple analogy with the structure of metal carbonyls'.

The authors recognized that this could well be evidence for dissociation but did not consider it 'proof of dissociation'. Subsequently there have been similar studies on nickel powder and more definitive and detailed kinetic work with tungsten.

Madey, Yates, and Stern (1965) using a mixture of $^{12}C^{18}O$ and $^{13}C^{16}O$ showed that the onset of rapid isotopic mixing occurred at ~873 K with tungsten which they associated with the desorption of the β_1 state. They consider three mechanisms.

(a) Via the bimolecular reaction

$$^{12}C^{18}O_{(ads)} + {}^{13}C^{16}O_{(ads)} \rightleftharpoons {}^{12}C^{16}O_{(ads)} + {}^{13}C^{18}O_{(ads)}.$$

The isotopically mixed CO then desorbs from the surface.

(b) Dissociation of CO into mobile adsorbed C and O species which subsequently recombine and desorb.

(c) Exchange of C impurity in the W surface with $CO_{(ads)}$.

Madey *et al.* (1965) ruled out (b) since no O_2 or CO_2 were desorbed at 2100 K when W was heated in CO at 1×10^{-6} torr, and ruled out (c) on the grounds that impurity carbon in bulk tungsten is known to diffuse to the surface only at temperatures above 1400 K. They also reported a field emission experiment where W was heated at 1600 K for 20 h at 10^{-10} torr. No pattern due to carbon was observed. Thus a bimolecular exchange reaction involving a four-centre complex

is postulated.

9.5. Electron-impact desorption

One of the first studies of the influence of electron impact on the CO + W system was by Menzel and Gomer (1964). The low-temperature, virgin, or γ state was dissociated by 80 eV electrons leaving a carbon residue; on thermally converting the γ state to the β state the cross-section for electron-stimulated desorption decreased by a hundred-fold, while for the α state (present together with the β state) the cross-section was some 10 times greater than for the γ state. It was suggested that in both the γ and α CO states the orientation of the molecule was such that the bonding to the surface was through the C atom while β-CO was in a horizontal position.

Studies by Redhead (1967) and Yates, Madey, and Payn (1967) of the ion-energy distribution of CO on W at 300 K indicated that two groups of positive ions having different kinetic energies were desorbed. The low-energy ions ($\sim$1 eV) were associated with $CO_{(ads)}$, having an activation energy of desorption of $\sim$76 kJ mol^{-1}, while the higher-energy ions ($\sim$7 eV) were associated with more strongly adsorbed CO. Yates *et al.* assigned the origin of these ions as the α-CO (1 eV, CO$^+$) and β_1-CO (7 eV, O$^+$) states, following the mass-spectrometric identification of the ions by Nishijma and Propst (1970). On the other hand Redhead tentatively assigned the two groups of ions to virgin (or γ) CO and α-CO. More recently Menzel (1968) and Coburn (1968) used mass-spectrometric studies and deduced that the α-CO state is in fact two states, α_1 yielding CO$^+$ and α_2 yielding O$^+$. The γ state did not yield an ion on electron impact.

One point that does emerge very clearly is that it is those CO molecules that are adsorbed at higher coverage that are most efficient in giving rise to positive ions during electron impact. These are also likely to be the most weakly adsorbed. Menzel (1975) has recently reviewed electron-stimulated desorption (ESD) in the context of surface bonding.

9.6. LEED studies

9.6.1. W (211) surface

CO adsorbed at room temperature produces a diffuse diffraction pattern with no well-defined extra spots and a high background intensity. On heating to progressively higher temperatures well defined c(6×4), p(2×4), and c(2×4) patterns were observed. In addition a 'complex pattern' with no obvious defined symmetry was reported. Chang (1968) associated each LEED pattern with a particular state of adsorbed CO, the latter being as defined by a flash desorption experiment, and Table 9.1 summarizes his results.

9.6.2. Pt (111), (100), and stepped Pt surfaces

Chemisorption of CO on (111) at room temperature has been reported to result in two different ordered structures, one with a c(4×2) unit cell and the other a (2×2) pattern. With the Pt (100) surface the clean (5×1) Pt pattern was replaced by a (1×1). On the other hand when stepped (111) and (100) surfaces were used no ordered structures were observed with CO, only an increase in the intensity of the scattered background. Lang, Joyner, and Somorjai (1972) suggest that lack of ordering with stepped surfaces may be a consequence of their inherently high activity

TABLE 9.1

CO structures on W (211) and their correlation with adsorbed states observed by flash desorption

Pattern	CO exposure (L)	Fractional coverage	Destruction temperature (K)	Associated state
c(6×4)	0·6	0·25	1200	β_2
p(2×4)	1·2	0·50	1000	β_1
c(2×4)	>1·4	$\frac{5}{8}$ to $\frac{10}{8}$	1000	β_1
Complex	>1·5	$\frac{5}{8}$	1200	β_2

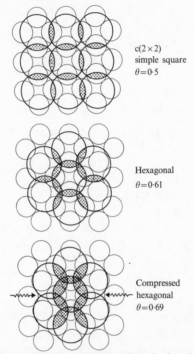

c(2×2)
simple square
$\theta = 0.5$

Hexagonal
$\theta = 0.61$

Compressed
hexagonal
$\theta = 0.69$

FIG. 9.7. Hard-sphere models of suggested unit cells for the observed long-range-ordered structures for CO on Ni (100). The small circles represent the substrate Ni atoms and the large circles the CO molecules. The choice of four-fold substrate sites for the CO molecules in the c(2×2) structure is based on the observation that all the integral-order beams remain sharp when the c(2×2) structure is disordered (see Section 3.7.2.). (From Tracy and Burkstrand 1974.)

leading to possible dissociation of the CO at the atomic surface steps. Heating the stepped surface to 473 K effects CO desorption but some carbon still remains on the surface as detected by Auger electron spectroscopy. This may be the carbon dissociated at the surface steps, although the authors are not clear as to what perturbing influence the electron beam has. What is absolutely clear is that stepped Pt surfaces (within 10° of the parent surface) are more reactive than flat surfaces; this is an important result and has more general implications in heterogeneous catalysis.

9.6.3. Ni (100) and (111) surfaces

Fig. 9.7 shows the hard-sphere-model structures derived from the LEED diffraction patterns for CO adsorbed on Ni (100) (Tracy and Burkstrand 1974). The lowest-coverage structure ($\theta = 0.5$) has a c(2×2) or ($\sqrt{2} \times \sqrt{2}$) R45° symmetry and, as drawn, the crystallographic positions of the admolecule are assumed to be at the fourfold symmetric sites. It should be remembered that the exact positions of surface species cannot be ascertained from LEED patterns, only their symmetry. When the exposure (and hence θ) is increased a hexagonal structure develops which on further adsorption becomes compressed.

In addition to these ordered structures there also occur disordered CO overlayers for all coverages at sufficiently high temperature. Tracy has shown this in terms of a T–θ phase diagram (Fig. 9.8); the solid curves indicate experimentally determined features while the broken lines represent the disordering temperature of the various ordered phases. There are present eutectic-like triple points at which there is equilibrium between the c(2×2), the hexagonal phase, and the disordered α_0 phase. This must mean that at such coverages the heats of adsorption of the three phases are similar.

Christmann, Schober, and Ertl (1974) have reported that with Ni(111) very diffuse extra spots of a ($\sqrt{3} \times \sqrt{3}$) R30° structure

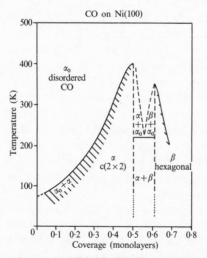

FIG. 9.8. T–θ phase diagram for Ni (100)–CO. The solid curves indicate experimentally determined features. The broken lines indicate possible locations for the remaining boundaries. (From Tracy 1974.)

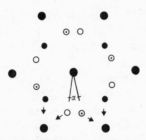

FIG. 9.9. Sketch of LEED patterns for CO adsorption on Ni (111) illustrating the superposition of three domain orientations and the continuous splitting of the spots. (From Christman, Schober, and Ertl 1974.)

formed at an exposure of 0·3 L. Split diffraction spots formed at 1·5 L exposure, and these (Fig. 9.9) are interpreted as being due to the superposition of three domain orientations rotated by 120° with respect to each other. An interesting feature of the data is that with continued exposure the angle α (Fig. 9.9) changed continuously, attaining a maximum value of 24° at saturation.

9.6.4. *Cu (100) surface*

Tracy (1972) reported that no CO adsorption occurred on Cu (100) at room temperature (as reflected by no observed changes in either the LEED pattern or work function) and that only on cooling to 77 K did the c(2×2) structure form. The pattern behaviour was, however, not straightforward, and beyond a CO exposure of 1·7 L the c(2×2) pattern changed continuously. Tracy interprets this smooth continuous change in structure as a smooth compression of a c(2×2) structure along the $\langle 110 \rangle$ direction which is the direction of the most prominent troughs in the surface. As the structure is compressed the molecules in adjacent rows maintain their relative positions along one of the two $\langle 100 \rangle$ directions in the surface. What gives rise to these anisotropic interactions is not clear, but the general phenomenon is also observed with CO adsorption on both Ni (100) and Pd (100).

Furthermore Chesters and Pritchard (1971) report virtually identical behaviour with CO on Cu (100) and analogous behaviour with Xe on Cu (100) at 77 K. These authors also followed simultaneously changes in work function and concluded that the c(2×2) structure at the maximum positive surface potential corresponded to twice the coverage previously suggested on the basis of comparative xenon adsorption measurements on poly-crystalline films. It is very significant that the 0·361 nm spacing of CO molecules in this surface structure is significantly less than that in solid carbon monoxide, and this structure can be further compressed and distorted by more CO adsorption. This 'compression-distortion' stage in adsorption involves weak CO binding and is not observed at 195 K.

One study (Joyner, McKee, and Roberts 1971) has however, reported diffraction patterns for CO adsorbed at room temperature on Cu(100). Initially a c(2×2) structure formed but at higher exposures to CO a p(2×2) structure formed. The clean surface was regenerated (as indicated by AES) on heating the crystal to 373 K in keeping with an adsorbate having a heat of adsorption less than about 80 kJ mol^{-1}, a value similar to isosteric heat values reported for CO adsorption on copper. The reason for the apparent stability of CO on Cu (100) at room temperature in this investigation compared with its desorption below room temperature reported both by Tracy and by Chesters and Pritchard is not known, but may be associated with undetected surface impurities or structural defects. More recent studies are reported elsewhere (Roberts 1977).

9.7. Field emission and field ion microscopy (FEM and FIM)

A characteristic feature of studies with field emission tips is the unusual nature of the surface, it being composed of a large number of crystallographic planes none of which is more than a few tens of nanometers across. Also there is a high proportion of surface atoms that are corner or edge sites. The high work function of the most closely packed planes means that it is not easy to observe adsorption on these.

Although Ehrlich, Hickmott, and Hudda (1958) concluded that CO is adsorbed non-dissociatively on a tungsten tip, more recently Kawasaki, Sensaki, and Sato (1968) have shown by means of a pulsed field emission technique that dissociation can occur at temperatures $\geqslant 923$ K and a CO pressure of $\sim 2 \times 10^{-7}$ torr. The power of field emission microscopy is no better illustrated than in studies of surface diffusion. Gomer (1958) reported that with tungsten at 20 K physically adsorbed molecules migrated over a chemisorbed layer and precipitated at its edges. This diffusion was crystallographically anisotropic and contrasts with the diffusion of chemisorbed CO which is not observed until above 700 K.

Three adsorbed states have been recognized for CO by FEM. The virgin state forms at very low temperatures (20–50 K), when the tungsten pattern is pseudo-clean, and an α state is believed to form on warming the virgin state to about 200 K. This α state is considered to lead to a decrease in the observed work function. At about the temperature that the α state is desorbed there occurs a further decrease in work function of about 0·5 eV. It is clear that rearrangement of the adsorbed layer as well as desorption is occurring since the desorption of 'electropositive' CO would lead to an increase in the work function of tungsten.

Above the α desorption temperature the FEM pattern is considered to reflect the β state; it appears that all planes are active in this adsorption. It should, however, be recalled that no comment can be made on the (110) plane since its high work function precludes observation in the FEM.

Ehrlich (1966), using FIM, reported the observation of individual CO molecules; disturbance of the surface layer by the field was considered to be only slight. Conclusions contrary to that of Ehrlich have been reported by Holscher and Sachtler (1966) who carried out combined FEM and FIM studies of CO on W. These authors give convincing arguments that the surface tungsten atoms are reorganized as a consequence of CO adsorption and refer to it as 'corrosive chemisorption'. They interpret their results as reflecting the penetration of CO into the subsurface at 300 K giving a two-dimensional structure with the tungsten atoms remaining at the surface. Further CO adsorption can occur on these atoms.

9.8. Infrared spectroscopy

There has been considerable effort put into the investigation of the adsorption of CO by infrared techniques starting with the classic work of Eischens and Pliskin (1958) who used metals dispersed on such materials as silica or alumina. Three bands were resolved at 2073, 1924, and 1870 cm^{-1} for CO on Ni; the intensity of the bands varied with CO pressure and in view of data with metal carbonyls it was suggested that CO stretch frequencies below 2000 cm^{-1} were due to CO acting as a bridging ligand. Bands with frequencies above $2000 \, cm^{-1}$ were attributed to monodentate carbonyls.

Subsequent studies have thrown some doubt on these assignments, monodentate CO ligands also showing stretching frequencies below $2000 \, cm^{-1}$, and Blyholder suggests that such low-frequency bands may be attributed to CO adsorption on edge or corner sites on the metal crystallites of the sample. He argues that sites of these types allow a greater degree of back-bonding by the $d\pi-p\pi^*$ mechanism leading to a further weakening of the C—O force constant. Blyholder therefore is of the view that all the bands observed in different investigations can be accounted for in terms of the linear CO structure (Blyholder 1964, Blyholder and Tanaka 1973).

The infrared absorption of chemisorbed carbon monoxide has also been studied directly with metal surfaces (as opposed to oxide-supported metals where support effects may be important). There have been in the main four variations: (a) evaporation of the metal into an oil film (Blyholder 1962); (b) evaporation of the metal in a high pressure of CO (Hayward 1963, Garland et al. 1965, Baker et al. 1968); (c) evaporation of the metal in an inert gas (Hayward 1963); (d) the use of the infrared reflection method where not only polycrystalline films formed in ultra-high-vacuum but also single crystal surfaces may be used (Greenler 1966, Pritchard and Sims 1970).

Although the film-in-oil technique is open to criticism, particularly on contamination grounds, the results of Blyholder have been most stimulating. The great advantage of the method is that unlike oxide-supported metals low-frequency vibrations are observable in view of the oil's transparency in this wavelength region.

Evaporating in a pressure of CO enables both high-area films to be prepared and the degree of surface contamination to be minimized; similarly, evaporating in an inert gas gives high areas. Hayward (1971) has summarized the infrared information available on CO adsorption on metal films. Before considering some specific systems it is worth referring to relevant information obtained from metal carbonyls. The nature of the metal–carbon bond in metal carbonyls is usually considered in terms of σ donation from the highest filled carbon 'lone-pair' molecular orbital of CO and π-back bonding into the $2\pi^*$ orbital of the ligand. This model, largely based on a decrease in the infrared stretching frequencies compared with free CO, has had very strong support from XPS studies of the metal carbonyls. The predominant bonding interaction has been shown to be donation from filled metal d orbitals to anti-bonding π^* orbitals on the ligand. Hillier and Saunders (1971) have carried out *ab initio* molecular-orbital calculations on $Ni(CO)_4$ and $Cr(CO)_6$ and compared their results with measured ionization potentials of both core and valence electrons. Their main conclusion is that the extent of back-bonding into the CO $2\pi^*$ orbital has probably been underestimated.

9.8.1. *CO adsorption on Cu*

Both Geenler and Pritchard have reported infrared data for the chemisorption of CO on Cu surfaces by the reflection technique. The theoretical background to the technique has been discussed in Chapter 5. Greenler and Tompkins (1971) and Pritchard and Sims (1970) observed a band at 2105 cm^{-1} for CO on Cu at room temperature. The intensity of the band increased with pressure in the range 0·02 to 2·3 torr. The results of Greenler and Tompkins (1971) and Pritchard and Sims (1970) were obtained with ostensibly polycrystalline copper films, whereas Chesters, Sims, and Pritchard (1970) defined the surface crystallography more precisely and used Cu (100) single crystals. They supplemented their spectroscopic data with surface-potential measurements and also extended their studies to a lower temperature (77 K). Two stages of adsorption were observed at 77 K; they were characterized using both the surface potential and infrared techniques (Fig. 9.10). No adsorption at low CO pressures was observed at room temperature. A particularly interesting and probably far-reaching conclusion is that, although the surface potential results with Cu (100) at 77 K were very similar to those observed with polycrystalline copper surfaces, in the infrared work (Fig. 9.10) the band at 2085 cm^{-1} may be compared with the sharp band at 2105 cm^{-1} observed with the polycrystalline surfaces. There was no evidence for a 2085 cm^{-1} band with the polycrystalline copper although, as the authors observe '... it seems very probable that (100) planes would have been present'. This is an important point since it raises the whole question of the general assumption that polycrystalline surfaces of f.c.c. metals are composed of equal

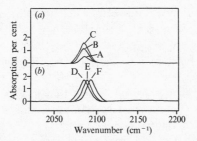

Fig. 9.10. Infrared spectra of CO on Cu (100) at 77 K: (*a*) during the adsorption of the species with a positive dipole at surface potentials of 116 mV (A), 185 mV (B), and 236 mV (C); (*b*) during the second stage of adsorption in which species with a negative dipole reduce the total surface potential from 236 mV (D) to 175 mV (E) and 120 mV (F). (From Chesters, Sims and Pritchard 1970.)

areas of the most close-packed planes (100), (111), and (110). Recent data by Pritchard *et al.* (1975) indicate that with high index and stepped Cu surfaces the infrared spectra of adsorbed CO is very similar to that observed with polycrystalline Cu surfaces while further studies by Horn and Pritchard (1976) relate different vibrational frequencies to specific surface structural features. Recent electron spectroscopic studies (Roberts 1977) emphasize the significance of the dual-electron transfer process (π^* and σ bonding) in determining the nature of the CO—Cu bond and how under certain conditions, adsorption of CO is observed on Cu (100) at room temperature. The bonding is, however, quite different from that for carbon monoxide adsorbed on copper at 77 K.

9.8.2. *CO adsorption on W and Pt*

King, Goymour, and Yates (1972) have used the reflection technique to study CO adsorption on a polycrystalline W ribbon. A band at about 2110 cm^{-1} is observed which is associated with the comparatively weakly held α-CO state. With increasing coverage in this state the band shifted to higher frequencies; at $\theta_{(\alpha \text{ state})} \simeq 0.42$ the band was centred at 2090 cm^{-1}, while at $\theta_{(\alpha \text{ state})} \simeq 1.0$ the band was nearly at 2120 cm^{-1}. It is interesting to recall (Hayward 1971) the rather ubiquitous occurrence of infrared absorption bands at 2100 ± 15 cm^{-1}; this for example is certainly the case for Ni, Cu, Fe, Ag, and Au. Also contamination effects are apparently not reflected in this absorption band which suggests that it is due quite clearly to a weakly held form of adsorbed CO. In gaseous metal carbonyls the CO stretching mode is observed at around 2120 cm^{-1} and some connection between the weakly held CO and the bonding of CO in the metal carbonyls is possible.

It is important also to emphasize that in general only a small fraction of the chemisorbed carbon monoxide is infrared active. For example McManus (see Hayward 1971) observed that the absorption band for CO on Fe was removed on evacuation at 300 K. Quite clearly only a comparatively small fraction of the adsorbed CO would be removed by evacuation since most of that adsorbed would have a high heat of chemisorption $(\sim 145 \text{ kJ mol}^{-1})$. Bradshaw and Pritchard (1970) have discussed this point and emphasize that with the transition metals much of the chemisorption is infrared inactive, while with the noble or sp metals adsorption, which of course is characterized by appreciably weaker bonding, is initially infrared active. The reason for the lack of infrared-active CO on the transition metals is as yet not clear, but evidence is gradually accumulating for the CO molecule not remaining intact in the adsorbed state.

One particular system that is worthy of further comment is the Pt + CO system studied by Blyholder and Sheets (1970). Bands were observed at 2045, 1815, and 480 cm^{-1} using the 'oil technique'. The band at 480 cm^{-1} had previously aroused considerable controversy in that it had been attributed to the Pt—C bond. The particular point at issue is whether meaningful infrared spectra can be obtained in the 500 cm^{-1} region using silica-supported platinum (Clarke, Farren, and Rubalcava 1968). The most intense band is at 2045 cm^{-1} and agrees with other work; the medium-intensity band at 1815 cm^{-1} is attributed to CO adsorption on metal atoms with relatively low co-ordination numbers. Changes in the 1815 cm^{-1} band intensity during sintering are compatible with this explanation. By studying the absorption spectrum of $C^{16}O$ on Pt Blyholder confirmed that the calculated isotopic shift in the infrared band at $\sim 480 \text{ cm}^{-1}$ is observed, thus confirming its assignment to the Pt—C stretching motion and thus supporting the earlier data of Clarke *et al.* (1968).

The advantages and problems associated with the use of infrared internal reflection spectroscopy for the study of adsorption on metal films have been considered recently by Rice and Haller (1975). They chose the adsorption of CO on Pd, and concluded that adsorbed species that gives rise to bands at $2000\ cm^{-1}$ are structure insensitive while those that absorb below $2000\ cm^{-1}$ are structure sensitive. They suggest that the latter involve bonding of both carbon and oxygen to Pd atoms on a low index plane.

9.9. Electron spectroscopic studies of CO adsorbed on Mo and W

9.9.1. *XPS studies of CO adsorption*

The adsorption of CO on polycrystalline molybdenum and tungsten films at both 300 K and temperatures approaching 77 K has been investigated by X-ray photoelectron spectroscopy (Atkinson, Brundle, and Roberts 1973, 1974*a*,*b*). The tungsten films were somewhat contaminated, as shown by the XPS spectra, and therefore we shall consider here only the data obtained with molybdenum films. There is clear evidence for the two CO states (the room-temperature $\beta + \alpha$ state and the low-temperature γ state) having distinct O(1s) binding energies (Fig. 9.11(a)). The γ states were formed by cooling a Mo film, which already had present the $\alpha + \beta$ states, to below 100 K and adsorbing CO to a pressure of $\sim 10^{-6}$ torr. The O(1s) binding energy moved during the filling of the γ states by about 4 eV to a lower binding energy. In terms of the simple charge-potential model we would conclude that there is a decrease in electron density at the O atom in CO (γ) compared with CO (β). It is known that the heat of CO adsorption on Mo films decreases linearly with coverage as the coverage increases in the temperature range 295–230 K, at a CO pressure of $\sim 10^{-2}$ torr. It is conceivable that the two effects go hand in hand; in other words that as the metal–CO bond strength decreases so also will the O(1s) binding energy increase, reflecting a decrease in the electron density on the O atom.

The XPS studies also allow some comments to be made on the nature of CO adsorbed in the state formed at 295 K by thermal conversion from the γ state. After exposure of a 'clean' Mo film to 6 L of CO at low temperature the O(1s) binding energy was 531·0 eV; on warming to 295 K the O(1s) binding energy decreased to 529·8 eV and this decrease was accompanied by a decrease in the peak width (Fig. 9.11(b)). These results are compatible with a mixture of states being formed at 77 K and the

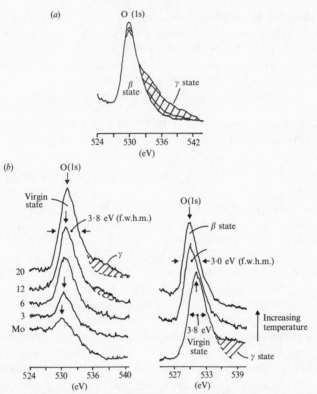

FIG. 9.11. (a) Interaction of CO with Mo first at 298 K (β state) and then at 77 K (γ state). (b) Oxygen O(1s) XPS peaks for CO adsorbed on Mo at 77 K (virgin + γ state) and subsequently on warming to 298 K (β state). The numbers on the left-hand curves give the CO exposure in langmuirs.

conversion of one of these (γ) to the room-temperature form which is characterized by a lower binding energy. There is no clear evidence for more than one state at room temperature.

XPS studies at 300 K and 100 K have been reported by Yates, Madey, and Erickson (1973) and Fig. 9.12 summarizes their results with a tungsten ribbon. They associate distinct O(1s) binding energies with the 'β state' and the α_1 and α_2 states at 300 K. At a lower temperature (100 K) the O(1s) profile is broader and at a higher binding (lower kinetic) energy. There are clearly obvious similarities with the results observed with evaporated films, although the α states if present on molybdenum were not resolved.

9.9.2. *UPS studies of CO adsorption*

Ultra-violet photoemission spectroscopy (UPS) has also provided some significant clues as to the nature of the surface species. Baker and Eastman (1973) reported EDC's for clean W (100) and after exposure to 1 L of CO at 300 K. The initial exposure to CO causes a disappearance of the surface-state peak at about 0·5 eV below the Fermi level and the appearance of a 2·5 eV wide CO-derived level at ~6 eV. This peak is characteristic of a β phase of CO adsorbed on tungsten characterized by a (1×1) LEED structure. On heating to 1100 K the CO converts to a new β state, with a c(2×2) LEED structure, but there is very little difference in the EDC's resulting from these two states. The main effect in converting the (1×1) CO structure to the c(2×2) structure is to increase the emission around 7·5 eV. Thus these photoemission data would lead us to the conclusion that changes in the electronic structure of adsorbed CO are minimal in converting from the (1×1) to the c(2×2). It should be recalled, however, that there is a significant decrease in the work function accompanying this structural change. Baker and Eastman, by determining an EDC at 300 K in the presence of CO at a pressure of $\sim 10^{-6}$ torr, investigated the $(\beta + \alpha)$ CO states. By subtracting the EDC for the β-CO state it is possible to compile a curve which to a first approximation represents the α-CO state. A peak centred at about -9 eV is considered to be characteristic of α-CO. Since the work function of clean W (100) is $\sim 5 \cdot 6$ eV and $\Delta\phi$ is $0 \cdot 6$ eV (decrease in ϕ), then the α-CO peak is centred

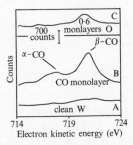

FIG. 9.12. O(1s) XPS signals from CO and oxygen adsorbed on polycrystalline tungsten. Curve A, background count level; curve B, O(1s) signal from adsorbed monolayer of CO on W; curve C, O(1s) signal from ~0·6 monolayer of oxygen adsorbed on tungsten. ($h\nu = 1254$ eV). (From Yates *et al.* 1973.)

at ~ 14 eV. This compares with the first ionization potential of CO(g), $14\cdot0$ eV, which corresponds to the 5σ molecular orbital.

By combining UPS with XPS studies over a range of substrate temperatures, information has been obtained on the binding energies of the C(1s) and O(1s) core electrons and also on the ionization potentials of the adsorbed CO molecular orbitals, if present, in the interaction of CO with molybdenum films (Atkinson *et al.* 1974*a*,*b*). Fig. 9.13 shows the He(I) spectrum of CO adsorbed on a molybdenum film (Atkinson *et al.* 1974*b*) at 85 K with increasing exposure to CO(g). The following features may be distinguished: (*a*) the development of a rather broad peak at about $9\cdot5$ eV below the Fermi level which disappears on warming to room temperature (curve E) and (*b*) the development of a peak at about $6\cdot5$ eV (~ 2 eV f.w.h.m.) which broadens and shifts to lower energy on warming to room temperature (curve E). These results may be compared with the X-ray-induced (Mg K α

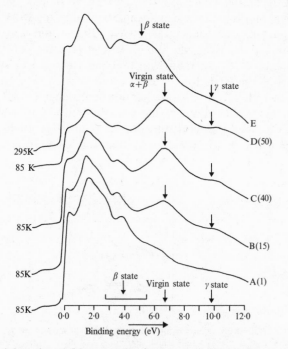

FIG. 9.13. Adsorption of CO on Mo film (HeI spectra). The numbers in brackets on the curves give the CO exposure in langmuirs.

radiation) spectra shown in Fig. 9.11. The peaks observed are described in terms of accepted terminology (γ, β, and virgin states) based on flash desorption studies.

9.10. XPS and UPS studies of the Fe+CO system

The adsorption of carbon monoxide on iron is particularly interesting in view of the variety of experimental approaches that have been used in its study (Ford 1970, Hayward 1971). They include isotopic exchange and infrared, both techniques indicating that under some circumstances CO is adsorbed molecularly and under others it may not be. These results prompted the use of XPS and UPS, particularly in view of the conclusions deduced from similar studies with W and Mo (Figs 9.11, 9.12, 9.13).

Ultra-violet spectra (He II) showed (Kishi and Roberts 1975) the presence of 'free CO-like' molecular orbitals at 80 K, positioned at about 7 eV and 11 eV below the Fermi level. On warming to, and leaving at, 295 K these peaks gradually diminished in intensity and were replaced by a single peak at about 5 eV below the Fermi level. The latter is not typical of molecular CO but can be reproduced by chemisorbing oxygen and is also observed when carbon (carbide) is present. Shifts in the X-ray-induced O(1s) spectra (Kishi and Roberts 1975) also support the view that at 290 K molecularly adsorbed CO dissociates slowly and that this surface carbide/oxide can further adsorb molecular carbon monoxide. Textor *et al.* (1977) have reported similar data for the Fe(111) surface.

9.11. Mechanism of CO bonding

We have considered a number of infrared studies of CO adsorption on metals. In general CO is infrared inactive at low coverage but active at high coverage with transition metals, while with sp metals such as copper, silver, and gold bands are observed at 2105, 2160, and $2110\ \mathrm{cm}^{-1}$ respectively (Ford 1970, Hayward 1971). Analogies have been drawn with the role of π bonding in metal carbonyls particularly for frequencies below $2100\ \mathrm{cm}^{-1}$; above $2100\ \mathrm{cm}^{-1}$ σ bonding is more important. Recent electron-spectroscopic studies provide general support for these views. Fig. 9.14 shows a correlation between the O(1s) binding energy and the heat of adsorption of CO on a number of different metals

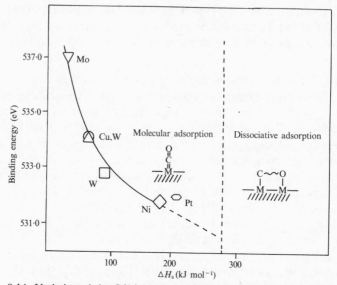

FIG. 9.14. Variation of the O(1s) binding energy in α and γ states of CO(ads) with ΔH_a on Cu, Ni, Pt, Mo (γ state only) and W (α states only). The molecular nature of the surface species is deduced from valence-level spectroscopy.

(Joyner and Roberts 1974). When the heat is greater than about $250\,kJ\,mol^{-1}$ the O(1s) value is always about $530\,eV$, i.e. close to the value observed with chemisorbed oxygen on metals. The suggestion is therefore made that there is a critical heat of chemisorption above which the CO dissociates at $290\,K$. This view is supported strongly by the presence of 'CO-like' molecular orbitals in the UPS data when the heats of chemisorption are less than $250\,kJ\,mol^{-1}$; they are absent when the heats are greater than this value. Iron is a borderline case (Kishi and Roberts 1975) in that dissociation occurs slowly at $290\,K$.

If we accept that the O(1s) binding energy reflects the electron density in the vicinity of the O atom then with increasing heat of adsorption the electron density also increases, presumably by back-bonding from the metal d-orbitals into the anti-bonding CO orbitals. This leads to weakening of the carbon–oxygen bond and ultimately to dissociation. Whether the carbon and oxygen atoms are then to be regarded as separate entities is another matter; all that can be concluded is that there is, on the basis of UPS, little bonding *between them*.

The detailed assignment of orbitals in the He(I) and He(II) spectra of adsorbed carbon monoxide has been discussed recently. A problem arises from the fact that in the adsorbed state only two bands are observed (Fig. 9.13), one at about 7 eV and the other at about 10 eV below the Fermi level. If 6 eV is added in order to compare with CO(g) these correspond to about 13 eV and 16 eV. The problem to be resolved is that three orbitals may be identified in CO(g) but only two in the adsorbed state. Lloyd (1974) drawing again on analogies with the spectra of metal carbonyls, suggests that the band with the lower ionization potential be assigned as $1\pi + 5\sigma$, i.e. due to composite orbitals, and the one of higher ionization potential be assigned as 4σ. Mason and his colleagues (Clarke, Gay, Law, and Mason 1975) in their studies of CO adsorption on platinum also support this point of view. Table 9.2 summarizes results of a number of studies with

TABLE 9.2

UPS studies of CO and related molecules on metals

System	Energy levels below E_F(eV)		Comments on surface species
	Observed	Plus 'correction' (6 eV)	
CO+Ni	7·5	13·5	Molecular chemisorption
Eastman & Cashion 1971	10·7	16·7	
CO+W (100)	6	12	Dissociative chemisorption
Baker & Eastman 1973	3	9	β phase CO: $C_{ads} + O_{ads}$:
	9	15	α phase CO: molecular adsorption
CO+Ti	5·7	11·7	
Eastman 1972	3·6	9·6	
CO$_2$+Ti	5·6	11·6	Dissociative chemisorption
Eastman 1972	3·5	9·5	
	8	14	
	(very weak band)		
O$_2$+Ti	6·7	12·7	Dissociative chemisorption
Eastman 1972	4·9	10·9	
Mo+CO (85 K)	10	16	Virgin, largely molecular (?)
↓ warm	6·5	12·5	
Mo+CO (290 K)	2·5–5·5	8·5–11·5	Dissociative chemisorption
Atkinson *et al.* 1974*a*	(no band at 10 eV)		

† These values may be compared with XPS and UPS studies of CO(g) where the following ionization potentials were observed: 15 eV (5σ), 17 eV (1π), 19·5 eV (4σ).

TABLE 9.3

O(1s) and C(1s) binding energies from various CO states and oxygen on W(110) and several other metal surfaces

Level	State of surface	W (110)	W (100)	W (poly)	Mo (poly)	Fe (poly)	Ru (001)	Ni (poly)
O(1s)	Virgin CO	531·6	531·7	531·5	531·2	531·3	531·7	531·4–531·8
	α CO	532·0	533·0	534·2 532·8	534·6			
	β CO	530·4	530·1	530·5	530·0	530·1	530·1	530·6
	Oxygen	530·1–530·4	530·1	530·3	530·0	530·0	529·8	529·7
C(1s)	Virgin CO	285·5	285·4	285·4	284·6	285·3		285·6–286·2
	α CO	285·8	286·2	287·3	287·3			
	β CO	283·1	283·0	283·1	282·7	282·9		283·5

All binding energies are referred to the Fermi level.
From Umbach *et al.* 1977.

comments on the likely state of surface species. These conclusions are supported by a detailed study by Umbach, Fuggle, and Menzel (1977) of CO adsorption on W (110). These authors combined X-ray and ultra-violet electron spectroscopy, core-level satellite spectroscopy, and Auger spectroscopy and Table 9.3 summarizes their X-ray data which are also compared with the results of a number of other systems. Ultra-violet photoelectron spectroscopy, with particular relation to gas adsorption on metal surfaces, has recently been reviewed by Bradshaw, Cederbaum, and Domcke (1975).

The correlation of infrared, surface-potential, and electron-spectroscopic data at different coverages is as yet not completely understood, but recent calculations by Miyazaki (1974) using a simple LCAO–MO approach indicate the direction that needs to be taken. Furthermore the CNDO calculations of Blyholder (1970) for CO chemisorbed on a cluster of 10 nickel atoms produce a pattern of valence orbital energy levels that reflect the photoelectron spectrum of nickel. It was deduced that the bonding of the carbon atom to the nickel involves mainly nickel s and p orbitals with little contribution from the d orbitals. On the other hand Grimley's (1962) quantum-mechanical calculations suggest that 84 per cent of the strength of the Ni—CO bond is contributed by back donation into the π^* orbitals (see also Ford 1970). This result is in qualitative agreement with the suggestion that back-bonding weakens the C—O bond accounting for the correlation between the O(1s) energy and the heat of adsorption (Fig. 9.14).

9.12. Conclusions

The complexities of the adsorption of carbon monoxide on metals have become more obvious as a result of recent LEED and electron-spectroscopic data (Roberts 1977). There is now overwhelming evidence for the dissociation of the β states on tungsten, molybdenum, and iron at room temperature, and the concept of 'states of adsorption' as deduced from kinetic data needs careful reappraisal. In this context the discussion of the possibility of corrosive chemisorption a decade or so ago based largely on field ion and field emission studies (Ehrlich 1966, Sachtler 1966) and also a number of spectroscopic papers where suggestions that the molecular integrity of carbon monoxide was in doubt should be recalled (Hayward 1971, McCoy and Smart 1973). Further support for the model for CO chemisorption discussed above comes from the results of energy-distribution data of field-emitted electrons (Young and Gomer 1974) and the isotopic exchange data of Anders and Hansen (1975). The assignment of orbitals in the spectrum of adsorbed carbon monoxide has received further confirmation from Gustafsson *et al.* (1975), who showed also that the $h\nu$ dependence of the photoionization cross-section is dominant in determining the surface sensitivity.

There are, however, still many facets of the chemisorption of CO on metals that are not resolved, and the interpretation of the sequence of LEED patterns observed is just one of them. A significant start has, however, been made with this problem in that Clarke, Gay, and Mason (1975) have shown how a molecular-dynamics calculation can account for the coverage dependence of the crystallography of CO adsorbed on Ni (100).

REFERENCES

ADAMS, D. L. (1974). *Surf. Sci.* **42,** 12.

ANDERS, L. W., and HANSEN, R. S. (1975). *J. chem. Phys.* **62** (12), 4652.

ANDERSON, J., and ESTRUP, P. J. (1967). *J. chem. Phys.* **46,** 563.

ATKINSON, S. J., BRUNDLE, C. R., and ROBERTS, M. W. (1973). *J. electron Spectrosc.* **2,** 105.

—— (1974a). *Faraday Disc. chem. Soc.* **58,** 62.

—— (1974b). *Chem. Phys. Lett.* **24** (2), 175.

BAKER, F. S., BRADSHAW, A. M., PRITCHARD, J., and SYKES, K. W. (1968). *Surf. Sci.* **12,** 426.

BAKER, J. M., and EASTMAN, D. E. (1973). *J. vac. Sci. Technol.* **10** (1), 223.

BEECK, O. (1940). *Proc. R. Soc. A* **177,** 62.

BELL, A. E., and GOMER, R. (1966). *J. chem. Phys.* **44,** 1065.

BICKLEY, R. I., ROBERTS, M. W., and STOREY, W. C. (1971). *J. chem. Soc. A* 2774.

BLYHOLDER, G. (1962). *J. chem. Phys.* **36,** 2036.

—— (1964). *J. phys. Chem.* **68,** 2772.

—— (1970). *J. vac. Sci. Technol.* **11** (5), 865.

BLYHOLDER, G., and SHEETS, R. (1970). *J. phys. Chem.* **74,** 4335.

BLYHOLDER, G., and TANAKA, M. (1973). *Bull. Chem. Soc. Japan* **46,** 1876.

BRADSHAW, A. M., CEDERBAUM, L. S., and DOMCKE, W. (1975). In *Structure and bonding*, vol. 24, 133. Springer-Verlag, Berlin.

BRADSHAW, A. M., and PRITCHARD, J. (1970). *Proc. R. Soc. A* **316,** 169.

BRENNAN, D., and GRAHAM, M. J. (1965). *Phil. Trans. R. Soc. A* **258,** 325.

BRENNAN, D., and HAYES, F. H. (1965). *Phil. Trans. R. Soc. A* **258,** 347.

CHANG, C. C. (1968). *J. electrochem. Soc.* **115,** 355.

CHESTERS, M. A., and PRITCHARD, J. (1971). *Surf. Sci.* **28,** 460.

CHESTERS, M. A., SIMS, M. L., and PRITCHARD, J. (1970). *Chem. Commun.* 1454.

CHRISTMANN, K., SCHOBER, O., and ERTL, G. (1974). *J. chem. Phys.* **60,** 4717.

CLARKE, J. K. A., FARREN, G. M., and RUBALCAVA, H. E. (1968). *J. phys. Chem.* **72,** 327.

CLARKE, T. A., GAY, I. D., and MASON, R. (1975). *Surf. Sci.* **50,** 137.

CLARKE, T. A., GAY, I. D., LAW, B., and MASON, R. (1975). *Chem. Phys. Lett.* **31,** 29.

COBURN, J. W. (1968). *Surf. Sci.* **11,** 61.

DEGRAS, D. A. (1967*a*). *Suppl. Nuovo Cim.* **5,** 408.

—— (1967*b*). *J. chem. Phys.* **64,** 405.

EASTMAN, D. E. (1972). *Solid State Commun.* **10,** 933.

EASTMAN, D. E., and CASHION, J. K. (1971). *Phys. Rev. Lett.* **27,** 1520.

EHRLICH, G. (1961). *J. chem. Phys.* **34,** 39.

—— (1966). *Disc. Faraday Soc.* **41,** 68.

EHRLICH, G., HICKMOTT, T. W., and HUDDA, F. G. (1958). *J. chem. Phys.* **28,** 506.

EISCHENS, R. P., and PLISKIN, W. A. (1958). *Adv. Catalysis* **10,** 1.

FORD, R. R. (1970). *Adv. Catalysis* **21,** 51.

GARLAND, C. W., LORD, R. C., and TROIANO, P. F. (1965). *J. phys. Chem.* **69,** 1188.

GASSER, R. P. H., THWAITES, R., and WILKINSON, J. (1967). *Trans. Faraday Soc.* **63,** 195.

GOMER, R. (1958). *J. chem. Phys.* **28,** 168.

—— (1966). *Disc. Faraday Soc.* **41,** 14.

GOMER, R. (1961). *Field emission and field ionization*, p. 179. Harvard Univ. Press, Cambridge, Mass.

GOYMOUR, C. G., and KING, D. A. (1973*a*). *J. chem. Soc. Faraday I* **69,** 736.

—— (1973*b*). *J. chem. Soc. Faraday I* **69,** 749.

GREENLER, R. G. (1966). *J. chem. Phys.* **44,** 310.

GREENLER, R. G., and TOMPKINS, H. G. (1971). *Surf. Sci.* **28,** 194.

GREGG, S. J., and LEACH, H. F. (1966). *J. Catalysis* **6,** 308.

GRIMLEY, T. B. (1962). *Proc. phys. Soc.* **79,** 1203.

—— (1969). In *Mol. Proc. Solid Surfaces Battelle Inst. Colloq.* (eds R. D. Gretz, E. Drauglis, and R. I. Jaffee), p. 299. McGraw–Hill, New York.

GUERRA, C. R. (1969). *J. Coll. interface Sci.* **29** (2), 229.

GUSTAFSSON, T., PLUMMER, E. W., EASTMAN, D. E., and FREEHOUF, J. L. (1975). *Solid St. Commun.* **17,** 391.

HAYWARD, D. O. (1963). *Gordon Conf. Paper*, see Hayward 1971.

—— (1971). In *Chemisorption and reactions on metallic films* (ed. J. R. Anderson). Academic Press, New York.

HILLIER, I. H., and SAUNDERS, V. R. (1971). *Mol. Phys.* **22,** 1025.

HOLSCHER, A. A., and SACHTLER, W. M. H. (1966). *Disc. Faraday Soc.* **41**, 1966.

HORGAN, A. M., and KING, D. A. (1969). In *The structure and chemistry of solid surfaces* (ed. G. A. Somorjai). John Wiley, New York.

HORN, K. and PRITCHARD, J. (1976). *Surf. Sci.* **55**, 701.

JOYNER, R. W., McKEE, C. S., and ROBERTS, M. W. (1971). *Surf. Sci.* **26**, 303.

JOYNER, R. W., and ROBERTS, M. W. (1974). *Chem. phys. Lett.* **29**, 447.

KAWASAKI, K., SENSAKI, K., and SATO, M. (1968). *Surf. Sci.* **11**, 143.

KING, D. A. (1975). *Surf. Sci.* **47**, 384.

KING, D. A., GOYMOUR, C. G., and YATES, J. T. (1972). *Proc. R. Soc. A* **331**, 361.

KISHI, K., and ROBERTS, M. W. (1975). *J. chem. Soc. Faraday Trans. I* **71**, 1715.

KOHRT, C., and GOMER, R. (1971). *Surf. Sci.* **24**, 77.

LANG, B., JOYNER, R. W., and SOMORJAI, G. A. (1972). *Surf. Sci.* **30**, 454.

LECANTE, J., RIWAN, R., and GUILLOT, C. (1973). *Surf. Sci.* **35**, 271.

LEIDHEISER, H., and GWATHMEY, A. T. (1948). *J. am. chem. Soc.* **70**, 1206.

LITTLE, J. G., QUINN, C. M., and ROBERTS, M. W. (1964). *J. Catalysis* **3**, 57.

LLOYD, D. R. (1974). *Disc. Faraday Soc.* **58**, 136.

McBAKER, M., and RIDEAL, E. K. (1955). *Trans. Faraday Soc.* **51**, 1597.

McCOY, E. F., and SMART, R. ST. C. (1973). *Surf. Sci.* **39**, 109.

McKEE, C. S., MURPHY, W. R., and ROBERTS, M. W. (1971). Unpublished work.

McMANUS, J. C. (1966). Unpublished work, see Hayward 1971.

MADEY, T. E., and YATES, J. T. (1967). *Nuovo Cim. Suppl.* **5**, 483.

MADEY, T. E., YATES, J. T., and STERN, R. C. (1965). *J. chem. Phys.* **42** (4), 1372.

MATHEWS, L. D. (1971). *Surf. Sci.* **24**, 248.

MENZEL, D. (1968). *Ber. BunsenGes. Phys. Chem.* **72**, 591.

—— (1975). *Surf. Sci.* **47** (1), 370.

MENZEL, D., and GOMER, R. (1964). *J. chem. Phys.* **41** (11), 3329.

MIYAZAKI, E. (1974). *J. Catalysis* **33** (1), 57.

NISHIJIMA, M., and PROPST, F. M. (1970). *J. vac. Sci. Technol.* **7**, 420.

ODA, Z. (1955). *Bull. Chem. Soc. Japan* **28**, 281.

—— (1956). *J. chem. Phys.* **25**, 592.

PITKETHLY, R. C. (1970). In *Chemisorption and catalysis* (ed. P. Hepple). Institute of Petroleum, London.

PRIMET, M., BASSET, J. M., MATHIEU, M. V., and PRETTRE, M. (1973). *J. Catalysis* **29**, 213.

PRITCHARD, J., and SIMS, M. L. (1970). *Trans. Faraday Soc.* **66**, 427.

PRITCHARD, J., CATTERICK, T., and GUPTA, R. K. (1975). *Surf. Sci.* **53**, 1.

QUINN, C. M. and ROBERTS, M. W. (1963). *Trans. Faraday Soc.* **59,** 985.

REDHEAD, P. A. (1961). *Trans. Faraday Soc.* **57,** 641.

—— (1967). *Nuovo Cim. Suppl.* **5,** 586.

RICE, R. W., and HALLER, G. L. (1975). *J. Catalysis* **40,** 249.

ROBERTS, M. W. (1977). *Chem. Revs.* **6,** 373, Chemical Society, London.

SACHTLER, W. M. H. (1966). *Disc. Faraday Soc.* **41,** 68.

TEXTOR, M., GAY, I. D. and MASON, R. (1977). *Proc. R. Soc. A* **356,** 37.

TRACY, J. C. (1972). *J. chem. Phys.* **56,** 2748.

TRACY, J. C., and BURKSTRAND, J. M. (1974). *CRC Crit. Rev. in Solid st. Sci.* **4,** Issue 3, 381.

UMBACH, E., FUGGLE, J. C., and MENZEL, D. (1977). *J. Electron. Spec.* **10,** 1.

WAGENER, S. (1957). *J. phys. Chem.* **61,** 267.

WEBB, A. N., and EISCHENS, R. P. (1955). *J. am. chem. Soc.* **77,** 4710.

YATES, J. T., and MADEY, T. E. (1969). *J. chem. Phys.* **51,** 334.

YATES, J. T., MADEY, T. E., and ERICKSON, N. E. (1973). *Chem. Phys. Lett.* **19,** 487.

YATES, J. T., MADEY, T. E., and PAYN, T. K. (1967). *Nuovo Cim. Suppl.* **5,** 558.

YOUNG, P. L., and GOMER, R. (1974). *J. chem. Phys.* **61** (12), 4955.

10

HYDROGEN AND NITROGEN

10.1. Introduction

HOMONUCLEAR diatomic gas-phase molecules might be expected, *a priori*, to provide relatively uncomplicated examples of adsorbate behaviour. Five elements exist in this form in their standard states at room temperature. Of these five, fluorine and chlorine are of such high reactivity that they may present practical problems in a vacuum system and relatively little work has been done with them. Oxygen is the subject of a succeeding chapter and we consider here the remaining pair, hydrogen and nitrogen.

Hydrogen, because of its electronic simplicity, is a suitable subject for theoretical considerations of adsorbate–surface bonding (see Chapter 12). It is important as a reactant in heterogeneously catalysed processes and also because exchange reactions of hydrogen and deuterium with organic molecules have been extensively investigated in relation to catalysis (Kemball 1959). Nitrogen, of course, is involved in the catalytic synthesis of ammonia, while its adsorption behaviour can be profitably compared with that of both hydrogen and carbon monoxide, the heteronuclear diatomic molecule with which it is isoelectronic.

The range of metals on which chemisorption occurs is found to be greater for hydrogen than for nitrogen (see Table 1.2) which, in part, is a reflection of their relative bond dissociation energies: 436 kJ mol^{-1} (H—H) and 945 kJ mol^{-1} (N≡N). In those cases in which adsorption does occur it will be found that the effect of crystallographic orientation on the process is fairly weak for f.c.c. metals but much more pronounced for b.c.c. and h.c.p. metals.

10.2. Hydrogen interaction with metals

10.2.1. *General aspects*

Owing to its position in the periodic table the general physical and chemical properties of hydrogen show a number of unique features, including very rapid mass transport, the largest possible isotope effects, and the occurrence of ions H^+ and H^- which are analogous on the one hand to alkali and on the other to halide ions. Of the compounds formed with metals, those of the simple binary type fall into three categories (see Shaw 1967, Mueller, Blackledge, and Libowitz 1968).

   (i) Ionic hydrides (M^+H^-) with the alkali and alkaline earth metals.

  (ii) Covalent hydrides, with copper and tin for example, and possibly lead (Mg and Al fall between classes (i) and (ii)).

 (iii) Metallic hydrides, with the lanthanides, actinides, the transition metals of groups IIIA, IVA, and VA, and also Cr, Ni, and Pd. These are metallic in appearance and have high electrical and thermal conductivities but the nature of the bonding involved is not well understood. The most intensively studied system has been $Pd-H_2$ (Lewis 1967), in which PdH is formed at 200 K and a non-stoichiometric compound at higher temperatures (e.g. $PdH_{0.6}$ at 373 K).

Formation of a binary hydride is an exothermic process, except in cases such as Ni and Cu where it is endothermic and Cr where it becomes endothermic at higher temperatures. For other metals which dissolve hydrogen without hydride formation, such as Fe, Co, Mo, Ag, and Pt, the process is endothermic, so that hydrogen uptake increases with increasing temperature. This may complicate the interpretation of thermal desorption spectra, since some hydrogen may enter the bulk rather than appear in the gas phase. The quantities of hydrogen which can be absorbed are very large in certain cases, including Pd, Ti, Zr, V, Ta, and U—zirconium hydride, for instance, contains twice as many H atoms per unit volume as liquid hydrogen! In the endothermic systems, however, the H to metal atom ratio is generally very small, in the range below 10^{-3} even at elevated temperatures; recent data for Pt (Ebisuzaki, Kass, and O'Keeffe 1968) indicates that at a hydrogen pressure of 1 atm and 300 K, the value may be as low as 10^{-11}.

Finally with metals such as W, Re, Os, and Ir no solubility is detectable by macroscopic techniques, although this does not rule out the possibility that amounts significant on the scale of surface processes may be dissolved; one monolayer of hydrogen atoms on a W(100) surface, if distributed through a $1\,cm^3$ tungsten specimen, would represent an H to W atom ratio of only 6×10^{-7}.

There is considerable metallurgical interest in the properties of hydrogen, stemming from problems of metal failure involving alloys of iron and more recently of titanium; these may suffer embrittlement in hydrogen atmospheres at relatively low temperature. Also, the interaction of hydrogen with transition metals, particularly nickel, palladium, and platinum, is of major importance for heterogeneous catalytic hydrogenation. In both the embrittlement and catalytic processes the dissociative chemisorption of hydrogen at the surface may be the rate-determining step, and the question immediately arises as to why transition metals, but not sp metals such as Al and Cu, are efficient in promoting this reaction. One possibility (Pearson 1971, Baetzold 1973) may be that electron transfer into the anti-bonding molecular orbital of the hydrogen molecule is required; d orbitals have the correct symmetry for this to occur, but s and p orbitals do not.

In the following sections we shall discuss the interactions of hydrogen with the group VIB metals Mo and W and the group VIII metals Ni, Pd, and Pt.

This choice is, of course, restricted; many other interesting systems have not been considered. In particular, hydrogen atoms, which may be conveniently produced from H_2 at a hot tungsten surface (see e.g. Anderson, Ritchie, and Roberts 1970), react with many metals which do not adsorb the molecular gas. It is possible, for example, to form in this way a hydride of lead ($PbH_{0.2}$) at 273 K (Roberts and Young 1970).

10.2.2. *Hydrogen adsorption on tungsten and molybdenum*

W(100). The adsorption of hydrogen on W(100) has been very extensively investigated, and a discussion of the accumulated observations provides a good example of the way in which a large range of modern techniques can be brought to bear on a single problem. At room temperature flash desorption reveals first a single state (β_2), desorbing with second-order kinetics (Fig. 10.1(a)). The activation energy for desorption decreases only slowly with increasing coverage, the initial value being $E_{d2}(\theta = 0) \approx 135$ kJ mol^{-1}. The β_2 state is complete at a fractional coverage estimated to be in the range $0.16 \leqslant \theta \leqslant 0.25$; at coverages greater than this a second state (β_1) appears in the flash desorption spectrum, for which $E_{d1}(\theta = 0) \approx 110$ kJ mol^{-1}. This state desorbs with first-order kinetics (Tamm and Schmidt 1969, 1970,

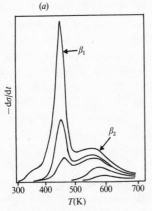

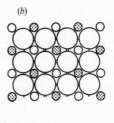

FIG. 10.1. W(100)–H_2. (a) Flash desorption sequence for successively larger amounts of H_2 adsorbed at 300 K. (b) Saturated surface: large circles W atoms, small open circles H atoms, small shaded circles H_2 molecules. $n_{\beta_1} = 5.0 \times 10^{14}$ molecules cm^{-2}, $n_{\beta_2} = 2.5 \times 10^{14}$ molecules cm^{-2}. (After Tamm and Schmidt 1969.)

1971a,b). Two values for the total adsorbate concentration at saturation have been reported, 1.5×10^{14} and 2.0×10^{14} molecules cm^{-2} (Tamm and Schmidt 1971b, Madey 1973, respectively); the latter, determined by a molecular beam method, is taken to be the more accurate. The relative saturation concentrations in the two states are $\sigma_1/\sigma_2 \approx 2$. There is some disagreement as to sticking-probability values and the form of their variation with coverage. Tamm and Schmidt (1970) and Ko, Steinbrüchel, and Schmidt (1974) found that S was constant for the β_2 state ($S_0 \approx 0.2$), while it decreased as $(1-\theta)$ for the β_1 state. In contrast Madey (1973), using a molecular beam method, and Barford and Rye (1974) reported a dependence of $(1-\theta)$ for S over the entire coverage range, with $S_0 \approx 0.5$. Diffraction results (Estrup and Anderson 1966, Adams and Germer 1970, Yonehara and Schmidt 1971) show that a $c(2 \times 2)$ pattern is formed initially, the half-order spots reaching maximum intensity at a coverage of ≈ 0.16, corresponding to the first appearance of β_1 in the desorption spectrum. Further exposure to hydrogen results in a rather complex sequence of pattern changes, which are accompanied by a general diminution in fractional-order intensities and culminate in the formation of a (1×1) pattern at $\theta = 1$. Evidence for a further state, β_3, has been obtained by Yonehara and Schmidt (1971); when a (100) crystal was cooled below 300 K after heating to 2500 K in vacuum, $\frac{11}{22}$ spots developed in the LEED pattern but a corresponding peak was not observed in the desorption spectrum. Hydrogen solution in tungsten is endothermic and it was suggested that as the crystal cools dissolved hydrogen diffuses from the bulk to the surface. On subsequent heating the adsorbate returns to the bulk owing to a rapid increase in its solubility with increasing temperature.

The observations relating to the β_1 and β_2 states were originally rationalized (Tamm and Schmidt 1969) in terms of a model involving initial dissociative adsorption into a $c(2 \times 2)$ structure with H atoms in sites of four-fold co-ordination (β_2 state) followed by molecular adsorption into the remaining four-fold sites (β_1 state, Fig. 10.1(b)). This accounts for the orders of the desorption processes and for the fact that $\sigma_1/\sigma_2 \simeq 2$, but a considerable amount of experimental evidence has now been accumulated which indicates that it cannot be correct with regard to the β_1 state; the points which follow illustrate the arguments.

(i) The co-adsorption of H_2 and D_2 leads to complete statistical mixing in both the β_1 and β_2 states, and also between the states (Tamm and Schmidt 1969).

(ii) If β_1 was a molecular state its activation energy for desorption would be expected to be appreciably different from that of the atomic β_2 state; this is not observed.

(iii) The work function is found (see e.g. Barford and Rye 1974) to vary linearly with coverage, indicating that the dipole moment per adsorbed particle is independent of coverage; this would be unlikely if two different states were occupied sequentially.

(iv) The cross-section for electron-stimulated desorption of H^+ passes through a maximum at a coverage $\theta \approx 0 \cdot 17$ which corresponds to completion of the β_2 state and also approximately to the occurrence of maximum intensity in the half-order diffraction spots. If two independent states filled sequentially the cross-section would not decrease at high coverage. Desorption of H_2^+ is never detected.

(v) Field emission energy-distribution (FEED) data reveal a local density of states due to hydrogen adsorption, which at coverages corresponding to the β_2 state is centred

~1·0 eV below the Fermi level ε_F in tungsten and has a full width at half maximum of 0·6 eV (Fig. 10.2, Plummer and Bell 1972). At higher coverages this level shifts upwards towards ε_F, diminishing in intensity until it has almost disappeared. Photoemission, which can probe more deeply below the Fermi level than can the FEED technique, shows similar behaviour, and in addition a level at ~5·4 eV below ε_F which also disappears on completion of the β_2 state (Fig. 10.3, Feuerbacher and Fitton 1973, Feuerbacher and Adriaens 1974). Again, these results suggest that at high coverages two distinct states do not coexist, but that a reorganization of the bonding electrons takes place which eliminates the spectral features associated initially with the β_2 state.

(vi) If β_1 was a molecular state, losses would be expected in the field emission energy distribution and in inelastic electron scattering spectra corresponding to excitation of H–H vibrations (at 0·55 eV). These are not detected (Plummer and Bell 1972, Propst and Piper 1967).

The observations listed avove, and others (see e.g. Yates and Madey 1971, Schmidt 1972), indicate that β_1 is not a distinct state, but that at higher coverages adatom—adatom interactions

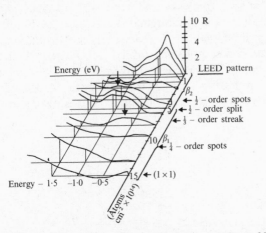

FIG. 10.2. W(100)–H_2. Field emission energy distribution results at 300 K. The enhancement factor R represents the contribution to the measured distribution of factors dependent on the electronic properties of the solid and its surface. (After Plummer and Bell 1972.)

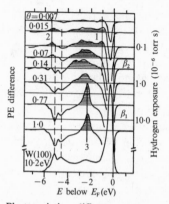

FIG. 10.3. W(100)–H_2. Photoemission difference spectra, that is, experimental spectra from which the spectrum of the clean surface (bottom curve) has been subtracted. The photon energy is 10.2 eV and θ is the fractional surface coverage. (After Feuerbacher and Fitton 1973.)

occur which produce modifications either in the character of the hydrogen–tungsten bonds, or in the physical position of the β_2 species, or both. As discussed in Chapter 6, nearest-neighbour repulsive interactions within a *single* adsorbed state can lead to a twin-peak desorption spectrum (see e.g. Adams 1974).

Electron-stimulated desorption ion–angular-distribution patterns (Madey, Czyzewski, and Yates 1975) suggest that in the β_2 state bonding is highly directional, so that the H atoms may be adsorbed directly above W atoms; this conclusion is supported by extended Hückel molecular orbital calculations (Anders, Hansen, and Bartell 1973). At higher coverages the ESD angular pattern changes, indicating a change in binding site. A possible model for a hydrogen monolayer consisting of a single state, in which

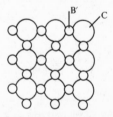

FIG. 10.4. Positions of H atoms (small circles) adsorbed on W(100) as suggested by Estrup and Anderson (1966). The saturation H-atom coverage is 1.0×10^{15} molecules cm^{-2}. (After Madey 1973.)

hydrogen atoms occupy 'bridge' sites, is shown in Fig. 10.4 (Estrup and Anderson 1966, Madey 1973).

Although this model can account in at least a qualitative way for many of the experimental facts, it still leaves a number of questions unanswered in detail. Why, for example, should EID from the β_2 state be much more probable than from the complete monolayer? Why should adjacent H atoms desorb (first-order kinetics) until an amount corresponding to σ_1 ($= 2\sigma_2$) is removed, etc? Perhaps the situation has been summarized best by Gomer (1975) in remarking that 'despite very careful and detailed experiments on what seems an extremely simple system nature can still baffle us'.

W(110). The flash desorption results for W(110) are quite similar to those for W(100). Following room-temperature adsorption two peaks, β_1 and β_2, are observed with binding energies of 111 and 137 kJ mol^{-1} respectively (Tamm and Schmidt 1971a). These have approximately equal populations at saturation and are atomically bound, exhibiting second-order desorption kinetics and complete isotopic mixing with D_2. RHEED results (Matysik 1972) show that a (2×1) structure is completed at $\theta = 0\cdot5$ as the β_2 state saturates. In contrast to W(100) the electron impact cross-section for desorption of H$^+$ increases at high coverage (Fig. 10.5(a)) (King and Menzel 1973). Also, the value for the

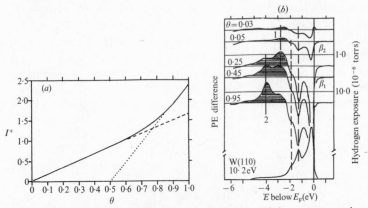

FIG. 10.5. W(110)–H$_2$. (a) Electron impact desorption H$^+$ ion current against fractional coverage θ derived assuming a linear fall of the sticking probability S with θ and $S_0 = 0\cdot025$. (After King and Menzel 1973.) (b) Photoemission difference spectra. (After Feuerbacher and Fitton 1973.)

more weakly bound β_1 state is greater than that for β_2, which is the expected behaviour. Adsorption causes a work-function decrease on this plane, but there is some disagreement as to the final magnitude of the change, values being quoted between $\sim 0 \cdot 0$ and $-0 \cdot 5$ eV (see Hopkins and Usami 1970, Barford and Rye 1974). Various features of the adsorption kinetics are also uncertain. The initial sticking probability S_0 lies somewhere between $0 \cdot 22$ and $0 \cdot 025$, while the coverage dependence of S is expressed either as $(1-\theta)$ or $(1-\theta)^2$ (Tamm and Schmidt 1971b, King and Menzel 1973, Barford and Rye 1974).

As with W(100) the thermal and electron-impact desorption data for (110) on first examination suggest the existence of two distinct adsorbed states, but the evidence is not conclusive. For example, the twin-peak thermal desorption spectrum may be approximately simulated on the assumption of a single state having a coverage-dependent activation energy (Tamm and Schmidt 1971a). Also, two states of equal population at high coverage would result in an intersection at $\theta \simeq 0 \cdot 5$ of the linear portions of the EID current–coverage curve (Fig. 10.5(a)); the observed intersection is at $\theta \simeq 0 \cdot 8$. Support for the idea of just one state is provided by FEED data (Plummer and Bell 1972) which show a single characteristic energy distribution building up with the coverage. Photoemission results (Fig. 10.5(b), Feuerbacher and Fitton 1973) are not so clear, however. A weak resonance initially present close to the Fermi level is extinguished as β_2 saturates (cf. W(100)), while a second level (2, Fig. 10.5(b)) at -4 eV appears as β_1 adsorption sets in. A third level (1) grows with the population of β_2 and persists at high coverage. The sequential filling of levels 1 and 2 has been taken as evidence for the existence of two adsorbed states, but level 1 also appears during adsorption on W(100) not only of H_2 but also of CO, N_2,

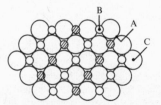

FIG. 10.6. W(110)–H_2. Possible adsorbate structures for the β_1 (open circles) and β_2 (shaded circles) states. (After Schmidt 1972.)

O_2, and C (Waclawski and Plummer 1972). Its nature is therefore in some doubt.

Possible structures for the hydrogen layer at half- and full-monolayer coverages are shown in Fig. 10.6 (Schmidt 1972). Note that in spite of the high symmetry of the substrate, anisotropic surface interactions must be present at $\theta = 0.5$ to produce the (2×1) structure. The complete layer would require a hydrogen density of 1.42×10^{15} atoms cm^{-2}; considering the inaccuracies involved in determining absolute coverages, the experimental values lying in the range $(1.4–1.8) \times 10^{15}$ are in good agreement with this value.

One cautionary note must be added in relation to W(110). Polizzotti and Ehrlich (1975), using a field emission microscope, failed to observe adsorption of H_2 directly from the gas phase onto this plane between 38 K and 298 K. Instead, the (110) was populated by surface diffusion of H atoms which had been formed by dissociative adsorption on neighbouring planes on the FEM tip. Similar observations for N_2 on W(110) and Re(0001) are discussed in Sections 10.3.2 and 10.3.5.

W(211). This is the simplest b.c.c. surface to contain two types of surface atom, those forming the close-packed rows which constitute the actual surface layer and those in the 'troughs' which, although they are in the second layer, are accessible none the less to small adsorbate molecules. This heterogeneity is reflected in the behaviour of W(211) towards hydrogen; two major desorption peaks, labelled β_1 and β_2, are observed to fill sequentially following adsorption at ~ 100 K (Rye, Barford, and Cartier 1973). These occur at 330 and 675 K; they appear to be due to atomically occupied states of approximately equal saturation populations and activation energies for desorption of 67 and 146 kJ mol^{-1}, respectively. An additional state desorbing at ~ 500 K, reported by Adams, Germer, and May (1970), was probably an experimental arteface (Rye and Barford 1971). Rye *et al.* (1973) also observe a high-temperature shoulder on their β_2 peak which 'may result from the (specimen) leads'. Its existence provides further illustration of the difficulties encountered when trying to establish quantitative adsorption and desorption relationships. If the main β_2 peak only is considered, the sticking probability for this state is found to remain constant at its initial

value of 0.57 over ~85 per cent of the β_2 coverage range. If, on
the other hand, the shoulder is included with the main peak, the
value of S_0 becomes ~1.0 and S varies as $(1-\theta)$ as reported by
Adams *et al.* (1970). Only one study of the β_1 state appears to
have been made (Rye *et al.* 1973); the value of S_0 was ~0.05 and
the coverage dependence was of the form $(1-\theta)$. No additional
features are observed in the LEED pattern as a result of adsorption, suggesting the formation of a (1×1) structure (Adams *et al.*
1970).

The work-function variation $\Delta\phi$ with hydrogen coverage on
W(211) contrasts quite sharply with that on (100) and on (110).
A distinct break in the plot (Fig. 10.7(a)) occurs at a point
corresponding to saturation of the β_2 state (Rye *et al.* 1973). This
provides strong evidence that two distinct adsorption states actually do exist in this system, one (β_2) leading to a work-function
increase of $\sim0.6\,\mathrm{eV}$ and the other (β_1) to a decrease of
$\sim0.27\,\mathrm{eV}$. Rye *et al.* (1973) attempted to obtain some insight as
to the nature of the sites involved by means of sequential
adsorption of deuterium and ethylene (D_2 was used to distinguish
between chemisorbed hydrogen and hydrogen resulting from
ethylene decomposition). If the β_2 state alone was saturated with
deuterium the amount of ethylene subsequently adsorbed was
similar to that adsorbed by the clean surface. If, however, both
the β_1 and β_2 states were initially filled, ethylene adsorption was
restricted to about one-third of the previous value. These results
can be rationalized by identifying β_1 sites with W atoms in the
close-packed surface rows and β_2 sites with atoms in the troughs.
Ethylene can easily bond to row atoms, either along one row in a
'cis' form in which the hydrogens are fully eclipsed or between
two rows in a 'trans' form (Fig. 10.7(b), Barford and Rye 1972).
It is also just possible that the σ–diadsorbed species could
fit into a trough, but impossible for the flat molecule approaching
from the gas phase to do so. Therefore the initial interaction must
be with the rows, and it seems reasonable to assume that these
also represent the sites for eventual ethylene adsorption. On the
basis of the mixed-adsorption results this means that the β_1
hydrogen sites, which are blocked by prior ethylene adsorption,
must also be identified with row atoms, leaving the troughs for β_2
adsorption. The signs of the work function changes then suggest
that the adatoms carry a partial positive charge; β_1 being above

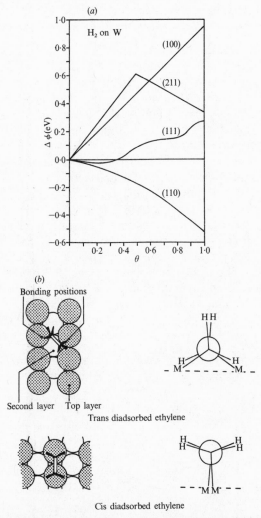

FIG. 10.7. (a) Work-function change *versus* coverage for H_2 on W(110), W(100), W(211) and W(111). (After Barford and Rye 1974.) (b) 'Cis' and 'trans' diadsorbed ethylene on W(211). The left-hand diagram depicts sites on W(211) expected to adsorb C_2H_4 in the given configuration. The projection on the right gives the configuration about the C—C axis. (After Barford and Rye 1972.)

and β_2 below the substrate surface plane would thus produce decreases and increases respectively in ϕ, as observed.

$W(111)$. Following (211) this surface represents a further stage in increasing structural complexity since it exposes both second- and third-layer atoms. Once again this change is qualitatively reflected in the thermal-desorption spectra of hydrogen adsorbed at 78 K (Tamm and Schmidt 1971a, see also Madey 1972, Barford and Rye 1974). In addition to molecularly adsorbed γ-species, four β-states are observed between 200 and 600 K. Because of the overlap of the desorption peaks kinetic analysis is difficult, but it appears that these states are atomic with activation energies for desorption ranging from 50 to 147 kJ mol^{-1}. The sticking-probability curve at 125 K (Fig. 10.8) shows a break at a coverage of $\sim 1\cdot 2 \times 10^{15}$ atoms cm^{-2} (Tamm and Schmidt 1971b), indicating the presence on the surface of at least two distinct states. Further evidence for the existence of separate states is provided by the work-function data (Fig. 10.7(a)), which exhibit a minimum at $\theta \sim 0\cdot 2$, a point of inflection at $\theta \sim 0\cdot 5$, and a relative minimum at $\theta \sim 0\cdot 75$, suggesting that the filling of each state may involve a change of slope or sign of $\Delta\phi(\theta)$ (Barford and Rye 1974).

It is possible to extract a limited degree of generalization from the foregoing data. For example, heterogeneity on (100) and

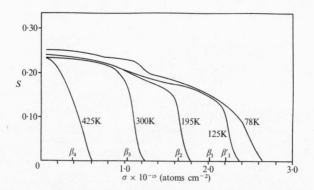

FIG. 10.8. $W(111)$–H_2. Sticking probability curves for different substrate temperatures. The saturation atom densities of the β_1, β_2, β_3, β_4 binding states are indicated. (After Tamm and Schmidt 1971b.)

(110) is probably induced, while that on (211) and (111) is intrinsic, being related to the presence of different binding sites. Also, Barford and Rye (1974) have pointed out that the amounts of hydrogen adsorbed may be correlated with the numbers of missing nearest-neighbour tungsten atoms in each plane; approximately one H atom is taken up for every two missing nearest neighbours.

Nevertheless, it will be apparent that many aspects of hydrogen adsorption on tungsten remain uncertain, in spite of the simplicity of the adsorbate. The complications may be due in part to the small size of hydrogen, which enables it to interact with surface sites inaccessible to larger species. Methane, for example, adsorbs on W(111) in only one state of low energy ($E_d = 34$ kJ mol^{-1}, Madey 1972), in sharp contrast to the five hydrogen states on this surface—four β and one γ. The complexity of the data, however, may reflect factors other than purely geometric ones. One example would be the adatom–adatom interactions already discussed, while a second would be provided by work-function changes. On W(110) the most likely site for adsorption may be the 'bridge' position B (Fig. 10.6). If we take the radius of an adsorbed atom to be approximately equal to the covalent radius of hydrogen, ~0·03 nm (Pauling 1960), then in this B position the centre of the adsorbate would lie about 0·055 nm above the line of centres of the top-layer tungsten atoms. The observed decrease in ϕ on (110) then indicates that hydrogen carries a partial positive charge. The same conclusion would be reached if instead the A or C sites (Fig. 10.6) were occupied; in fact it has just been argued that population of A-type sites on W(211) is responsible for a decrease in ϕ. On W(100) coverage data lead to the conclusion that at saturation the occupied sites must include those of either types C or B′ (Fig. 10.4) (B′ on (100) differs from B on (110) only in the absence of a second-layer W atom directly below the site). A positive hydrogen species in either of these positions would cause a decrease in ϕ, while ϕ actually increases continuously to saturation. We are therefore left with the following question: is the adsorbate charge on W(100) of a different sign to that on (110) and (211)? A final answer must await the location of the various adsorption sites with some precision.

Mo(100). There is general similarity between the interactions of hydrogen with Mo(100) and with W(100), as one would expect.

On molybdenum, however, the β_1 and β_2 states are less strongly bound than on tungsten. Since the lattice constants of the two metals are the same, it may be assumed that the adatom–substrate spacings are similar. The increased binding energy on W is therefore attributed to the greater radial extension of the 5d orbital, compared with the 4d in Mo, leading to greater overlap with H in the W case (Han and Schmidt 1971). The appearance in the molybdenum desorption spectrum of an additional β_3 state constitutes a second difference. As mentioned above, there is some evidence from LEED that a similar state exists on W(100) but that it dissolves endothermically in the bulk. If hydrogen is more strongly bound (less endothermic) in Mo the β_3 state may actually desorb in this case because its solubility does not increase rapidly enough with increasing temperature to permit bulk solution.

10.2.3. *Hydrogen adsorption on nickel, palladium, and platinum*

Adsorption of hydrogen on nickel surfaces, in particular, has been comprehensively investigated because of its importance in catalytic hydrogenation, but there has been some disagreement between data obtained by different workers. This may be in part the result of the strong influence which traces of carbon monoxide appear to exert on the hydrogen adlayer (see e.g. Yonehara and Schmidt 1971, Lapujoulade 1971) and in recent work efforts have been made to suppress CO partial pressures below $\sim 10^{-12}$ torr.

Ni(100) and Ni(111). These planes behave in very similar ways towards hydrogen (Lapujoulade and Neil 1972, 1973, Christmann *et al.* 1974). Flash desorption following adsorption at 258 K reveals two β states filling sequentially with activation energies for desorption of $\sim 81\,\mathrm{kJ\,mol^{-1}}$ (β_1) and 97 kJ mol^{-1} (β_2). There is uncertainty as to the value of the initial sticking probabilities in the range $S_0 \lesssim 10^{-2}$ (Horgan and Dalins 1974) to $S_0 \approx 0\cdot1$–$0\cdot25$ (Lapujoulade and Neil 1972, 1973, Christmann *et al.* 1974) and also as to the saturation coverage between $\sim 3 \times 10^{14}$ and $1\cdot5 \times 10^{15}$ atoms cm^{-2}, although the latter value is probably the more reliable. The work-function is increased at saturation (10^{-4} torr, 298 K) by $\sim 0\cdot2$ eV, $\Delta\phi$ being directly proportional to coverage up to the point corresponding to saturation of the β_2 state. The LEED patterns show no extra features on adsorption, only a decrease in intensity of the integral-order beams and a slight increase in the intensity of the background. The latter observation suggests that a disordered layer is formed at room temperature.

The adsorption of hydrogen on nickel is completely reversible above room temperature, and so adsorption isotherms can be readily constructed (Fig. 10.9(a)); $\Delta\phi$ provides a convenient measure of coverage (Christmann *et al.* 1974). The isosteric heat of adsorption may be then determined using the Clausius–Clapeyron equation

$$\left\{\frac{\partial \ln P}{\partial(1/T)}\right\}_\theta = \frac{-E_{ad}}{R}. \tag{10.1}$$

Results are shown in Fig. 10.9(b), from which it is seen that for both (100) and (111) E_{ad} is constant over the coverage range corresponding to the β_2 state, indicating the absence of long-range interactions between adatoms. At higher coverages the proportionality between $\Delta\phi$ and θ may no longer hold, but obviously (Fig. 10.9(b)) E_{ad} decreases as the β_1 state is filled.

There is some evidence, from the variation of the electrical resistance of thin nickel films on exposure to H_2 (Wedler and Bröcker 1971), that considerable electronic changes occur in the

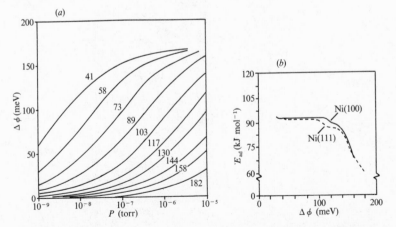

FIG. 10.9. Ni(100)–H_2. (a) Adsorption isotherms $\Delta\phi$ *versus* pressure. The numbers on the curves give the temperature in °C. (b) Isosteric heats of hydrogen adsorption on nickel as a function of $\Delta\phi$. (After Christmann *et al.* 1974.)

adsorbate as the β_1 state builds up and this suggests that, as in the case of W(100), β_1 and β_2 may not be distinct states but may represent a further example of induced heterogeneity.

Ni(110). In the case of the (110) plane of nickel, hydrogen adsorption involves two features of considerable interest. The first concerns the behaviour of the desorption spectra for different initial coverages (Christmann *et al.* 1974). Following adsorption at 300 K these spectra consist of a single peak, which at higher coverages exhibits first-order desorption behaviour with $E_d = 97\cdot8\ kJ\ mol^{-1}$. In the low-coverage range, however, the temperature at which the peak maximum occurs shifts slightly towards *higher* values as coverage increases. This could be the result of a zero-order desorption process, which is physically unlikely, or of an increasing adsorption energy (Lapujoulade and Neil 1973, Christmann *et al.* 1974). The latter explanation is confirmed by measurement of isosteric heats of adsorption using eqn (10.1); at very low coverage E_{ad} appears to be constant at $\sim90\ kJ\ mol^{-1}$, but it then *increases* to a second constant value of $97\cdot8\ kJ\ mol^{-1}$. An indication as to the nature of the surface process responsible for this change is provided by the adsorption isotherms (Fig. 10.10); the sudden increase in coverage above

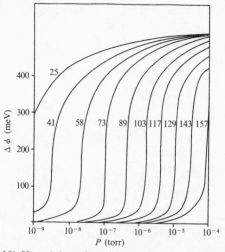

FIG. 10.10. Ni(110)–H$_2$. Adsorption isotherms $\Delta\phi$ *versus* pressure. The numbers on the curves give the temperature in °C. (After Christmann *et al.* 1974.)

some critical pressure is characteristic of two-dimensional condensation, which occurs because of attractive interactions between adatoms. This is quite common in physical adsorption (see e.g. Honig 1967), but Ertl and co-workers (Christmann *et al.* 1974) believe that Ni(110)—H_2 represents the first instance of its detection in a chemisorption system.

Initially hydrogen is randomly adsorbed, but at a coverage equivalent to $\Delta\phi \sim 60$ meV attractive interactions in the [110] direction, along the surface rows, produce a one-dimensional structure with periodicity equal to that of the nickel lattice, the transformation being accompanied by an increase in hydrogen binding energy. Considerably weaker adatom interactions occur in the [100] direction, across the rows. This picture is deduced from the appearance of a (1×2) LEED pattern at low coverage in which the half-order features are sharp streaks running in the [100] direction; these later contract into relatively sharp spots (May and Germer 1969, Taylor and Estrup 1974, Christmann *et al.* 1974).

Identification of the surface structures producing these diffraction patterns leads to the second point of interest concerning Ni(110)—H_2. This was one of the systems for which it was originally argued (Germer and MacRae 1962, May and Germer 1969) that fractional-order spots must be the result of surface reconstruction, since adsorbates such as hydrogen were not thought to have sufficient scattering power to generate new diffraction features (see Chapter 3). The alternative view—that the patterns were indeed due to hydrogen structures—was proposed by Bauer (1966) and supported by later theoretical calculations of scattering factors. The system has recently been re-examined by Taylor and Estrup (1974), who add weight to the reconstruction hypothesis. May and Germer (1969) pointed out that in a (1×2) hydrogen structure alternate [110] surface rows must be free of adsorbate, and at room temperature must remain so even at the highest exposures since a (1×1) pattern is never observed. This would be unexpected behaviour, and in fact Taylor and Estrup have shown that adsorption below 170 K produces initially a relatively weak (2×1) pattern, which is ascribed to a hydrogen structure with adsorbate on every surface row (Fig. 10.11(a)); theory indicates that the adsorption site directly above a surface Ni is preferred (Fassaert, Verbeek, and

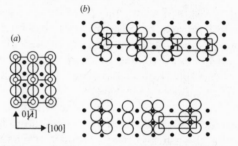

FIG. 10.11. Ni(110)–H$_2$. Surface structures. The large circles represent the top layer of Ni atoms. The solid circles represent the centres of the Ni atoms in the second layer. (a) Model of the (2×1) structure, containing half a monolayer of H atoms represented by small open circles. (b) The upper diagram represents a distorted substrate surface with (1×2) periodicity arising from paired displacement of the (100) rows; a mistake in the pairing is shown. The lower diagram illustrates an alternative model. Disorder arises from mistakes in the displacement along [100]. The H atoms are not shown. (After Taylor and Estrup 1974.)

van der Avoird 1972). At higher exposure the (1×2) pattern appears. The transformation involves a relatively small coverage increase but a large increase in diffracted intensity, which would be unlikely if a simple rearrangement of adsorbed hydrogen was occurring. It is suggested instead that [110] rows of nickel atoms are displaced to give a structure of the type shown in Fig. 10.11(b) (Tucker 1971, Taylor and Estrup 1974). In view of the fact that the process occurs at 170 K, this idea has the advantage that relatively minor atom movements are required, in contrast to the model proposed by May and Germer (1969) which involved movement of surface Ni atoms over distances of around 10 nm. As always, however, a final decision between the various models, both reconstructed and otherwise, must await a complete LEED intensity analysis.

Pd(111) and Pd(110). Although there have been numerous investigations of the remarkable bulk properties of the H$_2$/Pd system, relatively little attention has been paid to the *adsorption* of hydrogen on this metal. Roberts and Ross (1969) studied the interaction with films at pressures between 10^{-4} and 5×10^{-2} torr and obtained a value for the heat of formation of the hydride of 35 kJ mol^{-1}. A thermal-desorption study using polycrystalline wire has been carried out by Aldag and Schmidt (1971), while Ertl and co-workers (Conrad, Ertl, and Latta 1974) have investigated the behaviour of the (111) and (110) surfaces. In both cases thermal desorption following low exposures at room temperature produces a single peak at 373 K which is attributed to surface hydrogen. This has the characteristics required for a second-order desorption process with $E_d \approx 102 \, \text{kJ mol}^{-1}$ on (110) and 88 kJ mol^{-1} on (111), values which are confirmed and shown to be constant over a considerable range of surface coverage by measurement of the isosteric heats of adsorption. Measurements were also made using a stepped (111) surface consisting of terraces of (111) orientation nine atom rows wide separated by monatomic (111) steps. In general the adsorption energy was similar to that on the perfect (111) plane, but at low coverages it tended to a higher value comparable to that on (110) presumably owing to adsorption on sites at the steps. After higher exposures the desorption spectra from perfect (111) and (110) planes contain a second peak due to removal of hydrogen from the bulk. This becomes flatter and moves to higher temperatures as the time interval between exposure to the gas and desorption increases, that is, as the hydrogen is allowed to distribute itself more uniformly through the bulk.

The two crystal planes differ slightly in work-function behaviour, the increase on (111) of ~ 0.18 eV being only half as great as that on (110) (0·36 eV, $P_{H_2} \leq 10^{-5}$ torr). There is also a difference in the surface structures formed. On Pd(111) hydrogen produces no additional diffraction features but does cause variations in the intensity–voltage curves, indicating the existence of a (1×1) structure (Christmann, Ertl, and Schober 1973). In the case of Pd(110) a (1×2) pattern forms in which the half-order features are initially streaked but on further exposure contract to spots. This is similar to the behaviour of the Ni(110)—H$_2$ system, but in the present instance there is no particular evidence in favour of surface reconstruction. If this does not occur, observations such as the magnitude of the work-function change indicate that the relatively large (110) (1×2) unit mesh must contain more than a single hydrogen atom (Conrad *et al.* 1974).

In spite of the outstanding solubility of hydrogen in palladium, the results just outlined demonstrate that the adsorption properties of hydrogen on this metal are not appreciably different from those on nickel, the equivalent substrate in the first transition series. Differences in hydrogen binding energies between Ni and Pd, for example, are much smaller than those between Mo and W discussed previously. There does appear to be a kinetic distinction between the two systems, however; the initial sticking probability of hydrogen on evaporated palladium films is particularly high, ~0·9 (Atkinson *et al.* 1976), compared with that on films of nickel ($S_0 \approx 0·25$, Horgan and Dalins 1974) and other transition metals.

Pt(111). While it has been suggested in the past that hydrogen does not adsorb on Pt(111) at or above room temperature, there is now considerable evidence to the contrary (e.g. Baldwin and Hudson 1971, Weinberg and Merrill 1972, Lu and Rye 1974 1975, Christmann, Ertl, and Pignet 1976). Nevertheless, controversy remains as to the details of the process—the number of adsorbed states, activation energies for desorption, LEED patterns, and so on; for simplicity we shall outline the results of Christmann *et al.* since these were obtained under well-defined conditions of surface cleanliness and perfection of surface structure. Thermal desorption reveals two states, but at 300 K, because of their low binding energies, β_1 is absent and β_2 is present in low concentration only. This may explain previous failures to detect adsorption in certain cases. The β_1 and β_2 states appear to be atomically bound and to be of approximately equal population at saturation. The adsorption energy, determined from isotherm data using the Clausius–Clapeyron equation, is constant for the β_2 state at ~39 kJ mol^{-1} and then decreases as β_1 fills. Activation energies for desorption are similar since the activation energy for chemisorption is negligible.

Christmann *et al.* interpret their observations in terms of a model involving induced heterogeneity in which hydrogen atoms initially occupy a (2×2) array under the influence of indirect adatom–adatom repulsions. For $\theta > 0·5$ the vacant sites in this array are progressively filled, with an increase in repulsive interactions and a decrease in adsorption energy. The β_2 desorption peak corresponds to recombination of two H atoms from sites which have all nearest-neighbour sites vacant, while β_1 is the result of desorption from sites with occupied nearest neighbours. There are two problems with this picture, the first being the absence of any extra features in the LEED pattern—a (2×2) pattern would be expected at $\theta \lesssim 0·5$. The second problem is that the desorption kinetics cannot be quantitatively described by a simple model of induced heterogeneity involving a single type of binding site and lateral interactions. This assumes, however, that the adsorbate maintains configurational equilibrium during the desorption process. There is experimental evidence that at 150 K equilibrium may not be reached during adsorption and a non-equilibrium situation may therefore also exist in the lower-temperature reaches of the desorption programme. This could result in data which would not necessarily fit a simple model.

Sticking-probability, E_{ad} and work-function behaviour is shown in Figs 10.12 and 10.13. The $\Delta\phi(\theta)$ curve contrasts with that of the (111) surfaces of Ni and Pd which both undergo linear increases in θ with exposure to hydrogen; on Pt(111) the work function decreases. On all three surfaces, however, $\Delta\phi$ is rather small (~ 0.2 eV) and so the chemisorption bond is predominantly covalent. In the platinum case $\Delta\phi$ is proportional to $\theta^{1.3}$, a functional form which cannot be due to depolarization effects (see Chapter 12), but may be associated with indirect adatom–adatom interactions. There is no break in the $\Delta\phi(\theta)$ curve which might have corresponded to a second adsorption state.

The properties of surface atoms of low co-ordination number, that is atoms which are not constituents of a low index plane, are of particular interest since such atoms may possibly be associated with the 'active centres' of heterogeneous catalysis. Stepped surfaces provide a convenient means for study of sites of this type, and following the classical work of Lang, Joyner, and Somorjai (1972) several investigations have been carried out using

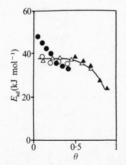

FIG. 10.12. Variation of the H_2 adsorption energy with coverage on a flat (open symbols) and a stepped (solid symbols) Pt(111) surface (Pt(997)). Triangles, isosteric heats as derived from $\Delta\phi$ data; circles, activation energies for desorption from thermal desorption spectra. (After Christmann and Ertl 1976.)

Pt substrates. For example, Bernasek and Somorjai (1975a,b), using a modulated molecular-beam scattering technique, have examined the hydrogen–deuterium exchange reaction on the stepped surfaces Pt-(S)[5(111)×(111)] and Pt-(S)[9(111)×(111)] which are characterized by terraces of (111) orientation five- and nine-atom spacings wide respectively, separated by monatomic (111) steps. In both cases the reaction probability, defined as the fraction of incident reactant molecules leaving the surface as product HD molecules at the specular angle, was found to be considerably greater than in the case of a Pt(111) surface. Therefore, the steps facilitate hydrogen dissociation. They also increase the efficiency of energy exchange between the incident gas molecules and the surface.

The interaction of hydrogen with the Pt-(S)[9(111)×(111)] surface (which in terms of Miller indices is Pt(997)), has also been investigated by Christmann and Ertl (1976); the results may be compared directly with those obtained by the same workers for Pt(111). The variations in adsorption-energy, sticking-probability, and work-function change as functions of coverage are of particular interest (see Figs. 10.12 and 10.13). For coverages greater than $\theta \approx 0.25$ the behaviour of all of these parameters is very similar on both planes. At lower coverages, however,

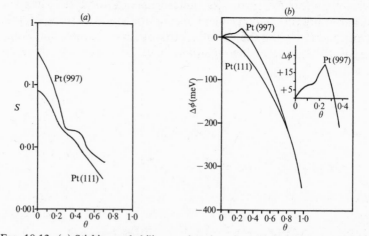

FIG. 10.13. (a) Sticking probability as a function of coverage at 120 K for H_2 on Pt(111) and on Pt(997). (b) Variation of the work function with coverage at 130 K for Pt(111) and Pt(997). The inset shows the region of small coverages with enlarged scales for Pt(997). (After Christmann and Ertl 1976.)

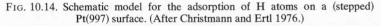

FIG. 10.14. Schematic model for the adsorption of H atoms on a (stepped) Pt(997) surface. (After Christmann and Ertl 1976.)

surface steps obviously provide sites at which hydrogen is more rapidly adsorbed, is held with a higher binding energy, and has a dipole of opposite sign compared with hydrogen on Pt(111). Further, there is evidence of a second discontinuity in the $S(\theta)$ and $\Delta\phi(\theta)$ curves at $\theta \approx 0.15$. Christmann and Ertl interpret these results in terms of nine hydrogen atoms adsorbed at saturation on each terrace (Fig. 10.14). The two sites at either end of a terrace are of higher binding energy than the remainder and so are occupied first. Also, the two sites are of different geometry and so hydrogen atoms adsorbed on them should have dipoles of different magnitudes, as observed (Fig. 10.13(b)). The difference in sign of the dipoles, compared with adsorption on (111), is ascribed to the altered electronic properties of the Pt edge atoms.

The four-fold increase in initial sticking probability caused by a step density of 11 per cent of the surface atoms cannot be due to a lowered activation energy for H—H bond fission, since dissociative hydrogen adsorption on Pt(111) is approximately temperature independent (Christmann *et al.* 1976, Norton and Richards 1974). Various possible explanations have been suggested; for example, gas molecules may be required to approach the surface with some particular orientation if they are not to be elastically reflected. The relatively high electric field in the vicinity of a step may sufficiently polarize an approaching molecule that the resulting torque influences the orientation. Alternatively, H_2 molecules in the precursor state may have an enhanced lifetime at a step, increasing the probability that the correct orientation is attained.

10.3. Kinetic and related studies of nitrogen interaction with metals

10.3.1. *General aspects*

Of the various parameters required for the characterization of an adsorption or catalytic system, gas pressure is among the most fundamental. Unfortunately, measurement of this quantity in the low-pressure region is not straightforward. The commonly used measuring devices employ either a hot filament as a source of electrons for gas ionization (Bayard–Alpert ionization gauge, mass spectrometer, etc.) or a high magnetic field (e.g. magnetron gauge), and so there may be quite strong perturbation of both the pressure and composition of the gas phase. In the case of the ubiquitous ion gauge, for example, two principal effects can be distinguished. The gauge may act as a pump by adsorption on the metal grid, by uptake of ions at the glass walls, and so on; secondly, gas molecules may be excited, ionized, or dissociated by electron bombardment or by collision with the hot filament and these species may find their way to the adsorbent surface. In addition, for all gauges absolute calibration presents a considerable problem.

Nitrogen suffers less from these difficulties than many other gases and with care accurate pressures can be determined; this fact accounts, in part, for the popularity of the gas as an adsorbate in kinetic experiments. Nevertheless, gauge effects have been observed in certain cases (see Yates and Madey 1969(b)), while nitrogen in various 'activated' forms can be highly reactive. If the molecular gas is deliberately brought into contact with a tungsten filament at high temperature, prior to adsorption, the species produced (ions, atoms, and excited molecules) adsorb on metals such as copper and nickel (Perdereau and Rhead 1971, Kirby et al. 1979, see also Chapter 12) which do not interact rapidly with N_2 molecules at room temperature and low pressure. Species containing a single nitrogen atom which occur on a surface as a result of ammonia adsorption can also produce unusual adsorbed nitrogen states (Peng and Dawson 1971).

The catalytic synthesis and decomposition of ammonia have been the subjects of many investigations over a long period of time. Such experiments represent a logical progression from work on hydrogen and nitrogen separately but also a progression to increased complexity, and so they are beyond the scope of this chapter. It may be of interest, however, to mention the LEED study by May, Szostak, and Germer (1969) and the kinetic investigations of Hansen (McAllister and Hansen 1973) and Dawson (Dawson and Peng 1973), since these serve as a convenient introduction to more recent work on the subject.

10.3.2. *Nitrogen adsorption on tungsten—high-energy binding states*

Polycrystalline tungsten. Nitrogen adsorbs rapidly on polycrystalline tungsten surfaces. At room temperature initial sticking probabilities have been reported in the range $0 \cdot 035 – 0 \cdot 6$, but the most reliable value appears to be $0 \cdot 61 \pm 0 \cdot 02$ determined by King and Wells (1972) using a molecular beam technique (see Section 6.4.2). The saturation coverage is between 2 and 3×10^{14} molecules cm^{-2} or about 40 per cent of a monolayer. This is noticeably different from the monolayer achieved on this substrate with N_2 at 77 K and with CO, for example, at room temperature. Comprehensive kinetic data in the form of sticking-probability curves for individual states of N_2 on tungsten as a function of both coverage and temperature (see Fig. 8.9(a)) were originally obtained by Ehrlich (1961) and have been discussed in Chapter 8.

Thermal desorption following adsorption at 298 K reveals both α and β states. The β state apparently consists of two substates β_1 and β_2 (Oguri 1963, Rigby 1965, Delchar and Ehrlich 1965, Madey and Yates 1966). The β_2 state appears immediately on exposure to nitrogen, becoming saturated at ~60 per cent of the total coverage (Fig. 10.15(a)) but the β_1 state starts to populate only at fractional coverages greater than ~$0 \cdot 2$ and continues to grow after saturation of β_2.

A twin peak in a thermal desorption spectrum is not necessarily indicative of two distinct adsorbed states on the surface. In the case of a single-crystal plane with one type of binding site only, it may be the result of adatom–adatom interactions within a single state (see Chapter 6) or of induced heterogeneity within the adsorbed layer produced by such interactions (Section 10.2.2). With polycrystalline surfaces other possible causes must also be considered. If the data for the W–N_2 system are analysed in the conventional way the fact that the temperature T_p of the maximum in the β_1 peak is independent of coverage suggests first-order kinetics; using a pre-exponential factor $\nu^{(1)} = 10^{13} \, s^{-1}$ the activation energy is calculated to be between 305 kJ mol^{-1} (Rigby 1965) and 343 kJ mol^{-1} (Yates and Madey 1965). It was originally assumed that the β_2 state desorbs with second-order kinetics since qualitatively T_p shifts to lower values with increasing

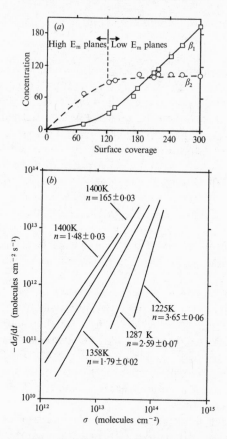

FIG. 10.15. Polycrystalline W–N$_2$. (a) Calculated concentrations of adatoms (broken curve) and molecular complexes (solid curve) as a function of coverage on a surface comprised of areas with high and low activation energies for diffusion. The plotted points are Rigby's (1965) experimental values of the concentrations desorbed from the β_1 (squares) and β_2 (circles) peaks. Concentrations and surface coverage are in units of 10^{12} molecules cm^{-2}. (After Robins *et al.* 1967.) (b) Analysis of desorption-rate measurements according to eqn 10.2. (After Madey and Yates 1966.)

coverage. On this basis Rigby (1965) obtained an activation energy for desorption of 313 kJ mol^{-1} from a plot of $\ln \sigma_0^{n-1} T_p^2$ *versus* $1/T_p$, where $n\ (=2)$ is the order of the process and σ_0 is the area under each desorption curve (see eqn (6.60)). A value of 352 kJ mol^{-1} was estimated independently by Yates and Madey (1965). These authors later pointed out, however, that decreasing T_p with increasing coverage would also result from a *first* order desorption with a varying activation energy (Madey and Yates 1966) and, further, that Rigby's data could equally well be fitted by assuming values of n other than 2. They adopted an alternative method of analysis in which the rate of desorption is considered as a function of coverage at a particular temperature. From eqn (6.51) it follows that, at constant temperature,

$$\left\{\frac{d(\ln(-d\sigma/dt))}{d(\ln \sigma)}\right\}_T = n + \frac{\sigma \, d(\ln k)}{d\sigma}. \tag{10.2}$$

Thus, if the rate constant k is coverage independent a plot of $\ln(-d\sigma/dt)$ versus $\ln \sigma$, at constant temperature, should be a straight line of slope n. Madey and Yates found this to be the case with their data at five temperatures between 1225 and 1440 K. The interesting point, however, was that the apparent desorption order decreased with increasing temperature (Fig. 10.15(b)).

Ehrlich (1966) has calculated that if only a single β_2 binding state is considered the above results imply that the desorption energy decreases with increasing coverage from about 390 kJ mol^{-1} at 2×10^{12} molecules cm^{-2} to 290 kJ mol^{-1} at 50×10^{12} molecules cm^{-2}. Such a change is more rapid than that observed in other experiments and, if correct, makes any distinction of the order difficult. This type of behaviour may also be reproduced, however, by considering a model involving competition between different binding states during desorption (Madey and Yates 1966). The surface is taken to consist of a number of patches, each of which is uniform throughout but different from the others. If during evaporation complete equilibrium is attained continuously owing to diffusion of adsorbed species between the patches, then a single integral desorption order should be found at all coverages except those so high that adatom interactions become important. The coverage-dependent non-integral order observed therefore suggests either that equilibrium between patches is not maintained or that the heats of adsorption and desorption vary considerably with coverage. The second possibility is considered to be the less likely on the basis of available experimental evidence. The question of equilibration between surface patches therefore assumes central importance, as it must for any model for desorption from a polycrystalline surface.

To account for the data shown in Fig. 10.15(b) Madey and

Yates (1966) assumed *complete lack* of equilibration between patches (their own isotopic mixing experiments showed this to be approximately correct, although conflicting evidence had been obtained by Rigby (1965)). It was postulated that two β_2 sub-states populate simultaneously and at equal rates up to 25 per cent of the total coverage. A single β_1 state then fills to saturation; this is in rough agreement with the experimental data (Fig. 10.15(a)). The three states occupy different regions on the surface, so that β_1 cannot be a 'gap-filling' phase located at vacant sites within the β_2 layer. It was assumed that β_1 desorbs with first-order kinetics ($E_d = 334$ kJ mol^{-1}, $\nu^{(1)} = 10^{13}$ s^{-1}), one β_2 state with second-order kinetics ($E_d = 343$ kJ mol^{-1}, $\nu^{(2)} = 0{\cdot}014$ cm^2 s^{-1}), and the other with first-order kinetics ($E_d = 372$ kJ mol^{-1}, $\nu^{(1)} = 10^{13}$ s^{-1}). No interchange between the sub-states is permitted during desorption. Other substate combinations might equally well have been chosen and no account was taken of factors such as coverage dependence of the pre-exponentials, but in spite of these and other approximations this model qualitatively reproduced the experimental results.

A further problem is presented by the β_1 state. Isotopic mixing experiments (e.g. Yates and Madey 1965) indicate that nitrogen in this state is dissociatively adsorbed—why then does it desorb with first-order kinetics? This question has been considered by Robins, Warburton, and Rhodin (1967) using a model basically similar to that just described. Its main novelty lies in the suggestion, also made by Rigby (1965), that essentially only *one* β adsorption phase exists. The detection of the β_1 and β_2 states then becomes a feature of the *desorption* process. Other assumptions made included those listed below.

(i) On dissociation of a nitrogen molecule the atoms initially carry a considerable fraction of the chemisorption energy and so must diffuse over the surface dissipating this energy to the lattice before they can become adsorbed (Ehrlich 1963). This means that, except at high coverage, first-order β_1 kinetics cannot be due to the concerted, single-step desorption of atom pairs adsorbed on adjacent sites.

(ii) The barriers E_m to adatom diffusion vary from one crystal plane to another. Field emission microscope results (see Chapter 8) suggest values of 146 kJ mol^{-1} on

W(100) and 84 kJ mol^{-1} on (111) and (110). The greater the depth of the potential wells encountered by a diffusing atom the more rapidly will it lose its energy to the lattice. Thus the rate of adsorption will initially be greater on high-barrier planes, since atoms formed on low-barrier areas will tend to diffuse out of these onto the high-barrier regions.

(iii) The surface is assumed to consist of two areas, one made up from high-barrier planes and the other from low-barrier planes, in the ratio 5:1.

(iv) The saturation coverage is taken to be 3×10^{14} molecules cm^{-2} on each area.

(v) The most radical assumption is that during desorption molecular complexes, composed of pairs of atoms, form on the surface through collision of diffusing atoms. These complexes are in dynamic equilibrium with the adatoms and may either dissociate or desorb. The desorption process will be *first order* and will contribute to the β_1 peak. By equating the rates of formation and dissociation of the complexes (rate constants k_1 and k_2 respectively) the following expression is obtained for their concentration σ_c:

$$\sigma_c = K\sigma_a^2 \qquad (10.3)$$

where σ_a is the adatom concentration and K is given by

$$K = \frac{k_1}{k_2} \exp\left\{\frac{-(E_m - E_c + E_G)}{kT}\right\}. \qquad (10.4)$$

The activation energy for surface diffusion is E_m. The difference in the activation energies for complex formation and dissociation $(E_c - E_G)$ (Fig. 10.16(a)) is equal to the difference in the β_1 and β_2 binding energies; experimentally this is found to be ~ 8 kJ mol^{-1} and so the term $(E_c - E_G)$ is neglected in comparison with E_m. If σ_t is the total concentration (molecules cm^{-2}) on a given plane then

$$\sigma_c + \tfrac{1}{2}\sigma_a = \sigma_t. \qquad (10.5)$$

(vi) During adsorption the low-barrier planes do not fill until coverage on the high-barrier planes is complete, taken to

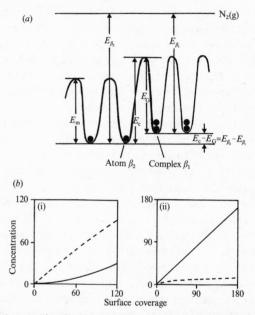

Fig. 10.16. Polycrystalline W–N_2. (a) Energy parameters relating to adsorbed nitrogen atoms and molecular complexes. (b) Typical equilibrium concentrations of adatoms (broken lines) and molecular complexes (solid lines) as a function of coverage on areas with (i) high ($K = 9 \cdot 25 \times 10^{-4}$) and (ii) low ($K = 0 \cdot 195$) activation energies for diffusion. Concentration and surface coverage are in units of 10^{12} molecules cm^{-2}. (After Robins *et al.* 1967.)

occur at ~40 per cent of saturation coverage (cf. Fig. 10.15(*a*)). (As will be seen below, this is *not* identical to the more artificial assumption in the model of Madey and Yates that the β_1 state does not fill until the β_2 state is complete.)

(vii) The number of molecules desorbed in the β_1 peak is equal to the number of molecular complexes in equilibrium with adatoms just below the desorption temperature, i.e. at ~1200 K.

(viii) There is a complete lack of equilibrium between the two areas on the surface during desorption, as assumed also in the model of Yates and Madey discussed above.

The relative concentrations in the β_1 and β_2 desorption peaks at 40 per cent saturation (only high-barrier areas filled) and at full coverage were taken from Rigby's data (Fig. 10.15(*a*)). Using eqn (10.3) and assumption (iv) values of K were then calculated for the high- and low-barrier areas. (The ratio K_h/K_1 was found to be $5 \cdot 0 \times 10^{-3}$. At this point in the argument use may be made of the diffusion-barrier data to check this calculation. From eqn (10.4) $K_h/K_1 \approx \exp\{(E_{m1} - E_{mh})/kT\}$. With $E_{m1} = 84$ kJ mol^{-1}, $E_{mh} = 146$ kJ mol^{-1}, $T = 1200$ K, this gives $K_h/K_1 = 2 \cdot 2 \times 10^{-3}$ in good agreement with the first calculation.)

These K values were used in conjunction with eqns (10.3) and (10.5) to calculate the concentrations of complexes and of adatoms on the two areas as functions of the coverage on each area (Fig. 10.16(b)). Finally, using assumptions (iii) and (vi), these concentrations, equivalent to the numbers of molecules in the β_1 and β_2 desorption peaks respectively, were calculated as functions of total surface coverage. The results are shown in Fig. 10.15(a) along with Rigby's experimental data. In spite of the many simplifying assumptions made the agreement between the two is good. Also, this model is seen to be consistent with the observed complexity in the order of the desorption kinetics. During a flash the molecular complexes desorb first into the β_1 peak. At higher temperature, when adatom desorption begins, an equilibrium number of complexes will still be maintained, however, and so the β_2 peak will have contributions from both first- and second-order processes, in qualitative agreement with the model of Madey and Yates.

This discussion emphasizes that interpretation of thermal desorption data may present considerable problems, especially for polycrystalline surfaces. The particular model just described is an approximate one only and takes no account of adatom–adatom interactions, for example. It remains, nevertheless, of considerable heuristic interest, as do many speculations of this kind.

$W(100)$. Nitrogen interacts rapidly with this crystal plane, and for exposures up to 100 L ($P \leqslant 3 \times 10^{-8}$ torr) at temperatures in excess of ~ 200 K a single β_2 state is observed in the thermal desorption spectrum (Clavenna and Schmidt 1970, Delchar and Ehrlich 1965). The desorption process appears to be second order, and at lower coverages ($\theta \leqslant 0 \cdot 2$) analysis in terms of eqn (6.55b) yields (Han and Schmidt 1971) $E_d = 334$ kJ mol^{-1} and $\nu^{(2)} = 0 \cdot 01$ cm^2 molecule^{-1} s^{-1}. At higher coverages there is deviation from this simple kinetic relationship. This could be due to coverage dependence of either E_d or $\nu^{(2)}$ but the latter is taken to be more likely.

A simple approach to the calculation of $\nu^{(2)}$ (Tamm and Schmidt 1969) assumes that the desorption rate may be taken as the rate of encounter between diffusing adatoms (r_{aa}) multiplied by the probability that on encounter a pair of adatoms will desorb. If n is the adatom concentration, then at low coverage the average distance between adatoms l is approximately $n^{-\frac{1}{2}} = (2\sigma)^{-\frac{1}{2}}$, where σ is the coverage in molecules cm^{-2}. The rate r_a at which a given adatom encounters other adatoms will be $r_a = (1/t)$, where t is the time taken to diffuse a distance l. The overall collision rate then is $r_{aa} = \frac{1}{2}(2\sigma r_a)$, the factor $\frac{1}{2}$ preventing collisions being counted twice. Finally, the desorption rate will be

$$\frac{-d\sigma}{dt} = r_{aa} \exp\left(\frac{-E'_d}{RT}\right) = \left(\frac{\sigma}{t}\right)\exp\left(\frac{-E'_d}{RT}\right) \qquad (10.6)$$

where E'_d is the activation energy for desorption of a pair of colliding adatoms. An expression for t may be obtained by equating l^2 with the mean-square displacement of a diffusing atom. Then (eqns (2.54) and (2.45))

$$\frac{1}{t} = l^{-2}\nu_0 a^2 \exp\left(\frac{-E_m}{RT}\right) = 2\sigma\nu_0 a^2 \exp\left(\frac{-E_m}{RT}\right) \qquad (10.7)$$

where ν_0 is the jump frequency and E_m is the activation energy for migration. Combining eqns (10.6) and (10.7) gives

$$\frac{-d\sigma}{dt} = 2\sigma^2 \nu_0 a^2 \exp\left(\frac{-E_d}{RT}\right) \qquad (10.8)$$

where $E_d = (E'_d + E_m)$ is the activation energy for desorption from the chemisorbed state. From eqn (10.8) it follows that the second-order pre-exponential factor is

$$\nu^{(2)} = 2\nu_0 a^2. \qquad (10.9)$$

For W(100), $a = 0.316$ nm; putting $\nu_0 = 10^{13}\,s^{-1}$ then gives $\nu^{(2)} \approx 2 \times 10^{-2}\,cm^2\,molecule^{-1}\,s^{-1}$. This value is in reasonable agreement with experimental results for hydrogen on W(100) (Tamm and Schmidt 1969). The above arguments break down at higher coverages. In that situation the finite size of the diffusing atoms must be taken into consideration, which can be done

(Clavenna and Schmidt 1970) by setting

$$l \approx n^{-\frac{1}{2}} - a_c \qquad (10.10)$$

where a_c is the minimum separation of an atom pair during collision. Then

$$\nu^{(2)} = 2a^2\nu_0\{1 - a_c(2\sigma)^{\frac{1}{2}}\}^{-2}. \qquad (10.11)$$

At low coverage this expression reduces to eqn (10.9), but at high coverage $\nu^{(2)}$ will be greater than $2a^2\nu_0$. By equating $2a^2\nu_0$ to the low-coverage experimental value of $\nu^{(2)}$ and using a_c as an adjustable parameter, Clavenna and Schmidt were able to obtain reasonable agreement between observed desorption data and eqn (6.55b) up to high coverage.

The sticking-probability curve for nitrogen on W(100) at 300 K is included in Fig. 10.20. The initial sticking probability is 0·58 (King and Wells 1974) and saturation coverage is 3×10^{14} molecules cm^{-2} (Clavenna and Schmidt (1970) quote $2·5 \times 10^{14}$ molecules cm^{-2}). The work function decreases continuously as the coverage increases (Fig. 10.21(a)). LEED results (Adams and Germer 1971a) indicate that a c(2×2) structure begins to form at low exposures. The half-order diffraction features are in the form of crosses or orthogonal streaks and so the adsorbed layer must consist of patches of nitrogen, each of which is elongated in a particular $\langle 11 \rangle$ direction on the surface (Fig. 10.17(a)). The fact that the integral-order spots are not streaked means that the patches are out of phase with each other in the $\langle 11 \rangle$ directions and also that they are not distinct islands, as suggested by Adams and Germer. Some patches must merge into others and there may be vacant sites within any one patch (see Chapter 3).

At coverages greater than $\theta \approx 0·2$ the streaked features in the pattern change to a roughly square shape indicating that patch growth does not occur in a preferred direction at this stage. For $0.4 < \theta \leqslant 0·5$ round, rather diffuse, half-order spots form at room temperature, but on annealing above 800 K these become quite sharp, indicating that the complete layer has an ordered c(2×2) structure. This correlates well with the observed saturation coverage.

The c(2×2) structure obtained with nitrogen and also with hydrogen on W(100) has been interpreted by Schmidt and co-workers (Tamm and Schmidt 1969, Clavenna and Schmidt 1970)

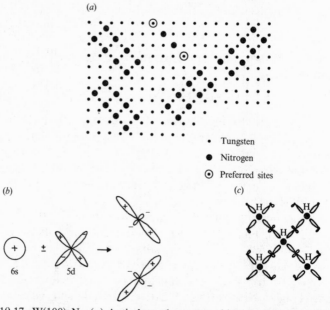

(a)

• Tungsten
● Nitrogen
⊙ Preferred sites

(b)   (c)

FIG. 10.17. W(100)–N$_2$. (a) Anti-phase placements of islands of nitrogen atoms. (After Adams and Germer 1971a.) (b) A schematic diagram of the orbitals formed by hybridization of a 5d orbital with a 6s orbital. (c) A possible structure of the bonding orbitals of W when the β_2 state of H$_2$ (filled circles) is formed on W(100). The other orbital in Fig. 10.17b is of higher energy and unoccupied. (After Tamm and Schmidt 1969.)

using a valence-bond picture which assumes that metal orbitals are important for the bonding of adsorbed species. In tungsten the 6s and 5d bands overlap in energy, and in the presence of an adsorbed atom the d electrons on four neighbouring W atoms will tend to form s-d$_{xy}$ hybrid orbitals, where x and y lie in the surface plane. Because of the even symmetry of the d orbitals the sum or difference in a linear combination of s and d orbitals produces a pair of hybrid orbitals which are essentially two lobed, as shown in Fig. 10.17(b). There is approximately one electron only in this state, and so when a bond is formed with an adatom by the orbital of lower energy in the hybrid pair the other, higher-energy, orbital will be unoccupied. Thus the sites adjacent to the occupied one will not be available for further adsorption and a c(2×2) layer will result (Fig. 10.17(c)).

When a σ-type bond is formed between the 2p$_x$ or 2p$_y$ orbital of an adsorbed nitrogen atom and a W atom, the W orbital will be extended in the direction of the nitrogen and also in the opposite direction, towards the adjacent adsorption site. Thus such adjacent sites will be preferred over isolated sites for further adsorption, leading to the formation of adsorbate patches. Further, the observed elongation of patches in $\langle 11 \rangle$ directions at low coverage implies that when two nitrogen atoms occupy adjacent sites the vacant sites at the ends of the adsorbate 'chain' are preferred to those to either side of it (Fig. 10.17(a)). Adams and Germer (1971a) have suggested that the long-range interaction required in this situation could be due to the formation of delocalized π-type bonds in $\langle 11 \rangle$ directions by overlap between nitrogen $2p_z$ orbitals and the d_z^2 orbitals on neighbouring W atoms.

W(*111*). Nitrogen interaction with this relatively 'rough' plane is much less rapid than with W(100); at room temperature the initial sticking probability is $\sim 0 \cdot 08$ and the saturation coverage $\sim 1 \cdot 25 \times 10^{14}$ molecules cm^{-2} or $\theta \approx 0 \cdot 11$ (King and Wells 1972). Flash desorption reveals an α peak evolving immediately the temperature is raised and a β_2 peak at temperatures greater than 1000 K (Delchar and Ehrlich 1965). At low concentrations (exposures up to ~ 10 L) the desorption kinetics are second order, with an activation energy of ~ 340 kJ mol^{-1}. At higher coverages the kinetics become complex; at an exposure of ~ 100 L the experimental curves indicate a second β-state with a concentration of about one-quarter of that of β_2 (Delchar and Ehrlich 1965). This deduction is supported by the observation of a break in the work-function exposure curve at the exposure at which the kinetics change.

There is some controversy as to the possibility of formation of β nitrogen on any plane, such as (111), which does not contain adsorption sites of four-fold co-ordination in the configuration found on b.c.c. {100} surfaces (see the discussion of W(110) below). It has been suggested (King and Wells 1972) that the observed β state on W(111) may result from the presence on the perfect surface of defect W adatoms; in suitable positions these would generate {100} sites.

W(*110*). The (110) is the most densely packed tungsten plane and is the least reactive towards nitrogen. For example on a field emission tip no work-function change on the (110) region was observed after an exposure of $2 \cdot 6 \times 10^4$ L (Van Oostrom 1967), while Delchar and Ehrlich (1965), using a single-crystal specimen, detected changes in work function on this plane only for gas pressures $\geqslant 10^{-2}$ torr; these were reversed on evacuation and corresponded with the properties of the low-binding-energy γ state. In two other studies, however (Madey and Yates 1967, Hopkins and Usami 1969) a small work-function increase ($+0 \cdot 03$–$0 \cdot 13$ eV) has been observed after an exposure of 200 L (at pressures up to 2×10^{-6} torr). Flash-filament kinetic measurements have also been reported (Madey and Yates 1967, Tamm and Schmidt 1971c) which indicate low but finite values for the rate of adsorption (initial sticking probability $0 \cdot 01$–$0 \cdot 05$) and saturation coverage ($1 \cdot 5 \times 10^{14}$–$0 \cdot 9 \times 10^{14}$ molecules cm^{-2}) at pressures below 3×10^{-5} torr (Fig. 10.18). On the other hand King and Wells (1972), using a molecular beam system capable of very accurate determination of sticking-probability values greater than $\sim 0 \cdot 01$, were unable to detect adsorption on this plane.

The possible presence of adsorbed impurities is critical in a system such as this where relatively long exposures are necessary. Considerable precautions have been taken by various workers

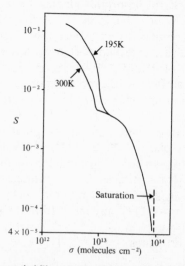

FIG. 10.18. Sticking probability *versus* nitrogen coverage on W(110). (After Tamm and Schmidt 1971c.)

(e.g. Madey and Yates 1967, Tamm and Schmidt 1971c) to ensure that the observed binding state consists predominantly of nitrogen but they do not entirely eliminate the possibility that impurity adsorption may influence the formation of this state. Also, the (110) represents another example of a surface devoid of {100} sites, which may be essential for nitrogen dissociation. It has been suggested (Singh-Boparai, Bowker, and King 1975) that the observed adsorption may be the result of a slight misorientation of the surface towards the (100) pole, which would produce a small number of sites of the required type. This possibility is supported by the apparent inability of a defect-free W(110) plane to adsorb H_2 directly from the gas phase (Polizzotti and Ehrlich 1975) and by the lack of reactivity of a perfect close-packed rhenium (0001) surface to nitrogen at 300 K (Liu and Ehrlich 1976, see Section 10.3.5.).

If adsorption *does* occur on W(110), the sticking-probability curve (Fig. 10.18) indicates the presence of two states, the minor one having an initial sticking probability of ~ 0.05 and saturating at $\sim 1 \times 10^{13}$ molecules cm^{-2}. The corresponding values for the major state are ~ 0.004 and 9×10^{13} molecules cm^{-2} respectively. These conclusions are supported by thermal desorption data.

High-index planes of tungsten. There are striking differences in the adsorption behaviour of nitrogen on the three low-index planes of tungsten and so it is of considerable interest to examine planes of higher index which combine structural features of the simpler planes. Since the greatest differences are between W(100) and W(110), the planes of the [001] zone are particularly appropriate for this purpose; those from the (100) pole to (310) are composed of (110) steps and (100) terraces, while those from (310) to (110) consist of (100) steps and (110) terraces (see Chapter 2). Adams and Germer (1971b) have investigated the (310) and (210) surfaces (Fig. 10.19(a)); the proposed structures of the corresponding saturated nitrogen overlayers, obtained from LEED experiments, are shown in Fig. 10.19(b). On the (310) plane two structures, which are mirror images of one another, are possible, and each contributes equally to the diffraction pattern (Fig. 10.19(c)). On both planes it is postulated that nitrogen adatoms occur on all (100) steps, double spaced along

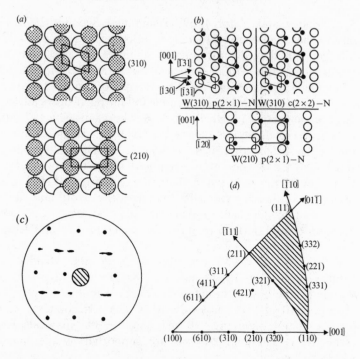

FIG. 10.19. (a) Sketches of (310) and (210) surfaces of a b.c.c. crystal projected upon a (100) surface. Atoms at step edges are shaded. (b) Sketches of proposed nitrogen surface structures. Substrate primitive unit meshes are indicated by light lines and overlayer meshes by heavy lines; ○ tungsten, ● nitrogen. Nitrogen on terraces not shown. (c) W(310)–N₂ LEED pattern. (d) Steriographic plot of the normals to various planes in a cubic structure. The shaded area indicates locations of tungsten planes which are predicted to be unreactive toward nitrogen at 300 K. (After Adams and Germer 1971b.)

the [001] direction, but that they do not occur on the (110) terraces.

Adams and Germer (1971b) generalized these observations to cover all planes of tungsten. Planes which contain surface atoms of co-ordination number 4, in the configuration occurring on W(100), are located in the stereographic triangle between the [001] and [$\bar{1}$11] zone axes (Fig. 10.19(d)). It is suggested that these surfaces should adsorb nitrogen with a high sticking probability, proportional to the (100) site density, and that they should have a saturation coverage equal to one-half the (100) site density; for planes $(hk0)$ in the $[001]$ zone this will be

$$\tfrac{1}{2}(h-k)/a_0^2(h^2+k^2)^{\frac{1}{2}}$$

atoms cm^{-2}. Surfaces on the $[\bar{1}11]$ zone have no four-fold sites and should therefore be inactive. Between the $[\bar{1}11]$ and $[\bar{1}10]$ axes such sites do occur, but in the configuration found on W(111), and so a relatively low reactivity would be expected in this region.

Accurate sticking-probability–coverage curves for nitrogen on the (310), (320), and (411) planes of tungsten have been measured by Singh-Boparai *et al.* (1975) and are reproduced in Fig. 10.20 which also includes the density of (100)-type sites on the various surfaces. Contrary to the suggestions of Adams and Germer it is apparent that there is not a direct correlation between (100) site density and either initial sticking probability or saturation coverage for the (100), (310), (411), and (320) planes. In fact, the final coverage on both (310) and (320) is significantly greater than the number of (100) sites. King and co-workers therefore suggest that such sites are essential for dissociation of the nitrogen molecule but that the adatoms so formed may diffuse out onto the (110) terraces. The energy for this migration is provided by the heat of adsorption which produces vibrationally excited adatoms capable of making a limited number of diffusive hops before their excess energy is dissipated

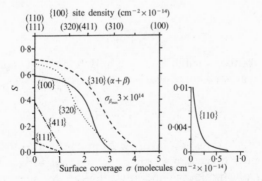

FIG. 10.20. A comparison of sticking probability profiles for nitrogen on tungsten single-crystal planes. In each case the gas temperature is 300 K; the result for the {320} plane refers to a crystal temperature of 360 K and the remainder to room temperature. (After Singh-Boparai *et al.* 1975.)

to the lattice (see Chapter 8). The presence of adatoms on the terraces would alter the contents of the unit mesh of the over-layer, but would not affect its size and shape. Therefore, this model does not conflict with the LEED results of Adams and Germer, since their work did not involve analysis of spot intensities.

The kinetic model used by King and Wells (1972) to describe the adsorption of nitrogen on W(100) has been discussed in Chapter 8. It may be readily extended (Singh-Boparai et al. 1975) to cover the (310) and (320) planes. The only new parameter which must be introduced is the probability ζ that a molecule in the physisorbed precursor state is in the vicinity of a site on a (100) step, whether filled or empty. It is assumed that the chemisorbed layer is equally distributed over steps and terraces and so the fractional coverage of step sites will be identical to θ, the fractional coverage of the entire surface. Under a lateral interaction energy ω the probability θ_{00} that a particular pair of nearest-neighbour step sites (required for N_2 dissociation) is empty is therefore given by eqn (8.55).

The sticking probability S is obtained from eqn (8.25) with $f(\theta) = \zeta\theta_{00}$ and $P'_d = P''_d$. Assuming the condensation coefficient c to be coverage independent, use of eqn (8.34) gives

$$S = c\left(1 + \frac{P_d}{P_a}\right)^{-1}\left\{1 + K\left(\frac{1}{\zeta\theta_{00}} - 1\right)\right\}^{-1} \qquad (10.12)$$

where

$$K = \frac{P'_d}{P_a + P_d} \qquad (10.13)$$

and the various probabilities P are as defined in Section 8.4. The zero-coverage sticking probability in this case is

$$S_0 = c\left(1 + \frac{P_d}{P_a}\right)^{-1}\left\{1 + K\left(\frac{1}{\zeta} - 1\right)\right\}^{-1}. \qquad (10.14)$$

Although this expression involves K, in contrast to the case of a non-stepped surface for which $\zeta = 1$, a value for c can still be obtained from the low-temperature limit of S_0, since both P_d/P_a and K tend to zero as the surface temperature tends to zero.

Consideration of the structures of the (320) and (310) planes (see Fig. 10.19(a)) indicates that reasonable values for ζ, the ratio of (100) step physisorption sites to total physisorption sites, are 0·33 and 1·0 respectively. Using these values Singh-Boparai *et al.* found that the variation of sticking probability with both temperature and coverage could be accurately described by eqns (10.12) and (10.14).

The kinetic data for the (411) plane (see Fig. 10.20(a)) cannot be accounted for by the above model, which requires nearest-neighbour {100} sites for N_2 dissociation. On b.c.c. (411) the {100} sites are in next-nearest-neighbour positions. A similar problem is presented by the (111) and (110) surfaces, as discussed previously. In all three cases, however, the reactivity is certainly considerably less than on (100), (310), and (320).

Work-function changes on planes in the [001] zone are shown in Fig. 10.21(a) (Adams and Germer 1971b). After adsorption at 300 K on the (210) and (310) surfaces, annealing up to 1200 K

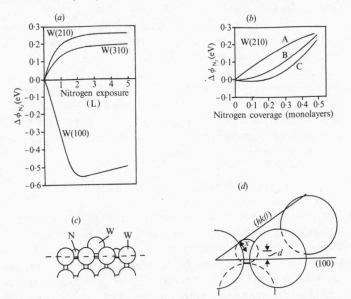

FIG. 10.21. Nitrogen on tungsten single crystal planes. (a) Changes in work function *versus* nitrogen exposure. (b) Effect of annealing upon the work-function changes on W(210). A, change in work function *versus* coverage for adsorption at 300 K; B, after anneal to 900 K; C, after anneal to 1200 K. (c) Section through a W(100) surface showing N adatoms and a defect W atom. (d) Projection of superficial tungsten atoms in a plane (hk0) onto the (100) plane, showing location of a nitrogen atom on a (100) site. Nitrogen atom radius $x = 0·059$ nm. ((a), (b), and (d) after Adams and Germer 1971b; (c) after Singh-Boparai and King 1976.)

reduced the work function (Fig. 10.21(b)). This was accompanied by a sharpening of features in the corresponding LEED patterns and so was due to an increase in the degree of order of the adlayer. The relatively large decreases in ϕ involved suggest that the establishment of long-range order may result in a considerable change in the electron-density profile at the surface.

Delchar and Ehrlich (1965) suggested that on W(100) a nitrogen atom carries a partial positive charge and lies above the plane of the surface—hence the observed decrease in work function. On planes which are rougher on the atomic scale the adatom may lie below the mean surface and produce a work-function increase. This model has been qualitatively tested by Singh-Boparai and King (1976) by evaporating tungsten onto a field emission microscope W tip to generate disordered W adatoms. Adsorption of nitrogen on this surface caused a decrease in emission from the {100} pole, indicating a surface dipole with the negative end outwards. Hence adatoms, such as that shown in Fig. 10.21(c) which are effectively 'inside' the surface, must be centres of positive charge in this situation also.

This model has been used by Adams and Germer (1971b) to derive a quantitative expression for $\Delta\phi$ for planes in the [001] zone. They assumed that on any such surface ($hk0$) the adatom lies at a fixed distance d above the (100) plane and that the dipole moment varies linearly with the distance x of the adatom from ($hk0$) (Fig. 10.21(d)); then

$$\Delta\phi_{(hk0)} = \Delta\phi_{(100)}(k-h)\left(\frac{ak}{2d}-h\right)(h^2+k^2)^{-1} \qquad (10.15)$$

where a is the tungsten lattice constant. No account was taken of adatoms on {110} regions of the surface, but these should make only a very small contribution to $\Delta\phi_{(hk0)}$. Eqn (10.15) does provide, in fact, a reasonable description of the experimental data, with a value of d corresponding to a nitrogen radius of 0·059 nm; this compares with the covalent double- and triple-bond radii of 0·062 and 0·055 nm, respectively.

It is interesting to note that nitrogen on W surfaces is positively charged, although nitrogen is the more electronegative of the two elements. Delchar and Ehrlich (1965) have emphasized that the dipole is determined by a number of effects, in addition to electronegativity differences; these include relative atomic sizes and hybridization of the orbitals involved, both bonding and non-bonding.

10.3.3. *Nitrogen adsorption on tungsten—states of intermediate binding energy*

In addition to β_2 nitrogen a second state of quite high binding energy may form on tungsten at high exposures. This has been designated β_1 (Clavenna and Schmidt 1970, Adams and Germer 1971b), although the choice is somewhat unfortunate since β_1 is established as a state of higher energy observed in desorption from polycrystalline samples. The evidence for this state is varied.

(i) The greatest concentration has been obtained on W(100) (Clavenna and Schmidt 1970) by repeatedly exposing a surface at 78 K to nitrogen at pressures up to 10^{-6} torr and periodically heating to 300 K to evaporate the low-binding-energy γ states. The maximum β_1 concentration observed (~40 per cent of the β_2 saturation coverage) appeared to be limited only by the maximum exposure (~2×10^3 L) which could be used without introducing serious contamination. This state desorbed at ~900 K with first-order kinetics and an activation energy of 203 kJ mol^{-1}.

(ii) When a clean (100) surface is exposed to nitrogen the work function first decreases to a minimum ($\Delta\phi \approx -0.4$ to -0.6 eV) at a coverage greater than 70 per cent of the β_2 saturation value. It then rises slowly at higher exposures (Fig. 10.21(a)) indicating the presence of an additional state with a negative dipole. After the work-function minimum a state additional to β_2 is observed in the desorption spectrum with a peak temperature of ~1000 K (Hopkins and Usami 1969).

(iii) Adams and Germer (1971b) find that on the (100), (210), and (310) planes of tungsten after a nitrogen exposure of 5 L no β_1 desorption peak is apparent, but that if the crystal is then exposed to carbon monoxide such a peak develops. The β_1 state was also detected without intentional exposure to CO, at exposures >30 L, but only in amounts proportional to the concentration of CO adsorbed from the background gases. In this work $^{15}N_2$ was used so that nitrogen species could be differentiated easily from CO by mass spectrometry and the β_1 state was identified as being due to nitrogen.

Adams and Germer suggest that β_1 nitrogen exists as nitrogen atoms at the vacant sites in the double-spaced β_2 nitrogen overlayers on W(100) and related surfaces. Carbon monoxide first adsorbs at these vacant sites and then undergoes place exchange with β_2 species. Also, it appears that the β_1 state is very similar to states produced by electron bombardment of molecularly adsorbed γ nitrogen (see, e.g., Yates and Madey 1969a) and to states resulting from the interaction of nitrogen atoms and ions, and even molecular nitrogen, with tungsten (Matsushita and Hansen 1970, Winters and Horne 1971, Dawson and Peng 1972).

The second state of intermediate binding energy is designated α; it is rather elusive. On polycrystalline surfaces it has been observed in some cases (Ehrlich 1956, Rigby 1965) but not in others (Yates and Madey 1965), while on those individual planes which have been investigated it has been detected on W(111) (Delchar and Ehrlich 1965) and on W(310) (Singh-Boparai et al. 1975).

On both polycrystalline wires and on the (111) plane the α and β states form simultaneously. Also, on both of these surfaces at room temperature the α saturation concentration is small, of the order of 10×10^{12} molecules cm^{-2} (10 per cent of the total saturation coverage), but on (310) it is remarkably high ($\sim$40 per cent of the total coverage). Polycrystalline specimens are distinguished by producing a maximum in the α concentration at 50–60 per cent of saturation. When the temperature is lowered the population on the polycrystalline surface is unaffected, except that the decrease at high coverage is no longer evident. On W(111), however, the saturation coverage increases with decreasing temperature; at 160 K it is a factor of $\sim$10 greater than at room temperature.

The desorption kinetics are difficult to determine precisely, but there is no mixing of nitrogen isotopes adsorbed into the α state and so it is concluded that it is molecularly bound. From the temperature range of its stability the activation energy for desorption is estimated to be $\sim$80 kJ mol^{-1} on polycrystalline samples and $\sim$60 kJ mol^{-1} on W(111).

The nature of the binding sites for the α species remains obscure. The concentration maximum observed on polycrystalline wires suggests that on these surfaces α adsorption occurs partially on sites from which the α molecules are subsequently displaced by β atoms (Rigby 1965). On the (111), however, no turning point in the coverage curve is observed and there is also direct evidence that α sites are not available to β species (Delchar and Ehrlich 1965). Delchar and Ehrlich have nevertheless surmised that α sites may be linked to the presence of occupied β sites, for example a surface defect under the influence of an adjacent β atom.

10.3.4. Nitrogen adsorption on tungsten—states of low binding energy

When nitrogen adsorbs on a polycrystalline tungsten surface at temperatures below ~ 170 K the total coverage at saturation ($\sim 6 \times 10^{14}$ molecules cm^{-2} at $\sim 5 \times 10^{-7}$ torr) is more than twice as large as the corresponding value at 300 K, and a third major state, in addition to the α and β states, becomes apparent in the flash desorption spectrum. This is γ nitrogen. It appears to form rapidly on all crystal planes, the population being greater than that in the β state. From the temperature and pressure ranges over which it is stable the heat of γ adsorption is estimated to be 40 kJ mol^{-1}. When compared with the similar heat of xenon adsorption on tungsten this value suggests that in the case of nitrogen the bonding forces are not simply of the van der Waals type since the mean value of the polarizability for nitrogen is only half that for xenon (Ehrlich 1961).

The γ species in general produces a decrease in work function, but on the (100) plane there is an indication of greater complexity. If the surface is saturated at room temperature and then cooled to 110 K further adsorption causes an additional decrease in work function of $0 \cdot 1$–$0 \cdot 2$ eV (Delchar and Ehrlich 1965, Van Oostrom 1967). Delchar and Ehrlich report that the curve first passes through a sharp minimum. If the surface is then warmed to 300 K the work function rises to a maximum and finally falls to a value similar to that achieved by room temperature adsorption. It has been suggested (Delchar and Ehrlich 1965) that the minimum and maximum are explicable in terms of a γ substate with a negative dipole forming only slowly on special sites. The minimum is due to the filling of this state at the end of adsorption at 110 K. On desorption it must be assumed that the γ_- state evaporates only after most of the positive dipole γ_+ species have been removed from the surface, hence producing the work-function maximum. The existence of these two substates has been confirmed by their resolution in a flash desorption spectrum (Clavenna and Schmidt 1970). They both desorb with first-order kinetics, the activation energies being $\gamma_+ = 38\cdot4$ and $\gamma_- = 43\cdot9\ kJ\ mol^{-1}$, in good agreement with the earlier estimates of Delchar and Ehrlich (1965). At saturation the coverages of γ_+, γ_-, and β_2 are each equal to $2\cdot5 \times 10^{14}$ molecules cm^{-2}.

10.3.5. *Nitrogen adsorption on rhenium*

In view of the inertness of the W(110) surface to H_2 and N_2 the behaviour of (0001) plane of an h.c.p. metal, such as rhenium, is of particular interest. It is known that nitrogen does adsorb on polycrystalline Re surfaces (see, e.g. Scheer and McKinley 1966, Yates and Madey 1969b) but at room temperature the rate of adsorption is low. On Re films the sticking probability is initially $0\cdot02$ and falls rapidly with coverage (McKee and Murphy, unpublished).

As discussed previously, a major difficulty arises when working with macroscopic single-crystal planes which are suspected to be inert; any observed activity may not be an inherent property of the perfect surface but may be due to a slight misorientation of the crystal or to surface defects. This problem has been overcome by Ehrlich and co-workers (see Liu and Ehrlich 1976) by the use of adsorbents in the form of field emission microscope tips, the perfection of which could be checked by field ion microscopy. Adsorption was followed by measuring changes in work function on individual planes using the probe-hole technique.

At 80 K nitrogen was found to adsorb non-dissociatively into a γ-state on all planes of Re, including (0001), with a binding energy of approximately $40\ kJ\ mol^{-1}$. At room temperature, however, the behaviour was quite different. Adsorption on atomically rough areas of the tip did occur, leading to an increase in the average work function, but no change in either work function or field emission current could be detected for the (0001) plane up to total exposures greater than 3×10^{19} molecules cm^{-2}. The sticking probability on the basal plane is therefore less than 10^{-5} at 300 K.

In contrast, if the emitter was saturated with nitrogen at 300 K and then heated in a vacuum to 500 K, the emission current from (0001) decreased considerably. Similarly, if the clean emitter was held at 500 K and exposed to gas at 300 K, adsorption was again detected, although the decrease in emission current from the basal plane initially lagged behind that from the tip as a whole. These results indicate that population of the (0001) plane does not occur directly from the gas phase. It must therefore involve surface transport of adatoms from neighbouring stepped planes which are capable of dissociating the N_2 molecule. This conclusion supports the suggestion that observed interaction of macroscopic W(110) specimens with nitrogen, and also hydrogen, is due to the presence of surface defects. It must not necessarily be concluded, however, that all close-packed planes are inert, since this is not found to be so with the (111) plane of f.c.c. rhodium (see Liu and Ehrlich 1976).

10.3.6. Nitrogen adsorption on nickel, palladium, and platinum

Nitrogen adsorbs rapidly on these metals at low temperatures, but at 300 K and pressures below 10^{-3} torr adsorption has been found, generally, to be negligible (on Pt there is evidence for very slow room-temperature adsorption, as discussed below). As a consequence, films of these metals can be used conveniently for thermal desorption experiments, and the Ni and Pd systems have been studied in this way (King 1968). On Ni three states were detected with the following binding energies: less than 29 kJ mol^{-1} (γ_1); 25–42 kJ mol^{-1} (γ_2); 38–59 kJ mol^{-1} (γ_3). Experimentally determined isotherms were well represented by the expression (Fowler and Guggenheim 1960)

$$P = \frac{\theta}{1-\theta} K \exp\left(\frac{-E_{d0}}{RT}\right)\exp\left(\frac{2\theta\omega}{KT}\right) \qquad (10.16)$$

where θ is the fractional coverage, K is a constant for the system at a given temperature, and E_{d0} is the zero-coverage desorption energy. This equation is identical to the Langmuir isotherm, except for the inclusion of the second exponential term which allows for a nearest-neighbour adsorbate–adsorbate interaction of magnitude $2\omega/z$, where z is the lateral co-ordination number. The best fit to the data was obtained with $E_{d0} = 46$ kJ mol^{-1} and $\omega = 3.7$ kJ mol^{-1}, giving a high-coverage desorption energy of

38 kJ mol^{-1}, in reasonable agreement with the thermal desorption results.

The sticking probability S was found to be constant at low coverage, θ, indicating adsorption by way of a precursor state, but it fell rapidly as the external surface of the film became saturated, indicating an immobile adsorbate. The $S(\theta)$ curve then flattened out again as sites in the pores of the film were occupied by gas-phase diffusion, before a final rapid fall corresponding to saturation of the internal surfaces. The initial sticking probability at 78 K on a smooth nickel surface was estimated to be 0·56.

On palladium films only γ_1 and γ_2 states were observed, the γ_3 state found on Ni being absent. Sticking-probability data lead to the conclusion that the nitrogen adlayer at 78 K is immobile on this metal.

Eischens and Jacknow (1965) reported a strong infra-red absorption band at 2202 cm^{-1} for nitrogen on silica-supported Ni; a similar band has been detected with thin Ni films (Bradshaw and Pritchard 1970) and with supported Pd and Pt (Van Hardeveld and Van Montfoort 1966). There has been considerable speculation as to the exact nature of the infra-red active species and its binding site. For example, it has been argued on the one hand that the nitrogen is weakly chemisorbed and on the other that it is physisorbed, but the weight of accumulated evidence appears to be in favour of the first of these claims. In nitrogen inorganic complexes, such as $[\text{Ru(NH}_3)_5\text{N}_2]^{2+}$, the M—N—N bond system is linear and an intense band occurs in the region 2010–2190 cm^{-1} due to N—N stretching. The bonding is believed to be π stabilized, involving donation from the mildly anti-bonding $3\sigma_\text{g}$ nitrogen orbital to the metal and back-bonding from occupied metal orbitals to the empty $1\pi_\text{g}$ orbital of nitrogen. A similar picture could account for the bonding of nitrogen to a metal surface (Bradshaw and Pritchard 1970, Wilf and Dawson 1976).

One of the most interesting observations has been that infra-red activity is detected only when the metal-particle size lies in the approximate range 1·5–7·0 nm (Van Hardeveld and Van Montfoort 1966). If metal crystallites adopt their equilibrium shapes, the number $n(B_x)$ of surface sites at which a metal adatom would have co-ordination (x), relative to the total number of atoms in the crystallite N, may be determined as a function of crystallite diameter d. For B_3 and B_4 sites (i.e. those occurring on the f.c.c. {111} and {100} planes respectively) $n(B_x)/N$ decreases monotonically as d increases. For B_5 sites, which occur for example on {110} and {113} planes (Fig. 10.22), the situation is quite different. When d is less than about 1·5 nm no such sites are to be expected, but for larger particles $n(B_5)/N$ rises steeply to a maximum in the range $1·8 \leqslant d \leqslant 2·5$ nm and then falls again to low values for $d \geqslant 7·0$ nm. It was therefore argued that infra-red-active nitrogen is associated with sites such as B_5. In the case of platinum it has been concluded (Nieuwenhuys and Sachtler 1973) that this material cannot be distinguished from other adsorbed nitrogen on the basis of heats of adsorption. The rate of change of the field emission current from various planes on a Pt emitter was determined as a function of temperature; the rate was found to be approximately equal for planes which included B_5 sites and for those which did not. This again suggests that only on certain crystal faces is adsorbed nitrogen able to absorb infra-red radiation. The fact that the work-function change due to nitrogen adsorption, and hence the dipole moment μ_1 of the adsorbate–metal complex, is greater on rough than on smooth planes was cited as supporting evidence. The optical properties of interest depend on the moment μ_2 between two adsorbed N atoms, and this quantity must be related to μ_1.

Several workers have reported that molecular nitrogen does not adsorb on platinum at 300 K and low pressure, but Wilf and Dawson (1976) have detected a slow uptake (sticking probability approximately 10^{-4}) at 7×10^{-8} torr on a polycrystalline filament. Thermal desorption revealed a single state obeying second-order kinetics with a coverage-independent activation energy of 80 kJ mol^{-1}. Also, the rate of formation of this state (but not the maximum coverage) was increased by dissociating the gas on a tungsten filament at ~2200 K prior to adsorption. Therefore it is concluded that the nitrogen is atomically adsorbed. It was

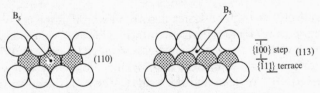

FIG. 10.22. B_5-type sites on f.c.c. (110) and (113) planes.

found that carbon monoxide competed with nitrogen for the surface sites involved in this state, but the influence, if any, of low concentrations of CO on the nitrogen adsorption mechanism was not established.

At low temperatures two additional states of nitrogen were detected, with activation energies for desorption varying from 31 kJ mol^{-1} at low coverage to ~15 kJ mol^{-1} at high coverage, in agreement with previous results (Nieuwenhuys and Sachtler 1973). The population of these states was very strongly influenced by the population of the higher-binding-energy state.

10.4. Electron spectroscopic studies of nitrogen and related molecules

There have been but a few investigations of the interaction of nitrogen and related molecules with metals by electron spectroscopy. Madey, Yates, and Erickson (1974) reported core-level data for the adsorption of both N_2 and NO on tungsten. They observed large chemical shifts, as much as 8 eV, between the N(1s) binding energies for weakly and strongly chemisorbed nitrogen, i.e. between the γ and β states. In this sense the results are clearly analogous to the shift in the O(1s) binding energies for strongly and weakly adsorbed carbon monoxide (Chapter 9). With nitric oxide the N(1s) values were interpreted by Madey *et al.* (1974) as reflecting essentially molecular adsorption at 100 K, but no final conclusion was drawn regarding its state at 300 K.

The interaction of nitrogen with iron has been studied by Kishi and Roberts (1976, 1977) who reported two N(1s) peaks, 400·2 eV and 405·3 eV, at 85 K. These were due to weakly adsorbed molecular species and tentatively assigned as

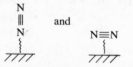

respectively. Both these states desorb on warming the adlayer to 295 K leaving a nitrogen-free surface. However, interaction of nitrogen with iron at 295 K gives rise to an N(1s) peak at 397 eV. This is a very slow process, the sticking probability being estimated as 10^{-6} or less, and the peak (397 eV) assigned to the N adatom. Kishi and Roberts (1976) suggest that the weakly held species are precursors to dissociation and that the low sticking probability into the chemisorbed state is attributed to the low concentration of the molecular nitrogen species present on the iron surface at 295 K. Gay, Textor, Mason and Iwasawa (1977) have observed very similar results for the Fe(111) surface.

By combining ultra-violet and X-ray photoelectron spectroscopy a detailed picture of the interaction of nitric oxide with iron, copper, and aluminium surfaces has been obtained (Kishi and Roberts 1977, Matloob and Roberts 1977, Carley and Roberts 1978). Fig. 10.23 shows the N(1s) data during the interaction of nitric oxide with iron at 85 K. Initially an N(1s) peak develops at 397 eV, i.e. indicating dissociative chemisorption with the formation of N adatoms. However, on further exposure to nitric oxide the N(1s) peak profile broadens to higher binding energy. Valence-level spectra indicate molecular adsorption of nitric oxide at this stage since the peaks in the He(II) spectrum can be correlated with peaks in NO(g).

On warming the adlayer from 85 to 295 K the N(1s) intensity shifts from the high-binding-energy region to peak at 397 eV. This is accompanied by a loss of orbital structure in the He(II) spectrum, the density-of-states curve now taking on the character

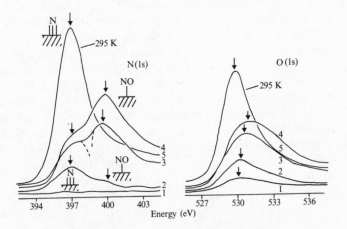

FIG. 10.23. Adsorption of NO on Fe at 85 K followed by warming the adlayer to 295 K. N(1s) and O(1s) spectra are recorded with increasing exposure (curve 2, 2×10^{-7} torr, 80 s; curve 4 (total), 2×10^{-7} torr, 480 s). Curves 1–4 at 85 K. (After Kishi and Roberts 1977.)

of an iron-oxide-type surface. Clearly the NO has dissociated between 85 and 295 K. Nitric oxide had previously been reported to dissociate on iron at high pressures at room temperature (Kishi and Ikeda 1974) but no information on the stability of the chemisorbed layer was available. Other studies of nitrogen containing molecules have been reported by Brundle and Carley (1975). With more detailed investigations there may well emerge a strong analogy between the surface chemistry of nitric oxide and carbon monoxide. Perhaps this should not be surprising in view of their similar electronic structures.

10.5. Conclusions

In the systems which have been considered hydrogen and nitrogen exhibit similar behaviour in many respects. Strong bonds are formed with surfaces by adatoms and in the atomic state both elements are highly reactive. The reactivities of the molecular gases, on the other hand, are restricted by the activation energy required for bond dissociation. The probable absence of dissociation on W(110), compared with the comparative ease of the process on W(100), is a good example of the influence of a 'geometric' factor in operation. Also, it is interesting that as non-dissociation of N_2 and H_2 on W(110) has been demonstrated, the dissociative adsorption of CO on W has been established (see Chapter 9). Thus the surface chemistries of these three molecules move some way towards integration.

The picture which emerges of the interactions of hydrogen and nitrogen with low-index metal planes is satisfying in the degree to which the kinetics, stereochemistry, and bonding are now understood. In the particular case of hydrogen, however, the difficulties involved in unravelling the nature of the bonding in detail are exacerbated by the unsuitability of most spectroscopic methods for studying hydrogen adatoms. Indeed, the unresolved questions and the general complexity of these simplest of systems clearly illustrate the inherent intricacy of chemisorption.

REFERENCES

ADAMS, D. L. (1974). *Surf. Sci.* **42**, 12.
ADAMS, D. L., and GERMER, L. H. (1970). *Surf. Sci.* **23**, 419.
—— (1971a). *Surf. Sci.* **26**, 109.
—— (1971b). *Surf. Sci.* **27**, 21.

ADAMS, D. L., GERMER, L. H., and MAY, J. W. (1970). *Surf. Sci.* **22**, 45.

ALDAG, A. W., and SCHMIDT, L. D. (1971). *J. Catalysis* **22**, 260.

ANDERS, L. W., HANSEN, R. S., and BARTELL, L. S. (1973). *J. chem. Phys.* **59**, 5277.

ANDERSON, J. R., RITCHIE, I. M., and ROBERTS, M. W. (1970). *Nature, Lond.* **227**, 704.

ATKINSON, G., COLDRICK, S., MURPHY, J. P., and TAYLOR, N. (1976). *J. less-common Metals* **49**, 439.

BALDWIN, V. H., and HUDSON, J. B. (1971). *J. vac. Sci. Technol.* **8**, 49.

BARFORD, B. D., and RYE, R. R. (1972). *J. vac. Sci. Technol.* **9**, 673.

—— (1974). *J. chem. Phys.* **60**, 1046.

BAUER, E. (1966). *Surf. Sci.* **5**, 152.

BAETZOLD, R. C. (1973). *J. Catalysis* **29**, 129.

BERNASEK, S. L., and SOMORJAI, G. A. (1975*a*). *J. chem. Phys.* **62**, 3149.

—— (1975*b*). *Surf. Sci.* **48**, 204.

BRADSHAW, A. M., and PRITCHARD, J. (1970). *Surf. Sci.* **19**, 198.

BRUNDLE, C. R., and CARLEY, A. F. (1975). *Faraday Disc. chem. Soc.* **60**, 51.

CARLEY, A. F., and ROBERTS, M. W. (1978). To be published.

CHRISTMANN, K., and ERTL, G. (1976). *Surf. Sci.* **60**, 365.

CHRISTMANN, K., ERTL, G., and PIGNET, T. (1976). *Surf. Sci.* **54**, 365.

CHRISTMANN, K., ERTL, G., and SCHOBER, O. (1973). *Surf. Sci.* **40**, 61.

CHRISTMANN, K., SCHOBER, O., ERTL, G., and NEUMANN, M. (1974). *J. chem. Phys.* **60**, 4528.

CLAVENNA, L. R., and SCHMIDT, L. D. (1970). *Surf. Sci.* **22**, 365.

CONRAD, H., ERTL, G., and LATTA, E. E. (1974). *Surf. Sci.* **41**, 435.

DAWSON, P. T., and PENG, Y. K. (1972). *Surf. Sci.* **33**, 565.

—— (1973). *J. phys. Chem.* **77**, 135.

DELCHAR, T. A., and EHRLICH, G. (1965). *J. chem. Phys.* **42**, 2686.

EBISUZAKI, Y., KASS, W. J., and O'KEEFFE, M. (1968). *J. chem. Phys.* **49**, 3329.

EHRLICH, G. (1956). *J. phys. Chem.* **60**, 1388.

—— (1961). *J. chem. Phys.* **34**, 29.

—— (1963). In *Metal surfaces: structure, energetics and kinetics.* American Society of Metals, Ohio.

—— (1966). *A. Rev. phys. Chem.* **17**, 295.

EISCHENS, R. P., and JACKNOW, J. (1965). *Proc. 3rd Int. Congr. on Catalysis*, p. 627. North-Holland, Amsterdam.

ESTRUP, P. J., and ANDERSON, J. (1966). *J. chem. Phys.* **45**, 2254.

FASSAERT, D. J. M., VERBEEK, H., and VAN DER AVOIRD, A. (1972). *Surf. Sci.* **29**, 501.

FEUERBACHER, B., and ADRIAENS, M. R. (1974). *Surf. Sci.* **45**, 553.

FEUERBACHER, B., and FITTON, B. (1973). *Phys. Rev.* **38**, 4890.

FOWLER, R. H., and GUGGENHEIM, E. A. (1960). *Statistical thermodynamics.* Cambridge University Press, London.

GAY, I. D., TEXTOR, M., MASON, R. and IWASAWA, Y. (1977). *Proc. R. Soc. A* **356**, 25.

GERMER, L. H., and MACRAE, A. U. (1962). *J. chem. Phys.* **37**, 1382.

GOMER, R. (1975). *Solid St. Phys.* **30**, 93.

HAN, H. R., and SCHMIDT, L. D. (1971). *J. phys. Chem.* **75**, 227.

HONIG, J. M. (1967). In *The solid–gas interface* (ed. E. A. Flood), Vol. 1, p. 371. Marcel Dekker, New York.

HOPKINS, B. J., and USAMI, S. (1969). In *The structure and chemistry of solid surfaces* (ed. G. A. Somorjai). John Wiley, New York.

—— (1970). *Surf. Sci.* **23**, 423.

HORGAN, A. M., and DALINS, I. (1974). *Surf. Sci.* **41**, 624,

KEMBALL, C. (1959). *Adv. Catalysis* **11**, 223.

KING, D. A. (1968). *Surf. Sci.* **9**, 375.

KING, D. A., and MENZEL, D. (1973). *Surf. Sci.* **40**, 399.

KING, D. A., and WELLS. M. G. (1972). *Surf. Sci.* **29**, 454.

—— (1974). *Proc. R. Soc. A* **339**, 245.

KIRBY, R. E., MCKEE, C. S., RENNY, L. V., and ROBERTS, M. W. (1979).

KISHI, K., and IKEDA, S. (1974). *Bull. chem. Soc. Japan* **47**, 2532.

KISHI, K., and ROBERTS, M. W. (1976). *Proc. R. Soc. A* **352**, 289.

—— (1977). *Surf. Sci.* **62**, 252.

KO, S. M., STEINBRÜCHEL, Ch., and SCHMIDT, L. D. (1974). *Surf. Sci.* **43**, 521.

LANG, B., JOYNER, R. W., and SOMORJAI, G. A. (1972). *Surf. Sci.* **30**, 454.

LAPUJOULADE, J. (1971). *J. Chim. phys.* **68**, 73.

LAPUJOULADE, J., and NEIL, K. S. (1972). *J. chem. Phys.* **57**, 3535.

—— (1973). *Surf. Sci.* **35**, 288.

LEWIS, F. A. (1967). *The palladium hydrogen system.* Academic Press, New York.

LIU, R., and EHRLICH, G. (1976). *J. vac. Sci. Technol.* **13**, 310.

LU, K. E., and RYE, R. R. (1974). *Surf. Sci.* **45**, 677.

—— (1975). *J. vac. Sci. Technol.* **12**, 334.

MCALLISTER, J., and HANSEN, R. S. (1973). *J. chem. Phys.* **59**, 414.

MADEY, T. E. (1972). *Surf. Sci.* **29**, 571.

—— (1973). *Surf. Sci.* **36**, 281.

MADEY, T. E., CZYZEWSKI, J. J., and YATES, J. T. (1975). *Surf. Sci.* **49**, 496.

MADEY, T. E., and YATES, J. T. (1966). *J. chem. Phys.* **44**, 1675.

—— (1967). *Nuovo Cim. Suppl.* **5**, 483.

MADEY, T. E., YATES, J. T., and ERICKSON, N. E. (1974). *Surf. Sci.* **43**, 526.

MATLOOB, M., and ROBERTS, M. W. (1977). *J. chem. Soc. Faraday Trans. I* **73**, 1393.

MATSUSHITA, K., and HANSEN, R. S. (1970). *J. chem. Phys.* **52**, 4877.

MATYSIK, K. J. (1972). *Surf. Sci.* **29**, 324.

MAY, J. W., and GERMER, L. H. (1969). In *The structure and chemistry of solid surfaces* (ed. G. A. Somorjai) paper 51. John Wiley, New York.

MAY, J. W., SZOSTAK, R. J., and GERMER, L. H. (1969). *Surf. Sci.* **15**, 37.

MUELLER, W. M., BLACKLEDGE, J. P., and LIBOWITZ, G. G. (1968). *Metal hydrides*. Academic Press, New York.

NIEUWENHUYS, B. E., and SACHTLER, W. H. M. (1973). *Surf. Sci.* **34**, 317.

NORTON, P. R., and RICHARDS, P. J. (1974). *Surf. Sci.* **41**, 293.

OGURI, T. (1963). *J. phys. Soc. Japan* **18**, 1280.

PAULING, L. C. (1960). *The nature of the chemical bond*. Cornell University Press, Ithaca, N.Y.

PEARSON, R. G. (1971). *Acct. chem. Res.* **4**, 152.

PENG, Y. K., and DAWSON, P. T. (1971). *J. chem. Phys.* **54**, 950.

PERDEREAU, J., and RHEAD, G. E. (1971). *Surf. Sci.* **24**, 555.

PLUMMER, E. W., and BELL, A. E. (1972). *J. vac. Sci. Technol.* **9**, 583.

POLIZZOTTI, R. S., and EHRLICH, G. (1975). *Bull. Am. phys. Soc. (II)* **20**, 857.

PROPST, F. M., and PIPER, T. C. (1967). *J. vac. Sci. Technol.* **4**, 53.

RIGBY, L. J. (1965). *Can. J. Phys.* **43**, 532.

ROBERTS, M. W., and ROSS, J. R. H. (1969). In *Reactivity of solids* (eds. J. W. Mitchell, R. C. Devries, R. W. Roberts, and P. Cannon), p. 411. John Wiley—Interscience, New York.

ROBERTS, M. W., and YOUNG, N. J. (1970). *Trans. Faraday Soc.* **66**, 2636.

ROBINS, J. L., WARBURTON, W. K., and RHODIN, T. N. (1967). *J. chem. Phys.* **46**, 665.

RYE, R. R., and BARFORD, B. D. (1971). *Surf. Sci.* **27**, 667.

RYE, R. R., BARFORD, B. D., and CARTIER, P. G. (1973). *J. chem. Phys.* **59**, 1693.

SCHEER, M. D., and MCKINLEY, J. D. (1966). *Surf. Sci.* **5**, 332.

SCHMIDT, L. D. (1972). *J. vac. Sci. Technol.* **9**, 882.

SHAW, B. L. (1967). *Inorganic hydrides*. Pergamon Press, Oxford.

SINGH-BOPARAI, S. P., BOWKER, M., and KING, D. A. (1975). *Surf. Sci.* **53**, 55.

SINGH-BOPARAI, S. P., and KING, D. A. (1976). *Surf. Sci.* **61**, 275.

TAMM, P. W., and SCHMIDT, L. D. (1969). *J. chem. Phys.* **51**, 5352.

—— (1970). *J. chem. Phys.* **52**, 1150.

—— (1971*a*). *J. chem. Phys.* **54**, 4775.

—— (1971*b*). *J. chem. Phys.* **55**, 4253.

—— (1971*c*). *Surf. Sci.* **26**, 286.

TAYLOR, T. N., and ESTRUP, P. J. (1974). *J. vac. Sci. Technol.* **11**, 244.

TUCKER, C. W. (1971). *Surf. Sci.* **26**, 311.

VAN HARDEVELD, R., and VAN MONTFOORT, A. (1966). *Surf. Sci.* **4**, 396.

VAN OOSTROM, A. (1967). *J. chem. Phys.* **47**, 761.

WACLAWSKI, B. J., and PLUMMER, E. W. (1972). *Phys. Rev. Lett.* **29**, 783.

WEDLER, G., and BRÖCKER, F. J. (1971). *Z. phys. Chem.* **75**, 299.

WEINBERG, W. H., and MERRILL, R. P. (1972). *Surf. Sci.* **33**, 493.

WILF, M., and DAWSON, P. T. (1976). *Surf. Sci.* **60**, 561.

WINTERS, H. F., and HORNE, D. E. (1971). *Surf. Sci.* **24**, 587.

YATES, J. T., and MADEY, T. E. (1965). *J. chem. Phys.* **43**, 1055.
—— (1969a). In *The structure and chemistry of solid surfaces* (ed. G. A. Somorjai), paper 59. John Wiley, New York.
—— (1969b). *J. chem. Phys.* **51**, 334.
—— (1971). *J. chem. Phys.* **54**, 4969.
YONEHARA, K., and SCHMIDT, L. D. (1971). *Surf. Sci.* **25**, 238.

11

THE INTERACTION OF OXYGEN WITH METAL SURFACES

11.1. Introduction

The standard free-energy charge for reactions of the kind

$$x\text{M(s)} + \tfrac{1}{2}\text{O}_2\text{(g)} \rightarrow \text{M}_x\text{O(s)}$$

is negative for all metals, recent thermodynamic studies having clarified the stability of the gold oxide. Some typical values are shown in Table 11.1 from which we can deduce, using the relationship

$$\Delta G^{\ominus}_{\text{f(oxide)}} = -RT \ln K_{\text{p}}$$

that for example in the case of Cu the equilibrium oxygen pressure for the

$$\text{Cu(s)} + \tfrac{1}{2}\text{O}_2\text{(g)} \rightleftharpoons \text{CuO(s)}$$

system is $\sim 10^{-10}$ torr at ~ 300 K. Thus from the thermodynamic viewpoint all metals should spontaneously form an oxide at low oxygen pressures, e.g. 10^{-6} torr at room temperature. Most clean metals form a chemisorbed layer of oxygen even at lower temperatures, and so the activation energy of oxygen chemisorption must be no more than about $10\,\text{kJ}\,\text{mol}^{-1}$. Further interaction leading to the growth of a distinct surface oxide phase usually requires a larger activation energy and is therefore kinetically controlled. The oxygen chemisorbed layer may therefore be considered as metastable with respect to the bulk oxide. The transformation from chemisorbed oxygen to bulk oxide has intrigued a number of investigators over the last decade. At what stage does lattice penetration occur? What is the mechanism of such a process? When can we refer to surface oxide? How does this oxide differ from the thermodynamically stable bulk oxide? These are the sorts of questions that have been posed and they have led to developments of experimental techniques specifically aimed to answer them. Simultaneously there has been an awareness of the theoretical implications with relation to the electronic structure of the surface and immediate subsurface regions.

We shall consider the metal + oxygen reaction as typifying all those metal–gas reactions that lead eventually to the growth of a new phase. Oxidation will therefore be treated in some detail, but clearly much of the mechanistic detail will be relevant to other analogous reactions such as sulphidation and halogenation, for which such detailed information is not yet available. Relevant reviews are those by Fehlner and Mott (1970), Lawless (1974), Musket (1970), Roberts (1962), Uhlig (1967), and Evans (1955).

11.2. Oxygen chemisorption and incorporation

In discussing oxygen chemisorption we shall bear in mind (a) the thermochemical evidence as reflected by the heat of chemisorption, (b) the kinetics of chemisorption, and (c) the role of surface structure and the nature of the adsorbed oxygen.

Heat of chemisorption. Most information on the heats of oxygen chemisorption ΔH_a on metals has been derived from direct calorimetric studies of the kind used by Brennan, Hayward, and Trapnell (1960). The variation of the heat of chemisorption with coverage (θ) for oxygen interaction with tungsten films at 288 K is shown in Fig. 11.1(a) and Table 11.1 summarizes the initial heats of chemisorption (ΔH_0) reported for some metals. In general there is a close correlation between ΔH_0 and the heat of formation of the corresponding stable oxide; this was recognized by Brennan *et al.* (1960) and extended by Roberts (1960) and Bond (1974) (Fig. 11.1(b)). The heat of chemisorption is clearly important since it is related to the energy of the metal-oxygen bonds formed. We could argue that it is possible to estimate the energy of the M—O bond from the equation

$$M + \tfrac{1}{2}O_2(g) \rightarrow M—O(ads) \tag{11.1}$$

and a knowledge of ΔH_0 and the dissociation energy of $O_2(g)$. For Ni (see Table 11.1 for ΔH_0) the M—O bond energy at 'zero coverage' at 300 K would be 225 kJ. Comparison shows that values calculated from the bond energies of the gaseous oxides are in reasonable agreement (Hayward 1971) with the initial heats of adsorption, suggesting that the bonds are similar in both cases. However, it is possibly unrealistic to pursue this approach any further in the light of recent developments in the elucidation

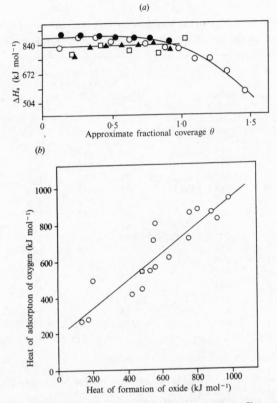

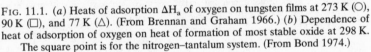

FIG. 11.1. (a) Heats of adsorption ΔH_a of oxygen on tungsten films at 273 K (○), 90 K (□), and 77 K (△). (From Brennan and Graham 1966.) (b) Dependence of heat of adsorption of oxygen on heat of formation of most stable oxide at 298 K. The square point is for the nitrogen–tantalum system. (From Bond 1974.)

of the surface crystallographic and electronic structure of chemisorbed oxygen.

11.2.1. Kinetics of oxygen chemisorption and incorporation

Much of the present-day thinking regarding the mechanism of metal oxidation stems from kinetic studies usually based on monitoring the oxygen pressure. These studies, although limited in the ultimate unambiguous information obtainable, have paved the way to the experimentally more sophisticated work that has emerged over the last few years. Fehlner and Mott (1970) discuss

Table 11.1

Comparison of heats of adsorption of oxygen on metal films with heats of oxidation and bond energies of gaseous oxides

Metal	Maximum coverage (mono-layers)	Initial heat of adsorption (kJ mol^{-1})	Stable solid oxide	Heat of formation of oxide (kJ mol^{-1})	Heats calculated from metal–oxygen bond energy of the following gaseous oxides (kJ mol^{-1})		
					MO	MO$_2$	MO$_3$
Mo	1·4	722	MoO$_2$	550	487±120	693	685
W	1·3	815	WO$_2$	563	802±80	798	810
Rh	1·0	441	Rh$_2$O	201	357	390	
Pd	0·7†	281	PdO	176	96	<164	
Pt	0·6†	281	PtO	143	281	424	
Ti	8·6	991	Ti$_3$O$_5$	987	827	831	
Mn	3–12	630	Mn$_2$O$_3$	643	307	432±100	
Fe	4·7	571	Fe$_3$O$_4$	563	<332		
Cr	2·5	731	Cr$_2$O$_3$	756	453±60	466±60	466±60
Co	2·3 / 3·0‡	420 / 353	Co$_3$O$_4$	428			
Ni	2·6 / 2·6‡	449 / 315	NiO	487	235		
Nb	3·8	873	Nb$_2$O$_5$	764		953	
Ta	3·2	890	Ta$_2$O$_5$	810	1040	970	

† Integral heat.

‡ Results of Brennan and Graham (1966) for adsorption at 77 K. All other data taken at 300 K.

Limits of experimental error are given for calculated heats in the last three columns when these errors exceed 40 kJ mol^{-1}.

Experimental heats of adsorption and surface coverages are taken from Brennan *et al.* (1960) except those marked with a double dagger.

From Hayward 1971.

the various kinetic laws observed and the models of oxidation compatible with these laws; a much earlier view is available in the article by Evans (1955). At high temperatures the process of oxidation is well understood in terms of the parabolic law where the oxide thickness x at time t is given by eqn (11.2) which follows directly from Fick's law of diffusion (eqn (11.3)):

$$x^2 = 2kt \tag{11.2}$$

$$\frac{dx}{dt} = \frac{k}{x}. \tag{11.3}$$

The diffusing species may be either the metal or oxygen and the rate constant k is a function of temperature and may or may not be pressure dependent according to the mechanism. At low temperatures, say at and below room temperature, we are intimately concerned with oxygen chemisorption and the stability of the adsorbed layer with respect to the surface oxide, which may or may not be the stoichiometric bulk oxide. In general, experience leads us to expect it to be distinctly different. The kinetic laws reported in this region of oxidation are usually of the logarithmic ($x \propto \log t$) or inverse logarithmic ($1/x \propto \log t$) type.

Oxygen chemisorption. All metals have been reported to interact with oxygen to form chemisorbed species and in this respect oxygen is unique in its reactivity. There are, however, some marked variations in the observed reactivity as reflected in the heat of chemisorption ΔH_a, the oxygen sticking probability S, and the tendency of the chemisorbed layer to rearrange to give an oxide phase. The latter is most easily reflected in terms of the coverage θ attained, but of course $\theta < 1\cdot0$ does not exclude surface rearrangement and other criteria based on such studies as changes in the work function, the observed electron diffraction pattern, and electron spectroscopy have to be used to obtain details of the molecular events occurring.

Nucleation. There have been a number of cases reported where kinetic information has led to the suggestion that oxygen chemisorption involves nucleation. Probably the most detailed studies of nucleation in oxidation are those with copper (Grönlund 1956, Bénard *et al.* 1959, and Mitchell and Lawless 1966). Fig. 11.2 shows the relationship between oxygen pressure, the time of oxidation, and the type of oxide film formed at 823 K. Three regions are present, in general agreement with the model of induction, followed by growth of nuclei leading to an oxide film. As anticipated on general grounds the induction period decreases with an increase in the oxygen pressure. The influence of temperature is complex and varies with the crystal face; the relative order of the length of induction period is (100), (111), (110), (311). Recently a detailed quantitative study of the nucleation and growth of Cu_2O on Cu(100) was made by Jardinier-Offergeld and Bouillon (1972); these authors made use of Rhead's (1965) theory to estimate values of the surface diffusion

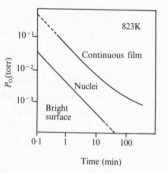

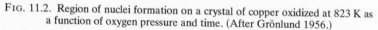

FIG. 11.2. Region of nuclei formation on a crystal of copper oxidized at 823 K as a function of oxygen pressure and time. (After Grönlund 1956.)

coefficient. The values were about 10^{-3} times smaller than the experimental values of Bradshaw *et al.* (1964). The fact that oxide particle density remains constant and that fast-moving particles are usually widely spaced has been explained by Rhead (1965) on the basis of Ostwald ripening. Experimental data supporting this idea have been obtained by Mitchell and Lawless (1966).

Although the main features of the high-temperature, low-pressure nucleation process are reasonably well established, details of the growth mechanism are not clearly defined. There is a need for coupling clean-surface approaches with electron microscope and electron diffraction techniques so that kinetic and structural data under well-defined experimental conditions are obtained.

If we consider, following Evans (1945), that nucleation is analogous to the sideways growth of a corrosion film on metals and that there are a fixed number of nucleii from which the oxygen monolayer grows laterally, the rate of oxygen uptake will clearly be proportional to the area at which oxygen adsorption can take place. Therefore the rate of decrease in oxygen pressure P_{O_2} will be given by

$$\frac{-dP_{O_2}}{dt} \propto 2\pi r \tag{11.4}$$

where r is the radius of the nucleus (or island). Now the actual number of oxygen molecules which are chemisorbed in unit time on unit area is constant at a given pressure so that the *area* of the layer will grow linearly with time. It therefore follows that

$$r \propto t^{\frac{1}{2}}$$

and

$$\frac{-dP_{O_2}}{dt} \propto t^{\frac{1}{2}} \tag{11.5}$$

where t is the time from the start of growth. This clearly would give rise to autocatalytic-type kinetics.

If instead of as above, where growth is assumed to be strictly two dimensional, the oxide layer builds up into hemispherical caps, then the rate can be shown to be dependent on $t^{\frac{2}{3}}$, since

$$\frac{-dP_{O_2}}{dt} \propto 2\pi r^2.$$

However, $r \propto t^{\frac{1}{3}}$ for a hemispherical cap so that

$$\frac{-dP_{O_2}}{dt} \propto t^{\frac{2}{3}}. \tag{11.6}$$

Clearly eqns (11.5) and (11.6) will only hold during initial growth if there are a very large number of nucleii, but once overlap occurs this simple model is inappropriate. Bloomer (1957) in his studies of oxygen chemisorption and oxidation of barium reported an initial acceleratory process, as also did Rhodin (1959) in his magnesium + oxygen system. In both cases the chemisorption was followed by measuring the sticking probability using the Wagener flow method, and the form of the data is shown in Fig. 11.3a. Values of n in the relationship rate $\propto t^{1/n}$ of between 0·5 and 0·8 were reported in keeping with the nucleation model discussed above. It is, however, conceivable that ion-gauge effects influenced the measured oxygen pressure giving rise to false kinetic data, but at least one other chemisorption kinetic study, $N_2 + Ca$ (Roberts and Tompkins 1959), where ion gauges were not used, supports (but does not establish!) the concept of nucleation during the chemisorption stage (Fig. 11.3(b)).

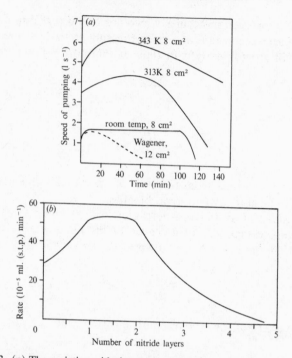

FIG. 11.3. (a) The variation with time and temperature of the speed of pumping of oxygen by barium (manifold pressure 2×10^{-5} torr). (From Bloomer 1957.) (b) The rate of nitridation of calcium, at a constant pressure of nitrogen of 9×10^{-3} torr and 295 K, as a function of the number of nitride layers present. (After Roberts and Tompkins 1959.)

More recently LEED and optical-simulation approaches to oxygen chemisorption give further support to island growth (nucleation). These are discussed in Chapter 3.

Incorporation at low temperature. It is now well established that in many metal + oxygen systems incorporation from the chemisorbed layer into the subsurface is facile, occurring at low temperature (77 K) in some cases. In the case of zinc and iron, for example, simple volumetric techniques were sufficiently sensitive for the conclusion to be reached that substantial oxidation (up to five equivalent monolayers) occurred at low temperature and low oxygen pressure. However, in many cases, of which nickel is a good example, more subtle techniques have had to be

employed. By monitoring changes in the work function (Quinn
and Roberts 1963, 1964) it became clear that substantial rearrangement of the chemisorbed layer formed at 77 K occurred
both on warming *in vacuo* to 300 K and at 300 K during subsequent oxygen interaction. Further evidence for these molecular
events is available from ultra-violet photoemission studies (Quinn
and Roberts 1965) where the energy distribution of photoelectrons was monitored with a view to observing 'chemical shifts'
during the transformation from metal to oxide. The work-
function approach was stimulated by (*a*) the place-exchange
model for oxygen incorporation proposed by Lanyon and Trapnell (1955), where a 'switch' between chemisorbed oxygen ($O^{\delta-}$)
and the 'underlying' metal atoms would be reflected by a change
in work function, and (*b*) the role of the surface potential of
chemisorbed oxygen and its associated field in lowering the
barrier for oxygen incorporation, the central theme in the
Cabrera–Mott (1948) theory of metal oxidation.

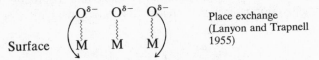

Surface — Place exchange (Lanyon and Trapnell 1955)

The behaviour of metals in general is typified by the three
examples (Quinn and Roberts 1964) shown in Fig. 11.4. With
molybdenum the work function ϕ increases by ~ 1.6 eV at 77 K
and an oxygen pressure of $\sim 10^{-3}$ torr; there is, however, very
little change on warming the adlayer to 295 K. With nickel the
work function increases by about 1.2 eV at 77 K; at 10^{-3} torr
there is a slight reversal of surface potential which was attributed
to molecularly adsorbed oxygen. On warming to 295 K ϕ returns
to the original clean-metal value; at 295 K each oxygen admission results in first an increase in ϕ and then after adsorption is
complete a decay in ϕ towards the clean-metal value. In the case
of aluminium decays in ϕ are observed at 90 K. There is therefore no evidence for place exchange with molybdenum in the
temperature range 77–295 K; chemisorbed oxygen is apparently
stable on nickel at 77 K, but place exchange is induced thermally
at about 190 K and increases in rate at higher temperature. With
aluminium place exchange is observed at 90 K but not at 77 K;
furthermore at 295 K little change in work function is observed

during the admission of oxygen doses, place exchange being rapid with the consequent dispersal of the negative dipole of the chemisorbed oxygen (Roberts and Wells 1969).

Copper is similar to nickel and iron to aluminium, except that with iron decays in work function are observed even at 77 K which is compatible with the fact that interaction of iron with oxygen is known to lead to the formation of more than one layer at this temperature. With aluminium changes in ϕ with time are observed only at 90 K. Uranium and lead are extreme examples of the aluminium-type behaviour in that the work function is *lowered* (Rivière 1964, Saleh, Wells, and Roberts 1964) during initial oxygen interaction at 295 K and only at a later stage does the work function increase.

One of the intriguing features of the interaction of oxygen with nickel is that the process of incorporation (reflected by a decrease in ϕ, Fig. 11.4) has a very low activation energy ($\sim 4 \text{ kJ mol}^{-1}$) yet the chemisorbed layer is stable at 77 K (Quinn and Roberts 1964, Roberts and Wells 1966a,b, Delchar and Tompkins 1967). The activation energy was measured by comparing rates of incorpora-

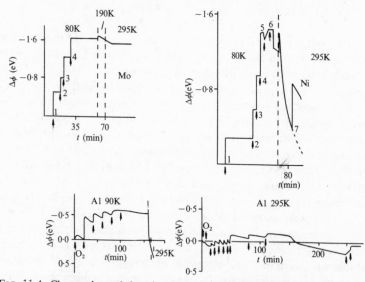

FIG. 11.4. Changes in work function during the adsorption of oxygen on Mo, Ni, and Al film surfaces in the temperature range 80–295 K (↑ indicates admission of a 'dose' of oxygen). (From Quinn and Roberts 1964.)

Fig. 11.5. Concerted vibration of four surface Ni(100) atoms on which an oxygen adatom is chemisorbed. (From Delchar and Tompkins 1967.)

tion at gradually increasing temperatures in the range 77–100 K, but at fixed extents of incorporation. Delchar and Tompkins (1967) made the interesting suggestion that if one considers the concerted vibration of four Ni atoms (Fig. 11.5) in the (100) surface then the probability of this event is calculated to be about 1 in 10^{12}. If the rate-determining step involves such a process leading to an oxygen adatom present at the centre of that arrangement being incorporated, then one can see quite clearly how it is the pre-exponential term in the rate expression which is of abnormally small magnitude, making it the controlling factor in the oxygen-incorporation process.

There is a very rough correlation (Roberts and Wells 1967), if we consider f.c.c. metals only, between the ease of oxygen incorporation and the interatomic spacing in the (100) plane of the surface layer (Table 11.2), and this is in keeping with the general concept proposed by Delchar and Tompkins and the contention that decreases in work function reflect lattice penetration. This is particularly striking for rather open metallic structures such as lead and uranium where the initial interaction leads to a decrease in ϕ. (Saleh et al. 1964, Rivière 1964).

TABLE 11.2
Size of adsorption sites on the (100) face

Element	Pb	U	Al	Cu	Ni
Diameter of site (nm)	0·147	0·136	0·119	0·105	0·088
Depth of site (nm)	0·073	—	0·059	0·052	0·044

The driving force for incorporation stems principally from the fact that chemisorbed oxygen is inherently unstable with respect to incorporated oxygen where co-ordination with the metal atoms is at a maximum. We can think of the force pulling the chemisorbed 'ion' into the subsurface as arising from the image force $q^2/4\ kx^2$ where q is the charge, k the dielectric constant, and x the distance from the metal surface. Therefore the activation energy E is $(W - q^2\ a/4\ kx^2)$ where W is the activation energy in the absence of the image force and a is the distance moved by the cation. When $x = 0.2$ nm the energy due to the image force is about 200 kJ mol^{-1}. If we think of W as being similar to the strength of the metal–metal bond, which can vary between 80 and 200 kJ mol^{-1}, then depending on the metal the observed activation energy E may be zero for an appreciable extent of the oxygen interaction process. As soon as E becomes positive the rate of reaction will become temperature dependent. Fehlner and Mott (1970) have drawn attention to correlations between the extent of oxide growth and bond strength. They consider not only metal–metal but also metal–oxygen values and distinguish between network-forming oxides of 'strong' metals and intermediate or modifying oxides of 'weak' metals. The distinction is, however, not always obvious. In the context of Fehlner and Mott's proposal it is relevant to draw attention to the Zachariasen rules (Zachariasen 1932) regarding the connection between the arrangment of the oxygen atoms around the neutral cation and the formation of a three-dimensional glassy network: MO_2 and M_2O_5 form network oxides if the oxygens form tetrahedra around the metal atoms; M_2O_3 forms a network if the oxygens are in a triangular arrangement around the metal; MO_3,

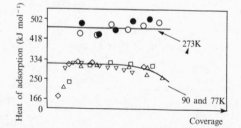

FIG. 11.6. The heats of adsorption of oxygen to saturation on nickel films at 273, 90, and 77 K. (From Brennan and Graham (1966).)

M_2O_7, and MO_4 are probable network formers; MO and M_2O should not form networks and are referred to as modifiers.

One investigation which deserves particular mention, since it puts forward an alternative model to the one discussed above for the $Ni + O_2$ system, is that of Horgan and King (1969, 1970) who reported sticking-probability–coverage data. The sticking probability exhibited a minimum value at about 4×10^{14} molecules cm^{-2}; this minimum was very obvious at 373 K, less pronounced at 290 K, and absent at 77 K. These authors, on the basis of the reactivity of low-coverage chemisorbed oxygen to $CO(g)$, concluded that initial oxygen interaction leads to molecularly adsorbed oxygen. They argue that, since the rate of $CO(g)$ interaction with low-coverage adsorbed oxygen is proportional to $[P_{CO(g)}]^2$, then the rate-determining step is likely to be given by eqn (11.7) rather than eqn (11.8):

$$2\,CO(g) + O_2(ads) \rightarrow 2\,CO_2(g) \tag{11.7}$$

$$CO(g) + O(ads) \rightarrow CO_2(g). \tag{11.8}$$

At higher coverages this molecularly adsorbed oxygen is not present. Calorimetric heats of adsorption are used as further support for the model since the heat has been reported to increase from about 170 kJ mol^{-1} at low coverage to a constant value of about 334 kJ mol^{-1} for the rest of the oxygen uptake at 77 K (Fig. 11.6) (Brennan and Graham 1966). At 273 K the heat is constant at about 462 kJ mol^{-1}, but unfortunately no data are reported at low coverage, corresponding to the region where the particularly low heat value (210 kJ mol^{-1}) was observed at 77 K.

Horgan and King associate the initial low heat of chemisorption (only observed at 77 K) and the minimum in the sticking-probability–coverage curve with molecularly adsorbed oxygen, which they suggest exists in the temperature range 195–373 K. The alternative model based on work-function and photoemission studies is that oxygen incorporation and oxide formation occurs above 195 K. This latter model is also compatible with the sticking-probability–coverage data at different temperatures and would explain the higher plateau heat observed at 273 K (462 kJ mol^{-1}) than at 77 K ($\sim$334 kJ mol^{-1}). It could not account, however for the three points in the ΔH_a–θ plot at low coverage (Fig. 11.6) which were in the range 170–252 kJ mol^{-1}.

It would be worthwhile reinvestigating ΔH_a at low θ and low temperature. At 290 K Klemperer and Stone (1957) also reported a ΔH_a which was invariant with θ but of appreciably greater exothermicity (630 kJ mol^{-1}). The model of Horgan and King therefore depends crucially on the second-order kinetics observed for the CO(g) and O(ads) reaction and the unusually low heat at low coverage for oxygen adsorption on Ni at one temperature (77 K).

Holloway and Hudson (1974) have pursued further the mechanism of oxygen interaction with nickel using single crystals and LEED, AES, and work-function measurements to follow the surface process. They recognize three stages with Ni(100): (1) a fast dissociative chemisorption leading to surface structures based on the initial nickel interatomic spacing, (2) a rapid oxidation leading to epitaxial NiO two layers thick, and (3) a slow growth of bulk NiO. Holloway and Hudson therefore suggest that initial oxygen interaction is dissociative; furthermore they conclude that beyond the minimum in the sticking-probability–coverage curve the formation of NiO occurs (stage 2 above), rather than penetration of the oxygen into the nickel. Obviously one cannot tell the difference between 'oxygen penetration' and 'oxidation' from kinetic studies alone. In this context it is worth recalling, however, that photoemission results (Quinn and Roberts 1965, Roberts and Wells 1966a,b, Roberts 1972) were interpreted in terms of place-exchange and oxide formation. This was based on both electron yield and electron energy distribution studies as a function of the temperature of the nickel and oxygen pressure. Smith and Anderson (1976) have recently also considered critically the suggestions of Holloway and Hudson and reinforce the above views by drawing particular attention to the extensive experimental evidence for oxygen incorporation at exposures well below those at which NiO diffraction spots first appear.

Ni + O$_2$ is probably the most thoroughly studied metal–oxide system, and, except for some minor points, the model that has emerged from the various techniques employed with both polycrystalline and single-crystal surfaces is a consistent one. Any differences that have been noted are frequently a function of the *precise* experimental conditions. For example Holloway and Hudson (1974) attribute differences in the work-function behaviour of films and single crystals to differences in oxygen pressure. Probably a more important factor is the exact temperature of the nickel; for example Holloway and Hudson observed first an increase and then a decrease in work function when Ni(100) at 302 K was exposed to oxygen at $\sim 10^{-6}$ torr. Wells and Roberts (1966a,b) observed similar behaviour with polycrystalline films in a flow system but with the additional observation of a recovery of the work function at high oxygen exposure. The important difference between the two experiments was undoubtedly that the films were maintained at much higher temperature, 440 K; the pressure in the film work was in the range 8×10^{-6}–2×10^{-4} torr. Ni(210) surfaces were shown (Kirby, McKee, and Roberts 1976) to facet to give (100) type planes when exposed to oxygen at 85 K and the crystal warmed to 295 K. This emphasizes the need to combine surface-structure (LEED) studies with other methods of investigation.

11.3. Oxide-film growth or oxidation

A simplified model (Roberts 1962) of the situation during oxide growth is shown in Fig. 11.7. There are two limiting cases. (1) positively charged cations and electrons migrate in the same direction from the metal–oxide interface to the outer surface; (2) negatively charged anions migrate inwards and electron movement occurs in the opposite direction. We shall, however, start with the ideas put forward by Cabrera and Mott (1948) since they are a good basis for discussing the growth of an oxide on a metal surface (Kruger and Yolken 1964, Roberts 1962). They made the following assumptions.

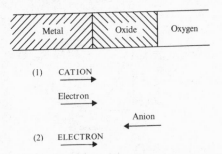

FIG. 11.7. Model of oxide growth. (From Roberts 1962.)

(1) The slow step in oxide growth is at the metal–oxide interface and involves entry of the metal 'ion' from the metal into the oxide lattice; there are N' special sites at this interface and their vibrational frequency is v. The barrier to entry, W, is the activation energy of the oxidation process. Subsequent diffusion through the oxide to the oxide–gas interface is fast.

(2) Oxygen molecules from the gas phase dissociate at the oxide surface giving rise to chemisorbed oxygen adatoms. These adatoms are negatively polarized, leading to a surface potential V across the oxide of thickness X. Consequently a field $F = V/X$ acts across the oxide.

(3) The field F effectively lowers the barrier W to a value $(W - qaF/2)$ where q is the charge on the ion and $\frac{1}{2}a$ the jump distance a being the interatomic spacing.

Fig. 11.8 shows the lowering of the energy barrier to incorporation for the iron + oxygen system as a function of oxide thickness. In this case we would anticipate multilayer oxide growth at 77 K and this is observed even at an oxygen pressure of $\leqslant 10^{-6}$ torr. The growth rate has thus the form of eqn (11.9):

$$\frac{dX}{dt} = N'aqv \exp\left\{ -\frac{W - qaV/2X}{RT} \right\} \tag{11.9}$$

which leads to the inverse logarithmic law

$$\frac{1}{X} = A^* - B^* \ln t.$$

On the other hand, if the Cabrera–Mott model does not apply and the growth process has an activation energy that increases

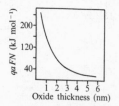

FIG. 11.8. The field effect as a function of oxide thickness; N is the Avogadro constant. (From Roberts 1962.)

linearly with oxide thickness then the growth law is of the form

$$\frac{dX}{dt} \propto \exp\left(-\frac{W+X}{RT}\right)$$

which integrates to give a logarithmic law of the kind

$$X = A \ln(1 + Bt).$$

Now whether the activation energy E conforms either to

$$E = W + f\left(\frac{1}{X}\right) \quad \text{or} \quad E = W + f(X)$$

is difficult to ascertain unambiguously and consequently logarithmic and inverse logarithmic laws are notoriously difficult to distinguish, particularly when data are available only over a restricted range of oxide thickness X (Roberts 1962).

Ghez (1972) has recently shown, however, that the inverse logarithmic law is not a strictly accurate solution of the original Cabrera–Mott equation. A more correct test of the rate law would be a plot of $1/X$ *versus* $\ln(t/X^2)$. Therefore the debate as to whether the kinetics of oxide growth obey inverse or direct logarithmic laws may not have been particularly valuable.

Crucial to the Cabrera–Mott theory is the field F which is assumed to vary with oxide thickness, $F = V/X$. Now the variation of the rate of oxide growth with oxygen pressure provides further information on the relevance of the term V, since this will clearly be dependent on the concentration of oxygen adatoms (adions). The original Cabrera–Mott theory assumed that the surface would be saturated with oxygen 'adions' at a pressure of $\sim 10^{-4}$ torr, but there is now strong evidence for the rate to be pressure dependent. This can be accommodated by the theory since surface potential data indicate a dependence of the surface potential (i.e. V) on the oxygen pressure.

An interesting suggestion has recently been put forward by Fehlner and Mott (1970) that the activation energy E is better described by

$$E = W - \tfrac{1}{2}qaF + \beta X \tag{11.10}$$

where βX is a term analogous to $f(X)$ above but reflects a structural change in the oxide during growth. Use of eqn (11.10) leads to a logarithmic law.

There are at the moment no data available from which the relative magnitudes of $qaF/2$ and βX can be ascertained. In fact all that is known is that E increases with X (Roberts and Wells 1966a,b, Eley and Wilkinson 1960, Kirk and Huber 1968, Crossland and Roettgers 1966). Direct structural evidence on changes in oxide morphology is sparse but what is available points to rearrangements in some cases.

Some elegant and mechanistically significant results have recently been obtained by Hunt and Ritchie (1972a,b) and Ritchie and Tandon (1972) who studied the influence of doping the gaseous phase (O_2) with iodine on the oxidation of aluminium. At 373 K the oxidation rate, monitored by resistance change of an Al film, increased by up to a factor of 7 (Fig. 11.9); the magnitude of the effect diminished as oxidation progressed.

The formation of the iodide in preference to the oxide is thermodynamically very unlikely; moreover the rate readjusted itself to the lower value on removal of the iodine. We have here clear evidence for the rate being influenced by the nature of the chemisorbed species, i.e. whether $I^{\delta-}$ or $O^{\delta-}$. Clearly in terms of the Cabrera-Mott model, which is considered to apply for the $Al + O_2$ system, the field F increased owing to further chemisorption of $I^{\delta-}$ species. As oxidation proceeds to greater thicknesses the effect would obviously diminish and this is observed (Fig. 11.9). Also if the mechanism is correct it implies that iodine chemisorbs on amorphous aluminium oxide much more readily than oxygen, which is in keeping with the known electron affinities of the two atoms.

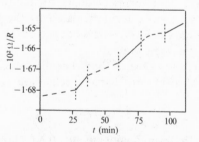

FIG. 11.9. The effect of traces of iodine on the oxidation rate of aluminium at 373 K: broken lines oxidation in oxygen; solid lines oxidation in oxygen which contained iodine. (From Hunt and Ritchie 1972b.)

At very high temperatures (~1200 K) parabolic growth (eqn (11.2)) is generally observed with Ni. At moderate temperatures (up to 773 K) a variety of kinetic laws have been reported (Evans 1955) and recently Graham and Cohen (1972), using UHV techniques and clean surfaces, have attempted to rule out surface impurities as determining the various kinetic laws. During the formation of the first 3 nm a direct logarithmic law is obeyed. For greater oxide thickness (3–4 nm) the growth is parabolic. The logarithmic equation $X = A \ln t + B$ is independent of the temperature range but clearly at higher temperatures it is obeyed only for a short initial period of the reaction. The dependence of rate on oxygen pressure is $\frac{1}{6}$.

In a separate study Graham et al. (1972) investigated the influence of surface preparation on the oxidation kinetics and, not surprisingly, important effects were observed. The authors on the basis of their results distinguish between oxidation by lattice diffusion with an activation energy of ~218 kJ mol^{-1} and oxidation by leakage path diffusion with an activation energy of ~155·4 kJ mol^{-1}

We have not considered how electrons transfer during oxidation or whether this process can be rate-limiting. Fromhold and Cook (1966, 1967) have used the earlier theories of Mott and Cabrera, and Mott, to calculate the oxide thickness at which rate control by ion migration shifts to rate control by electron tunnelling. They quote a value of about 3 nm. This of course is much larger than the limiting thicknesses reported for most metals at 295 K, so that tunnelling is unlikely to be rate limiting. Fehlner and Mott (1970) are of the opinion that even if the maximum tunnelling distance of 3 nm is exceeded, localized states in the band gap of an amorphous oxide can lead to impurity or 'hopping' conduction. In this process, familiar in doped semiconductors such as silicon or germanium, the electron tunnels from one centre to another, the activation energy for the process being quite small. Fehlner and Mott therefore dismiss electron tunnelling as a possible slow step and favour ionic movement.

The possibility of anion movement was not considered in the original Cabrera–Mott theory, but recent anodic-oxidation work by Davies et al. (1965) at Chalk River, Canada, has shown that anion mobility at low temperature is not only possible but dominates during the anodization of Hf, Ta, Nb, W, and Zr. All these oxides are glass formers and the ability of anions to move through these oxide lattices can be best understood in terms of the glassy structures. Network formers also have channels through which a large anion can move while the much smaller and highly charged cations are more rigidly bound within the network structure.

If therefore we accept that in general anionic transport must be a distinct possibility, it is important to consider how an oxygen ion formed at the oxide–oxygen interface is transported through the oxide to the metal–oxide interface. It is likely that a large barrier exists at the oxide–oxygen interface for anion movement leading to anion incorporation within the oxide. This is analogous to the barrier for cation entry at the metal–oxide interface which is crucial to the Cabrera–Mott theory. Once within the oxide there is little likelihood of the ion being trapped en route to the metal–oxide interface. Fehlner and Mott suggest that the field lowers the activation energy barrier for anion entry and is therefore strictly analogous to the cation-entry model.

Rate control at the oxide-oxygen interface. Evidence for oxidation being controlled by molecular events occurring at the oxide–oxygen interface has become more prevalent during recent years. For example Boggio (1972) has shown that in the case of the low-temperature oxidation of copper a surface reaction similar to that proposed by Grimley (1955) is in keeping with the experimental evidence based on the measurement of work-function changes on oxidizing Cu(111), (110), and (100) surfaces.

In the analogous metal–sulphide system there are a number of clearly defined cases where the kinetics of sulphidation by $H_2S(g)$ are compatible with the rate-determining step being the dissociation of H_2S at the sulphide surface. Both the sulphidation of lead and iron films conform to first-order kinetics (Saleh *et al.* 1964, Roberts and Ross 1966) and in the case of Fe, sulphidation beyond the monolayer occurs according to the stoichiometry of eqn (11.11):

$$Fe_{(s)} + H_2S_{(g)} \rightarrow FeS_{(s)} + H_{2(g)} \tag{11.11}$$

(Sulphidation of lead is intriguing in that it results in the recognition of a stable hydride of lead.) The kinetics of the $Fe + H_2S$

reaction (eqn (11.11)) obey the relationship

$$-\left(\frac{dP_{H_2S}}{dt}\right)_{\theta,T} = A(\theta)\frac{P_{H_2S}}{(2\pi mkT)^{\frac{1}{2}}}\exp\left(\frac{-E(\theta)}{RT}\right)$$

where $A(\theta)$ and $E(\theta)$ were related by

$$\log A(\theta) = 0\cdot44\, E(\theta) - 6\cdot9. \qquad (11.12)$$

Clearly first-order plots ($\log P_{H_2S}$ versus time) were only linear under conditions where $A(\theta)$ and $E(\theta)$ did not vary appreciably. This did not obtain until θ was greater than about four equivalent monolayers, when $E(\theta)$ was approaching a constant value of about $72\ kJ\ mol^{-1}$. It is also pertinent to mention that the Cabrera–Mott model could not account for the observed kinetics, although one of the prerequisites, namely that E versus $1/\theta$ was linear, was obeyed. The first-order pressure dependence and the variation of the pre-exponential factor with θ argued against the simple model. When the observed kinetics are considered in terms of the absolute rate theory the variation in $A(\theta)$ with the extent of sulphidation (θ) can be interpreted in a gradual transition from immobile to mobile adsorption as θ (and the temperature) increase.

11.4. Flash desorption and the nature of surface oxygen

Studies combining flash desorption and mass spectrometry have largely been confined to the $W + O_2$ system for obvious reasons. This system has been of considerable interest ever since Langmuir's observation of the production of volatile tungsten oxides above 1200 K, an observation that stems from the strength of the tungsten–oxygen surface bond. The essence of the experimental approach is to expose a tungsten ribbon to oxygen and then flash the ribbon to a high temperature, analysing the desorbed species by mass spectrometry. The work of McCarroll (1967) and Ptushinskii and Chuikov (1967) showed that both oxygen atoms and a complex series of oxides (W_3O_9, W_2O_6, WO_3, WO_2, and WO) were desorbed, the former at low oxygen exposures and the latter only after high exposures. There has been some debate (McCarroll 1967, Ptushinskii and Chuikov 1967) as to whether surface oxides are formed at room temperature. King, Madey, and Yates (1971) have more recently provided experimental evidence for the various adsorption states and have proposed four stages in the formation of the tungsten oxide products observed in the flash desorption spectra (Fig. 11.10). Stages 1, 2, and 3 are all reported to occur at 300 K, but with increasing oxygen exposure of 5×10^{-6}, 10^{-4}, and 10^{-2} torr s respectively. The limiting thickness of the oxide film was equivalent to about 3 monolayers at 300 K. Stage 4 occurs between 500 and 1000 K and an oxygen exposure of 1 torr s; it is of unlimited thickness. The rate of formation of this oxide film was shown to be directly proportional to the oxygen pressure for pressures $\sim 10^{-4}$ torr. It is interesting to note in passing that unlimited oxide-film growth above a critical temperature is a feature inherent in the Cabrera–Mott model. If we rearrange eqn (11.9), putting $F = V/X_L$, then the limiting thickness X_L is given by

$$\frac{1}{X_L} = \frac{2W(1 - \alpha T)}{VaqN} \tag{11.13}$$

where N is the Avogadro constant,

$$\alpha = -\frac{R}{W} \ln\left(\frac{dX_L}{dt} \frac{1}{N' aqv}\right)$$

and dX_L/dt is the experimentally defined limiting rate (say 0·01 monolayer min^{-1}). Therefore when $T = 1/\alpha$ the limiting oxide thickness is infinite. In the case of the $Ba + O_2$ system (Bloomer 1957) the critical temperature is 350 K, while in the $Ca + N_2$ system (Roberts and Tompkins 1959) it is 435 K. For $W + O_2$ the Cabrera–Mott model is not considered applicable to stage 4. In particular this stage is considered to be controlled by the rate of oxide desorption and to be independent of the concentration of oxygen on the surface, i.e. it is zero order:

$$-\frac{d\sigma_{WO_2}}{dt} = k \exp\left(-\frac{E_d}{RT}\right). \tag{11.14}$$

From plots of rates at various temperatures (Arrhenius plots) E_d,

Stage 4	WO_2
Stage 3	WO_2, WO_3, W_2O_6
Stage 2	WO, WO_2, WO_3
Stage 1	Atomic Oxygen

FIG. 11.10. Stages in the oxidation of tungsten. (From King *et al.* 1971.)

the activation energy for desorption of WO_2, was found to be 418 kJ mol^{-1} and k estimated to be 10^{34} molecules cm^{-2} sec^{-1}:

11.5. Electron spectroscopy

The impact of the development of electron-spectroscopic techniques has already been significant in the field of metal oxidation. Studies have largely been concerned with 'thick oxide' films with very little having been reported for oxygen chemisorption and the reactions immediately following chemisorption leading to oxide growth. This is understandable in view of the dearth of equipment available, at least until very recently, with ultra-high-vacuum facilities. Where there has been quite informative progress is in the field of alloy oxidation, and the particular advantage of ion etching has been highlighted. Ion etching enables a layer-by-layer removal ('onion peeling') of the oxide with analysis at each stage leading to a chemical-composition–depth profile. Whether ion etching perturbs in some way or other the chemical composition of the surface has not as yet been clearly established; caution is therefore necessary in interpreting such data.

11.5.1. *Interaction of oxygen with metals*

Molybdenum. Both volumetric-adsorption and work-function data support strongly the view that for oxygen exposures of $\sim 10^3$ L interaction is confined to the chemisorbed layer in the temperature range 77 on 295 K (Quinn and Roberts 1964). X-ray photoelectron spectroscopy indicates only one O(1s) peak, centred at 530·4 eV (Atkinson, Brundle, and Roberts 1974).

Increasing the severity of the oxygen-interaction conditions to 10^{-1} torr of oxygen at 420 K resulted in broadening of the Mo(3d$_{\frac{3}{2}}$) peak; the Mo(3d$_{\frac{5}{2}}$) peak was unchanged as also was the O(1s). At 520 K and 10^{-1} torr of oxygen the main Mo peaks decreased in intensity, the Mo(3d$_{\frac{3}{2}}$) broadened further, and the O(1s) increased in intensity. A new and quite distinct O(1s) feature also developed at higher binding energy, centred at about 234 eV. There is therefore clear evidence for XPS being able to distinguish between oxygen atoms in differing electronic environments and this is supported by the fact that reversibly adsorbed water on molybdenum exhibits quite a distinct O(1s) peak from that observed with irreversibly chemisorbed water which is considered to be dissociated.

Tungsten. Oxygen adsorption at 300 K gives rise to a single X-ray induced O(1s) peak at all coverages (which is invariant in energy) (Yates, Madey, and Erickson 1974). Furthermore, oxygen interaction at 950 K and an exposure of 800 L results in the same O(1s) peak. This is particularly interesting in that under these experimental conditions Yates, Madey, and Erickson have reported that multilayer oxide growth occurred. Also, if we accept that WO, WO_2, WO_3, and W_2O_6 species do exist at 300 K, as observed by flash desorption line-of-sight mass-spectrometry, it follows that XPS (or ESCA) cannot distinguish between oxygen atoms bound to tungsten atoms in different oxidation states. An alternative possibility is of course that the oxides observed by flash desorption may be formed during the flash process itself, i.e. that they reflect the technique used rather than the molecular situation on the surface. The results of the $Mo + O_2$ system tend to support the contention that the flash-filament data should be viewed with caution until they are proved not to be influenced by the technique.

A recent development has been the observation that ions liberated from an adsorbed oxygen layer on tungsten by electron-stimulated desorption show sharply peaked symmetric angular distributions which are in registry with the substrate. The full potentialities of these results are yet to be assessed (Czyzewski, Madey, and Yates 1974).

Nickel. Photoemission studies of the $Ni + O_2$ system have covered a wide range of photon energies (6 eV up to 1486 eV). The aim of the earlier work using low energy ultra-violet radiation (Quinn and Roberts 1965) was to ascertain whether there was any evidence for perturbation of the energy distribution of the photoelectrons when from other studies (e.g. work function, Fig. 11.4) place exchange or incorporation had been indicated.

Photoelectron-yield studies at 295 K indicated clearly the occurrence of two simultaneous processes, one leading to a decrease in photocurrent and the other to a recovery in the yield. The latter was activated in the sense that there was no evidence for it occurring at lower temperature (Fig. 11.11) These results with polycrystalline Ni ribbons paralleled accurately work-function results with Ni films and strengthened the model for 'place exchange'. The photoelectron energy distribution data

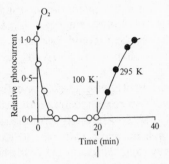

Fig. 11.11. Variation of photocurrent during oxygen interaction with a polycrystalline nickel ribbon maintained at about 100 K (○) followed by warming *in vacuo* ($P < 10^{-7}$ mm) to 295 K (●). $\lambda = 215$ nm. (From Roberts and Wells 1966*a*.)

extended the model in that they suggested a distinct change in electron energy which was interpreted as reflecting a surface 'oxide'. This 'oxide', on the basis of chemical reactivity to H_2 and CO, was not considered to be NiO (Roberts and Wells 1966*a,b*).

Schön and Lundin (1972) have investigated oxidized nickel by ESCA. The spectrum of a nickel foil at 298 K, mechanically scratched prior to the investigation, exhibited Ni 2p signals which not surprisingly reflected a mixture of metallic Ni and nickel oxide. The narrow Ni $2p_{\frac{3}{2}}$ peak at low binding energy is due to metallic nickel, whereas the broad peak at 2 eV higher binding energy and the accompanying satellite peak 7 eV apart, are attributed to nickel oxide. The satellite peaks have been observed by Novakov and Prins (1972) and Castle and Epler (1974) in nickel oxide and attributed to shake-up processes. Jørgensen (1971) has interpreted the satellite peaks in nickel oxide as being due to exchange interaction (multiplet splitting). If unpaired electrons are present (i.e. the spin quantum number is non-zero), the binding energy for spin-up and spin-down inner electrons will be different. A Hartree–Fock calculation by Watson and Freeman (1960) gives an energy difference of 1·9 eV between spin-up and spin-down 2p electrons in Ni^{2+} which is appreciably smaller than the observed splitting of 7 eV in NiO. On heating to 573 K *in vacuo* the complex satellite structure, reflecting the oxide, disappears. Schön and Lundin also followed the O(1s) spectrum; at 298 K two peaks are present, separated by 1·8 eV.

The behaviour of these two peaks on heating etc. was investigated, but no clear conclusion emerged, largely because of the possibility that the surface could become contaminated with oxygen species other than those derived from oxidized nickel (e.g. H_2O(ads)).

Recently Norton and Tapping (1975) and Brundle and Carley (1975) have investigated oxygen interaction with nickel between 77 K and 300 K by XPS. Although Brundle and Carley invoke the participation of a nucleation process to explain their results (following Holloway and Hudson (1974)) they conclude that

'interpretation of the XPS results substantially bears out some of the early conclusions from work-function measurements'.

Carley, Joyner, and Roberts (1974) used the large variations in work function known to occur in the $Ni + O_2 + H_2$ system to explore the question of reference levels in XPS. They concluded that it was not appropriate just to add the work function, as is frequently the practice. Hagstrum (1976) has considered the reference-level problem in some detail and has concluded that

'for simple adsorption systems the measurable macroscopic parameter that comes closest to giving the ionization energy of a surface orbital is the binding energy of the orbital with respect to the Fermi level plus the work function of the saturated, uniformly covered surface'.

For non-uniformly covered surfaces (patches) the problem is more complex and the $Ni + O_2 + H_2$ system is one example.

Chromium. Appearance-potential spectra have been reported (Houston and Park 1971) for atomically clean and oxygen-exposed chromium surfaces. Apart from additional structure due to the oxygen K-shell excitation the exposure alters both the positions and amplitudes of the Cr L_3, and L_2 transitions. Fig. 11.12(a) shows the derivative spectra for the clean and oxygen-exposed surfaces with the O (K-shell) excitation clearly visible. The shifts in the thresholds of the L_3 and L_2 transitions on oxidation are shown in Fig. 11.12(b). There is a shift of about $0 \cdot 6$ eV in the L_3 level which is also accompanied by an increase in the width from an apparent $2 \cdot 5$ to about $3 \cdot 3$ eV as oxygen removes electrons from the chromium; the L_3 peak width is related to the unfilled d band. Little information can be deduced from changes in the peaks since there is overlapping of the L_2

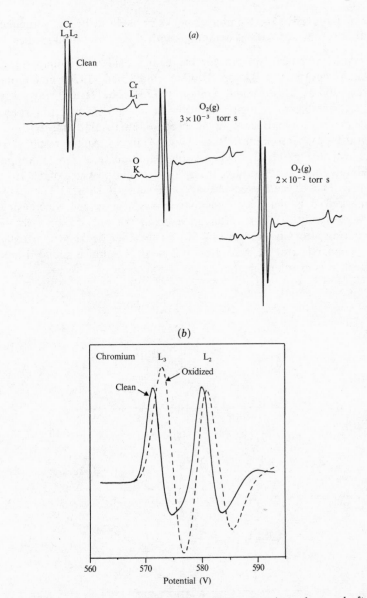

FIG. 11.12. (a) Appearance potential spectra for chromium, clean and after exposure to oxygen at room temperature. (b) Shifts in the L_3 and L_2 levels after oxidation. (From Houston and Park 1971.)

and L_3 levels. Overlapping of peaks is likely to be the principal disadvantage of APS as compared with X-ray excitation studies.

Titanium. Auger spectra for clean and oxidized titanium have been reported by Bishop, Rivière, and Coad (1971) and more recently by Bassett and Gallon (1973). The latter study was performed with a high-resolution hemispherical electron spectrometer, while Bishop *et al.* used a low-resolution retarding-grid-type spectrometer. Bassett and Gallon's Auger results are shown in Fig. 11.13, oxygen being admitted to a clean titanium film at room temperature, up to a maximum pressure of 1×10^{-4} torr. The oxygen exposure at each stage is shown. The peak marked 5 is the most sensitive to surface oxygen, decreasing substantially in height after an exposure of only 5L and disappearing after an exposure of 50L. The other peaks are virtually unchanged after 5L but after 50L peak 7 is much reduced and peak 6 is changing shape.

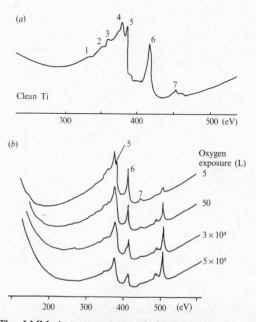

FIG. 11.13. The LMM Auger spectrum of titanium during interaction with oxygen. (*a*) Clean surface. (*b*) After various exposures as indicated. (After Bassett and Gallon 1973.)

TABLE 11.3
Major peaks in the LMM Auger spectrum of clean titanium

Peak no.	Assignment	Energy (eV)	
		Calculated	Observed
1	$L_3M_1M_1$	330	332
2	$L_3M_1M_{2,3}$ ^{1}p ⎫	358	⎧352
3	$L_3M_1M_{2,3}$ ^{3}p ⎭		⎩361
4	$L_3M_{2,3}M_{2,3}$	383	380
5	Possibly a shake-up peak associated with $L_3M_{2,3}M_{4,5}$	—	385
6	$L_3M_{2,3}M_{4,5}$	418	416
7	$L_3M_{4,5}M_{4,5}$	449	448

The assignments of the various peaks are shown in Table 11.3; peak 7, involving two electrons in the conduction band, is reduced as electrons are removed from the band on forming oxide. Peak 6, involving one electron in the conduction band, changes considerably and in the final spectrum resembles the corresponding peak from vacuum-cleaved TiO_2. The behaviour of peak 5 is consistent with either assignment, although the fact that it is more quickly affected by oxidation than either peak 6 or peak 7 suggests that it may be sensitive not only to the *number* of electrons in the conduction band but also to the number of states available at the Fermi surface. Bassett and Gallon suggest that it may reflect a shake-up process, in other words that it is produced by a $L_3M_{2,3}M_{4,5}$ transition together with excitation of a 3p electron into a vacant band state above the Fermi level. The observed energy difference of 31 eV is of the correct magnitude for such a process since the 3p level in Ti lies 34 eV below the Fermi level.

Eastman (1972), using He(I) radiation and what he considered to be hydrogen-contaminated titanium, reported that initial exposure to oxygen resulted in a peak at −5·6 eV and attenuation of the d band emission. He suggests that with further oxygen exposure an insulator forms, probably n-type TiO_2, with oxygen-derived p bands about 5 eV wide with an edge some 3 eV below the Fermi level.

Copper. The Cu+oxygen system has probably been the subject of more investigations than any other. They include the work of Schön (1973) with bulk CuO and (ostensibly) Cu_2O where observations of Cu 2p satellites were reported in X-ray induced spectra. These satellites had also been observed previously by Novakov (1971). Unfortunately in both these studies the chemical stoichiometry of the two oxide surfaces within the escape depth of the photoelectrons was not known. This was partly due to the non-UHV conditions of the spectrometer. Therefore the assignment of satellite structure to the paramagnetic Cu^+ ion could not be regarded as unequivocal. In chemically better-defined systems shake-up satellites had been accepted as reflecting paramagnetic ions, diamagnetic species not generally giving rise to such peaks and this led Castle and Epler (1974), Evans, Evans, Parry, Tricker, Walters and Thomas (1974), and Braithwaite, Joyner, and Roberts (1975) to explore whether Cu^+ was indeed formed during the oxidation of copper. Although quantitative aspects of the satellite structures observed with copper films are as yet unclear, there is little doubt that a *prima facie* case was established that Cu^+ was formed after long exposures to oxygen at 290 K. Studies with single crystals, Cu(100), were more definitive in that the assignment was based on data obtained over a range of temperatures and oxygen pressures using Auger spectroscopy, XPS, LEED, and thermodynamic arguments. More detailed comments on the oxidation of Cu(100) are left until the LEED data are presented in Section 11.6.

Zinc. There has been little work on the oxidation of 'clean' zinc surfaces although zinc was central to much of the earlier work on metal oxidation. There is no doubt from simple volumetric studies (Roberts 1975) that zinc oxidizes to an average thickness of some 1 or 1·2 nm at 77 K and an oxygen pressure of 10^{-6} torr. Briggs (1975), using XPS, has proposed a model invoking a patchwise growth of oxide based on observations of the angular variation of the Zn (2p) and O (1s) signals.

Alkaline-earth metals. Photoemission (u.v.) studies have been reported of Sr, Ca, and Ba. Helms and Spicer (1972) showed with Sr that oxygen states formed at about 5 eV below the Fermi level were very sensitive to oxygen, and that after an exposure of 40 L had a f.w.h.m. value comparable to that expected of bulk SrO.

An interesting point which the authors emphasize is that the emission from the Sr remains unchanged, and they very reasonably interpret this as being due to penetration of the lattice by O^{2-} ions maintaining a metal-rich layer at the surface. Another way of putting this is that the oxide formed is soluble in the metal at room temperature. There are analogies here with the predictions of the Cabrera–Mott theory and the concept of a critical temperature above which the oxide thickness is infinite. The results with Ba are rather similar (Kress and Lapeyre 1972).

11.5.2. *Interaction of water vapour with metals*

There are very few systematic studies of the kinetics of the interaction of water vapour with clean metal surfaces. This is understandable in view of the difficulty of handling water quantitatively at low pressures and temperatures. There are in general two experimental approaches: either the surface is monitored directly as in Suhrmann's (1955) photoelectric work or the gas phase is monitored as in for example Kemball's mass-spectrometric study of the $W + H_2O$ interaction (Imai and Kemball 1968).

Suhrmann and Wedler (Suhrmann 1955) combined measurements of film resistance during adsorption with photoelectric emission studies; the latter enabled the work function to be calculated by the Fowler method. In the case of nickel films they reported an increase in the temperature range 77–90 K. At 193 K and particularly at 273 K the work function increased and there was a further increase in the electrical resistance. The authors interpret the changes as reflecting first molecular adsorption at 77 and 90 K, the positive dipole of the adorbed water molecule leading to a reduction in work function, and then at 273 K dissociation of the adsorbed molecule.

Imai and Kemball (1968) investigated the interaction of water vapour with tungsten in the temperature range 273–573 K by mass-spectrometric analysis of the gas phase paying particular attention to the evolution of hydrogen. At 273 K the interaction was apparently straightforward since it conformed to the stoichiometric relationship

$$W + H_2O_{(g)} = WOH + \tfrac{1}{2}H_{2(g)} \qquad (11.15)$$

470   THE INTERACTION OF OXYGEN

At much higher temperatures (573 K) water vapour reacts with tungsten leading to bulk oxidation of the metal and the formation of hydrogen, while at a lower temperature (473 K) the surface attains a composition intermediate between WO_2H and WO_2.

X-ray photoelectron spectroscopy has more recently provided unambiguous information (Atkinson *et al.* 1974) on the molecular nature of the interaction of water vapour with molybdenum. The crux of the argument is that the O(1s) binding energy of the oxygen atom in molecularly adsorbed water is nearly 3 eV higher than that for dissociated water when chemisorbed oxygen is formed. Fig. 11.14 shows the O(1s) spectrum for a clean molybdenum film which has interacted with water vapour at a pressure of about 10^{-7} torr (*a*) at 295 K and (*b*) after cooling to 77 K followed by further adsorption of water vapour. At 295 K the major peak is at 530 eV, and this is typical of that expected for chemisorbed oxygen on metals, but an incipient peak or shoulder is clear at a binding-energy value of about 533 eV. If this latter peak is associated with molecularly adsorbed water, and there is strong evidence for this from electron-spectroscopic studies of multilayers of water on gold surfaces at 77 K (binding energy, 532·7 eV), then we can conclude that at 295 K dissociation giving chemisorbed oxygen is also accompanied by some molecularly adsorbed water even at a pressure of about 10^{-6} torr. It is estimated that approximately 20 per cent of the monolayer at 295 K involves molecular water. Central to an interesting discussion (Blyholder and Sheets 1975) of the factors that determine whether water is dissociatively or molecularly adsorbed is the

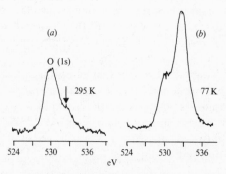

FIG. 11.14. O(1s) spectrum for interaction of water with a molybdenum film at (*a*) 295 K and (*b*) 77 K.

electron density available at the metal atom and how this may be modified by such electron-withdrawing groups as BF_3, a Lewis acid. (The analogy with the model for CO adsorption on metals is clear, see Chapter 9.)

Electron-spectroscopic studies have been reported (Fuggle *et al.* 1975) for a number of metals (Mg, Al, Cr, and Mn) with the claim that it is possible to distinguish hydroxide species from both chemisorbed oxygen and molecularly adsorbed water, while Castle and Epler (1975) have indicated a correlation between XPS analysis and electrochemical history in the aqueous corrosion of aluminium.

11.6. Low-energy electron diffraction

It is probably unfortunate that some of the first metal + gas systems to be investigated by LEED were concerned with oxygen interaction. The development of new LEED patterns during oxygen exposure was attributed to reconstruction, i.e. the premise was accepted that scattering by light atoms such as oxygen did not contribute to the diffraction process. Therefore in order to interpret oxygen-interaction patterns, which differed from the original clean-metal substrate pattern, the movement of metal atoms (reconstruction) had to be invoked. This was not difficult to accept for the $Ni + O_2$ system (Germer and Hartman 1960) but unfortunately the concept was more widely applied to include other adsorbates such as hydrogen, carbon monoxide, and nitrogen. Some time later it was accepted that light atoms could scatter electrons and produce their own patterns. It should be emphasized that spot positions give information on the unit cell structure but provide no information on the composition of the unit cell and this clearly is a serious disadvantage particularly in systems where reconstruction does in fact occur. Combining LEED information with chemical-composition data, say from AES or XPS, offers considerably more scope in surface chemistry. One other note of caution is the considerable evidence now available that LEED can hide a high degree of structural anarchy (McKee, Perry, and Roberts 1973).

May (1969) has recently suggested that in addition to place exchange another possible model for reconstruction is the formation of excited oxide molecules which have high kinetic energy parallel to the surface. This energy may enable metal atoms to move over considerable distances and lead to reconstruction. For example May suggests that a (2×1) reconstructed layer could form when chemisorption at a defect site leads to the formation of two hot molecules which then move away from the site of dissociation before coming to rest. Some caution is, however, needed in interpreting LEED structures, since ordering may be the result of interaction energies in the adsorbed layer and the act of 'coming to rest' is likely to be more complex than merely a matter of losing kinetic energy. Some of the fundamental factors influencing the occurrence of ordered LEED patterns are discussed in Chapter 3.

Tungsten. Chang and Germer (1967) and Tracy and Blakely (1969 a,b) studied the interaction of oxygen with the W(211) plane; one reason for interest in this particular system is its similarity with the Ni(110)+O$_2$ system which Germer and his colleagues had studied previously. Fig. 11.15 shows the sequence of LEED patterns observed at room temperature and also a marble model of the (211) plane. It should be noted that the close-packed rows of top-layer atoms are separated by open channels or 'troughs'.

There first appears a p(2×1) pattern which subsequently fades away to give a p(1×2) structure; the p(1×2) pattern remains unaltered up to an oxygen exposure of 9L. The p(2×1) pattern, with new fractional order spots $h + \frac{1}{2}$, k, develops gradually; when the intensity of the fractional-order spots is at a maximum the surface is covered by a structure that is doubly spaced in the direction along the close-packed rows.

An interesting feature during the development of the p(2×1) pattern is the streaking of the $h + \frac{1}{2}$, k diffraction spots in a direction which is normal to the densely packed rows of atoms. With increasing oxygen exposure these streaks coalesce into sharp spots. At low oxygen coverage Chang and Germer, and Tracy and Blakely, interpret these streaks as islands of the double-spaced structure, long in the direction of the rows but narrow at right angles to the rows. This would be in keeping with high mobility and therefore growth along the rows. It has been shown however (McKee, Perry, and Roberts 1973) that this is an oversimplified picture and that the actual structure is rather more

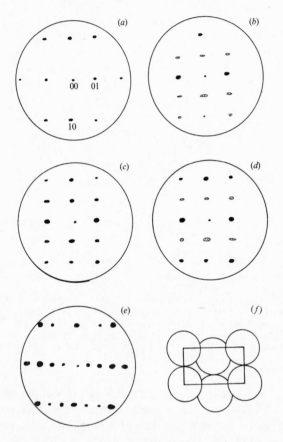

FIG. 11.15. W(211)+O$_2$ LEED patterns at 86 eV. (a) clean surface (b) oxygen exposure 0·5 L (c) 1 L (d) 1·5 L (e) 9 L. The unit mesh of the clean surface is shown in (f). (After Chang and Germer 1967.)

complicated (see Section 3.7.2). With continued exposure the $h+\frac{1}{2}, k$ spots gradually fade away with the appearance of $h, k+\frac{1}{2}$ diffraction spots which become sharper and more intense with exposure. At an exposure of 9 langmuir (Fig. 11.15(e)) the pattern is well developed and is a p(1×2) structure. This is unchanged on further exposure at room temperature. Atomistic models for the formation of this sequence of structures have been proposed by McKee, Perry, and Roberts (1973) and by Ertl and Plancher (1975).

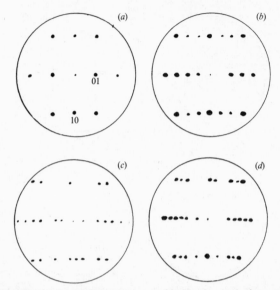

FIG. 11.16. W(211)+O_2 LEED patterns at 92 eV, observed after exposure of the clean surface to ~100 L of oxygen and momentarily heating in vacuum to the following temperatures: (a) p(1×1) pattern, 2000 K (b) p(1×2), 1760 K (c) p(1×3), 1650 K and (d) p(1×4), 1810 K. There is about one monolayer of oxygen in the p(1×1) structure, and more in the others. (After Chang and Germer 1967.)

If the p(1×2) pattern is heated, p(1×3), p(1×4) patterns and a p(1×1) which is not the original clean surface since oxygen is present, can be formed (Fig. 11.16). Chang and Germer do not attempt to interpret these patterns except to draw attention to their similarity to faceted surfaces produced when the (211) crystal is heated to between 1300 and 1700 K during exposure to oxygen. The p(1×n) patterns with $n \geqslant 3$ are distinctly different to the p(1×n) observed on heating surfaces with adsorbed oxygen *in vacuo* and Chang and Germer attribute them to (110) facets.

Riwan, Guillot, and Paigne (1975) have reported recently a LEED study of oxygen adsorption on Mo(100) while Kennett and Lee (1975) have made a thorough investigation of faceting during molybdenum oxidation.

Iron. LEED data have been reported by a number of investigators for the Fe(100)+O_2 system. The early results are summarized in Table 11.4 taken from the paper by Melmed and

Carroll (1973). It can be seen that there is agreement that eventually a (111) FeO layer is formed, but there is some disagreement on oxygen exposures necessary for obtaining inter-mediate patterns and in some cases intermediate patterns are not reported. Clearly different exposures for observing a particular pattern can lead to different assessments of surface oxygen cover-age and therefore to different interpretations of the pattern. The aim of Melmed and Carroll's work was to eliminate, or at least minimize, the area of conflict, and to achieve this they added ellipsometry and work-function facilities to their LEED equip-ment; it is worth noting, however, that they had no electron-spectroscopic facilities for determining surface composition and their substrate was an oriented evaporated film.

Melmed and Carroll concluded that (i) a $c(2 \times 2)$ pattern was first formed, (ii) this was usually followed by a $p(2 \times 1)$ but sometimes a $c(3 \times 1)$ pattern, (iii) $c(1 \times 1)$ Fe and (111) 'FeO' patterns coexisted, and (iv) finally only the (111) 'FeO' pattern was observed. There are, however, notes of caution in their work regarding the sensitivity of some diffraction structures to the electron beam, for example the $p(2 \times 1)$ structure faded away in the presence of the beam. Moving the beam to another part of

TABLE 11.4

Summary of LEED results for the system $Fe(100) + O_2$

Surface structure (surface-mesh nomenclature and/ or authors' interpretation)	O_2 exposure necessary for LEED pattern visibility[†]			O_2 exposure necessary for LEED pattern visibility[‡]		
	Initial (L)	Maximum (L)	Final (L)	Initial (L)	Maximum (L)	Final (L)
(1) $c(2 \times 2)$ $\frac{1}{4}$ monolayer	(1) 0·2	1·7	—	(1) 0·5	1	>2
(2) $c(2 \times 2) + c(3 \times 1)$	(2) not reported			(2) 2		
(3) $c(3 \times 1)$	(3) 40	60 ($\frac{2}{3}$ monolayer)	150	(3) 2	3–5 ($\frac{1}{3}$ monolayer)	
(4) $c(3 \times 1)$ splitting	(4) not reported			(4) after $c(3 \times 1)$		8–10
(5) $c(1 \times 1)$	(5) 150			(5) 10		80
(6) $c(1 \times 1) + (111)$ 'FeO'	(6) not reported			(6) 80		
(7) (111) 'FeO' only	(7) 150–300			(7) >80		

From Melmed and Carroll 1973.

† Pignocco and Pellissier 1967.

‡ Molière and Portele 1969; Portele 1969.

the surface restored the pattern. The authors conclude that the p(2×1) pattern (and the c(3×1) when present) is likely to be due to 'weakly bound adsorbate (oxygen) on top of the first monolayer'. It is interesting that a consistent way of inducing the c(3×1) structure to form was to anneal the p(2×1) at about 800 K *in vacuo* ($<10^{-9}$ torr) to give the c(2×2) and then expose this structure to oxygen at room temperature. It would appear that the LEED patterns do not reveal the whole story regarding the detailed structure of the surface since two similar LEED patterns (the p(2×1) in this case) behave differently to further oxygen. Clearly the ability of the p(2×1) structure to chemisorb oxygen is dependent on the detailed atomic nature of the surface. Since it is now well recognized that substantial surface disorder may well be undetected by LEED, patterns alone are insufficient to deduce the details of structure. It may well be that the insensitivity of LEED to structural imperfections could lead to over-facile interpretations of surface reactivity. In this context it is worth recalling that the work-function changes observed (Quinn and Roberts 1963, 1964) when oxygen reacted with metal surfaces were very dependent on the details (e.g. substrate temperature, oxygen pressure) of the interaction.

Other LEED studies of iron have been reported by Kobayashi and Kato (1969), Kato and Kobayashi (1971), and Ueda and Shimizu (1974), while Ertl and Wandelt (1975) described electron-spectroscopic results of high-temperature oxidation. Tricker, Thomas, and Winterbottom (1974) indicate the possible potential of Mössbauer spectroscopy.

Copper. There have been a number of investigations by LEED of the interaction of oxygen with single crystals of copper. Many of them have emphasized high-temperature interaction but little attention has been paid to the region of temperature between 77 K and 290 K. The earlier studies, summarized recently (Braithwaite *et al.* 1975), did not have the advantage of electron spectroscopy (AES, XPS, or UPS) and therefore we confine ourselves to the most recent investigation. These results with Cu(100) (Braithwaite *et al.* 1975) show many of the features of earlier studies and are summarized in the two schemes shown (Figs. 11.17 and 11.18). Information is given on the LEED structures observed and the associated O(1s) data (peak position, full

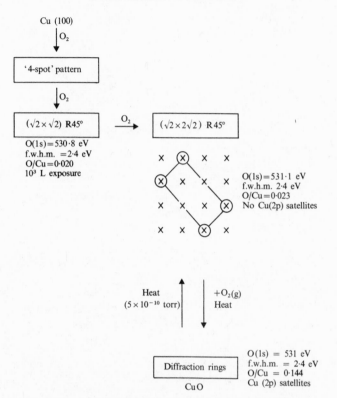

FIG. 11.17. Cu(100) + O₂ (290 K). LEED structure and XPS data.

width at half-maximum height (f.w.h.m.), the O(1s)/Cu (2p) ratio, and whether these are associated with the Cu (2p) peak shake-off satellites).

From the totality of data and supporting thermodynamic argument it is concluded that following the formation of the $(\sqrt{2} \times \sqrt{2})$ R45° pattern at 290 K, which is interpreted as reflecting the chemisorption of oxygen in the positions of four-fold co-ordination, incorporation occurs, without change of pattern, to form Cu_2O as reflected by XPS. This structure can be converted by heating in oxygen at 600 K to CuO, as indicated by appropriate diffraction rings and Cu (2p) satellites characteristic of Cu^{2+}. The $(\sqrt{2} \times 2\sqrt{2})$ R45° pattern could be reformed by heating *in vacuo*, consistent with the thermodynamics of the decomposition reaction $2CuO \rightarrow Cu_2O + \frac{1}{2}O_2$.

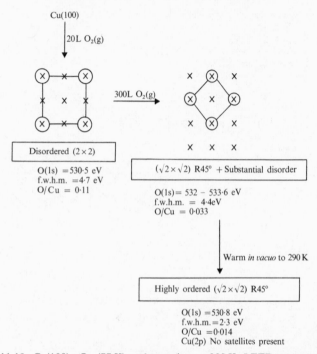

Fig. 11.18. Cu(100)+O$_2$ (77 K) and warming to 290 K. LEED structures and XPS data.

At 77 K a disordered (2 × 2) pattern forms first (Fig. 11.18) but at higher exposure to oxygen transforms to the ($\sqrt{2} \times \sqrt{2}$) R45° structure. The O(1s) data indicate a range of metal–oxygen bond strengths in that the f.w.h.m. value was large, 4·4 eV, and two distinct O(1s) peaks were evident, one at about 531 eV and the other at nearly 534 eV. The higher-binding-energy peak disappeared as a result of oxygen desorption on warming from 77 to 290 K.

Ultra-violet and X-ray-induced spectroscopic studies have been reported for oxygen interaction with polycrystalline copper by Evans *et al.* (1974).

11.7. Conclusions

The oxidation of metals is dependent on a number of parameters encompassing chemical, physical, and structural properties of both the metal and the oxide formed. The various stages in oxide growth have been isolated for discussion, as also have the different approaches of interpretation. Emphasis has been given to the early stages of oxygen interaction in keeping with other aspects discussed in this book. This should not be taken to mean that such areas as the structure of thick oxide films, the understanding of breakaway phenomena, and the role of alloying elements in controlling oxidation are of any less significance. The reader will find the following references useful for a wider view of oxidation than presented here: Evans (1955), Lawless (1974), Hauffe (1965), and Kubaschewski and Hopkins (1962). The detailed mechanism of oxygen chemisorption, incorporation, and nucleation leading to oxide growth is not completely understood but there is no question that progress has been dramatic over the last decade. Development of ideas based on the classical experimental approaches of the 1950's has been possible largely through LEED and electron spectroscopy (XPS, UPS, and AES). One such example is the $Cu(100)$+oxygen system (Braithwaite *et al.* 1975). There has been considerable interest in the relative insensitivity of the $O(1s)$ binding energy for oxygen in different environments, e.g. chemisorbed oxygen compared with oxide, and this has stimulated theoretical calculations. We have not considered the silver–oxygen system, but a recent study (Engelhardt and Menzel 1976) gives a good critical summary of previous data and also new LEED and electron-spectroscopic results.

REFERENCES

ATKINSON, S. J., BRUNDLE, C. R., and ROBERTS, M.W. (1974). *Disc. Faraday Soc.* **58.** 62.

BASSETT, P. J., and GALLON, T. E. (1973). *J. electron Spectrosc.* **2,** 101.

BÉNARD, J., GRØNLUND, F., OUDAR, J., and DURET, M. (1959). *Z. Elektrochem.* **63,** 799.

BISHOP, H. E., RIVIÈRE, J. C., and COAD, J. P. (1971). *Surf. Sci.* **24,** 1.

BLOOMER, R. N. (1957). *Brit. J. appl. Phys.* **8,** 321.

BLYHOLDER, G., and SHEETS, R. W. (1975). *J. Catalysis* **39,** 152.

BOND, G. C. (1974). *Heterogeneous catalysis: principles and applications.* Clarendon Press, Oxford.

BOGGIO, J. E. (1972). *J. Chem. Phys.* **57,** 4738.

BRADSHAW, F. J., BRANDON R. H. and WHEELER, C. (1964). *Acta. Met.* **12,** 1057.

BRAITHWAITE, M. J., JOYNER, R. W., and ROBERTS, M. W. (1975). *Disc. Faraday, Soc.* **60,** 89.

BRENNAN, D., and GRAHAM, M. J. (1966). *Disc. Faraday Soc.* **41**, 95.

BRENNAN, D., HAYWARD, D. O., and TRAPNELL, B. M. W. (1960). *Proc. R. Soc. A* **256**, 81.

BRIGGS, D. (1975). *Disc. Faraday Soc.* **60**, 81.

BRUNDLE, C. R., and CARLEY, A. F. (1975). *Chem. Phys. Lett.* **31**, 423.

CABRERA, N., and MOTT, N. F. (1948). *Rep. Prog. Phys.* **12**, 163.

CARLEY, A. F., JOYNER, R. W., and ROBERTS, M. W. (1974). *Chem. Phys. Lett.* **27**, 580.

CASTLE, J. E., and EPLER, D. C. (1974). *Proc. R. Soc. A* **339**, 49.

—— (1975). *Surf. Sci.* **53**, 286.

CHANG, C. C., and GERMER, L. H. (1967). *Surf. Sci.* **8**, 115.

CROSSLAND, W. A., and ROETTGERS, H. T. (1966). *Phys. Failure Electron.* **5**, 158.

CZYZEWSKI, J. J., MADEY, T. E., and YATES, J. T. (1974). *Phys. Rev. Lett.* **33**, 777.

DAVIES, J. A., DOMEIJ, B., PRINGLE, J. P. S., and BROWN, F. (1965). *J. Electrochem. Soc.* **112**, 675.

DELCHAR, T. A., and TOMPKINS, F. C. (1967). *Proc R. Soc. A* **300**, 141.

EASTMAN, D. E. (1972). *Solid St. Commun.* **10**, 933.

ELEY, D. D., and WILKINSON, P. R. (1960). *Proc. R. Soc. A* **254**, 327.

ENGELHARDT, H. A., and MENZEL, D. (1976). *Surf. Sci.* **57**, 591.

ERTL, G., and PLANCHER, M. (1975). *Surf. Sci.* **48**, 364.

ERTL, G. and WANDELT, G. (1975). *Surf. Sci.* **50**, 479.

EVANS, S., EVANS, E. L., PARRY, D. E., TRICKER, M. J., WALTERS, M. J., and THOMAS, J. M. (1974). *Disc. Faraday Soc.* **58**, 97.

EVANS, U. R. (1945). *Trans. Faraday Soc.* **41**, 365.

—— (1955). *Rev. pure appl. Chem. R. Aust. chem. Inst.* **5**, (1), 1.

FEHLNER, F. P., and MOTT, N. F. (1970). *Oxid. Metals* **2**, 59.

FROMHOLD, A. T., and COOK, E. L. (1966). *J. chem. Phys.* **44**, 4564.

—— (1967). *Phys. Rev.* **158**, 600.

FUGGLE, J. C., WATSON, L. M., FABIAN, D. J., and AFFROSSMAN, S. (1975). *Surf. Sci.* **49**, 61.

GERMER, L. H., and HARTMAN, C. D. (1960). *J. appl. Phys.* **31**, 2085.

GHEZ, R. (1972). *J. chem. Phys.* **58**, 1838.

GRAHAM, M. J., and COHEN, M. (1972). *J. electrochem. Soc.* **119**, 879.

GRAHAM, M. J., SPROULE, G. I., CAPLAN, D., and COHEN, M. (1972). *J. electrochem. Soc.* **119**, 883.

GRIMLEY, T. B. (1955). In *Chemistry of the Solid State*, ed W. E. Garner. Butterworth, London.

GRÖNLUND, F. (1956). *J. Chim. Phys.* **53**, 60.

HAGSTRUM, H. D. (1976). *Surf. Sci.* **54**, 197.

HAUFFE, K. (1965). *Oxidation of metals.* Plenum Press, New York.

HAYWARD, D. O. (1971). In *Chemisorption and reactions on metallic films* (ed. J. R. Anderson). Academic Press, London, New York.

HELMS, C. R., and SPICER, W. E. (1972). *Phys. Rev. Lett.* **28**, 565; *Appl. Phys. Lett.* **21**, 237.

HOLLOWAY, P. H., and HUDSON, J. B. (1974). *Surf. Sci.* **43**, 123.

HORGAN, A. M., and KING, D. A. (1969). In *The structure and chemistry of solid surfaces* (ed. G. A. Somorjai). John Wiley, New York.
—— (1970). *Surf. Sci.* **23**, 259.
HOUSTON, J. E., and PARK, R. L. (1971). *J. chem. Phys.* **55**, 4601.
HUNT, G. L., and RITCHIE, I. M. (1972*a*). *J. chem. Soc. Faraday Trans. I* **68**, 1413.
—— (1972*b*) *Surf. Sci.* **30**, 475.
IMAI, H., and KEMBALL, C. (1968). *Proc. R. Soc. A* **302**, 399.
JARDINIER-OFFERGELD, M., and BOUILLON, F. (1972). *J. vac. Sci. Technol.* **9**, 770.
JØRGENSEN, C. K. (1971). *Chimia*, **25**, 213.
KATO, S., and KOBAYASHI, H. (1971). *Surf. Sci.* **27**, 625.
KENNETT, H. M., and LEE, A. E. (1975). *Surf. Sci.* **48**, 591.
KING, D. A., MADEY, T. E., and YATES, J. T. (1971). *J. chem. Phys.* **55**, 3236, 3247.
KIRBY, R. E., McKEE, C. S., and ROBERTS, M. W. (1976). *Surf. Sci.* **55**, 725.
KIRK, C. T., and HUBER, E. E. (1968). *Surf. Sci.* **9**, 217.
KLEMPERER, D. F., and STONE, F. S. (1957). *Proc. R. Soc. A* **243**, 375.
KOBAYASHI, H., and KATO, S. (1969). *Surf. Sci.* **18**, 341.
KRESS, K. A., and LAPEYRE, G. J. (1972). *Phys. Rev. Lett.* **28**, 1639.
KRUGER, J., and YOLKEN, H. T. (1964). *Corrosion*, **20**, (1), 29.
KUBASCHEWSKI, O., and HOPKINS, B. E. (1962). *Oxidation of metals and alloys.* Butterworth, London.
LANYON, M. A. H., and TRAPNELL, B. M. W. (1955). *Proc. R. Soc. A* **227**, 387.
LAWLESS, K. R. (1974). *Rep. Prog. Phys.* **37**, 231.
McCARROLL, B. (1967). *Surf. Sci.* **7**, 499.
McKEE, C. S., PERRY, D., and ROBERTS, M. W. (1973). *Surf. Sci.* **39**, 176.
MAY, J. W. (1969). *Surf. Sci.* **18**, 431.
MELMED, A. J., and CARROLL, J. J. (1973). *J. vac. Sci. Technol.* **10**, 164.
MITCHELL, D. F., and LAWLESS, K. R. (1966). *J. paint Technol.* **38**, 575.
MOLIÈRE, K., and PORTELE, F. (1969). In *The structure and chemistry of solid surfaces* (ed. G. A. Somorjai). John Wiley, New York.
MUSKET, R. G. (1970). *J. less-common Metals* **22**, 175.
NORTON, P. R. and TAPPING, R. L. (1975). *Disc. Faraday Soc.* **60**, 71.
NOVAKOV, T. (1971). *Phys. Rev.* **B3**, 2693.
NOVAKOV, T., and PRINS, R. A. (1972). In *Electron Spectroscopy* (ed. D. A. Shirley). North-Holland, Amsterdam.
PIGNOCCO, A. J., and PELLISSIER, G. E. (1967). *Surf. Sci.* **7**, 261.
PORTELE, F. (1969). *Z. Naturforsch. A* **24**, 1268.
PTUSHINSKII, YU. G., and CHUIKOV, B. A. (1967). *Surf. Sci.* **7**, 507.
QUINN, C. M., and ROBERTS, M. W. (1963). *Nature, Lond.* **200**, 648.
—— (1964). *Trans. Faraday Soc.* **60**, 899.
—— (1965). *Trans. Faraday Soc.* **61**, 1775.
RHEAD, G. E. (1965). *Trans. Faraday Soc.* **61**, 797.
RHODIN, T. N. (1959). *Structure and properties of thin films* (ed. C. A.

Neugebauer, J. B. Newkirk, and D. A. Vermilyea). John Wiley, New York.

RITCHIE, I. M., and TANDON, R. K. (1972). *Surf. Sci.* **22,** 199.

RIVIÈRE, J. C. (1964). *Brit. J. appl. Phys.* **15,** 1341.

RIWAN, R., GUILLOT, C., and PAIGNE, J. (1975). *Surf. Sci.* **47,** 183.

ROBERTS, M. W. (1960). *Nature, Lond.* **188,** 1020.

—— (1962). *Q. Rev.* **16,** (1), 71.

—— (1972). In *Specialist periodical reports: surface and defect properties of solids* (eds M. W. Roberts and J. M. Thomas), vol. 1. Chemical Society, London.

—— (1975). *Disc. Faraday Soc.* **60,** 157.

ROBERTS, M. W., and ROSS, J. R. H. (1966). *Trans. Faraday Soc.* **62,** 2301.

ROBERTS, M. W., and TOMPKINS, F. C. (1959). *Proc. R. Soc. A* **251,** 369.

ROBERTS, M. W., and WELLS, B. R. (1966a). *Disc. Faraday Soc.* **41,** 162.

—— (1966b). *Trans. Faraday Soc.* **62,** 1608.

—— (1967). *Surf. Sci.* **8,** 453.

—— (1969). *Surf. Sci.* **15,** 235.

SALEH, J. M., WELLS, B. R., and ROBERTS, M. W. (1964). *Trans. Faraday Soc.* **60,** 1865.

SCHÖN, G. (1973). *Surf. Sci.* **35,** 96.

SCHÖN, G. and LUNDIN, S. T. (1972). *J. electron Spectrosc.* **1,** 105.

SMITH, G. D. W., and ANDERSON, J. S. (1976). *J. chem. Soc. Faraday Trans. I* **72,** 1231.

SUHRMANN, R. (1955). *Adv. Catalysis,* **7,** 303. Academic Press, New York, London.

TRACY, J. C., and BLAKELY, J. M. (1969a). In *The structure and chemistry of solid surfaces* (ed. G. A. Samorjai). John Wiley, New York. Paper 65.

—— (1969b). *Surf. Sci.* **15,** 257.

TRICKER, M. J., THOMAS, J. M., and WINTERBOTTOM, A. P. (1974). *Surf. Sci.* **45,** 601.

UEDA, K., and SHIMIZU, R. (1974). *Surf. Sci.* **43,** 77.

UHLIG, H. H. (1967). *Corrosion Sci.* **7,** 325.

WATSON, R. E., and FREEMAN, A. J. (1960). *Phys. Rev.* **120,** 1125.

YATES, J. T., MADEY, T. E., and ERICKSON, N. E. (1974). *Surf. Sci.* **43,** 257.

ZACHARIASEN, W. H. (1932). *J. Am. chem. Soc.* **54,** 3841.

12

METAL ADSORPTION ON METALS

12.1. Introduction

IN 1933 Taylor and Langmuir described the interaction of caesium with polycrystalline tungsten in a paper which remains as one of the landmarks in surface chemistry, and leading on from that work metallic adsorption has been extensively studied during the intervening years. The reasons for this interest are many, but they fall into three general categories.

(i) Metal atoms, particularly the alkalis, represent one of the simplest types of adsorbate. Therefore they should offer reasonable prospects for theoretical treatment.

(ii) The systems are experimentally convenient; for example, transition-metal adsorbates can be directly imaged in the field ion microscope, while adsorption of alkali metals is accompanied by large work-function decreases which make them ideal candidates for study by low-energy electron emission techniques.

(iii) Metallic adsorption is of importance in relation to technological operations such as fabrication of thin-film devices. Also, alkali and mixed alkali–oxygen adsorbates find applications in thermionic and photoelectric cathodes and in energy conversion and ion-propulsion systems (see, for example, Wilson 1967, Sommer 1968, Chen and Papageorgopoulos 1971).

In this chapter we shall first consider some aspects of metal-surface diffusion and nucleation processes and then discuss electronic properties, adsorption energies, and surface geometry. This leads naturally to a description of quantum-mechanical theories of metallic adsorption. Finally, although the subject is not encompassed by the title of the chapter, it would seem to be a further natural progression to consider current theories of non-metallic adsorption.

12.2. Surface diffusion and nucleation

12.2.1. *Single-atom mobilities*

A considerable amount of information on surface diffusion has been accumulated by the study of macroscopic mass-transfer processes such as grain-boundary grooving and thermal faceting, but the field ion microscope alone offers the possibility of direct observation of diffusion at the atomic level. As we have seen in Chapter 6 only metal adatoms can be successfully imaged at present and so the systems investigated have been mainly restricted to transition metals on refractory substrates. In general the technique involves evaporation of a small number of adatoms onto an FIM tip held at low temperature (20–80 K). Helium is then introduced, the field applied, and the image photographed. Next the temperature is raised to a predetermined value for some fixed period of the order of a few minutes in the absence of the field and changes in position of adatoms at this higher temperature are recorded in another image upon re-cooling to 20–80 K.

As a result of an ideal diffusion experiment one should be able to define precisely not only parameters such as the overall rate and the activation energy of the process and their variation with crystal plane, but also the co-ordinates of the diffusing atom and its substrate neighbours before and after each diffusive hop. The latter information is difficult to obtain from FIM micrographs for a number of reasons including the low visibility of substrate atoms in the central regions of low-index crystal planes, local enhancement of magnification at adatoms, and possible distortion of adatoms in the imaging field. Information of this type has been obtained, however, for a single hop of a tungsten adatom on a W(111) plane by Graham and Ehrlich (1973, 1974). On this plane there are two likely sites for an adatom, both of trigonal symmetry (Fig. 12.1); the 'lattice' site (site 3) would be the one occupied in adding a further layer to the b.c.c. lattice, while the 'fault' site (site 2) would be occupied in a layer which constituted a stacking fault in the lattice. In a total of 175 observations it was found that following deposition, adatoms never occurred at positions on the surface other than lattice or fault sites; in fact the latter were occupied in 10 cases only. Moreover, heating to 480 K had no effect on 'lattice' adatoms, but caused 'fault' adatoms to move to lattice sites. Thus the lattice site represents

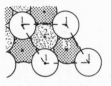

FIG. 12.1. Atomic arrangement of (111) plane of tungsten: layer1, surface layer;
layer $2 = -a/(2\sqrt{3})$; layer 3, $-a/\sqrt{3}$ (a, lattice spacing).

the most stable adatom position, although the barrier to migration into it from a fault site is high, $\sim97\ kJ\ mol^{-1}$.

Diffusion of adatoms from lattice sites was first detected at
$\sim600\ K$. At this temperature observations were complicated by
movement of impurities from the shank of the specimen onto the
tip, but nevertheless it was established that an adatom at a lattice
site always jumped to another adjacent *lattice* site, the activation-
energy barrier being of the order of $17\cdot4\ kJ\ mol^{-1}$. Such a jump is
depicted in Plate 6, which must represent one of the most
detailed observations yet obtained of the most elementary of
kinetic events.

The FIM technique was first exploited for observation of
surface diffusion by Ehrlich and Hudda (1966) in an investigation
of tungsten adsorption on tungsten. Suppose the potential between a migrating atom and any substrate surface atom is a
function only of their separation and is independent of the presence of neighbouring surface atoms, that is, total adatom–surface
interaction is pairwise additive. Then surface diffusion should
occur most readily over the close-packed, atomically smooth
planes, since on these the diffusing atom is in contact with the
minimum number of substrate atoms. Even a qualitative examination of the field ion micrographs reveals that, at least for W on
W, this is not the case. After heating to 288 K no movement of
adatoms occurs on the (110) plane, the most densely packed
plane of the b.c.c. lattice, nor on (310), but movement does occur
on the rougher (211) plane. On further heating to 319 K migration takes place on (110), but on (310) and also on (111) the
adatoms still remain immobile. Again, from similarity of atomic
structures, behaviour on the (321) face might be expected to
parallel that on (211), but in fact diffusion on (321) is more
closely related to that on (110). The sequence of mobilities is

therefore

$$(211) > (321) \approx (110) > (111) \approx (310)$$

and this is confirmed when the activation energies for surface migration E_m are determined from the temperature dependence of the diffusion rates. If the diffusive jumps are uncorrelated and f is the jump frequency, the mean square displacement of an atom in time t will be (see Section 2.10)

$$\langle x^2 \rangle = ft \langle l^2 \rangle \tag{12.1}$$

where l^2 is the mean-square distance covered in a single jump. We can also write

$$f \approx \frac{kT}{h} \exp\left(\frac{\Delta S}{R}\right) \exp\left(-\frac{E_m}{kT}\right). \tag{12.2}$$

Then, if the temperature dependence of l can be assumed to be small, E_m may be determined from a plot of $\log\langle x^2 \rangle$ against $1/T$.

The activation energies for atom migration are found to be 86, 83, and 53 kJ mol^{-1} for movement over the planes (110), (321), and (211) respectively (Table 12.1). In the latter two cases the values refer to diffusion *along* the surface troughs; movement at 90° to this, over the surface steps, is considerably slower, the mean-square displacement being lower by a factor of 10 or more. Further data, obtained by Ehrlich and by Bassett and co-workers, are collected in Table 12.1. The sequence of E_m values confirms

TABLE 12.1

Surface self-diffusion parameters for tungsten and rhodium

System	D_0[†] (m^2 s^{-1})	E_m[†] (kJ mol^{-1})	$E_{m\,calc}$[‡] (kJ mol^{-1})
W/W(110)	$2 \cdot 6 \times 10^{-7}$	86	52
W/W(211)	$3 \cdot 0 \times 10^{-12}$	53	42
W/W(321)	$3 \cdot 7 \times 10^{-8}$	83	—
Rh/Rh(111)	$2 \cdot 0 \times 10^{-8}$	15	—
Rh/Rh(110)	$3 \cdot 0 \times 10^{-5}$	58	—
Rh/Rh(331)	$1 \cdot 0 \times 10^{-6}$	62	—

[†] W data from Ehrlich and Hudda (1966) and Bassett and Parsley (1970). Rh data from Ayrault and Ehrlich (1974).
[‡] Wynblatt and Gjostein (1970).

the earlier qualitative observations on ease of diffusion on various planes, but presents the problem of explaining why this sequence differs from that predicted (Drechsler 1954) by the pairwise interaction model, namely

$$E_m(111) > (310) > (321) \approx (211) > (110).$$

The main discrepancies are that the theoretical E_m value on (211) is too high and that on (110) too low relative to that on (321).

A number of factors which are not taken into account in the simple pairwise interaction model, but which may influence the value of E_m, have been analysed by various authors. Ehrlich and Kirk (1968), recalling that diffusion in the bulk involves the co-ordinated motion of the lattice as a whole, considered the effects on surface diffusion of distortions of the lattice around the diffusing adatom. In a purely qualitative way it can be seen that for an adatom moving along a trough in a W(211) surface the activation energy E_m would be considerably reduced if the two protruding rows of surface atoms with which the adatom is in contact simultaneously moved outwards. In contrast, on W(321) only one side of the diffusion channel could 'open up' in this way since the second row of boundary atoms is embedded in the (110)-type structure of a terrace (Fig. 12.2) and is not free to undergo a large displacement. This effect may therefore contribute to the relatively low observed value of E_m on (211) compared with (321). It has been investigated in more detail by Wynblatt and Gjostein (1970). These authors calculated the interaction energy $\phi(x_{ij})$ of a hypothetical isolated pair of atoms i and j at separation x_{ij} using a modified Morse function which had been found (Girifalco and Weizer 1959) to lead to theoretical vacancy migration energies in a number of cubic metals which were in good agreement with experiment. If x_0 is the equilibrium

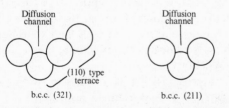

FIG. 12.2. Diffusion channels on b.c.c. (321) and (211) surfaces.

separation of the atom pair and α is a constant then

$$\phi(x_{ij}) = E_{ij}[\exp\{-2\alpha(x_{ij} - x_0)\} - 2\exp\{-\alpha(x_{ij} - x_0)\}]. \quad (12.3)$$

Putting $x_{ij} = x_0$ gives $\phi_0 = -E_{ij}$ showing that E_{ij} is the dissociation energy of the pair. The energy of the ith atom in a crystal is given by the pairwise sum of $\phi(x_{ij})$ over all other atoms, i.e.

$$\Phi_i = \sum_{j \neq i} \phi(x_{ij}). \quad (12.4)$$

The co-ordinates of an atom in a relaxed state are then determined by minimizing Φ_i, and the method is applied successively to each atom in the vicinity of an adatom (out to sixth nearest neighbours) until the energy of the configuration changes by less than some predetermined value between iterations. This procedure is carried out with an adatom in its equilibrium position on the surface and also at the saddle point in the diffusion path; hence E_m is determined. The results of the calculation for W migration on W(110) and (211) are shown in Table 12.1. It is seen that although quantitative agreement with experiment is not particularly good, the order of the values is at least correct.

Lattice fluctuations may also lead to unusually low values of the pre-exponential factor D_0 for surface diffusion (Ehrlich and Kirk 1968). If D is the diffusion coefficient we may write (cf. eqns (12.1) and (12.2), and Chapter 2)

$$D = Pf\langle l^2 \rangle$$

$$\simeq PA\langle l^2 \rangle \exp\left(-\frac{E_m}{kT}\right)$$

$$= D_0 \exp\left(-\frac{E_m}{kT}\right) \quad (12.5)$$

where P is the probability that the diffusion channel expands just as the diffusing atom approaches and $A = (kT/h)\exp(\Delta S/R)$. For an expansion of $\sim 0{\cdot}02$ nm it is found that $P \simeq 3 \times 10^{-3}$ at 300 K, which is in reasonable agreement with the experimental estimate that D_0 is about four orders of magnitude lower than normal for diffusion on W(211).

While the majority of FIM data relate to surface self-diffusion on b.c.c. tungsten, it is important to note that for f.c.c. rhodium the results (Table 12.1) are completely different (Ayrault and

Ehrlich 1974). On the most closely packed plane, Rh(111), diffusion is extremely rapid compared with the relatively low rate on W(110). Also, the considerable difference in rate between W(211) and W(321) is not observed in the case of the comparable rhodium planes, (110) and (331). Rhodium, in fact, conforms with the intuitive expectation that the diffusion rate should be directly related to the atom density of a given plane and this suggests that the tungsten results may be anomalous, a suspicion which is enhanced by a very significant observation made with rhodium. On the (110) plane the motions of pairs of Rh adatoms were found to be correlated; adatoms in adjacent channels in the surface (closest approach 0·38 nm) tended to remain near each other during diffusion, rather than to move independently. This also occurred even when the atoms were separated by one surface channel (closest approach 0·76 nm) and it resulted in a *decrease* in both the activation energy and pre-exponential factor for diffusion, compared with the case of movement of a single adatom. Rapid diffusion on W(211) may therefore be related to the presence of other W adatoms on the plane rather than to any feature of the tungsten lattice. We might also note that correlated movement provides direct evidence for adatom–adatom interactions, theoretical aspects of which are discussed in Section 12.5.

Up to this point we have discussed the diffusion of atoms over perfect crystal planes, but the behaviour of adatoms at surface steps is also of importance since movement over steps must occur during crystal growth, faceting, and other mass-transport processes. Field ion microscope evidence indicates that in many cases migrating atoms are efficiently reflected at the edges of planes, the process being particularly effective on W(110) where the probability of escape over the plane edge is $\sim 1 \times 10^{-4}$. It was originally suggested (Ehrlich and Hudda 1966) that at this position the energy barrier to diffusion is higher than in the middle of a plane. A difference of no more than 5–20 kJ mol^{-1} would be required to account for the observed reflections of transition metals on tungsten at temperatures up to ~ 400 K (Bassett and Parsley 1970, Tsong 1972); Bassett and co-workers (Bassett 1973), however, have examined diffusional loss of a number of adsorbates from W(110) at various temperatures and in every case have found that the activation energy for loss was equal to that for diffusion over the plane. This suggests that reflection may be the manifestation, not of a high activation energy, but of a low pre-exponential factor which could arise if the process required a correlated displacement or relaxation of substrate atoms.

12.2.2. *Clustering*

Whatever its origin, reflection at the edge of a plane provides a convenient barricade behind which adatoms may be confined on a particular plane and it has been exploited by Bassett and by Tsong in an elegant series of FIM investigations into adatom–adatom interactions and cluster formation. Two or more atoms are deposited on the plane of interest at low temperature ($\leqslant 77$ K) and the tip is then heated to a temperature sufficiently high to permit movement over the plane, yet low enough to prevent adatom loss over the step at the plane's edge. Since the surface is normally cooled to 20–80 K for imaging, adatoms are only observed at positions corresponding to potential–energy minima; the effects of interactions on processes occurring at higher temperatures must therefore be deduced indirectly. Image interpretation also presents some technical difficulties—adatoms in close proximity to one another generally can not be resolved, the location of an adatom relative to the substrate lattice may be difficult to define exactly, and field distortion of adatoms may occur. In spite of these problems a considerable amount of information has been accumulated on the most basic of chemical reactions, those between small numbers of individual atoms.

For the series of transition metals Ta, Mo, W, Re, Ir, Pt two types of behaviour have been distinguished. Adatoms having less than seven valence electrons (i.e. Ta, Mo, W) produce stable dimers on both W(110) and W(211) surfaces and clusters of several adatoms take the form of two-dimensional islands. In contrast, for adatoms in which the d-shell is more than half-filled the bonding is relatively restricted and directional; on W(211) Ir and Pt behave in a similar way to W, but on W(110) stable dimers are not easily formed and larger clusters consist of linear chains of adatoms (Bassett and Parsley 1969, 1970, Tsong 1972). Rhenium represents an intermediate case in that the trimer may be either triangular or linear. Taking tungsten as an example of the first type, it is found that, when more than one tungsten atom is deposited on W(110), clusters rapidly form at room temperature. It may be noted, however, that for $T < 390$ K two adatoms are always separated by the substrate nearest-neighbour distance of 0·274 nm and never (in 100 observations) by the next-nearest-neighbour distance 0·314 nm (Tsong 1972), indicating that the difference in the interaction potential energies at these two separations is of the order of $kT \ln(100) = 15$ kJ mol^{-1} at 390 K. This compares with a dimer binding energy of $\sim$100 kJ mol^{-1}, estimated from the fact that dimer dissociation sets in at 390 K. The W–W interaction is relatively short ranged on (110) but on (211) two tungsten atoms in neighbouring troughs have been observed to diffuse together at a separation of 0·447 nm (Tsong 1972); hence the interatomic potential is dependent on substrate structure. Rhenium provides an example of the second type of behaviour in that the interaction between two Re atoms on W(211) is similar to that between two W atoms, but on W(110) no Re–Re pairs are observed in the temperature range 320–390 K. Above 390 K the thermal energy is sufficient to surmount the repulsive potential between Re atoms and clusters of three or more atoms form. Dimers of bond length 0·274 nm also exist but are unstable, suggesting that the bond strength is less than 100 kJ mol^{-1}. At temperatures below 320 K weakly bound dimers with a bond length of $\sim$0·7 nm are observed. These data indicate that the Re–Re interaction potential on W(110) is oscillatory, with minima of $\sim$100 kJ mol^{-1} at 0·274 nm separation, and $\leqslant$75 kJ mol^{-1} at $\sim$0·7 nm (Tsong 1972).

Like rhenium, iridium and platinum do not produce stable dimers on W(110); instead interesting structures in the form of linear chains of adatoms are observed (Bassett 1973), often with several chains growing parallel to each other at a separation of $\sim$1·5 nm (Plate 8). In some ways this behaviour resembles that of a covalently bonded rather than a metallically bonded system; it may reflect the filling of anti-bonding adatom–metal orbitals once the d-shell is half-filled. The iridium chains appear to be aligned with the close-packed [111] direction in the W(110) surface, but in the cases of Ta, W, and Re clusters there is some evidence that the adatom–adatom bond axis may lie up to $\sim$10° away from [111] (Bassett 1970, 1973).

Several properties of these transition-metal clusters indicate that they cannot be described in terms of a simple pair-bond model in which nearest-neighbour bonds are assumed to be comparable in strength to similar bonds in the bulk metal. Thus, for dimers on W(110) the bonds are weak, that of W–W being $\sim$30 kJ mol^{-1} compared with a value of 140 kJ mol^{-1} in bulk tungsten. Also, the cluster interactions may be highly non-additive. The stability of Ir chains on W(110) increases with increasing chain length and the corresponding dissociation temperatures indicate that the energy of the bond attaching the terminal atom to the chain increases from 7 kJ mol^{-1} in the dimer to 50 kJ mol^{-1} for chains of 5 or more atoms (Bassett 1970). Finally, cluster mobility again demonstrates that pairwise additivity of adatom bonds is not applicable. While the mobilities of dimers are smaller than those of single atoms by factors of 2–10, mobility does not decrease rapidly as cluster size increases. For example, clusters of 9–10 Re atoms on W(110) are found to have a mobility $\sim$1 per cent of that of a single rhenium adatom, while the activation energies for diffusion are approximately the same for the two cases. These observations may indicate either that inter-adatom bonding in the clusters is weak or that the formation of inter-adatom bonds weakens the adatom–surface bonding (Bassett 1970).

12.2.3. *The effects of impurities and lattice defects*

The surface diffusion process may be considerably influenced by the presence of impurities and defects. Perhaps one of the most delicate examples is the observation (Tsong 1972) that during experiments involving single-atom diffusion on an FIM tip the adatom often became trapped at a particular point on the surface from which it did not move at the original surface temperature. In addition it was occasionally found that a migrating atom would visit some small region on the surface more often than would be expected in the case of an unrestricted random walk. The additional adatom binding energy at these surface 'traps' was estimated to be $\sim$14 kJ mol^{-1}, which is relatively small compared with the normal binding energy of several hundred kJ mol^{-1} or to the activation energy for surface diffusion ($\sim$100 kJ mol^{-1}). The traps were not distinguishable as separate features in the field ion images, but they were presumably associated with defects such as impurity atoms, vacancies, or dislocations present in or just below the surface.

Both increases and decreases in surface self-diffusion rates have been observed in the presence of impurities at concentrations in the monolayer range. It was suggested by Blakely (1963) and demonstrated by Gjostein (1967) that if the surface self-diffusion coefficient D_s $(= D_0 \exp - E_s/kT)$ derived from mass-transport experiments is plotted as a function of the reciprocal reduced temperature T_f/T, where T_f is the melting point, the results for a number of pure metals fall on a common curve leading to a single value of D_s at the melting point of

$$D_s(T_f) \approx 3 \times 10^{-8} \quad \text{to} \quad 1 \times 10^{-7} \, \text{m}^2 \, \text{s}^{-1}.$$

This observation enables one to detect the presence of impurities since this will result in D_s values which do not fall on the base curve. In certain cases, for example a silver surface with sulphur as impurity or a copper surface with halogen impurity, quite remarkable increases in D_s have been observed (Delamare and Rhead 1971). These are particularly large at high temperature where D_s may be greater than that for the clean metal by factors of up to 10^4. It was pointed out by Rhead (1969) that such values can actually be made to fit the base curve if the corresponding reduced temperature is calculated in terms of a surface melting point $(T_{f,s})$ which is lower than that of the clean substrate. An estimate of $T_{f,s}$ is obtained by assuming it to be the temperature at which D_s for the impurity system is approximately equal to $3 \times 10^{-8} \, \text{m}^2 \, \text{s}^{-1}$. Values of D_s greater than this are taken to be characteristic of a two-dimensional liquid. LEED studies of sulphur adsorption on Fe, Ni, Cu, and Ag (see Delamare and Rhead 1971) show that two-dimensional surface sulphides are formed and these might be expected to have melting points similar to those of the corresponding three-dimensional compounds. In the case of Ag/S for example, the D_s *versus* T_f/T curve shows a discontinuity at ~ 1060 K (Perdereau and Rhead 1967) while the melting point of silver sulphide is 1115 K. Further support for these ideas has come from measurement of surface diffusivities in copper–halogen systems (Delamare and Rhead 1971). The copper halides have low melting points (e.g. CuCl, 703 K) and the observed values of D_s at any given temperature are among the highest recorded.

In the experiments just discussed the diffusion coefficient was derived by observing the rate at which grooves form at grain

TABLE 12.2

Activation energies for surface self-diffusion E_s of W, Ta, and Mo in the presence of C, Si, and O impurities

	E_s, clean[†] (kJ mol^{-1})	E_s, carbon[†] (kJ mol^{-1}) Low coverage–high coverage	E_s, silicon[†] (kJ mol^{-1}) Low coverage–high coverage	E_s, oxygen[‡] (kJ mol^{-1})
W	290	290–830	290–680	405–434
Ta	280	280–475	—	
Mo	238	242–435	242–270§	

[†] Bettler *et al.* 1974.
[‡] Pichaud and Dreschler 1972.
§ This value is probably not maximum for Si on Mo.

boundaries on the surface of a polycrystalline specimen. Measurements have also been made in the field emission microscope (Melmed 1967); for example, the effects of oxygen, carbon, and silicon on the surface self-diffusion of W, Ta, and Mo have been investigated by Pichaud and Dreschler (1972) and Bettler, Bennum, and Case (1974). In all cases the activation energy E_s was significantly increased by the presence of impurity (see Table 12.2), in contrast to the effects of halogens and sulphur on copper and silver discussed previously. The behaviour of oxygen is also interesting in that on copper (Bradshaw, Brandon, and Wheeler 1964) and silver (Rhead 1965) it produced marked increases in diffusion coefficients, although in these experiments the surface conditions were not well defined—a general problem with much earlier work.

A number of different atomistic models for the surface self-diffusion process have been proposed (see Delamare and Rhead 1971), but none has been finally substantiated owing to lack of experimental information on factors such as the role of surface defects or the possible importance of the collective movement of groups of atoms as opposed to migration of single atoms. While increases in D_s may be rationalized in terms of two-dimensional compound formation, decreases in this parameter may be due to impurity adsorption at kink sites on the surface (Bettler *et al.* 1974).

The activation energies on a tungsten surface for removal of a

W atom from one kink site to another, and from a ledge site to a terrace site, are respectively 290 and 145 kJ mol^{-1}; the activation energy for W atom diffusion over terrace sites (i.e. on the (110) plane) is ~100 kJ mol^{-1}. The value of E_s determined by the techniques described above will include contributions from all three of these processes, but obviously the effect of kink sites will predominate. Thus preferential impurity adsorption at kinks might be expected to hinder the transfer of substrate atoms from the kink to a neighbouring ledge or terrace site. The fact that the C—W and Si—W bonds are stronger than the W—W bond (respective energies 280, 184, and 145 kJ per W atom) then correlates with the observed decreases in diffusivities on W in the presence of these impurities.

12.2.4. Faceting

In the last section we mentioned the use of grain-boundary grooving for investigation of surface self-diffusion and this phenomenon represents one example of a variety of changes in surface structure which can occur in the presence of an impurity. Another would be surface roughening, or etching, and its reverse; thus carbon or silicon on a tungsten field emission tip cause dark areas to appear in the (433) regions of the pattern at temperatures below a critical temperature T_c in the range 1250–1575 K (Bettler et al. 1974). On heating through less than 5 K at T_c these areas disappear 'almost as if a shutter had been closed'; on cooling they reappear just as sharply.

The most striking impurity effect, however, is faceting, the rearrangement of a flat surface into a hill-and-valley structure. This generally occurs when a metal (or non-metal) is in contact with a mildly reactive atmosphere at temperatures approaching the melting point, but it has also been observed at low temperature. The earlier work on this subject has been reviewed by Moore (1963). The main experimental technique employed was optical microscopy, but more recently information on the atomic scale has been obtained by the application of LEED, RHEED, and Auger electron spectroscopy. One of the classic systems to be studied was that of silver heated in air at around 1200 K; this treatment typically produces a striated surface consisting of linear facets of {100} or {111} orientation separated by planes of more complex orientation (Fig. 12.3). This system was studied in detail

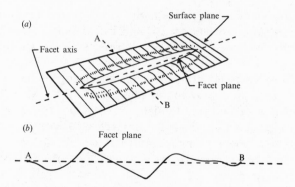

FIG. 12.3. (a) Surface with a linear facet. (After Mullins (1961). (b) Section through (a).

by Chalmers, King, and Shuttleworth (1948) who found that the facets could be removed by heating in nitrogen and that for faceting to be complete temperatures above 1100 K were required.

Chalmers *et al.* suggested that the faceted structure represents a position of minimum specific surface work and that the planes generated are those of low energy. This has been confirmed for the Ag/O$_2$ system, for example, by Rhead and McLean (1964). The measured γ-plot (i.e. variation of specific surface work with surface orientation, see Chapter 2) for orientations near the (111) pole of silver, in atmospheres of both hydrogen and air, is shown in Fig. 12.4. The driving force for faceting is the considerable decrease in γ as the structure changes towards the (1̄11) pole. Oxygen adsorption causes a large reduction in the average value of γ and there is an additional reduction at (111), accentuating the 'cusp' in the γ plot at this point, owing to the fact that the concentration of adsorbed oxygen is slightly higher ($\sim$4 per cent) on the low-index plane than on surfaces of other orientation (Rhead and Mykura 1962).

The γ plot, and hence the stability of a given surface, may be quite sensitive to the nature and extent of adsorption and it may be noted that faceting does not always produce low-index planes. For example, if copper is heated to high temperatures in oxygen, {111} and {100} facets are indeed frequently observed, but certain orientations, on cooling to a critical temperature between 800

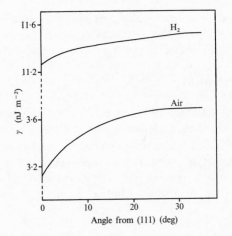

Fig. 12.4. γ plots for orientations of silver near the (111) pole in atmospheres of air and hydrogen (After Rhead and McLean 1964).

and 1000 K, produce {410} and {530} facets (Fig. 12.5; Legrand-Bonnyns and Ponslet 1975). Perdereau and Rhead (1971), using LEED, have also observed formation of {410} facets on exposure of copper single crystals to 10^{-6} torr of oxygen at room temperature; the clean surfaces in this case had a stepped (100) structure and were oriented 10° and 20° to (100) along the [001] zone (Fig.

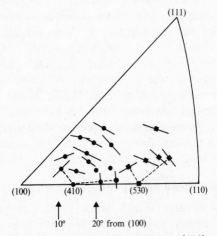

Fig. 12.5. Orientations of copper observed to produce {410} and {530} facets in oxygen. (After Legrand-Bonnyns and Ponslet 1975.)

12.5). Possible 'chemical' influences on the γ plot may be illustrated by comparing the interactions of copper surfaces with
oxygen and with nitrogen which has been 'activated' in the gas
phase either by electron bombardment or by contact with a
tungsten filament at ~2600 K. With oxygen Cu(210) does not
facet at room temperature, but gives (410) and (530) facets at
~900 K; with activated nitrogen this surface produces (100) and
(110) facets at 300 K (Kirby et al. 1979). The 10° and 20°
surfaces in contrast are readily faceted by oxygen at room temperature, but not by activated nitrogen (Perdereau and Rhead
1971). The interaction of Ni(210) with activated nitrogen is
identical to that of copper, whereas in the presence of iodine
Ni(210) forms (540) facets and a second set lying at a very small
angle to (100) (Tucker 1969).

Obviously the factors regulating facet formation may be quite
complex in any given case. While the process often occurs only at
high temperatures it is also observed at room temperature and
even, in the case of activated nitrogen faceting of Ni(210), at
temperatures as low as 150 K (Kirby, McKee, and Roberts 1976).
An elegant macroscopic theory of the kinetics of facet growth
according to a number of different mechanisms has been developed by Mullins (1961), but on the atomic scale no detailed
kinetic models are available. Certainly facet formation requires
the transport of very considerable quantities of metal; Tucker
(1967) has estimated that in the case of oxygen faceting of
Rh(210) an amount equivalent to 50 monolayers may be involved. In contrast, only a relatively small concentration of
adsorbed material may be required to initiate the process, less
than a monolayer for the O_2/Rh(210) and activated–N_2/Cu(210)
systems (Tucker 1967, Kirby et al. 1979). In other cases much
larger quantities of adsorbate may be necessary, however; for
example it has been suggested (Legrand-Bonnyns and Ponslet
1975) that oxygen can form (410) and (530) facets on copper
only when the bulk of the solid is saturated with oxygen.

12.3. Transition-metal adsorption on tungsten

Diffusion measurements enable one to investigate the variation in
potential experienced by an adatom as it moves over the surface,
but the binding energy of the atom at equilibrium on an adsorption 'site' is also of fundamental interest. (Binding energy in this

context is effectively synonomous with activation energy for desorption and, assuming adsorption to be non-activated, with heat of adsorption. It is not to be confused, of course, with the electron binding energy referred to in electron spectroscopy.) It is difficult to determine this quantity using macroscopic specimens because metal adatoms tend to move to kinks, steps, or other defects on the surface and it is displacement from these sites which is rate determining in a normal evaporation experiment, for example.

The field ion microscope provides a means of avoiding this difficulty since adatoms may be field desorbed at temperatures which are sufficiently low to prevent thermal diffusion. The field required to remove a selected adatom (in the form of an ion) from a region of crystalline perfection may be measured and related to the binding energy. The technique has the additional advantage of being applicable to adsorbates which are bound to the surface as strongly as the substrate atoms and this fact, together with the high visibility of transition-metal adsorbates in the microscope, makes these atoms ideal candidates for study.

Two possible mechanisms are available for field desorption, one involving movement of the desorbing species over the potential barrier tending to restrain it at the surface and the other tunnelling through the barrier; the latter predominates at low temperature ($\leqslant 70$ K for W on W (Müller and Tsong 1969)). Unfortunately the theory is not sufficiently advanced to enable precise conclusions to be drawn (Müller 1969, Bassett 1973), but qualitative trends in the binding energies of different adatoms on a given surface plane can be established. Results obtained by Ehrlich and Kirk (1968) and by Plummer and Rhodin (1968) for the binding energies of single W atoms on various planes of tungsten are shown in Table 12.3. The differences between the two sets of data illustrate some of the uncertainties just mentioned; for example, the large discrepancy between the values for W(110) arises principally because of the choice of the work function for this plane: $5 \cdot 5$ eV (Ehrlich and Kirk) or $8 \cdot 2$ eV (Plummer and Rhodin). Qualitatively, however, the conclusions to be drawn from these data are clear. One might have expected energies to be highest on high index planes since the adatoms in these situations can make contact with a relatively large number of substrate atoms; this is not observed. A tungsten atom is less

TABLE 12.3

Binding energies of W adatoms on W

Plane	Binding energy (kJ mol^{-1})			
	Experimental		Theoretical	
	(a)†	(b)‡	(c)†	(d)§
(110)	513	800	560	447
(211)	680	670	750	738
(100)	—	775	—	758
(310)	650	—	758	—
(111)	580	650	767	883
(321)	640	720	785	748
(411)	600	—	835	—

† Ehrlich and Kirk 1968.
‡ Plummer and Rhodin 1968.
§ Dreschler 1954.

strongly bound on (411) than on the smoother (211) plane, for example. Also, the experimental results do not correlate well with values calculated assuming a pairwise additive potential (column (d), Table 12.3) or using a Morse potential summed over all lattice points to a distance of 2·5 lattice spacings from the adatom (column (c), Table 12.3). This is particularly true for the higher index faces.

A factor which is not considered in the pairwise interaction model is redistribution of electron density at the surface (Smoluchowsky 1941). On high-index planes the contours of the electron cloud will not follow those of the atom cores, the electron density being drained from protruding atoms to some extent and concentrated in the surface troughs. Experimental support for this idea has been obtained from work function measurements during the deposition of W atoms on a W(110) substrate (Besocke and Wagner 1973). The result of redistribution is that protruding atoms will form weaker bonds with an adatom than will substrate atoms in a low-index plane, on which little charge smoothing occurs. Detailed calculations of this effect would be extremely difficult, but by using a very simple estimate of the extent of charge depletion Plummer and Rhodin (1968) were able to show that when smoothing was taken into account the pairwise interaction model gave reasonable agreement with experiment, at least on the lower-index planes (110), (100), (211), and (111).

Binding energies for atoms of metals in the sixth period, on the four low-index planes of tungsten, are shown in Fig. 12.6(a) (Plummer and Rhodin 1968). The order of the energies for any given adatom on the four faces is (110)>(100)>(112)>(111); the two apparent exceptions of Ir on (100) and (110) and Pt on (111) and (112) are not considered significant because of experimental problems with these particular measurements. We may now reach the simple but important conclusion that the binding energy is determined by two factors: (i) the electronic configuration of the adatom; (ii) the crystallography of the surface.

The variation of the binding energy with the electronic properties of the adatom, rising to a maximum with rhenium, is similar

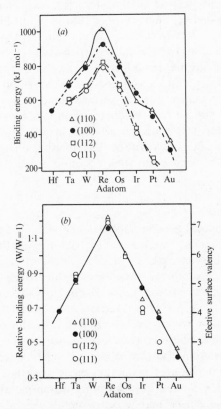

FIG. 12.6. (a) Binding energies of 5d transition-metal adatoms on the four low-index planes of tungsten. (b) Relative binding energies, with respect to tungsten, and effective surface valency. (After Plummer and Rhodin 1968.)

to the variation of the cohesive energies of the 5d transition metals. When a transition-metal atom approaches the surface of a metal the atomic levels will broaden into bands, the widths of which are determined by the strength of the interaction with the surface (see Section 12.5). It may be assumed that the degenerate atomic d levels will split into bonding and anti-bonding bands. The binding energy of the adatom is then expected to increase across the periodic table from hafnium, with the configuration $5d^2 6s^2$ (i.e. effectively four bonding electrons), to rhenium $(5d^5 6s^2)$ with seven bonding electrons. Thereafter the energy will decrease since the additional d electrons will be anti-bonding. The point is emphasized (Fig. 12.6(b)) by normalizing the binding energies with respect to tungsten, for each plane, and calculating the effective number of electrons contributing to the bonding, or the 'effective surface valency', on the assumption that for tungsten $(5d^4 6s^2)$ *six* equally participating electrons are involved. It turns out that the effective surface valency is integral in almost all cases; platinum and iridium each exhibit two values, with the higher occurring on the surface planes of highest binding energy. Also, the surface valency corresponds to one of the known oxidation states of the adatom in all cases except the values of 3 for Pt and 5 for Ir. The binding energy E_d of an adatom A on a plane i can thus be represented by the simple relationship

$$E_d \propto v_A F_i$$

where v_A is the effective surface valency and F_i is a crystallographically dependent function which determines the extent of charge smoothing on plane i.

A number of attempts have been made to calculate these binding energies from theory, for example by Cyrot–Lackmann and Ducastelle (1971). These authors used a tight-binding approach, and their results are compared with the experimental data in Fig. 12.7. There is agreement as to the general form of the energy variation, but the observed dependence on crystal plane and the position of maximum energy are not reproduced theoretically. The binding energies of transition-metal adatoms on surfaces of other 3d, 4d, and 5d series metals were also calculated and showed the same features as the tungsten system. A similar theoretical study has been carried out by Thorpe (1972).

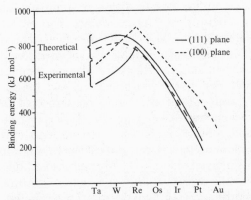

FIG. 12.7. Comparison of theoretical and experimental binding energies of 5d transition-metal adatoms on W(111) and W(100). (After Cyrot-Lackmann and Ducastelle 1971.)

12.4. Alkali and alkaline-earth adsorption on transition metals

As mentioned in the introduction to this chapter these adsorption systems have been investigated in considerable detail because of relative ease of experimentation and because of their scientific and technological interest. In this section we shall discuss some of the results obtained in relation to surface structures, dipole moments, and binding energies.

12.4.1. Surface structures

A particularly comprehensive study of the interactions of Na, K, and Cs with the three low-index faces of Ni has been made by Gerlach and Rhodin (1968, 1969, 1970) using LEED, work-function, and thermal desorption measurements. We shall consider their work with Na in some detail since it provides a well-documented example of the influence of coverage on structure. Similar investigations have been carried out using tungsten substrates by groups in Kiev (see Medvedev and Yakivchuk 1974) and in Pennsylvania (see Papageorgopoulos and Chen 1973); the results are, in general terms, similar to those obtained with nickel.

A. *Ni(100)–Na*. When Na adsorbs on this surface at room temperature the LEED pattern at low coverage $(0 < \theta < 0.25)$ is similar to that for clean Ni(100), except that the 00 spot is

surrounded by a diffuse ring of diameter approximately proportional to the square root of the coverage (Fig. 12.8). Rings of
much lower intensity are also detected around other spots. As
discussed in Chapter 3 such patterns may be interpreted in terms
of a 'semi-random' surface structure which cannot be described
by a single unit mesh but in which, nevertheless, the adatoms are
more or less evenly spaced, the standard deviation of the average
distance between neighbours being small (Fig. 12.8(a)). It is
assumed that the overlayer is coherent with the substrate, with
each sodium atom occupying an identical site between four nickel
atoms. Laser simulation (Fig. 12.8(b)) confirms that this structure
would produce the observed diffraction pattern. As the coverage
is increased beyond $\theta = 0.25$ the degree of order in the overlayer
increases until a perfect c(2×2) pattern is formed at $\theta = 0.5$; still
further adsorption produces sodium multilayers.

B. Ni(111)–Na. In this system also, for $0 < \theta < 0.33$, a diffuse
ring is observed around the 00 spot in the diffraction pattern
indicating a limited range of Na–Na spacings centred on a mean
value. At low coverage the adatom–substrate interaction should
represent the strongest force in the surface region and it may be
assumed that the overlayer is coherent. For $\theta > 0.33$, however,
adatom–adatom interactions begin to predominate and the overlayer rearranges into a hexagonal structure to maximize packing
(Fig. 12.9(a)). The sodium atoms are now forced out of the sites
of high co-ordination and their positions are no longer related to
the Ni–Ni spacing; in other words an incoherent structure

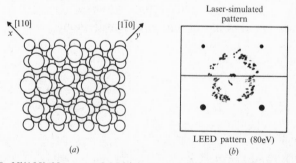

(a)   (b)

Fig. 12.8. Ni(100)–Na at $\theta = 0.25$: (a) proposed structure; (b) schematic comparison of laser-simulated diffraction pattern from this structure with LEED
pattern. (After Gerlach and Rhodin 1969.)

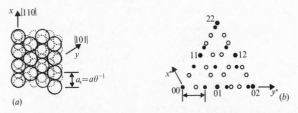

(a)   (b)

FIG. 12.9. Ni(111)–Na for $0.33 < \theta < 0.49$: (a) proposed incoherent hexagonal structure; (b) predicted diffraction pattern. The open circles represent double diffraction beams. (After Gerlach and Rhodin 1969.)

forms. The original diffuse ring has at this stage broken up into a hexagonal array of six spots which continue to move outwards from 00 as coverage increases, their separation from 00 being $a^*\theta^{\frac{1}{2}}$ where a^* is the reciprocal lattice vector for clean Ni(111) (Fig. 12.9(b)). This indicates that $\mathbf{a}_s$, the Na–Na spacing in real space, is inversely proportional to $\theta^{\frac{1}{2}}$ (Fig. 12.9(a)). There is also an inner hexagonal array of spots (Fig. 12.9(b)) which are the result of double diffraction between the overlayer and the substrate. The reciprocal vectors $\mathbf{a}_d^*$ are given (see Chapter 3) by the difference (or sum) of the reciprocal vectors for the two layers

$$\mathbf{a}_d^* = \mathbf{a}^* - \mathbf{a}_s^* = \mathbf{a}^*(1 - \theta^{\frac{1}{2}})$$

and so they move *inwards* towards the origin as coverage increases.

For coverages greater than $\theta = 0.45$ a second Na layer begins to form, although the first layer is not complete until $\theta \simeq 0.6$.

C. *Ni(110)–Na.* In contrast to (100) and (111), well-ordered overlayers do not form at room temperature on (110); heating to several hundred degrees centigrade is required. At low coverages ($\theta < 0.25$) coherent structures are produced in which rows of Na atoms sit in the troughs in the Ni surface which run in the [110] direction. The diffraction pattern observed at $\theta = 0.11$ and the proposed structure are shown in Fig. 12.10. The third-order spots in the y^* direction are due to the fact that only every third trough contains sodium. Third-order features also occur in the x^* direction since the Na atoms are triple spaced along each trough. They appear as streaks because there is no correlation between the positions of adatoms in the different troughs. We can therefore think of 'lines of disorder' running in the [110] direction, between

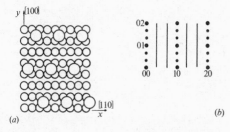

FIG. 12.10. Ni(110)–Na for $\theta = 0 \cdot 11$: (a) proposed structure; (b) calculated
diffraction pattern. (After Gerlach and Rhodin 1969).

the rows of sodium atoms, and hence we expect streaks in the
diffraction pattern in the direction normal to this (see Section
3.7.2). The fact that the third-order, but not the integral-order,
features are streaked indicates that the adsorbate structure does
not occur in small islands separated by clean surface; if it is
present in patches these must have dimensions greater than the
coherence width of the incident electron beam, i.e. greater than
say 10 nm (see Section 3.7).

At $\theta = 0 \cdot 25$ alternate troughs are occupied by adatoms which
are double spaced within any trough; again the rows are randomly shifted relative to each other in the [110] direction. For
coverages between $\theta = 0 \cdot 11$ and $0 \cdot 25$ the troughs will contain
mixtures of units consisting of double-spaced and triple-spaced
sodium atoms (Fig. 12.11(a)), the proportion of the former
increasing with increasing coverage. This causes the third-order
streaks present at $\theta = 0 \cdot 11$ (Fig. 12.10(a)) to move continuously
towards each other (Fig. 12.11(b)) until they coalesce into a
single half-order streak at $\theta = 0 \cdot 25$. The phenomenon is similar
to the continuous spot splitting observed during oxygen adsorption on Ni(110) (see Chapter 11). Note that these structural
changes involve considerable movement of sodium on the surface, both from one site to another within any given trough and
also from trough to trough.

Increasing coverage between $\theta = 0 \cdot 25$ and $0 \cdot 32$ causes the
half-order streaks to move away from the 00 spot, while other,
less intense, double-diffraction streaks move inwards towards this
spot. We have encountered similar behaviour in discussing hexagonal spot arrays produced by Na on Ni(111), and as in that
case we may assume that for $\theta > 0 \cdot 25$ the structure becomes

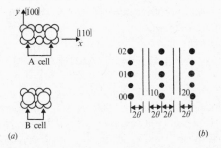

FIG. 12.11. Ni(110)–Na for $0\cdot11<\theta<0\cdot25$: (a) two types of cell from which proposed structure is composed; (b) calculated diffraction pattern. (After Gerlach and Rhodin 1969.)

incoherent (in the [110] direction). The sodium spacings within each trough are determined not by the potential of the nickel surface, but by adatom–adatom interactions. Since the spacing of the streaks is found to be 2θ it follows that the sodium spacing is inversely proportional to coverage in this range (Fig. 12.12(a)).

For coverages between $0\cdot32$ and $0\cdot64$ monolayers the adatom spacing in the [110] direction remains constant at $\sim1/(2\times0\cdot32)=1\cdot56$ times the Ni–Ni spacing and the streaks remain stationary. The additional sodium occupies the troughs which were vacant at $\theta=0\cdot32$ until all troughs are filled at $\theta=0\cdot64$. Once again there must be considerable rearrangement of sodium on the surface as coverage increases. At $\theta=0\cdot64$ the streaks in the y^* direction have been replaced by spots since the sodium positions are now correlated from trough to trough (Fig. 12.12(b)). Further sodium up to a coverage of $\theta=0\cdot71$ is accommodated by compression of the structure in the [110] direction, while beyond this coverage a second adlayer forms.

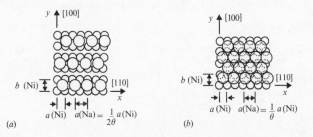

FIG. 12.12. Ni(110)–Na: (a) proposed structure for $0\cdot25<\theta<0\cdot31$; (b) proposed structure for $0\cdot64<\theta<0\cdot71$. (After Gerlach and Rhodin 1969.)

From the results presented above we see that the structures of alkali-metal layers on nickel are known in considerable detail; in particular, coverages have been established with sufficient accuracy to enable a reasonably unambiguous analysis of the LEED patterns to be carried out. An interesting feature appears when the minimum radii (r_m) of the alkali adatoms are calculated from the surface adatom densities corresponding to maximum coverage in the first layer. Values of r_m are shown in Table 12.4 (Gerlach and Rhodin 1969) along with 'reduced effective radii' $R = r_m/r_a$, where r_a is the atomic radius in the bulk alkali metal. It is seen that in all cases the adatoms are *smaller* than the corresponding bulk atoms, the differences for potassium and caesium being considerable. A similar observation has been made for Na on W(211) (Chen and Papageorgopoulos 1970), so the effect appears to be independent of the substrate. This is quite contrary to expectation; previous observations of adlayer densities greater than those predicted on the basis of bulk atomic radii had been rationalized (Taylor and Langmuir 1933) by assuming that the substrate surface was rough and therefore had an effective area greater than the geometric area used in the calculation.

The results also show that there must be repulsive adatom–adatom interactions at all coverages and, further, that on the two-fold symmetric (110) surface these forces must be anisotropic. For example in the sodium structures at $\theta = 0.11$ and 0.25 the adatom spacing in the [110] direction is only 70.7 per cent of the spacing in the [100] direction across the troughs, indicating

TABLE 12.4
Minimum radii of alkali-metal adatoms on nickel

System	Minimum effective radius r_m (nm)	Atomic radius r_a (nm)	Reduced effective radius $R = r_m/r_a$
Ni(111)–Na	0·177	0·190	0·93±0·02
Ni(100)–Na	0·175	0·190	0·925
Ni(110)–Na	0·175	0·190	0·92±0·01
Ni(110)–K	0·197	0·235	0·84±0·02
Ni(110)–Cs	0·207	0·267	0·78±0·02

From Gerlach and Rhodin 1969.

higher repulsions in this latter direction. These prevent the occupation of adjacent troughs. Again, in the coverage range $0.64 < \theta < 0.71$ the Na structure is found to be compressible in the [110] but not in the [100] direction. These observations are compatible with Grimley's predictions of indirect adatom–adatom interactions by way of the metal conduction band, which would be anisotropic on a surface of two-fold symmetry (Grimley 1967a,b, Grimley and Walker 1969).

12.4.2. Binding energies

Binding energies of alkali adsorbates on transition-metal substrates have generally been determined using the field and thermal desorption techniques (see Todd and Rhodin (1974) for a guide to earlier work). Thermal desorption rate and temperature–time plots for Ni(110)/Na are shown in Fig. 12.13(a). The large peak on the left represents desorption from second and higher layers. Activation energies E_d and frequency factors $\nu^{(1)}$ for desorption at a given absolute coverage (σ) were determined by measuring the desorption rate R_d at various temperatures at this coverage. Assuming that the process is first order then

$$R_d(\sigma, T) = \sigma \nu^{(1)}(\sigma) \exp\left(-\frac{E_d}{kT}\right). \tag{12.6}$$

Hence a plot of $\ln R_d$ *versus* $1/T$ at constant σ yields the required quantities. The results for Na on Ni(111) and Ni(110) are shown in Fig. 12.13(b); those for K and Cs on Ni(110) are qualitatively similar.

The desorption-energy curves for Ni(111) involve relatively little structure for coverages at which the first sodium layer is incomplete, and this correlates with the observation that the surface structure varies on this plane in a smooth and continuous way; the same is true of the (100) plane. In contrast the desorption-rate curves for Ni(110) consist of multiple features, in accordance with the greater structural complexity of the adlayers in this case. Some of these features may be identified, at least tentatively, with particular structures. For example, the desorption rate is rather high for coverages up to $\theta \simeq 0.25$ (peak I, Fig. 12.13(a)) and E_d and $\nu^{(1)}$ are approximately constant (Fig. 12.13(b)). Between $\theta \simeq 0.25$ and 0.3, however, R_d is low (minimum II, Fig. 12.13(a)) and E_d and $\nu^{(1)}$ fall rapidly; this is

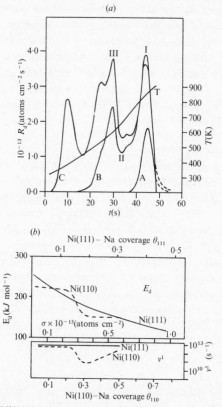

FIG. 12.13. (a) Ni(110)–Na: temperature–time plot T and thermal desorption rate plots at the following coverages: $\theta_A = 0.10$; $\theta_B = 0.46$; $\theta_c = 0.86$. (b) Activation energy for desorption E_d and frequency factor $\nu^{(1)}$ for desorption *versus* coverage for Na on Ni(110) and Ni(111). (From Gerlach and Rhodin 1970.)

just the coverage range in which the structure becomes incoherent in the [110] direction along the surface troughs. For $\theta > 0.32$ compression within occupied troughs ceases and unoccupied troughs begin to fill, while E_d remains constant and $\nu^{(1)}$ rises. Also, the desorption rate peak at 610 K (peak III, Fig. 12.13(a)) probably corresponds to the structure in which two-thirds of the troughs contain incoherent rows of Na atoms.

For ease of comparison with theory it is important to determine binding energies at zero coverage. Values may be obtained by extrapolation from higher coverages, but these may involve uncertainties due to difficulties in measuring coverage and to possible changes in surface structure with coverage. Ideally, one should like to determine the binding energy of a single adatom directly, as is done for transition-metal adsorbates in the field ion microscope (see Section 12.3). Unfortunately, for alkali atoms the ion imaging field is greater than the desorption field, but this is not the case in the field emission microscope. Here, the problem is one of insufficient resolution to observe individual atoms directly. It has been shown, however (Plummer and Young 1970, Todd and Rhodin 1974), that the arrival of a single atom at the tip, or its desorption from the tip, can be detected by a positive or negative step in the field-emitted current. Todd and Rhodin (1974) utilized this phenomenon to determine the field necessary to desorb single atoms; in turn the field can be related to the zero-coverage binding energy E_0, although this step involves a number of approximations. Values for various alkali-metal–tungsten-plane combinations are shown in Fig. 12.14 relative to the value for Cs on W(100) (300 ± 5 kJ mol^{-1}). Two trends emerge: for a given substrate plane

$$E_{Na} < E_K < E_{Cs}$$

and for a given alkali atom

$$E_{(111)} < E_{(112)} < E_{(110)}.$$

These results are in general agreement with the E_0 values obtained by extrapolation from higher coverages. In contrast to

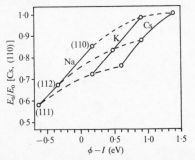

FIG. 12.14. The relative binding energies within the alkali system: one atom of Na, K, and Cs on the (110), (112), and (111) planes of tungsten. Solid lines connect points for the same alkali; broken lines connect points for the same substrate plane; ϕ is the work function of given tungsten plane and I is the alkali ionization potential. (After Todd and Rhodin 1974.)

transition-metal adsorption on tungsten, where the binding energy is determined principally by the nature of the adatom (Section 12.3), in the present case substrate crystallography is obviously of equal importance.

12.4.3. *Work-function changes and dipole moments*

In many cases adsorption of the various alkali and alkaline-earth metals on both polycrystalline and single-crystal transition-metal surfaces leads to work-function–coverage curves of the general type shown for Na, K, and Cs on Ni(110) in Fig. 12.15(*a*)

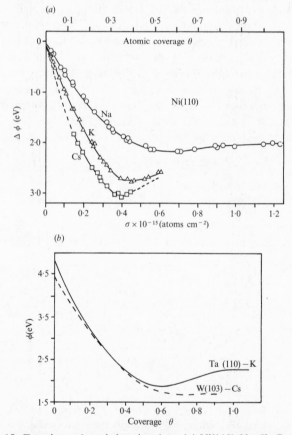

FIG. 12.15. Experimental work-function data: (*a*) Ni(110)–Na, K, Cs, capacitor method (Gerlach and Rhodin 1970); (*b*) Ta(110)–K, electron-beam retarding-potential method (Fehrs and Stickney 1971); W(103)–Cs, field emission method (Sidorski *et al.* 1969.)

(Gerlach and Rhodin 1970); these were measured by the capacitor technique. Representative curves determined by the field emission method for tungsten (Sidorski, Pelly, and Gomer 1969) and by the electron-beam retarding-potential technique for tantalum (Fehrs and Stickney 1971) are shown in Fig. 12.15(b). Ideally the experimental work-function data should relate to individual crystal planes and should include values of absolute surface coverage since these are necessary for calculation of dipole moments. Although many investigations have been carried out using a variety of techniques, these requirements have been satisfied in relatively few instances and even then some discrepancies remain between the results obtained by various workers studying the same system. Some of the difficulties encountered in the field emission experiments have been summarized by Fehrs and Stickney (1971). In the case of macroscopic samples at 300 K complications may result from the formation of adsorbate islands, since the alkalis are known to diffuse below this temperature.

Curves such as those in Fig. 12.15 show three principal features of interest, the first being the initial slope which is directly related to the zero-coverage dipole moment (μ_0) of the adsorbate–substrate bond. Since an adsorbate coverage of σ atoms cm^{-2} produces a work-function change $\Delta\phi$ given by

$$\Delta\phi = 2\pi\sigma\mu \tag{12.7}$$

it follows that

$$\mu_0 = \frac{c}{2\pi}\left[\frac{\partial(\Delta\phi)}{\partial\sigma}\right]_{\sigma\to0}. \tag{12.8}$$

In eqn (12.8) μ_0 is expressed in debyes (1 debye = 1×10^{-8} e.s.u.) and $\Delta\phi$ in electron volts; c ($=\frac{1}{300}$) is then the conversion factor from volts to e.s.u. In Section 12.4.2 mention was made of difficulties encountered in estimating zero-coverage binding energies by extrapolation from higher coverages; the same difficulties arise when determining μ_0 values and this is reflected in the uncertainties associated with the data shown in Table 12.5. Todd and Rhodin (1974) have therefore used the field emission method, outlined in the last section, to determine single-atom dipole moments; the results are lower than those obtained by other techniques but considerable uncertainty as to absolute μ_0

TABLE 12.5

Work-function changes $\Delta\phi$ and zero-coverage dipole moments μ_0 for alkali-metal adsorption on Ni and W

System	Plane	$-\Delta\phi$(eV)	μ_0 (debye)	Method[†]
Ni–Na	(111)		7·4±0·5	C
	(110)		3·2±0·3	C
W–Na	(110)	0·23–0·29	6·0–7·6	F
	(112)	0·14–0·18	3·6–4.8	F
	(111)	0·13–0·17	3·4–4·4	F
Ni–K	(110)		5·3±0·5	C
W–K	(112)	0·19–0·29	5·0–7·6	F
Ni–Cs	(110)		7·0±0·7	C
W–Cs	(112)	0·16–0·27	4·2–7·0	F

[†] C, capacitor, using eqn 12.8, data from Gerlach and Rhodin (1970); F, field emission probe hole, data from Todd and Rhodin (1974).

values arose from the necessity of estimating the effective area of an adatom during treatment of the experimental data.

In spite of these various problems, two general conclusions emerge regarding the zero-coverage dipole moment (Table 12.5): (i) for a given adatom on different planes of a given metal μ_0 decreases as the work function (or atom density) of the plane decreases, e.g. for Na on Ni the order is $(110) < (100) \leqslant (111)$; (ii) for a given crystal plane μ_0 increases in the order $Na < K < Cs$. Data of this type have provided a point of comparison with experiment for quantum-mechanical theories of adatom–surface bonding and will be further discussed in Section 12.5.

From Fig. 12.15 we see that as coverage increases the work function generally passes through a minimum value (ϕ_m) at some coverage σ_m, although this is not always the case (see, for example, the behaviour of Cs on W(103), Fig. 12.15(b)). The magnitudes of ϕ_m and σ_m represent the second point of interest in these curves, while the third is the approximately constant value which the work function assumes at some high coverage. Values of dipole moment μ as a function of Na, K, and Cs coverage on Ni(110), obtained by application of the Helmholtz equation (12.7) to $\Delta\phi$ data, are shown in Fig. 12.16. Comparison with Fig. 12.15(a) shows that, although ϕ passes through a minimum in each case, there are no discontinuities in μ at the corresponding coverages σ_m. Neither are there any significant changes at these points in the structures of the adlayers as determined by LEED. Evidence in the case of lithium adsorption on W(211) (Medvedev, Naumovets, and Smereka 1973) also suggests that the work function behaviour depends more on short-range order within the adlayer than on the long-range order detected by LEED.

The Helmholtz equation (12.7) predicts a linear change in ϕ with surface coverage σ. The non-linear decrease actually observed during alkali adsorption can be qualitatively ascribed to 'depolarization' effects arising because of various adsorbate–adsorbate interactions, for example the long-range Coulomb potential at a given adparticle due to all other adsorbate dipoles. Discussions of the details of the classical approach to this problem have been given by a number of authors (see, for example, MacDonald and Barlow 1963, Sidorski and Wojciechowski 1971). In outline, the argument runs as follows. The electrostatic field at a distance x from a permanent dipole, in a direction normal to the dipole axis, is $2\mu/x^3$. In an adsorbed layer of N particles the total field at a given dipole due to all other dipoles is

$$F = -\tfrac{1}{2} \sum_{i=1}^{N} 2\mu x_i^{-3} = -\mu \sum_{i=1}^{N} x_i^{-3}. \qquad (12.9)$$

This summation was evaluated by Topping (1927). The result, which appears to be approximately independent of the geometry

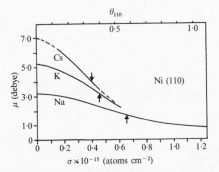

FIG. 12.16. Dipole moment *versus* coverage for Na, K, and Cs on Ni(110). Arrows indicate coverages at which work-function minima are observed. (After Gerlach and Rhodin 1970.)

of adsorption sites, is

$$F \approx -9\mu/r^3 \tag{12.10}$$

where r is the adatom separation. Consider mobile adsorption on a square lattice with unit mesh side a. At any fractional coverage θ we may assume a single value of r. Then

$$\theta = \frac{\text{area occupied per adatom at } \theta = 1}{\text{area occupied per adatom}} \approx \frac{a^2}{r^2}$$

and therefore

$$r \approx a\theta^{-\frac{1}{2}} \tag{12.11}$$

A relationship of similar form should apply in the case of immobile adsorption (Sidorski and Wojciechowski 1971).

When a polarizable species is placed in a field F it may be assumed that the dipole induced is proportional to the field strength, i.e.

$$\mu = \alpha F \tag{12.12}$$

where α is the polarizability. If the dipole moment at coverage θ can be written in the form (Miller 1946)

$$\mu = \mu_0 + \alpha F \tag{12.13}$$

where μ_0 is the moment of an isolated adsorbed species, then from eqns (12.10), (12.11), and (12.13) it follows that

$$\mu = \mu_0 - 9\alpha\mu\theta^{\frac{3}{2}}a^{-3}. \tag{12.14}$$

Hence, using eqn (12.7), we find that

$$\Delta\phi = \frac{2\pi\mu_0\sigma}{1 + 9\alpha\theta^{\frac{3}{2}}a^{-3}}. \tag{12.15}$$

Putting $\sigma = \theta/a^2 = \theta\sigma_1$, where σ_1 is the adsorbate density at monolayer coverage, this rearranges to

$$\frac{\theta}{\Delta\phi} = \frac{1}{c_1} + \frac{c_2}{c_1}\theta^{\frac{3}{2}} \tag{12.16}$$

where (eqn (12.8))

$$c_1 = 2\pi\mu_0\sigma_1 = c\left(\frac{d\Delta\phi}{d\theta}\right)_{\theta=0}$$

and

$$c_2 = 9\alpha\sigma_1^{\frac{3}{2}}.$$

Eqn (12.16) is often referred to as the Topping relationship. Unlike the Helmholtz equation (12.7) it predicts a minimum in $\phi(\theta)$ at $\theta_m = (2/c_2)^{\frac{2}{3}}$ and it has been found (Schmidt and Gomer 1966, Swanson and Strayer 1968, Sidorski et al. 1969) to describe the behaviour of a number of systems adequately at coverages below the minimum. In spite of this agreement, however, the model may not accurately represent the physical processes in the adlayer (Sidorski and Wojciechowski 1971). The polarizabilities derived from the constant c_2 (eqn (12.16)) are high, for example 0.067 nm^3 and 0.050 nm^3 for Cs on W(211) and W(100) respectively (Sidorski et al. 1969); this should mean that the work-function values observed by a technique involving a low applied field, such as thermionic emission, should differ from those obtained from field emission, a high-field method. No significant difference is detected in practice.

At coverages above σ_m the Topping approach definitely breaks down (Swanson and Strayer 1968). In an attempt to rationalize the behaviour of $\Delta\phi$ in the monolayer region Sidorski and Wojciechowski (1968) have suggested that the alkali layer at this point may be of sufficient density to have developed a band structure. The work function would then be determined by the energies, relative to the bottom of the conduction band, of the Fermi level (ε_F) and of the vacuum level (ε_{vac}), e.g. for Cs

$$\phi = \varepsilon_{vac(Cs)} - \varepsilon_{F(Cs)}.$$

Both ε_{vac} and ε_F will be functions of the lattice constant of the 'metal'. From our discussion of the geometry of alkali layers, Ni(110)–Na for example (Section 12.4.1), we have seen that, as the monolayer is approached ($\theta = 0.64$–0.71), the structural effect of increasing coverage is a compression of the adsorbate in the [110] direction. According to the above argument this would lead to a variation in ϕ. Further, if the alkali layer did possess a band structure the monolayer value of ϕ should be approximately independent of the nature of the substrate; this is observed experimentally (Swanson and Strayer 1968).

A number of other approaches to the high-coverage problem have been explored, for example a phenomenological study by

Gyftopoulos and Levine (1962). Also, Lang (1971) has used the theory of the inhomogeneous electron gas, in which the ion-core lattice of the substrate is replaced by a semi-infinite uniform positive background—this is descriptively referred to as the jellium model of a metal. The adsorbate ions are also replaced by a positive slab adjoining the metal. The electron density associated with this system of positive charge is then obtained from a self-consistent wave-mechanical calculation and from this density the work function is derived. Changes in coverage are introduced by changing the density of the adsorbate slab. Lang thus takes a model which is relatively simple and attempts to treat it with a high degree of precision. The results obtained are in reasonable agreement with experimental data in the cases of quantities which are not sensitive to the substrate work function, such as the values of $\Delta\phi$ at $\theta = 1$ and at θ_m.

12.5. Quantum-mechanical treatment of chemisorption

12.5.1. *General principles*

The alkali metals occupy a rather special position in the field of chemisorption theory since, along with hydrogen, they constitute the least complex class of adsorbate. A considerable effort has been directed towards a description of their adsorption behaviour, and one of the early applications of quantum-mechanical ideas to surface phenomena was an attempt (Gurney 1935) to account for the effect of alkali and alkaline-earth adsorption on the work function of tungsten. It was originally thought that the reduction in ϕ which accompanies adsorption of K, Rb, and Cs could be explained in the following way. The ionization potentials I of the these atoms (e.g. K 4·32 eV, Cs 3·89 eV) are less than the work function of tungsten (polycrystalline $\phi_W = 4·5$ eV) and so the adsorbate valence level will lie above the Fermi level in the metal (Fig. 12.17(a)). Therefore a complete transfer of an electron from adsorbate to metal might be expected, creating an electrical double layer at the surface with the positive side outwards; this, of course, would produce a considerable reduction in work function. Such a scheme may be approximately correct for K, Rb, and Cs, but a problem arises when it is observed that Na, and also Ca, Sr, and Ba, affect ϕ in essentially the same way; their ionization potentials, however, are greater than ϕ_W (e.g.

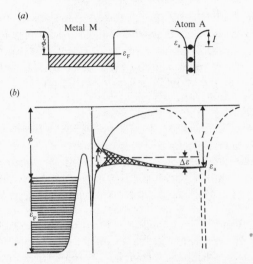

FIG. 12.17. (a) Energy levels in a metal and an isolated atom. (b) Changes in an adatom valence level on approaching a metal surface. (After Gadzuk 1969.)

$I_{Ba} = 5·19$ eV) and therefore electron transfer to the metal apparently should not occur.

The first satisfactory explanation of this problem was presented in a classic four-page paper by Gurney (1935), remarkable not only for its brevity and its clarity but also for the fact that it still provides the basis for a major area of discussion in chemisorption theory. One considers first an alkali atom A at some distance from a metal M (Fig. 12.17(a)), the point of interest being the potential energy of an electron along a line normal to the surface of M and passing through the core of A. As the atom approaches the surface any atomic levels lying below the bottom of the metal conduction band will remain discrete and localized on the atom since the amplitude of the corresponding wavefunction will fall off exponentially within the metal; these levels need not be considered further in the present context. We focus attention instead on a valence level in the atom (of energy ε_a), which we suppose lies above the bottom of the metal conduction band. The Schrödinger equation for the system metal plus adatom is then solved, using methods to be discussed presently, and hence we obtain a set of allowed levels each of which belongs jointly to the metal and the atom. On examination of the amplitude of the wavefunction in the vicinity of the adatom core we find that when the atom is some distance from the surface, $|\psi|^2$ is appreciable only for energies close to ε_a (Fig. 12.17(b). Effectively, we have a discrete atomic level, slightly broadened by the presence of the metal. When the atom lies closer to the surface, however, two important effects are detected. First, there is a shift $\Delta\varepsilon$ in the position of the adatom valence level to $\varepsilon'_a = \varepsilon_a + \Delta\varepsilon$. This can be understood in terms of the image forces acting on the valence electron and is further discussed below.

The second effect concerns the value of $|\psi|^2$ around the adatom core. For small separations this shows appreciable magnitude over a range of energies on either side of ε'_a (Fig. 12.17(b)), Gurney 1935). In formal terms one can picture the valence electron at the shifted level as resonating between the metal and the adatom. According to the uncertainty principle, if energy is to be measured within an accuracy 2Γ the measurement must be extended over a period of time $\Delta t = h/2\Gamma$, where h is Planck's constant. If the valence electron resided permanently on the adatom, $\Delta t = \infty$ and $\Gamma = 0$, so that the valence level would be well defined as at large adatom–surface separations. When the electron has a finite lifetime on the adatom, however, its energy becomes less certain and we can think in terms of a broadened level of half-width $\Gamma \simeq h/2 \Delta t$, centred at ε'_a. Of course, once the atom–metal coupling is 'switched on' all levels strictly belong to the complete system, atom plus metal, and the original atomic state decays to zero, but it is nevertheless convenient to consider this discrete level as persisting in the combined system in the form of a broadened 'virtual' level.

To make the foregoing conclusions quantitative we must return to the solution of the Schrödinger equation. Since the wavefunctions extend throughout the complete system we are faced with a many-electron problem involving all the metal conduction-band electrons in addition to the adatom valence electron.

The most common approach has employed self-consistent field (SCF) molecular orbital theory in the Hartree or Hartree–Fock approximations, but before considering the application to surfaces it may be useful to recall the essential ingredients of these methods as originally constituted for treatment of molecules. (Richards and Horsley (1970) present a particularly concise summary, as does Atkins (1974); see in addition any general text, e.g.

Pilar (1968), Dewar (1969), Raimes (1972)). The complete Hamiltonian H for an N-electron system may be written in the form

$$H = H_0 + H'$$

$$= \sum_{i=1}^{N} H_i + H' \qquad (12.17)$$

where

$$H_i = \frac{-\hbar^2}{2m} \nabla_i^2 - \sum_{m=1}^{N} \frac{Z_m e^2}{\mathbf{r}_{im}} \qquad (12.18)$$

and

$$H' = \sum_{i<j}^{N} \sum_{i<j}^{N} \frac{e^2}{\mathbf{r}_{ij}}. \qquad (12.19)$$

Here H_0 is a one-electron operator, involving the co-ordinates of a single electron and relating to the motion of this electron in a potential due to the attraction of N nuclei of atomic numbers Z_m. It would be the appropriate Hamiltonian if there were no interactions between the various electrons. The operator H' relates to the potential generated by such interactions. In the hypothetical absence of H' the Schrödinger equation

$$H_0 \Psi = E \Psi \qquad (12.20)$$

can be solved by assuming that each electron i has a separate wavefunction, or orbital ψ_i. The total wavefunction Ψ, which involves the co-ordinates of all N electrons, can then be written as the product of functions ψ_i, each of which involves the co-ordinates of one electron only, i.e.

$$\Psi(\mathbf{r}_1, \mathbf{r}_2, \ldots, \mathbf{r}_N) = \prod_{i=1}^{N} \psi_i(\mathbf{r}_i). \qquad (12.21)$$

(We might note that to be a correct wavefunction Ψ should include spin terms, and it should be antisymmetric in the electron co-ordinates; these modifications can be included without altering our outline of the method). Combining eqns (12.20) and (12.21) and remembering the partial differential nature of ∇_i^2, we obtain

$$\sum_{i=1}^{N} \frac{1}{\psi_i} \left(-\frac{\hbar^2}{2m} \nabla_i^2 - \sum_{m=1}^{N} \frac{Z_m e^2}{\mathbf{r}_{im}} \right) \psi_i = E. \qquad (12.22)$$

Thus the many-electron equation (12.20) can be separated into N one-electron equations of the form

$$H_i \psi_i = \varepsilon_i \psi_i. \tag{12.23}$$

The total energy of the system E will be

$$E = \sum_{i=1}^{N} \varepsilon_i. \tag{12.24}$$

The electron interactions must now be considered. The problem is that each term in H' (eqn (12.19)) involves the co-ordinates of two different electrons and so the Schrödinger equation

$$H\Psi = E\Psi \tag{12.25}$$

is not separable. It was suggested first by Hartree (1928) that an approximate solution could be obtained if the repulsion between any two electrons was averaged over all positions of one of them, thus giving a term $F(\mathbf{r}_i)$ which is a function of the co-ordinates of the second electron only. When this is done the Schrödinger equation becomes

$$H\Psi = \left\{ \sum_{i=1}^{N} H_i + \sum_{i=1}^{N} F(\mathbf{r}_i) \right\} \prod_{i=1}^{N} \psi_i(\mathbf{r}_i) = E \prod_{i=1}^{N} \psi_i(\mathbf{r}_i). \tag{12.26}$$

As was the case with the non-interacting system (eqns (12.20) and (12.22)) we now have an equation which is separable into N one-electron equations of the type

$$(H_i + F_i)\psi_i = \varepsilon_i \psi_i. \tag{12.27}$$

When the solutions to these equations, ψ_i and ε_i, are inserted in (12.24) and (12.26) we obtain values for the overall system wavefunction Ψ and energy E, within of course the approximations involved in the introduction of the terms F_i.

The actual form of F is chosen to be such that the total system energy E is minimized. In the Hartree–Fock version of the SCF method the one-electron eqns (12.27) become

$$\left\{ H_i + \sum_{j=1}^{N} (J_j - K_j) \right\} \psi_i = \varepsilon_i \psi_i \tag{12.28}$$

where

$$J_j \psi_i(1) = \left\{ \int \psi_j^2(2) \frac{e^2}{r_{12}} \, dV_2 \right\} \psi_i(1) \qquad (12.29)$$

and

$$K_j \psi_i(1) = \left\{ \int \psi_j(2) \psi_i(2) \frac{e^2}{r_{12}} \, dV_2 \right\} \psi_j(1). \qquad (12.30)$$

The Coulomb operators J_j relate to the coulombic repulsions between, for example, electron 2 in orbital j and electron 1 in orbital i. Now, $\psi_j^2(2)\,dV_2$ represents the electron density in volume element dV_2 due to electron 2 and e^2/r_{12} represents the repulsive force between two electrons. Thus, it follows that the integral in eqn (12.29) gives the repulsion between electron 2, averaged over all positions in orbital j, and electron 1. This is just an expression of the original Hartree picture. The exhange operators K_j were introduced by Fock to allow for the fact that the formulation of J_j ignores spin correlation, or the tendency for electrons of the same spin to stay apart, which is a consequence of the Pauli principle. The interaction between a pair of electrons with parallel spins will be overestimated in $J_j\psi_i$; to an extent which might be expected to depend on the overlap between orbitals j and i, hence the presence of the terms $\psi_j\psi_i$ in eqn (12.30). It may be noted that, while spin correlation is taken into account, Hartree–Fock theory does not consider Coulomb correlation, the tendency for electrons (irrespective of spin) to remain apart because of electrostatic repulsion. If exchange is ignored the treatment is known as the Hartree approximation.

At this stage one problem remains—the J_j and K_j involve ψ_j, which we are trying to determine. An iterative procedure must therefore be used in which we guess a set of orbital functions ψ_j for all electrons in the system and use these to obtain an approximation to $\sum\limits^{N} (J_j - K_j)$. This is then inserted in eqn (12.28) to calculate a first improved wavefunction ψ_i for one of the electrons i; ψ_i is next used to calculate an improved interaction potential for some other electron k, and hence from (12.28) we derive an improved ψ_k. The procedure is repeated until further iterations cause negligible change in the wavefunctions. at which stage a self-consistent solution to the problem has been achieved.

In general it is not possible to solve the Hartree–Fock equations (12.28) numerically for molecules; instead the molecular orbitals ψ_j are represented as linear combinations of atomic orbitals (LCAO's) X_k

$$\psi_j = \sum_k c_{jk} X_k \qquad (12.31)$$

and the equations are solved numerically. The overall problem thus reduces to determination of the coefficients c_{jk} for each electron in the system. The X_k are referred to as the basis set and the larger this is the more accurate will be the solution obtained for the Hartree–Fock equations.

To outline the way in which the theory may be applied to adsorption problems we shall examine some calculations in the Hartree approximation by Bennett and Falicov (1966) relating to alkali adsorption on tungsten. We consider one-electron functions $\psi(\varepsilon, l, m)$ which extend throughout the metal and the atom; each state is characterized by three quantum numbers ε (energy), l, and m. Let the energy density of these states be $n(\varepsilon)$ and let V be the adatom volume. Then we may define the local density of states $\rho(\varepsilon)$ at the adsorbate by the relationship

$$\rho(\varepsilon) = \sum_{l,m} \int_V n_{lm}(\varepsilon) |\psi(\varepsilon, l, m; \mathbf{r})|^2 \, d^3r. \qquad (12.32)$$

The term $n(\varepsilon)$ gives the number of system states with quantum numbers l, m in the energy range $d\varepsilon$, while the integral $\psi^2 d^3r$ over V determines the contribution of each of these states in the particular volume corresponding to the adatom. Thus $\rho(\varepsilon)$ may be interpreted as the weight with which the adsorbate state is represented in all system states of a given energy. The negative charge q^- carried by the adatom is obtained by populating the states $\rho(\varepsilon)$ with electrons according to the Fermi distribution $f(\varepsilon)$ (see Chapter 2). If we limit consideration to a zero-temperature approximation, then $f(\varepsilon) = 1$ for $\varepsilon \leqslant \varepsilon_F$ and $f(\varepsilon) = 0$ for $\varepsilon > \varepsilon_F$, where ε_F is the Fermi energy in the metal. Thus we may write

$$q^- = \int_{-\infty}^{+\infty} f(\varepsilon)\rho(\varepsilon) \, d\varepsilon \xrightarrow[(T \to 0)]{} \int_{\infty}^{\varepsilon_F} \rho(\varepsilon) \, d\varepsilon. \qquad (12.33)$$

It turns out that $\rho(\varepsilon)$ approximates to a Lorentzian distribution, of half-width Γ, centred at the position ε'_a of the shifted level, so we may also write

$$\rho(\varepsilon) = \frac{\Gamma/\pi}{(\varepsilon - \varepsilon'_a)^2 + \Gamma^2}. \qquad (12.34)$$

Combining the last two equations (remembering that the integral of $(x^2 + a^2)^{-1}$ is $(1/a)\tan^{-1}(x/a)$) gives

$$q^- = \frac{1}{\pi}\left\{\tan^{-1}\left(\frac{\varepsilon_F - \varepsilon'_a}{\Gamma}\right) + \frac{\pi}{2}\right\}. \qquad (12.35)$$

This expression offers a quantitative basis for Gurney's (1935) model for work-function changes due to alkali and alkaline-earth adsorption. The charge carried by the adatom will depend on the extent to which the virtual state is occupied, and this in turn will be determined by the location of the state with respect to the Fermi level in the metal. If the virtual level ε'_A is situated below the Fermi level in the metal there will be a considerable electronic population in the region of the adatom. This will neutralize the adatom ion core, so that there will be no net charge on the adsorbate and an essentially metallic bond will be formed between adatom and metal. In the second case the virtual level lies in the vicinity of the Fermi level and will be only partially occupied by electrons. The result is a bond showing both ionic and metallic characteristics. Finally, if the normally occupied portion of the virtual level lies completely above ε_F the adparticle will be completely ionized. In the case of an alkali atom this requires that the entire broadened level must lie above ε_F. In the case of an alkaline-earth atom, however, only half of the originally doubly occupied s level need lie above ε_F for the adsorbed particle to carry a single positive charge and so to make a contribution to the electrical double layer at the surface similar to that made by an alkali atom.

Preparatory to setting up the Schrödinger equation for the adsorbate–metal system we choose a basis set consisting of the following: (i) an atomic s-like wavefunction Φ; (ii) a continuum of metallic wavefunctions $X(\varepsilon, l, u)$, where l and u are momentum quantum numbers. It is assumed that all functions are normalized and that Φ is orthogonal to all X's. The one-electron wavefunctions for the complete system are then given by the linear combinations

$$\psi(\varepsilon, l, m) = a(\varepsilon, l, m)\Phi + \sum_{l'} \int \mathrm{d}u \int b(\varepsilon, l, m; \varepsilon', l', u) X N_{l'} \, \mathrm{d}\varepsilon'$$

$$(12.36)$$

where $N_{l'}$ is the density of metallic states with angular momentum l' in the plane perpendicular to the metal surface and the a's and b's are constants to be determined. The form of the full-system wavefunctions ψ is shown in Fig. 12.18 (Gadzuk 1967). At an energy not lying within the range of the broadened atomic state, ψ oscillates within the metal and decays exponentially outside the surface, i.e. it resembles the metallic function X. For energies which do fall within this range, however, the wavefunction has considerable amplitude in the region of the adatom ion core, where it resembles the atomic wavefunction Φ, and reduced amplitude in the metal. The broken lines indicate how ψ is formed by combination of slightly overlapping unperturbed atom and metal wavefunctions.

A considerable simplification is possible when we remember that we are primarily interested in quantities ε_a' and Γ (eqn 12.34) which are properties of the virtual adsorbate level; we are not immediately concerned with the wavefunctions within the metal, nor with the continuum-level energies. Therefore attention may be confined to the region outside the metal. Since the wave functions X decay exponentially here, the predominant contribution to ψ will be from the atomic level as modified by interaction

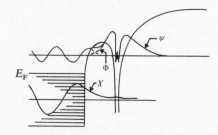

FIG. 12.18. Wavefunctions for electrons at various energies: X is a wavefunction for an electron not within the energy of the broadened band; ψ is a possible wavefunction for an electron within the band; Φ is an atomic wavefunction. (After Gadzuk 1967.)

with the metal, i.e. from $a(\varepsilon, l, m)\Phi$. Selection rules show that

$$a(\varepsilon, l, m) = 0 \quad \text{for} \quad l \neq 0, m \neq 0$$

and so we need consider only $a(\varepsilon, 0, 0)$. Eqn (12.32) for the local density of states $\rho(\varepsilon)$ now becomes

$$\rho(\varepsilon) = \int n_{0,0}(\varepsilon) \, |a(\varepsilon, 0, 0)|^2 \, |\Phi|^2 \, d^3r. \qquad (12.37)$$

Since only one system state can have identical quantum numbers, $n_{0,0}(\varepsilon) = 1$. Also, Φ is normalized and so the integral over Φ^2 is unity. Thus we arrive at the very simple relationship

$$\rho(\varepsilon) \equiv |a(\varepsilon, 0, 0)|^2. \qquad (12.38)$$

From eqn (12.36) we need, therefore, determine only the single constant a; this can be done by straightforward perturbation methods to give

$$|a(\varepsilon, 0, 0)|^2 = \frac{|V_\varepsilon|^2}{(\varepsilon - \varepsilon_{a'}) + \pi V_\varepsilon^4} \qquad (12.39)$$

where

$$|V_\varepsilon|^2 = \int |\langle X(\varepsilon, 0, \mu) \,|H'| \, \Phi\rangle|^2 \, N_0 \, d\mu \qquad (12.40)$$

$$\varepsilon_{a'} = \langle \Phi \,|H'| \, \Phi\rangle \qquad (12.41)$$

and $H' = H_i + F_i$ (cf. eqn (12.27)) is the one-electron Hamiltonian. Comparison of eqns (12.34) and (12.39) shows that

$$\Gamma = \pi \, |V_\varepsilon|^2. \qquad (12.42)$$

As one might have anticipated $\varepsilon_{a'}$ is just the energy of the atomic level Φ existing in a system with total energy operator H', while the virtual level width Γ involves the strength of the interaction $\langle X \,|H'| \, \Phi\rangle$ between Φ and the metal states X.

To evaluate finally $\varepsilon_{a'}$ and Γ an expression is required for the Hamiltonian H' for the adatom valence electron in the region outside the metal. It will be of the form

$$H' = H_i + F_i$$

$$= \left(-\frac{\hbar^2}{2m}\nabla^2 - \frac{e^2}{r}\right) + \left(\frac{-e^2}{4x} + \frac{e^2}{R}\right). \qquad (12.43)$$

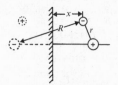

F<small>IG</small>. 12.19. Classical picture of an adatom and the image charges it induces in the metal. (After Gadzuk 1967.)

The potential-energy terms are generated by averaging the interactions of the valence electron with the remaining electrons in the system in the following way.

(i) The coulombic attraction between the valence electron and the spatially averaged charge of the adsorbate positive ion core, i.e. $-e^2/r$ (see Fig. 12.19).

(ii) The interaction of the electron with the unperturbed metal surface, represented approximately by the classical image potential $-e^2/4x$, where x is the distance from the electron to the surface.

(iii) The interaction of the electron with the metal, as perturbed by the presence of adsorbate positive ion core; classically this would be a repulsion $+e^2/R$ ($\approx +e^2/2x$, (cf. Fig. 12.19) between the electron and the image of the ion core.

The combined image terms (ii) and (iii) ($\approx +e^2/4x$) account for the upward shift of the adatom level from ε_a to $\varepsilon_{a'}$ (Fig. 12.17(b)) which increases the degree of ionization and plays a central role in stabilizing the cationically adsorbed species. The inclusion of the term (ii) assumes that as the electron tunnels from the metal to the adatom its image in the metal is generated sufficiently rapidly for the interaction to reach $-e^2/4x$ before the electron tunnels back to the metal. This appears to be the case for systems such as Cs/W (Hewson and Newns 1974) and has been assumed by various workers (Gadzuk 1967, Remy 1970); Bennett and Falicov (1966) have, on the other hand, taken the total image interaction to be $e^2(1-q^-)/2x$.

Having established a suitable Hamiltonian, eqns (12.40) and (12.41) are used to calculate ε_a' and Γ; hence we obtain $\rho(\varepsilon)$ and q^- by way of (12.34) and (12.35). For alkali adsorption on low-index faces of transition metals Γ is of the order of 0·3–1·5 eV; the level shift is $\Delta\varepsilon \approx 0\cdot5$–$1\cdot0$ eV, and the effective charge on the alkali adatom is 0·65–0·9 (Bennett and Falicov

1966, Gadzuk 1969, Remy 1970, Gadzuk, Hartman, and Rhodin 1971).

12.5.2. *The Anderson model*

A calculation in the Hartree approximation is adequate for treatment of alkali adsorption since the adsorbate carries a high positive charge; only a single electron need ever be considered in the adsorbate orbital Φ. If we now move on to examine hydrogen adsorption, for example, the same is no longer the case. Some degree of electron transfer from metal to adatom may occur and we must allow for the possibility of two electrons being on the adatom simultaneously; this requires a Hartree–Fock approach. Following Grimley (1971) the problem may be illustrated by taking an isolated hydrogen atom and forming a hydride ion. The second electron enters the 1s level, but because of the Coulomb repulsion between the two electrons there is a large upward self-consistent shift in the 1 s energy from ε_a to ε_{a^-} (Fig. 12.20). This intra-atomic repulsion term U is given by

$$U = (I - A) = 1250 \text{ kJ mol}^{-1}$$

for hydrogen, where I is the ionization potential and A the electron affinity of the 1s level. If the system in the form H^- approaches a metal surface the energy of the affinity level ε_{a^-} will be shifted downwards by an amount $e^2/4x$ due to the interaction of the negative ion with its image in the metal. At the surface the effective repulsion term is therefore (see Fig. 12.21(a))

$$U_e = U - e^2/2x \ (\approx 580 \text{ kJ mol}^{-1} \text{ for hydrogen}).$$

In a formal way U_e includes both Coulomb repulsion and exchange interactions (Anderson 1961, Thorpe 1972). An adequate

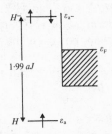

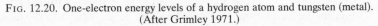

FIG. 12.20. One-electron energy levels of a hydrogen atom and tungsten (metal).
(After Grimley 1971.)

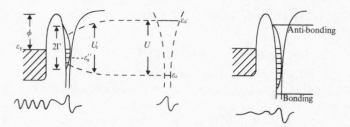

FIG. 12.21. (a) Origin of the effective intra-atomic repulsion term U_e in the Anderson model. (b) Localized states above and below the metal conduction band. (After Gomer 1975a.)

treatment of this term represents one of the main difficulties of the calculation. Next we consider transfer of an electron from the adsorbate to *extended* electron states in the metal. In this case there is no level shift comparable to that on the adsorbate, principally because electrons in the metal are very efficiently screened from one another, and so a second repulsion term of type U is not required.

The Hamiltonian for the complete system will be similar to that considered previously (eqn (12.43)) in connection with alkali adsorption, except for the addition of a term $U_e\langle n_\downarrow\rangle$ representing the interaction of an electron in the spin-up state (say) with the average population $\langle n_\downarrow\rangle$ of the spin-down state on the adsorbate. A Hamiltonian of this form was first introduced by Anderson (1961) when considering magnetic impurities in solids, and it has been applied to chemisorption problems by various workers including Grimley (1967) and Newns (1969). The calculations are in principle very similar to those of Bennett and Falicov considered previously. The main parameter of interest remains the local density-of-states function (eqn (12.34)) and, provided the adatom–surface interaction is not very strong, it retains the form shown in Fig. 12.17. When, however, the strength of the interaction is high it becomes possible for discrete localized states to separate out, an empty one above and a filled one below the metal band (Fig. 12.21(b)).

12.5.3. *Surface molecules*

If an adatom interacted strongly with only a limited number of atoms in the surface of a metal it would be possible to consider the system in terms of a 'surface molecule' weakly coupled to the remainder of the solid (Fig. 12.22, Grimley 1971, Thorpe 1972). We might then think of the split-off states mentioned above as bonding and anti-bonding levels in this surface molecule. If the picture was realistic it should lead to a considerable simplification in chemisorption theory, and observations such as the desorption of nickel carbonyl from a CO adlayer on Ni suggest that it may indeed have substance (Thorpe 1972).

The principal interaction to be considered is between the adatom and metal orbitals, which are restricted within the region of a few surface atoms rather than extending throughout the metal. Coulomb and exchange interactions between these metal electrons may now be important, just as they are between electrons on the adatom, and a second term of similar form to $U_e \langle n_\downarrow \rangle$ should be included in the Hamiltonian to take account of this effect. This generates what is referred to as a Hubbard Hamiltonian (Hubbard 1964); using this, calculations proceed essentially as before.

The concept of the surface molecule in the Hubbard limit has been applied by Thorpe (1972) to calculate the binding energies of 5d transition atoms on tungsten. The experimental trends (Plummer and Rhodin 1968) were well reproduced (although the absolute values were not). On the basis of relatively simple considerations it has been suggested (Kelly 1974) that the surface-molecule picture may be of limited applicability, but more detailed calculations by Grimley and Pisani (1974) show that this is not necessarily correct.

12.5.4. *Indirect adatom–adatom interactions*

If an electron moves in the field of two identical nuclei which are separated in a vacuum, its wavefunction will fall rapidly to zero in the intervening space (Fig. 12.23(a)). If the nuclei are at the

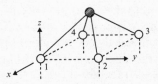

FIG. 12.22. W_4X pyramidal surface molecule on b.c.c. W(100). (After Thorpe 1972.)

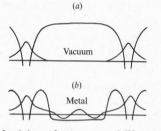

FIG. 12.23. Wavefunctions for (a) two free atoms and (b) two atoms adsorbed on a metal. (After Grimley 1967a.)

surface of a metal, however, the non-localized substrate orbitals provide a path for coupling between them. The wavefunction now oscillates in the metal (Fig. 12.23(b)) rather than falling to zero and so produces an indirect interaction between the adsorbed atoms. The existence of such an interaction was originally suggested by Koutecky (1958) and first investigated quantitatively, using the Anderson model, by Grimley (Grimley 1967a,b, Grimley and Walker 1969, see also Einstein and Schrieffer 1973). The interaction is found to be oscillatory (alternately attractive and repulsive) and of long range, falling off as the inverse cube of distance. It may occur with alkali and hydrogen atoms on transition metals, but not with many molecules. It provides an appealing explanation for a number of experimental observations, including ordered diffraction patterns produced by adsorbate layers at low coverage.

12.5.5. *Alternative approaches to chemisorption theory*

In addition to the Anderson and related approaches already outlined, a number of other variations on self-consistent-field molecular-orbital theory have been investigated, for example extended Hückel theory (see e.g. Anders, Hansen, and Bartell 1973). Also, it is hardly surprising that the second major method for the description of molecular bonding, valence-bond theory, has also been applied to the adsorption problem (see Paulson and Schrieffer 1975). Other approaches which have been investigated include density-functional theory, to which reference has already been made during the discussion of work-function variations in adsorbed alkali layers (Section 12.4.3). This has been applied (Ying, Smith, and Kohn 1975) to describe various aspects of hydrogen adsorption.

Comprehensive information on chemisorption theory can be found in a number of reviews, for example Gomer (1974, 1975a,b), Grimley (1975), and Schrieffer and Soven (1975).

REFERENCES

ANDERS, L. W., HANSEN, R. S., and BARTELL, L. S. (1973). *J. chem. Phys.* **10**, 5277.

ANDERSON, P. W. (1961). *Phys. Rev.* **124**, 41.

ATKINS, P. W. (1974). *Quanta, a handbook of concepts.* Clarendon Press, Oxford.

AYRAULT, G., and EHRLICH, G. (1974). *J. chem. Phys.* **60**, 281.

BASSETT, D. W. (1970). *Surf. Sci.* **23**, 240.

—— (1973). In *Specialist periodical reports: surface and defect properties of solids* (eds M. W. Roberts and J. M. Thomas), vol. 2, p. 34. Chemical Society, London.

BASSETT, D. W., and PARSLEY, M. J. (1969). *Nature, Lond.* **221**, 1046.

—— (1970). *J. Phys. D Appl. Phys.* **3**, 707.

BENNETT, A. J., and FALICOV, L. M. (1966). *Phys. Rev.* **151**, 512.

BESOCKE, K. and WAGNER, H. (1973). *Phys. Rev.* B **8**, 4597.

BETTLER, P. C., BENNUM, D. H., and CASE, C. M. (1974). *Surf. Sci.* **44**, 360.

BLAKELY, J. M. (1963). *Prog. mater. Sci.* **10**, 395.

BRADSHAW, F. J., BRANDON, R. H., and WHEELER, C. (1964). *Acta Metall.* **12**, 1057.

CHALMERS, B., KING, R., and SHUTTLEWORTH, R. (1948). *Proc. R. Soc.* A **193**, 465.

CHEN, J. M., and PAPAGEORGOPOULOS, C. A. (1970). *Surf. Sci.* **21**, 377.

—— (1971). *Surf. Sci.* **26**, 499.

CYROT-LACKMANN, F., and DUCASTELLE, F. (1971). *Phys. Rev.* B **4**, 2406.

DELAMARE, F., and RHEAD, G. E. (1971). *Surf. Sci.* **21**, 267.

DEWAR, M. J. S. (1969). *The molecular orbital theory of organic chemistry.* McGraw-Hill, New York.

DRECHSLER, M. (1954). *Z. Elektrochem.* **58**, 327.

EHRLICH, G., and HUDDA, F. G. (1966). *J. chem. Phys.* **44**, 1039.

EHRLICH, G., and KIRK, C. F. (1968). *J. chem. Phys.* **48**, 1465.

EINSTEIN, T. L., and SCHRIEFFER, J. R. (1973). *Phys. Rev.* B **7**, 3629.

FEHRS, D. L., and STICKNEY, R. E. (1971). *Surf. Sci.* **24**, 309.

GADZUK, J. W. (1967). *Surf. Sci.* **6**, 133.

—— (1969). In *The structure and chemistry of solid surfaces* (ed. G. A. Somorjai), paper 43. John Wiley, New York.

GADZUK, J. W., HARTMAN, J. K., and RHODIN, T. N. (1971). *Phys. Rev.* B **4**, 241.

GERLACH, R. L., and RHODIN, T. N. (1968). *Surf. Sci.* **10**, 446.

—— (1969). *Surf. Sci.* **17**, 32.

—— (1970). *Surf. Sci.* **19**, 403.

GIRIFALCO, L. A., and WEIZER, V. G. (1959). *Phys. Rev.* **114**, 687.

GJOSTEIN, N. A. (1967). In *Surfaces and interfaces* (eds. J. J. Burke, N. L. Reed, and V. Weiss), vol. 1, p. 271. Syracuse University Press, Syracuse, New York.

GOMER, R. (1974). *Adv. chem. Phys.* **27,** 211.

—— (1975*a*). *Solid State Phys.* **30,** 93.

—— (1975*b*). *Acct. chem. Res.* **8,** 420.

GRAHAM, W. R., and EHRLICH, G. (1973). *J. chem. Phys.* **59,** 3417.

—— (1974). *Surf. Sci.* **45,** 530.

GRIMLEY, T. B., (1967*a*). *Proc. phys. Soc.* **90,** 751.

—— (1967*b*). *Proc. phys. Soc.* **92,** 776.

—— (1971). *Ber. BunsenGes. phys. Chem.* **75,** 1003.

—— (1975). In *Progress in surface and membrane science* (eds. J. F. Danielli and M. D. Rosenberg), vol, 9, p. 71. Academic Press, New York.

GRIMLEY, T. B., and PISANI, C. (1974). *J. Phys. C Solid State Phys.* **7,** 2831.

GRIMLEY, T. B., and WALKER, S. M. (1969). *Surf. Sci.* **14,** 395.

GURNEY, R. W. (1935). *Phys. Rev.* **47,** 479.

GYFTOPOULOS, E. P., and LEVINE, J. D. (1962). *J. appl. Phys.* **33,** 67.

HARTREE, D. R. (1928). *Proc. Cambridge Phil. Soc.* **24,** 89, 111, 426.

HEWSON, A. C., and NEWNS, D. M. (1974). *Jap. J. appl. Phys. Suppl.* **2,** (2), 121.

HUBBARD, J. (1964). *Proc. R. Soc. A* **281,** 401.

KELLY, M. J. (1974). *Surf. Sci.* **43,** 587.

KIRBY, R. E., McKEE, C. S., and ROBERTS, M. W. (1976). *Surf. Sci.* **55,** 725.

KIRBY, R. E., McKEE, C. S., RENNY, L. V., and ROBERTS, M. W. (1979). To be published.

KOUTECKY, J. (1958). *Trans. Faraday Soc.* **54,** 1038.

LANG, N. E. (1971). *Phys. Rev. B* **4,** 4234.

LEGRAND-BONNYNS, E., and PONSLET, A. (1975). *Surf. Sci.* **53,** 675.

MACDONALD, J. R., and BARLOW, C. A., JR. (1963). *J. chem. Phys.* **39,** 412.

MEDVEDEV, V. K., and YAKIVCHUK, A. I. (1974). *Sov. Phys.—Solid State* **16,** 634.

MEDVEDEV, V. K., NAUMOVETS, A. G., and SMEREKA, T. P. (1973). *Surf. Sci.* **34,** 368.

MELMED, A. J. (1967). *J. appl. Phys.* **38,** 1885.

MILLER, A. R. (1946). *Proc. Camb. phil. Soc.* **42,** 292.

MOORE, A. J. W. (1963). In *Metal surfaces: structure, energetics and kinetics* (eds. W. D. Robertson and N. A. Gjostein), p. 173. American Society of Metals, Ohio.

MÜLLER, E. W. (1969). In *Molecular processes on solid surfaces* (eds. E. Drauglis, R. Gretz, and R. I. Jaffee), p. 400. McGraw-Hill, New York.

MÜLLER, E. W., and TSONG, T. T. (1969). *Field ion microscopy: principles and applications.* Elsevier, Amsterdam.

MULLINS, W. W. (1961). *Phil. Mag.* **6,** 1313.

NEWNS, D. M. (1969). *Phys. Rev.* **178,** 1123.

PAPAGEORGOPOULOS, C. A., and CHEN, J. M. (1973). *Surf. Sci.* **39,** 283.

PAULSON, R. H., and SCHRIEFFER, J. R. (1975). *Surf. Sci.* **48,** 329.

PERDEREAU, J., and RHEAD, G. E. (1967). *Surf. Sci.* **7,** 175.

—— (1971). *Surf. Sci.* **24,** 555.

PICHAUD, M., and DRESCHLER, M. (1972). *Surf. Sci.* **32,** 341.

PILAR, F. L. (1968). *Elementary quantum chemistry.* McGraw-Hill, New York.

PLUMMER, E. W., and RHODIN, T. N. (1968). *J. chem. Phys.* **49,** 3479.

PLUMMER, E. W., and YOUNG, R. D. (1970). *Phys. Rev. B* **1,** 2088.

RAIMES, S. (1972). *Many-electron theory.* North-Holland, Amsterdam.

REMY, M. (1970). *J. chem. Phys.* **53,** 2487.

RHEAD, G. E. (1965). *Acta Metall.* **13,** 223.

—— (1969). *Surf. Sci.* **15,** 353.

RHEAD, G. E., and McLEAN, M. (1964). *Acta Metall.* **12,** 401.

RHEAD, G. E., and MYKURA, H. (1962). *Acta Metall.* **10,** 843.

RICHARDS, W. G., and HORSLEY, J. A. (1970). *Ab initio molecular orbital calculations for chemists.* Clarendon Press, Oxford.

SCHMIDT, L. D., and GOMER, R. (1966). *J. chem. Phys.* **45,** 1605.

SCHRIEFFER, J. R., and SOVEN, P. (1975). *Phys. Today* April, p. 24.

SIDORSKI, Z., PELLY, I., and GOMER, R. (1969). *J. chem. Phys.* **50,** 2382.

SIDORSKI, Z., and WOJCIECHOWSKI, K. F. (1971). *Acta Phys. Polon. A* **40,** 661.

SMOLUCHOWSKI, R. (1941). *Phys. Rev.* **60,** 661.

SOMMER, A. H. (1968). *Photoemissive materials.* John Wiley, New York.

SWANSON, L. W., and STRAYER, R. W. (1968). *J. chem. Phys.* **48,** 2421.

TAYLOR, J. B., and LANGMUIR, I. (1933). *Phys. Rev.* **44,** 423.

THORPE, B. J. (1972). *Surf. Sci.* **33,** 306.

TODD, C. J., and RHODIN, T. N. (1974). *Surf. Sci.* **42,** 109.

TOPPING, J. (1927). *Proc. R. Soc. A* **114,** 67.

TSONG, T. T. (1972). *Phys. Rev. B* **6,** 417.

TUCKER, C. W. (1967). *Acta Metall.* **15,** 1465.

—— (1969). In *The structure and chemistry of solid surfaces* (ed. G. A. Somorjai), paper 58. John Wiley, New York.

WILSON, R. G. (1967). *Surf. Sci.* **7,** 157.

WYNBLATT, P., and GJOSTEIN, N. A. (1970). *Surf. Sci.* **22,** 125.

YING, S. C., SMITH, J. R., and KOHN, W. (1975). *Phys. Rev. B* **11,** 1483.

13

THE IMPACT OF NEW EXPERIMENTAL
DEVELOPMENTS ON HETEROGENEOUS
CATALYSIS

13.1. Introduction

THE detailed understanding of a heterogeneously catalysed reaction necessitates pinpointing the 'active centre' which is clearly a difficult task when it is appreciated that the claim is frequently made that 'such centres' constitute only some 1 in 10^3 or even less of the surface atoms. The investigator has frequently had to resort to the asterisk to describe the 'active site', but with the advent of experimental methods which probe the surface directly there is now optimism that, having defined the surface in terms of its chemical composition, its crystallography and chemical bonding, at least the probability of isolating the active site itself will be substantially enhanced. The point is sometimes made that the newer techniques in surface chemistry enable us to find out more and more about less and less, but maybe that is what is at the heart of understanding the mechanistic details of a catalytic reaction!

In this chapter we consider a number of examples where impact on current thinking in heterogeneous catalysis is already evident. We draw particular attention to details of bonding of surface species, the role of defective-surface crystallography in influencing chemical reactivity, and the surface enrichment that occurs with alloys. The examples chosen illustrate the future potential and are in no way intended to be comprehensive.

13.2. Catalytic behaviour of tungsten carbide

Tungsten carbide is known to be a good catalyst (Levy and Boudart 1973) for a number of chemical reactions, including the isomerization of 2, 2-dimethyl propane, which are also catalysed by platinum, an expensive catalyst! The question that immediately follows is: Why should tungsten carbide behave like platinum?

Bennet *et al.* (1974), taking note of the possible correlation between catalytic activity and the density of d-like electron states at the Fermi level of the metal, investigated the band structure of WC and compared it with W and Pt. Using XPS they showed that tungsten carbide has a 'density-of-states curve' more like Pt than W. Fig. 13.1 compares the valence-band spectrum of WC with those of clean Pt and the (111) plane of W. The density of states at the Fermi level, $N(\varepsilon_F)$, of WC is intermediate in character between that of W, where the Fermi level falls in a region of low density of d-like states, and that of Pt, where the Fermi level falls in a region of high density of d-like states. The authors suggest that an important factor leading to the increase in $N(\varepsilon_F)$ is the change from a body-centred structure (where ε_F falls in a broad region of low density of states) to, in tungsten carbide, a hexagonal structure where ε_F falls in a region of high density of states. The high $N(\varepsilon_F)$ value is therefore concluded to be an important parameter in determining catalytic activity. However, it is not the only one, since on this argument alone vanadium would be expected to be a substitute for platinum whereas in fact it is not. Ross and Stonehart (1975) on the other hand, have shown recently, using Auger electron spectroscopy, that the material used by Levy and Boudart (1973) was an atypical

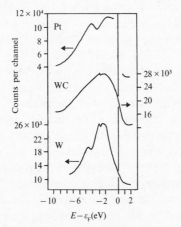

FIG. 13.1. Valence-band XPS spectra of WC, Pt, and W (the (111) plane). The Fermi level ε_F has been placed so that a standard line Au $4f_{\frac{5}{2}}$ lies 87·6 eV below it. (From Bennett *et al.* 1974.)

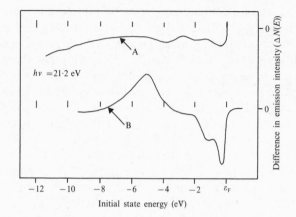

FIG. 13.2. UPS difference curves; curve A, Pt+5 langmuir O_2; curve B, Ni(111)+ 10 langmuir O_2 (Eastman Demuth 1974.)

tungsten carbide. In particular its surface composition differed from that of the bulk in that it was carbon deficient.

In an analogous way Spicer and his colleagues (Spicer *et al.* 1976), prompted by the unusual catalytic activity of platinum, posed the question as to 'how the bonding of the reactants to platinum differs from their bonding to other d metals'. Fig. 13.2. shows the UPS difference curves obtained (Eastman and Demuth 1974) at a photon energy of 21·2 eV for oxygen adsorbed on platinum and nickel. These difference curves are obtained by subtracting the energy-distribution curves for the clean metal surface from those for the oxygen-covered surface. Negative portions of the difference curves therefore correspond to decreases in electron emission, while positive portions correspond to an increase as a result of oxygen adsorption. It is obvious that platinum is very different to nickel in that the 'density of states' (difference curve) shows a relatively uniform decrease in emission right across the valence band with no evidence for emission from electron states associated with oxygen. The conclusion that the bonding of oxygen to platinum is different to that with nickel follows naturally; whether this also accounts for platinum's unusual catalytic activity is, however, speculative at this point in time.

13.3. Nature of a supported molybdenum hexacarbonyl olefin disproportionation catalyst

Whan, Barber, and Swift (1972) investigated the changes which occurred during heating of molybdenum hexacarbonyl supported on γ-alumina. Such thermal treatment is known to induce catalytic activity for the disproportionation of propylene at 298 K. Loss in intensity of the C(1s) peak on heating *in vacuo* to 383 K was attributed to removal of some of the CO ligands around the central molybdenum atom. There was only a small change in the Mo peak. However, on exposure to air there was a drastic change in the Mo 3d spectra towards higher binding energy and the overall profile suggested strongly that several different species of molybdenum in several different chemical environments were present. The authors conclude that the active species in disproportionation catalysts of this type are molybdenum entities which have an oxidation number probably greater than zero but less than 6, attached to electron-withdrawing sites on the alumina.

13.4. Unsuspected surface reactions

The interaction between hydrogen sulphide and lead surfaces which have chemisorbed oxygen showed two interesting features (Roberts, Saleh, and Wells 1964): (*a*) such surfaces were more reactive to hydrogen sulphide than were the clean lead surfaces; and (*b*) a one-to-one correlation was reported to exist between the hydrogen sulphide uptake and the number of oxygen adatoms present on the surface, which led not unnaturally to the suggestion that S=O bonds were formed even at room temperature. If this in fact was the case then electron spectroscopy offered a unique method of exploring the molecular processes since distinct chemical shifts would be anticipated in the S(2p) levels in going from $S_{(ads)}$ to $S{=}O_{(ads)}$. Fig. 13.3 shows the O(1s) and S(2p) spectra for a lead surface first exposed to oxygen and then to hydrogen sulphide at room temperature (Kishi and Roberts 1975a). The O(1s) peak intensity was found somewhat surprisingly to decrease with continued exposure, and there was a concurrent development of an S(2p) peak. Clearly the earlier suggestion that S=O bonds were formed was incorrect and dissociation of hydrogen sulphide followed by desorption of water was the likely mechanism. In the earlier studies the water

(a)

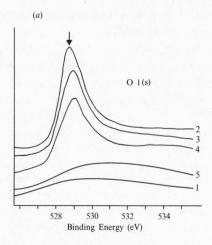

(b)

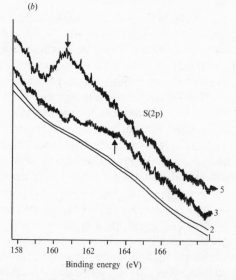

FIG. 13.3. XPS study of H_2S interaction with $Pb + O_{(ads)}$ at 290 K. (a) O(1s) peak. (b) S(2p) peak. Curve 1, Pb; curve 2, O_2, 2×10^{-5} torr, 100 s; curve 3, H_2S, 10^{-6} torr, 100 s; curve 4, H_2S, 5×10^{-6} torr, 100 s; curve 5, H_2S, 2×10^{-5} torr, 150 s.

would have been condensed out in a trap maintained at 193 K. This therefore explains the one-to-one correlation between oxygen pre-adsorbed and 'sulphur' adsorbed. These views were amply confirmed by investigating the reaction at lower temperature (77 K) and gradually warming to room temperature where the decrease in the O(1s) intensity at 530 eV was replaced first by a wide O(1s) peak at about 532 eV, typical of condensed water; then, as the temperature approached 273 K the large peak at 532 eV also disappeared reflecting desorption of the 'adsorbed' water.

We have therefore an example of a surface reaction where an oxide is converted into a sulphide through an intermediate reduction process involving dissociation of the hydrogen sulphide molecule and desorption of water. Moreover, all the molecular events occur below room temperature and with what must be a comparatively small activation energy.

13.5. Role of minor elements

The addition of minor elements, sometimes referred to as promotors, is a common occurrence in the industrial practice of heterogeneous catalysis. The exact role of such promoters is not always clear. We give two examples here where minor elements present at the surface of a solid have been shown to determine in an unequivocal manner (a) the surface structure and (b) the chemical reactivity of the surface.

Leysen, Hopkins, and Taylor (1975) have recently investigated the way in which electropositive elements such as caesium, calcium, and potassium can induce structural ordering of zinc oxide surfaces. Using a combination of LEED and AES they showed that on the polar (000$\bar{1}$)-O surface adsorption of caesium led to a well-resolved ($\sqrt{3} \times \sqrt{3}$) superstructure. The Cs ions were very strongly bound and could only be removed with difficulty by ion bombardment. Clearly the adsorbed caesium is capable of inducing the clean (1×1) surface to transform to a very stable ($\sqrt{3} \times \sqrt{3}$) structure. The possible analogy with real catalysts is obvious and is reminiscent of the suggestions of McCarroll Edmonds, and Pitkethly (1969) and Edmonds, McCarroll, and Pitkethly (1970) for the role of 'surface reorientation' in heterogeneous catalysis and also the much earlier work of Cimino, Boudart, and Taylor

(1954) who suggested that the addition of potassium oxide to iron reduces the extent of dissociation in the adsorbed species during the hydrogenolysis of ethane. In the latter context it is pertinent to recall that sulphur pre-adsorbed on an evaporated iron surface (Kishi and Roberts 1975b) reduces the activity of that surface for the molecular breakdown of adsorbed carbon monoxide. This conclusion, thought to reflect the role of back-bonding into the anti-bonding orbitals of CO from the metal d band, was based on the fact that peaks in the ultra-violet-induced photoelectron spectrum, associated with molecular CO, were not replaced slowly by the peaks typical of atomic-like carbon and oxygen which occurred with clean iron. Further confirmation was obtained from carbon and oxygen core-level spectra.

Electron spectroscopy has also had important industrial implications, and one example where surface analysis by AES was useful was in unravelling the role of lead as a poison of some commercial copper catalysts (Bahsin 1974). Examination of the catalysts by more conventional physical and chemical analysis and by scanning electron microscopy had failed to reveal any significant differences.

13.6. Crystal orientation in metal surfaces

There are very few studies where unequivocal correlations have been drawn between catalytic activity and crystal orientation. Furthermore it is frequently assumed in calculations of surface-atom densities that ostensibly polycrystalline metal surfaces are composed of equal concentrations of the more closely packed planes, i.e. in the case of an f.c.c. structure (100), (111), and (110) planes.

In a discussion (Roberts and Stewart 1970) of the catalytic activity of polycrystalline gold surfaces for the decomposition, and also oxidation, of methanol the possibility of thermally created active sites was considered. More recently Somorjai and his colleagues (Lang, Joyner, and Somorjai 1972a,b) showed in some beautiful experiments that stepped surfaces were more active than flat surfaces. One might argue that this was not unexpected, but that would be to do less than justice to work which proved that at least for hydrogen dissociation surface steps are important. Bernasek and Somorjai (1975) and Kahn,

Petersen, and Somorjai (1974) have more recently shown that steps are relevant to other reactions including those of hydrocarbons. These, taken together with the gold reactivity data, led to experiments being performed on *polycrystalline* gold surfaces using LEED and AES (Isa, Joyner, and Roberts 1975). The strategy adopted was largely speculative in the sense that LEED studies are by the normal requirements of the method only carried out with *single crystals*. The results (Plate 9) were interesting and revealing in two ways: (*a*) with continued treatment, discrete simple LEED patterns emerged and (*b*) the interpretation of these patterns led to the conclusion that the surface was composed predominantly of stepped (111) planes. There is therefore direct experimental evidence for a thermally induced ordered structure on polycrystalline gold and moreover that the structure is of a kind (stepped (111)) which is likely to show enhanced chemical reactivity. No experimental method other than LEED would provide this obviously important information. Whether the steps are *the active sites* remains to be established, however. Ibach (1975) has recently shown that the enhancement of reactivity at steps may well be a consequence of the lowering of the activation energy arising from the ease of charge transfer from the solid to the adsorbate. This is reminiscent of some early ideas on the role of promotors in catalysis, the induction effect of Boudart (1952), and is also analogous to the activation of photocathodes by alkali-metal atoms. Recently Somorjai and Blakely (1975) have refined the role of steps in catalysis by showing that steps, and kink sites within steps, are distinctly different catalytic sites on a platinum surface where C—C, C—H, and H—H bond scissions occur during hydrocarbon reactions.

Pritchard, Catterick, and Gupta (1975) have approached the problem of the crystallographic composition of polycrystalline surfaces in a different way. They have used the adsorption of carbon monoxide as a probe of the surface structure of polycrystalline copper surfaces and conclude that they 'often consist mainly of high index facets...'.

13.7. Atomic composition of alloy surfaces

Alloys have, as emphasized elsewhere, played an important role in the development of our current views on heterogeneous catalysis, correlations of catalytic activity being made with their bulk characteristics. The question remained, however, as to whether there was a direct connection between surface and bulk properties. During the last few years considerable effort has been expended to try to answer this question. Initially the approach was to use the so-called surface-titration method, which is the application (Van der Plank and Sachtler 1968) of chemisorption to determine the concentration of an active metal in the surface of an alloy with an inactive component. With the advent of electron spectroscopy (AES and XPS), however, more definitive information has been possible, although ultra-violet photoemission data had already paved the way in that strong evidence for surface enrichment of a Pt–Au alloy by Pt had been obtained by Bouwman and Sachtler (1970); later Bouwman and Biloen (1974), studying the Pt–Sn alloys, explored in detail the question of surface enrichment by combining AES with XPS. After oxidation there is a marked decrease in the Pt $4f_{\frac{7}{2}}$ signal measured relative to the Sn $3d_{\frac{5}{2}}$ signal. On reduction the situation is reversed. It is concluded that tin enrichment occurs on oxidation and that after reduction the surface returns to a composition nearer that of the bulk but still richer in tin.

The driving force for surface segregation or enrichment is the decrease in specific surface work, the latter being determined by the mutual interactions between substrate atoms and adatoms (see Chapter 2). Since large mutual interactions between substrate atoms are reflected by a large heat of sublimation or high specific surface work, surface enrichment with the component, with the smallest heat of sublimation is favoured. Although calculations based on regular solution theory (Burton, Hyman, and Fedak 1975) suggest that segregation of the more volatile constituent to the surface of a bulk alloy or small alloy particle, a microcluster alloy, may occur for all alloy systems regardless of temperature, more recent studies (Burton and Machlin, 1976) indicate that a rule based on the melting curve of the alloy is preferable.

Christmann and Ertl (1972) made use of the fact that if there was a difference between the surface composition of, say, an alloy and the bulk then this would show up in angular electron spectroscopic studies. They investigated two cases by AES, one a homogeneous Ag–Pd alloy and the other where enrichment of

Ag had occurred at the surface. In the latter case the ratio of the intensities of the Pd (331 eV) to the Ag (354 eV) signals varied with the angle of the primary electron beam. With the homogeneous alloy the ratio was constant and independent of angle.

The catalytic selectivity of alloys has recently been reviewed (Clarke 1975).

13.8. Surface reactions of adsorbed hydrocarbons

Bond (1966) proposed that in such reactions as hydrogenation, dehydrogenation, and dimerization catalysed by transition metals, π-d bonding plays an important role. Demuth and Eastman (1974), with this very much in mind, studied the valence orbitals for physically adsorbed and chemisorbed acetylene, ethylene, and benzene. Since π-orbital ionization energies and several σ-orbital energies are observed, σ-and π-orbital relaxation shifts (non-chemical bonding) and π-orbital bonding shifts can be approximately separated to estimate π–d interaction strengths and chemisorption energies (Fig. 13.4). The philosophy adopted in the Demuth–Eastman approach is to assume that shifts in non-bonding molecular orbitals are due entirely to 'relaxation effects', thus giving a measure of these effects and enabling due cognizance of them to be taken in the shifts observed in the bonding orbitals. This therefore enables the shift in energies due

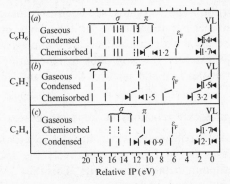

FIG. 13.4. Vertical ionization energies and Fermi (ε_F) and vacuum levels (VL) for the gaseous, condensed and chemisorbed phases of (a) benzene, (b) acetylene, and (c) ethylene, all plotted relative to σ-orbital ionization potentials (IP) for the gas phase. Relaxation shifts are given by the vacuum-level shifts while bonding shifts are given by relevant π-orbital shifts. (From Demuth and Eastman 1974.)

only to chemisorption or surface reaction to be ascertained. These authors showed that reactions which were endothermic in the gas phase were exothermic surface reactions when allowance was made for the changes in orbital energies. In particular, by studying C_2H_4 and C_2H_2 comment could be made on the dehydrogenation of ethylene. The generality of the approach will be crucially dependent on whether the assumption that relaxation effects are the same for all orbitals turns out to be correct.

Plummer, Waclawski, and Vorburger (1974) have also carried out photoelectron studies, supplemented by LEED and workfunction measurements. They investigated the adsorption of ethylene on W(110). The conclusion, perhaps not surprising in view of the well-established surface chemistry of ethylene, was that self-hydrogenation occurred to form C_2H_2-type surface species. These are the 'acetylenic residues' proposed by Jenkins and Rideal (1955) some 20 years previously, solely on the basis of kinetic studies with nickel. Plummer *et al.* contrast their conclusions with those reported earlier by Barford and Rye (1972).

Mason and his co-workers (Clarke, Gay, and Mason 1975, Clarke, Gay, Law, and Mason 1975) have used the powerful combined XPS–UPS–LEED approach to make a thorough study of the role of surface structure in determining the reactivity of platinum surfaces for the adsorption of various olefins and fluorinated olefins. They distinguish between associative and dissociative chemisorption and recognize the conditions necessary to give rise to intramolecular elimination of HF. They emphasize the analogies that exist with organometallic complex chemistry and have recently reviewed this approach to elucidating the bonding of simple molecules to metal surfaces (Mason and Textor 1976).

13.9. Small molecule reactions at stepped surfaces

Somorjai and his colleagues were the first to attempt to model catalyst surfaces by studying quantitatively the role of surface steps in determining catalytic reactivity. This they did by coupling LEED, which enabled surface structure to be defined, with a variety of other experimental techniques (AES, massspectrometry and angular-distribution studies (Lang, Joyner and Somorjai, 1972; Bernasek and Somorjai, 1975)). These authors chose to study the interaction of H_2 with stepped and atomically flat surfaces of platinum and reported a profound influence of surface steps. More recent work by Christmann and Ertl (1976) confirmed the importance of surface steps in the kinetics of hydrogen interaction with platinum but suggested that their influence is somewhat less than that observed by Somorjai. Other studies, such as those of Bonzel and Ku (1973) of the adsorption of oxygen on Pt(111), were less clear in delineating a precise role for surface steps while Engel et al. (1977) believe that steps play a significant part in determining the kinetics of desorption of oxygen from stepped tungsten surfaces. A recent review (Somorjai, 1975) emphasises the role of steps and the concept of 'active sites' in reactions of larger molecules such as hydrocarbons.

Angular-distribution studies (Bernasek and Somorjai 1975) of the product and scattered reactants from surfaces also provide a clue to the role of the steps in promoting the dissociation of hydrogen on a platinum surface, and Fig. 13.5 shows the angular distributions of H_2, D_2, and HD from the (111) and two stepped surfaces Pt-(S)[5(111)×(111)] and Pt-(S)[9(111)×(111)]. The angle of incidence, indicated by the arrow, is at the same angle that corresponds to the maximum intensity in the scattered beam, therefore indicating the specular nature of the scattering process. It can be seen that the width of the scattering pattern increases and the specular intensity decreases as the step density is increased from (111) to [9(111)×(111)] to [5(111)×(111)]. This is a clear indication of more efficient energy transfer between the incident gas molecules and the surface as the step density is increased. The angular distribution of the desorbed product HD is cosine (peaked at the surface normal), indicating complete thermal accommodation with the surface prior to desorption.

Analogous data were obtained with the decomposition of methylamine and formic acid. It is interesting to note that in the case of CO_2 formation from formic acid the angular-distribution data were better fitted to a $\cos^2\theta$ rather than a $\cos\theta$ function. The mechanistic implication is not clear but the authors suggest that it might be connected with the possible 'autocatalytic' decomposition mechanism proposed by Madix, Falconer, and McCarty (1974) for formic acid decomposition on nickel. More general aspects of molecular beam scattering are discussed elsewhere (Chapters 6 and 8).

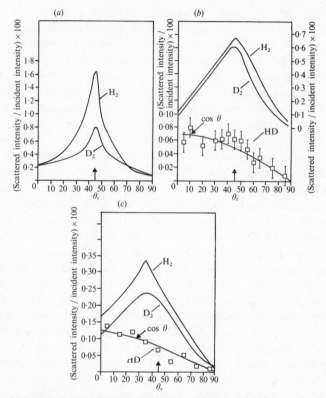

FIG. 13.5. Angular distribution of H_2, D_2, and HD from (a) Pt(111), (b) Pt–(S)[9(111)×(111)], and (c) Pt–(S)[5(111)×(111)]; $T_s = 1000$ K, $T_g = 300$ K. θ_r is the angle from the surface normal; incident angle 45°. (From Bernasek and Somorjai 1975.)

13.10. Theoretical studies

Why do such metals as palladium, iron, and nickel appear so frequently as the active component in commercial catalysts? Similarly why do some enzymes and proteins also appear to depend on such transition metals for their activity? One of the more promising theoretical developments, the self-consistent-field X-alpha (SCF-Xα) scattered-wave approach to quantum chemistry, is already sufficiently advanced for the quantitative investigation of catalytic and biocatalytic systems. Certain crucial factors (already recognized experimentally!) are becoming clear. One point, for example, is the important role played by the d electrons and the particular advantage of the SCF-Xα calculations is that they can cope with metal clusters which can be considered to model small catalyst particles. Messmer *et al.* (1974), Quinn (1974), and Rösch and Rhodin (1974) have used this approach for probing the electronic structure of metal clusters and chemisorption by them. Since all the atoms of a Cu_8 or Ni_8 cluster are surface atoms, then we have a measure of the respective 'surface density of states'. The orbitals mapped out therefore illustrate the types of 'surface chemical bonds', and of particular interest are the directed d lobes at the cube corners which may be significant for the catalytic activity of small transition metal particles. Jones, Jennings, and Painter (1975) have also examined two aspects of the electronic structure of metals: first the differences in orbital structure and hence bonding properties of surfaces of transition metals in a series, and secondly the differences between the bonding orbitals on stepped and flat surfaces of the same metal. Extended orbital lobes present only at steps suggest increased activity compared with flat surfaces and offer a possible explanation for the observed enhanced chemical reactivity of stepped surfaces. When we turn our attention to catalysis perhaps the simplest case to consider is the dissociation of molecular hydrogen by a transition metal such as palladium. The SCF-Xα electronic energy levels for a Pd_4 cluster at the centre of which is a hydrogen atom are shown in Fig. 13.6. Slater and Johnson (1974) draw particular attention to the 'splitting off'

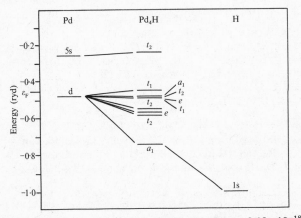

FIG. 13.6. SCF–Xα calculation for Pd_4H. (1 rydberg $= 2.18 \times 10^{-18}$ J).

of a hydrogen-bonding a_1 level from the bottom of the Pd_4 d band. This splitting apparently coincides with experimental photoemission data for the two-phase mixture of β-PdH with Pd. The exact role of d-electrons in chemisorption is, however, still to be resolved, an alternative view being recently put forward by Melius (1976). Inherent in this model is the participation of back-bonding, a concept currently favoured to explain the surface chemistry of carbon monoxide.

Perhaps the most important aspect of recent theoretical approaches to the surface chemistry of metals is that we now have available experimental techniques which enable these theoretical ideas to be scrutinized. It is probably true to say that it was the experimental developments of the last decade that enticed the theoreticians to explore further this general area of science. It is now possible to define experimentally surface structure and surface bonding, both unprecedented in the field of interfacial science; and this has undoubtedly been a stimulus to theoretical work. The impact of surface physics on fundamental aspects of catalysis was reviewed recently by Yates (1974), McCarroll (1975), Roberts (1977), Tompkins (1977) and Thomson (1977).

'Chemistry without catalysis would be a sword without a handle, a light without brilliance, a bell without sound.'

A. MITTASCH.

REFERENCES

BAHSIN, M. W. (1974) *J. Catalysis* **34,** 356.

BARFORD, B. D., and RYE, R. R. (1972). *J. vac. Sci Technol.* **9,** 673.

BENNETT, L. H., CUTHILL, J. R., MCALISTER, A. J., ERICKSON, N. E., and WATSON, R. E. (1974). *Science* **184,** 563.

BERNASEK, S. L., and SOMORJAI, G. A. (1975). *Surf. Sci.* **48,** 204.

BOND, G. C. (1966). *Disc. Faraday Soc.* **41,** 200.

BONZEL, H. P. and KU, R. (1973). *Surf. Sci.* **40,** 85.

BOUDART, M. (1952). *J. am. chem. Soc.* **74,** 3556.

BOUWMAN, R. and BILOEN, P. (1974). *Analyt. Chem.* **46,** 136.

BOUWMAN, R., and SACHTLER, W. M. H. (1970). *J. Catalysis* **19,** 127.

BURTON, J. J., HYMAN, E., and FEDAK, D. G. (1975). *J. Catalysis* **37,** 106.

BURTON, J. J. and MACHLIN, E. S. (1976). *Phys. Rev. Lett.* **37,** 1433.

CHRISTMANN, K., and ERTL, G. (1972). *Surf. Sci.* **33,** 254.

——— (1976). *Surf. Sci.* **60,** 365.

CIMINO, A., BOUDART, M., and TAYLOR, H. S. (1954). *J. phys. Chem.* **58**, 796.

CLARKE, J. K. A. (1975). *Chem. Rev.* **75**, 291.

CLARKE, T. A., GAY, I. D., LAW, B., and MASON, R. (1975). *Disc. Faraday Soc.* **60**, 119.

CLARKE, T. A., GAY, I. D., and MASON, R. (1975). *Physical Basis for Heterogeneous Catalysis* (eds. E. Drauglis, and R. I. Jaffee). Plenum Press, New York.

DEMUTH, J. E., and EASTMAN, D. E. (1974). *Phys. Rev. Lett.* **32**, 1123.

EASTMAN, D. E., and DEMUTH, J. E. (1974). *Jap. J. app. Phys. Suppl.* **2**, Part 2, 827.

EDMONDS, T., McCARROLL, J., and PITKETHLY, R. C. (1970). *Ned. Tijdsch. Vacuumtech.* **8**, 162.

ENGEL, T., VON DEM HAGEN, T. and BAUER, E. (1977). *Surf. Sci.* **62**, 361.

IBACH, H. (1975). *Surf. Sci.* **53**, 444.

ISA, S., JOYNER, R. W., and ROBERTS, M. W. (1975). *J. Chem. Soc. Faraday Trans. I* **72**, 540.

JENKINS, G. I., and RIDEAL, E. K. (1955). *J. chem. Soc.* 2490.

JONES, R. O., JENNINGS, P. J., and PAINTER, G. S. (1975). *Surf. Sci.* **53**, 409.

KAHN, D. R., PETERSEN, E. E., and SOMORJAI, G. A. (1974). *J. Catalysis* **34**, 291.

KISHI. K., and ROBERTS, M. W. (1975a). *J. chem. Soc. Faraday Trans. I* **71**, 1721.

—— (1975b). *J. chem. Soc. Faraday Trans. I* **71**, 1715.

LANG, B., JOYNER, R. W., and SOMORJAI, G. A. (1972a) *Surf. Sci.* **30**, 454.

—— (1972b). *J. Catalysis* **27**, 405.

LEVY, R. B., and BOUDART, M. (1973). *Science* **181**, 547.

LEYSEN, R., HOPKINS, B. J., and TAYLOR, P. A. (1975). *J. Phys. C. solid St. Phys.* **8**, 907.

McCARROLL, J. J. (1975). *Surf. Sci.* **53**, 297.

McCARROLL, J. J., EDMONDS, T., and PITKETHLY, R. C. (1969). *Nature, Lond.* **223**, 1260.

MADIX, R. J., FALCONER, J. and McCARTY, J. (1974). *J. vac. Sci. Technol.* **11** (1), 266.

MASON, R., and TEXTOR, M. (1976). In *Specialist periodical reports: surface and defect properties of solids* (eds. M. W. Roberts and J. M. Thomas), vol. 5. Chemical Society, London.

MELIUS, C. F. (1976). *Chem. Phys. Lett.* **39**, 287.

MESSMER, R. P., JOHNSON, K. H., DIAMOND, J. B., and KNUDSEN, S. K. (1974). General Electric Tech. Rep. No. 74 CRDO52, General Electric Co. Schenectady, N.Y.

PLUMMER, E. W., WACLAWSKI, B. J., and VORBURGER, T. V. (1974). *Chem. Phys. Lett.* **28**, 510.

PRITCHARD, J., CATTERICK, T., and GUPTA, R. K. (1975). *Surf. Sci.* **53**, 1.

QUINN, C. M. (1974). *Disc. Faraday Soc.* **58,** 94.

ROBERTS, M. W. (1977). *Chem. Revs.* **6,** 373. Chemical Society, London.

ROBERTS, M. W., SALEH, J. M., and WELLS, B. R. (1964). *Trans. Faraday Soc.* **60,** 1865.

ROBERTS, M. W., and STEWART, T. I. (1970). In *Chemisorption and Catalysis* (ed. P. Hepple). Institute of Petroleum. London.

RÖSCH, N., and RHODIN, T. N. (1974). *Disc. Faraday Soc.* **58,** 28.

ROSS, P. N., and STONEHART, P. (1975). *J. Catalysis,* **39,** 298.

SLATER, J. C., and JOHNSON, K. H. (1974). *Phys. Today,* **27** (10), 34.

SOMORJAI, G. A., and BLAKELY, D. W. (1975). *Nature, Lond.* **258,** 580.

SOMORJAI, G. A. (1975). In *The Physical Basis for heterogeneous catalysis* (eds: E. Drauglis, and R. I. Jaffee). Plenum Press, New York.

SPICER, W. E., YN, K. Y., LINDAU, I., PIANETTA, P., and COLLINS, D. M. (1976). In *Specialist periodical reports: surface and defect properties of solids* (eds. M. W. Roberts and J. M. Thomas), vol. 5, p. 103. Chemical Society, London.

THOMSON, S. J. (1977). In *Specialist periodical reports: catalysis* (ed. C. Kemball), vol. 1. Chemical Society, London.

TOMPKINS, F. C. (1977). In *Proc. 6th Int. Congress on Catalysis* (ed G. C. Bond, P. B. Wells, and F. C. Tompkins), p. 32. The Chemical Society, London.

VAN DER PLANK, P. and SACHTLER, W. M. H. (1968). *J. Catalysis* **12,** 35.

WHAN, D. A., BARBER, M., and SWIFT, P. (1972). *Chem. Commun.* 198.

YATES, J. T. (1974) *Chem. and Eng. News* **52,** part 34, 19.

APPENDIX: WAVE THEORY

Sinusoidal waves are of considerable importance in the theory of electromagnetic radiation, and also in many other aspects of physics, since any wave motion can be analysed in terms of sinusoidal waves by the methods of Fourier analysis (see for example Stone (1963) for a lucid account of Fourier methods). Consider the wave shown in Fig. A.1(a), at time $t = 0$. The magnitude ψ of the disturbance is given as a function of distance x from the origin by

$$\psi = A \cos\left(\frac{2\pi x}{\lambda} + \delta\right). \tag{A.1}$$

(Since one cycle of the disturbance corresponds to a change in x equal to the wavelength λ and also to a change in the cosine function of 2π rad, any value of x is expressed in radians as 2π (x/λ)). The amplitude A represents the maximum displacement, while the phase angle δ (rad) determines the position of the cosine function on the x axis. Since δ can be changed by any integral multiple of 2π without affecting ψ, it is convenient for it always to be located in the range $0 \leqslant \delta > 2\pi$. From eqn (A.1) a maximum in the disturbance will thus occur at a distance $x_m = -(\delta/2\pi)\lambda$ from the origin. When only one wave is considered the origin may always be chosen so that $\delta = 0$, but when two or more waves are involved their relative phases are of primary importance.

If the wave propagates in the positive x direction with constant amplitude and constant velocity c, then at some later time $t = t$

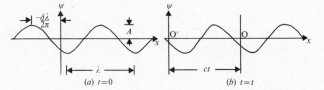

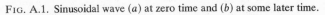

(a) $t = 0$ (b) $t = t$

FIG. A.1. Sinusoidal wave (a) at zero time and (b) at some later time.

the situation will be as shown in Fig. A.1(*b*), and so the disturbance, as a function of both *x* and time, is given by

$$\psi = A \cos\left\{\frac{2\pi}{\lambda}(x - ct) + \delta\right\}. \tag{A.2}$$

The frequency is $\nu = c/\lambda$ Hz, and again, since one cycle corresponds to a change in the cosine term of 2π rad, it is convenient to define an angular frequency $\omega = 2\pi\nu$ rad s^{-1}. Eqn (A.2) may then be rewritten

$$\psi = A \cos\left(2\pi\frac{x}{\lambda} + \delta - \omega t\right). \tag{A.3}$$

It is often convenient to represent a wave by a complex number

$$z = x + i\,y$$

where *x* is the real and *y* the imaginary part of *z*, and $i = \sqrt{-1}$. In turn *z* may be represented by a vector lying in a plane which is referred to 'real' and 'imaginary' axes, the vector components in the real and imaginary directions being equal to the corresponding parts of *z* (Fig. A.2). Then

$$z = r(\cos\theta + i\sin\theta) \tag{A.4}$$

where $r = |z|$ is the magnitude or modulus of the vector, and θ represents its orientation with respect to the real axis. It will be seen at a later stage that the vector form is useful when waves are to be added.

Eqn (A.4) leads to an alternative representation of the complex

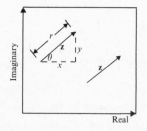

FIG. A.2. Representation of a wave by a complex number.

number by way of the Euler relationship

$$e^{i\theta} = \cos\theta + i\sin\theta.$$

Then

$$z = r(\cos\theta + i\sin\theta) = re^{i\theta}. \tag{A.5}$$

By comparing eqns (A.3) and (A.5) it is seen that a wave of amplitude A may be represented as the *real* part of a complex number of modulus $|z| = A$:

$$\psi = A\cos\theta = \text{Real}\,(A\,e^{i\theta}). \tag{A.6}$$

The complex conjugate of z is defined as

$$z^* = Ae^{-i\theta}$$

and the modulus of z may be obtained from the relationship

$$|z| = (zz^*)^{\frac{1}{2}} = (A^2 e^{i\theta} e^{-i\theta})^{\frac{1}{2}} = A.$$

This exponential form is not only concise but also enables multiplication and division of wave expressions to be carried out easily:

$$\psi_1\psi_2 = A_1 A_2 \exp\{i(\theta_1 + \theta_2)\} \tag{A.7}$$

$$\psi_1/\psi_2 = A_1/A_2 \exp\{i(\theta_1 - \theta_2)\}. \tag{A.8}$$

It is understood that the desired result is given by the *real* parts only of eqns (A.7) and (A.8). Eqn (A.6) may further be used to separate the time-dependent factor in the wave expression (eqn. (A.3)):

$$\psi = A\exp\left\{i\left(\frac{2\pi x}{\lambda} + \delta\right)\right\}\exp(-i\omega t)$$
$$\psi = A_c \exp(-i\omega t) \tag{A.9}$$

The quantity A_c ($= A\exp\{i(2\pi x/\lambda + \delta)\}$) is known as the *complex amplitude* of the motion and its value completely specifies any oscillation of a given frequency.

For some types of wave, e.g. water waves, it is possible experimentally to determine the amplitude, but in diffraction of electromagnetic radiation only the intensity I is accessible to measurement. This is the time average of the energy falling on a unit area lying normal to the direction of propagation, and for a harmonic wave it is proportional to the square of the amplitude.

Plane and spherical progressive waves of constant amplitude and velocity

Consider a point P in the path of the disturbance and let **r** be the vector from the origin to P (Fig. A.3(a)). Suppose that the displacement at P is given by

$$\psi(\mathbf{r}, t) = A \cos(\mathbf{k}.\mathbf{r} + \delta - \omega t) \tag{A.10}$$

where **k**, a constant, is the *wavevector*. At a given time the *wavefronts*, i.e. the *surfaces on which ψ is constant*, will correspond to those surfaces over which **k**.**r** is constant. From the definition of the vector dot product this requirement may be written as

$$\mathbf{k}.\mathbf{r} = kr \cos \sigma = \text{constant} \tag{A.11}$$

where k, r are the magnitudes of **k** and **r**, and σ is the angle between these two vectors. From Fig. A.3(a) it is seen that eqn (A.11) is satisfied for all points lying on any plane normal to **k**, and hence eqn (A.10) represents a plane harmonic progressive wave.

For the particular case in which **r** is parallel to **k**, **k**.**r** = kx and

$$\psi_x = A \cos(kx + \delta - \omega t) \tag{A.12}$$

where $k = |\mathbf{k}|$ is the *wavenumber*. At a given instant t_0 successive maxima will occur at x and $(x + \lambda)$ and so

$$\{k(x + \lambda) + \delta - \omega t_0\} - \{kx + \delta - \omega t_0\} = 2\pi$$

Hence $\qquad\qquad\qquad k = 2\pi/\lambda. \tag{A.13}$

Substitution of eqn (A.13) in eqn (A.12) demonstrates that eqns

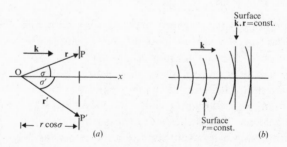

FIG. A.3. (a) Plane wave and (b) spherical wave.

(A.3) and (A.12) are alternative expressions for the profile of a plane wave in a direction parallel to the wavevector **k**.

From eqn (A.12) any wavefront, e.g. that corresponding to maximum displacement, always satisfies the relation

$$kx + \delta - \omega t = 2n\pi$$

where n is an integer. Thus the *wave velocity* c is given by

$$c = \frac{dx}{dt} = \frac{\omega}{k}. \tag{A.14}$$

Hence the direction of the wavevector **k** represents the direction of travel of the plane wave, and its modulus, the wavenumber k, is related to the wavelength and to the angular frequency through eqns (A.13) and (A.14).

In the case of a spherical wave the wavefronts are the surfaces $r = \text{constant}$, and a harmonic spherical wave is represented by

$$\psi = \frac{A}{r} \cos(kr + \delta - \omega t). \tag{A.15}$$

For large values of r the variation of A/r will be much less rapid than that of the cosine term and so over a limited region the spherical wave will approximate to a plane wave (Fig. A.3(b)).

The superposition of waves

The wave equations in both Maxwell's theory and the scalar model are of a type such that, if $\psi_0, \psi_1, \ldots, \psi_{(N-1)}$ are possible solutions, any linear combination of these is also a solution. When considering interference between a number of waves N the resultant disturbance ψ may thus be obtained as the sum of the individual disturbances. Consider sinusoidal waves of the same frequency but different amplitudes and phase angles:

$$\psi_n = A_{cn} \exp(-i\omega t).$$

Then the resultant is

$$\psi = \left(\sum_0^{N-1} A_{cn} \right) \exp(-i\omega t) = A_c \exp(-i\omega t) \tag{A.16}$$

where the complex amplitude A_c is the sum of the individual

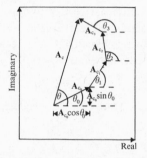

FIG. A.4. Addition of the complex amplitudes of 4 waves to give the complex amplitude A_c of the resultant wave.

complex amplitudes A_{cn}. Note that the resultant is itself sinusoidal. The addition of the complex amplitudes may be considered in terms of vector addition (Fig. A.4). The location of a given vector in the complex plane with respect to the origin is unimportant and so it is permissible to draw the vectors end to end. If we write (Fig. A.4)

$$\mathbf{A}_{cn} = A_{cn}(\cos \theta_n + \mathrm{i} \sin \theta_n)$$

then

$$\theta = \tan^{-1}\left[\frac{\sum A_{cn} \sin \theta_n}{\sum A_{cn} \cos \theta_n}\right] \qquad (A.17)$$

and

$$A_c = \frac{\sum A_{cn} \sin \theta_n}{\sin \theta} = \left\{\left(\sum A_{cn} \sin \theta_n\right)^2 + \left(\sum A_{cn} \cos \theta_n\right)^2\right\}^{\frac{1}{2}}$$

$$(A.18)$$

where θ and A_c are the phase angle and amplitude respectively of the resultant disturbance. Since rotation of Fig. A.4 has no effect on $\mathbf{A}_c$ it is usual to set one of the phase angles θ_n (usually θ_0) equal to zero. In general the resultant phase angle θ is not of interest unless it is required to combine the resultant wave with a further wave. Also the time factor in eqn (A.16) may be omitted since it is of no interest in diffraction problems, remaining unchanged throughout the calculations; the amplitude A_c and its square, the intensity I, are the important quantities.

REFERENCES

STONE, J. M. (1963). *Radiation and optics: an introduction to the classical theory.* McGraw-Hill, New York.

AUTHOR INDEX

SUBJECT INDEX

Absolute rate theory, 41 *et seq.*, 266, 267, 459
absorption, 17
accommodation, 243, 284, 305
  coefficient, thermal, 22, 235, 236, 307, 308, 314 *et seq.*
acetic acid, 213
acetone, 179
acetylene, 26, 545, 546
  adsorbed species, 1, 215
activated complex, 42, 43, 266
activation energy, *see* adsorption; chemisorption; diffusion; etc.
active site, 536, 543, 547
activity patterns, 28
adatom, *see also* adsorbate
  –adatom collision rate, 414
  alkali, minimum radius, 508
  charge, 524, 525, 528
  clusters, 490 *et seq.*
  cross-sectional area, 295, 296, 349
  dimers, 490, 491
  electronic structure, 500
  linear chains, 490, 491
  packing, 292, 294
  polar, 282
  spacing, modulation, 158
  surface bond, 146
  trimers, 490
  tunnelling, 499
  vacancy pair, 91
adlayer, *see also* adsorbate; adatom
  coherent, 504
  defects in, 25
  degree of disorder, 159, 424
  displacement relative to substrate, 134
  distortion, 158, 360, 361, 507, 517

induced heterogeneity, 395, 398, 402, 403, 407
intrinsic heterogeneity, 395
adsorbate, *see also* adlayer; adatom
  compression of, 360, 361, 507, 517
  co-ordination, 147
  coverage, 80, 85, 135, 160, 234, 260, 268
  domains, 120
  electronegative, 248, 249
  electronic properties, 25, 147, 242, 397, 405
  equilibrium population, 275
  fractional coverage, 118, 250, 516
  highly charged, 120
  islands, 160, 164, 472, 506, 513
  minimum energy configuration, 275
  patches, 415, 417
  site symmetry, 131
  structures
    nomenclature, 117
    prediction of, 119, 120
    rotational symmetry, 120
  translational freedom, 267
adsorbate–adsorbate interactions, 38, 119, 120, 150, 151, 158, 249, 266, 275, 307, 326, 349, 387, 395, 397, 407, 409, 413, 417, 422, 429, 489, 490 *et seq.*, 504, 507 *et seq.*, 515, 531, 532
  and adsorption kinetics, 337 *et seq.*, 422
  and thermal desorption, 275 *et seq.*
  attractive, 399
  indirect, 402, 403, 509, 531, 532